de Gruyter Textbook
Pfanzagl · Parametric Statistical Theory

Johann Pfanzagl

Parametric Statistical Theory

With the assistance of R. Hamböker

W
DE
G

Walter de Gruyter
Berlin · New York 1994

Author

Johann Pfanzagl
Mathematisches Institut
Universität Köln
Weyertal 86—90
D-50931 Köln
Germany

1991 Mathematics Subject Classification: 62-01

Keywords: Theory of estimation, testing hypothesis, confidence intervals, asymptotic theory, asymptotic optimality

∞ Printed on acid-free paper which falls within the guidelines of the ANSI to ensure permanence and durability.

Library of Congress Cataloging-in-Publication Data

Pfanzagl, J. (Johann)
    Parametric statistical theory / Johann Pfanzagl.
        p.   cm. — (De Gruyter textbook)
    Includes bibliographical references and index.
    ISBN 3-11-014030-6 (acid-free). — ISBN 3-11-013863-8
(pbk. ; acid-free)
        1. Mathematical statistics. I. Title. II. Series.
    QA276.PA77   1994
    519.5′4—dc20   94-21850
                                                                      CIP

Die Deutsche Bibliothek — Cataloging-in-Publication Data

Pfanzagl, Johann:
Parametric statistical theory / Johann Pfanzagl. — Berlin ; New York :
de Gruyter, 1994
    (De Gruyter textbook)
    ISBN 3-11-013863-8 Pb.
    ISBN 3-11-014030-6 Gewebe

Printed in Germany.
Printing: Arthur Collignon GmbH, Berlin. Binding: Lüderitz & Bauer GmbH, Berlin.

Preface

Scire tuum nihil est, nisi te scire hoc sciat alter.

(Persius, I, 27)

This book presents a survey of advanced parametric statistical theory. It requires a basic knowledge of measure theory and probability theory, at about the level of Ash (1972) or Dudley (1989). For the German speaking reader, familiarity with the material in Bauer's (1990a,b) "Maß– und Integrationstheorie" and about half of his "Wahrscheinlichkeitstheorie" (with an English edition forthcoming) can be considered as an excellent preparation. Auxiliary results from measure theory and probability theory which are not easily found in textbooks (like uniform versions of limit theorems, or measurable selection theorems) are summarized in the last section of each chapter.

Students with these mathematical prerequisites and some knowledge of elementary statistics will have no problem in mastering the material in this book. It corresponds to a two–term course, totalling about 150 hours (including exercises).

The book is a result of classes held over many years for students of mathematics at the University of Cologne. It owes much to some of my former assistants who cooperated with me in the supervision of exercises and diploma theses. Among these are C. Hipp, U. Einmahl, L. Schröder, and R. Hamböker.

I am particularly obliged to R. Hamböker. Without his unresting help I would have been unable to bring the manuscript into its final shape.

The TEX version was done by E. Lorenz.

Cologne, January 31, 1994   J. Pfanzagl

Introduction

Mathematical statistics is a tool for drawing conclusions from realizations of a random variable to the underlying probability measure. A basic knowledge of the physical background of the random phenomenon usually gives some information about the general nature of this probability measure: Prior to the observation we know that the probability measure belongs to some family of probability measures, say $\mathfrak{P}$. Statistical theory becomes particularly simple if this basic family is parametric, i.e. if $\mathfrak{P} = \{P_\vartheta : \vartheta \in \Theta\}$, where Θ is a subset of a Euclidean space. In asymptotic theory we consider in this book parametric families $\{P_\vartheta^n : \vartheta \in \Theta\}$, where the observations are independent and identically distributed.

Our approach to the statistical procedures is simple–minded and earth-bound, in a sense. Starting from the possible uses of an estimator, we try to develop concepts for the evaluation of estimators, and methods for the construction of "good" estimators. To consider estimation as a special decision procedure seems to be of no help. To squeeze every statistical problem into the Procrustean–bed of decision theory prevents the statistician from considering the possible solutions according to criteria inherent to the particular problem.

Paraphrasing a statement of Wittgenstein's in one of his letters to Russell, a book consists of two equally important parts: The part which has been written, and the part which has been omitted. In accordance with our simple–minded approach we have omitted some of the traditional contents of textbooks on mathematical statistics when they appeared to us neither practically useful nor mathematically interesting. As an example we mention the celebrated Cramér–Rao bound. This bound is not attainable, in general, thus lacking any clear operational significance for finite sample sizes, and it plays no role in asymptotic theory since it is based on the assumption of unbiasedness for every sample size. Other concepts which are omitted as irrelevant for applications can be characterized by the catchwords "admissible" and "minimax".

Robust methods and Bayesian theory fell victim to space limitations. Throughout the book, the considerations are restricted to dominated families of probability measures, to avoid complications of purely mathematical nature, without relevance for applications. Most results are presented under the more restrictive assumption of mutual absolute continuity. We think that this is no serious shortcoming for the statistician interested in applications. (Thereby, we also avoid the discussion about the interpretation of statistical procedures the results of which are a composite of probabilistic and "exact" assertions.)

The book contains some historical remarks. They reflect the predilections of the author, and do not cover the whole field in a systematic way.

Parameters and functionals

The inference based on the realization of a random variable usually aims at a particular feature of the underlying probability measure, expressed by some functional κ, defined on $\mathfrak{P}$. The considerations of this book are mainly restricted to finite dimensional functionals $\kappa : \mathfrak{P} \to \mathbb{R}^q$.

Basically it is the functionals (like means, or quantiles, or failure rates) which have a meaningful interpretation. Parameters which are used for a description of the basic family $\mathfrak{P}$ may have a meaningful interpretation if they play a role as a functional. For the purpose of illustration, let $\{E_a : a > 0\}$ be the family of exponential distributions. If E_a describes a life distribution, a is expected duration of life, $a \log 2$ the median life time etc. If one is uncertain whether the exponential distribution fits reality with sufficient accuracy, one may be inclined to consider the larger family of gamma distributions $\{\Gamma_{a,b} : a, b > 0\}$ (which contains E_a as $\Gamma_{a,1}$) as a more realistic model. But what is the use of an estimate for the parameter a under the assumption that the observations are distributed as $\Gamma_{a,b}$? In this case, the mean life time is ab, the median life time $aT(b)$ (with $T(b)$ defined by $\int_0^{T(b)} x^{b-1}e^{-x}dx = \Gamma(b)/2$). Embedding $\{E_a : a > 0\}$ in the larger family modifies the meaning of the parameter a, and we have to get clear about which functional we wish to estimate. Some statisticians might be inclined to consider the true distribution $\Gamma_{a,b}$ as a "distorted" or "contaminated" exponential distribution, and to find the way back to the original "undistorted" distribution by projecting $\Gamma_{a,b}$ into the family $\{E_\alpha : \alpha > 0\}$, i.e. to assign to each pair (a, b) the value $\alpha(a, b)$ for which a certain distance between E_α and $\Gamma_{a,b}$ is minimal. The functional $\kappa(\Gamma_{a,b}) = \alpha(a, b)$ thus defined has the advantage of admitting robust estimators. It has the shortcoming of lacking any meaningful interpretation. To answer a meaningful question one has to decide which functional is to be estimated, and whether the interest is in a functional of the true (if distorted) distribution, or in a functional of the original distribution (in case this functional is identifiable from observations originating from the distorted distribution).

"Exact" versus asymptotic theory

Chapters 3, 4 and 5 contain results of "exact", i.e. nonasymptotic, nature about confidence coefficients, significance levels, unbiasedness and optimality. These results are exact only within the model. Since models are hardly ever accurate images of reality, the results of the "exact" theory hold approximately only as soon as we turn to applications. Therefore, nothing is lost by the application of asymptotic results, and much can be gained. Roughly speaking, "exact" solutions in parametric theory are feasible for exponential families only. Even in this restricted area, certain problems remain unsolved. This is typically the case for "curved" exponential families. The most popular example for the failure of the exact theory is the Behrens–Fisher problem.

Asymptotic theory takes advantage of the regularity brought in by many independent repetitions of the same random experiment, a regularity which finds its simplest expression in the Central Limit Theorem. Whereas distributions of different estimators are incomparable in general, and "optimal" estimators a rare exception, limit distributions are normal in regular cases, and this makes a comparison feasible.

Asymptotic theory is "general" in the sense that it applies to arbitrary parametric families, subject to some regularity conditions. If only results of the exact theory were available, many practical problems would necessarily remain unsolved or — more probably — they would be treated under unrealistic assumptions (e.g., an omnipresence of normal distributions), thus risking a model bias the amount of which is unknown. In contrast to this, asymptotic theory offers the possibility of representing reality by a more accurate model, thus reducing the danger of a model bias.

The crucial drawback of asymptotic theory: What we expect from asymptotic theory are results which hold approximately (like estimators which are approximately median unbiased and the concentration of which is approximately maximal). What asymptotic theory has to offer are limit theorems. By taking a limit theorem as being approximately true for large sample sizes, we commit an error the size of which is unknown. To obtain bounds for such errors is not impossible. (As an example think of the Berry–Esséen bound for the accuracy of the normal approximation to the distribution of sample means.) However: Such error bounds will usually grossly overestimate the true errors. Edgeworth expansions are an efficient tool for reducing the errors of asymptotic results and for clarifying their general structure. Realistic information about the remaining errors may be obtained by simulations.

The asymptotic considerations in this book are mainly restricted to the i.i.d. case, i.e. to independent, identically distributed observations. Moreover, we confine ourselves to an asymptotic theory of first order, i.e. to an approximation by limit distributions. We hope that the reader who thoroughly understands the problems connected with the simplest case will have a solid base for the study of more general and more refined asymptotic methods.

Contents

Chapter 1
Sufficiency and completeness

1.1 Introduction

Motivates the concepts of "exhaustive" and "sufficient" statistics which preserve all "information" in the sample.

Let $(X, \mathcal{A})$ be a measurable space, and $S : (X, \mathcal{A}) \to (Y, \mathcal{B})$ a measurable map. Moreover, let $\mathfrak{P}$ denote a family of p–measures (probability measures) $P|\mathcal{A}$. If, instead of the observation x, only $S(x)$ is known, this will, in general, result in some loss of "information": It will be impossible to match each statistical procedure based on x by a statistical procedure based on $S(x)$. This is, however, not necessarily so. It may occur, that for each statistical procedure based on x there is a statistical procedure based on $S(x)$ which is at least as good. If this is the case, we call the function S "exhaustive".

Turning this intuitive idea into a mathematical concept meets with the difficulty that there is no general concept for comparing statistical procedures, varied as tests, confidence procedures or estimators.

As far as tests are concerned, one could think of requiring that for any critical function $\varphi : X \to [0, 1]$ there is a critical function $\psi \circ S : X \to [0, 1]$ with the same power function on $\mathfrak{P}$ (i.e. with $P(\varphi) = P(\psi \circ S)$ for all $P \in \mathfrak{P}$). The situation is almost hopeless if one tries to describe in mathematical terms that one estimator is "at least as good" as another. (See the discussion in Chapter 2.) Even if one would succeed in each instance: It is not clear, in advance, whether "exhaustive for tests" is the same as "exhaustive for estimators", say.

One does not depend on special criteria for the evaluation of various statistical procedures if it is possible to obtain, starting from $S(x)$, a random variable $\hat{x}$ with the same distribution as the original observation x. Any statistical procedure, applied to $\hat{x}$, has then — for every $P \in \mathfrak{P}$ — exactly the same performance as the statistical procedure applied to x. With $P \in \mathfrak{P}$ unknown, the device leading from $S(x)$ to $\hat{x}$ has to be independent of P, of course.

In technical terms the transition from $S(x)$ to $\hat{x}$ is accomplished by a Markov kernel $M|Y \times \mathcal{A}$.

Definition 1.1.1. A statistic $S : (X, \mathcal{A}) \to (Y, \mathcal{B})$ is *exhaustive* for $\mathfrak{P}$ if there exists a Markov kernel $M|Y \times \mathcal{A}$ such that

$$\int M(y, A) P \circ S(dy) = P(A) \qquad \text{for } A \in \mathcal{A} \text{ and } P \in \mathfrak{P}. \tag{1.1.2}$$

Recall Theorem 1.10.33 according to which, for Polish spaces $(X, \mathcal{A})$, for any Markov kernel $M|Y \times \mathcal{A}$ there exists a function $m : Y \times (0, 1) \to X$ such that, for every $y \in Y$, the induced distribution of $u \to m(y, u)$ under U (the uniform distribution over $(0, 1)$) is identical to $M(y, \cdot)$, i.e. $U \circ (u \to m(y, u)) = M(y, \cdot)$. If S is exhaustive for $\mathfrak{P}$, i.e. $\int M(y, A) P \circ S(dy) = P(A)$ for $A \in \mathcal{A}$ and $P \in \mathfrak{P}$, we obtain $(P \times U) \circ ((x, u) \to m(S(x), u)) = P$. If we know $S(x)$ and determine a realization u from the uniform distribution over $(0, 1)$, we obtain with $m(S(x), u)$ a random variable which has exactly the same distribution as the original x, for every $P \in \mathfrak{P}$.

There is a convenient alternative if the sufficient statistic S is boundedly complete and if there exists an ancillary statistic $T : (X, \mathcal{A}) \to (Z, \mathcal{C})$ such that $x \equiv f(S(x), T(x))$ for $x \in X$. Then S and T are stochastically independent for $P \in \mathfrak{P}$ by Basu's Theorem 1.8.2, and, with $P_0 \in \mathfrak{P}$ arbitrarily fixed, $(P \times P_0) \circ ((x, u) \to f(S(x), T(u))) = P$ for every $P \in \mathfrak{P}$ by the Addendum to Proposition 1.8.11.

Example 1.1.3. For $\vartheta > 0$ let E_ϑ denote the exponential distribution, given by its Lebesgue dentity $x \to \vartheta^{-1} \exp[\vartheta^{-1} x]$, $x > 0$. The function $S_n(x_1, \ldots, x_n) = \sum_1^n x_\nu$ is exhaustive for the family $\{E_\vartheta^n : \vartheta > 0\}$.
$T_i(x_1, \ldots, x_n) := x_i / \sum_1^n x_\nu$, $i = 1, \ldots, n$, is ancillary, and

$$x_i = S_n(x_1, \ldots, x_n) T_i(x_1, \ldots, x_n).$$

Hence the condition $x = f(S(x), T(x))$ is fulfilled with $f(y, z) = yz$. If we determine a realization $(u_1, \ldots, u_n)$ from E_1^n, then

$$\left(\sum_1^n x_\nu u_1 / \sum_1^n u_i, \ldots, \sum_1^n x_\nu u_n / \sum_1^n u_i \right)$$

has for every $\vartheta > 0$ the same distribution E_ϑ^n as $(x_1, \ldots, x_n)$.

Definition 1.1.4. A statistic $S : (X, \mathcal{A}) \to (Y, \mathcal{B})$ is *sufficient* for $\mathfrak{P}$ if for every $A \in \mathcal{A}$ there exists a conditional expectation, given S, say φ_A, which does not depend on $P \in \mathfrak{P}$, i.e.

$$\int \varphi_A(y) 1_B(y) P \circ S(dy) = P(A \cap S^{-1} B) \qquad \text{for } B \in \mathcal{B}, \ P \in \mathfrak{P}.$$

If $\mathfrak{P}$ is dominated, sufficiency of S implies (see Proposition 1.3.1) for every measurable map from X into a Polish space the existence of a conditional distribution, given S, which does not depend on $P \in \mathfrak{P}$.

Historical Remark 1.1.5. The concept of an "exhaustive" statistic goes back to Blackwell (1951). With estimators in mind, Fisher developed the idea that a statistic S is sufficient if any statistic T has a conditional distribution, given S, which is independent of $P \in \mathfrak{P}$. Knowing $T(x)$, in addition to $S(x)$, can, therefore, contribute nothing to our knowledge about the "true" P. Fisher's idea of a sufficient statistic goes back to (1920, p. 768). He illustrates this idea by showing — for the sample size $n=4$ — that the conditional distribution of $n^{-1}\sum_1^n |x_\nu - \bar{x}_n|$, given $(n^{-1}\sum_1^n (x_\nu - \bar{x}_n)^2)^{1/2}$, with respect to $N^n_{(\mu,\sigma^2)}$, does not depend on μ and σ^2. The basic idea is repeated in a number of succeeding papers, but the corresponding formulations in mathematical terms (connecting sufficiency with some kind of factorization of the densities) remains rather vague, even as late as Fisher (1934, pp. 288ff.).

Remark 1.1.6. Some authors try to define a more subtle concept of *partial sufficiency*. Given a parametrized family of p–measures $\{P_{\vartheta,\tau} : \vartheta \in \Theta, \tau \in T\}$, a statistic $S|X$ is "partially sufficient" in some intuitive sense if it contains all information needed for the inference about ϑ (in the presence of the unknown parameter τ). Though it is certainly a meaningful problem to find an appropriate mathematical concept for this intuitive idea, a satisfactory solution is still missing. Some authors say that S is sufficient for ϑ if S is sufficient for $\{P_{\vartheta,\tau} : \vartheta \in \Theta\}$, for each $\tau \in T$, and if $P_{\vartheta,\tau} \circ S$ depends on ϑ only. It is, however, hard to understand how this corresponds to the intuitive idea of "sufficiency for ϑ" outlined above.

The reader interested in various definitions of "partial sufficiency" is referred to D. Basu (1978).

The relation between exhaustivity, sufficiency, and the factorization of densities will be investigated in Sections 1.2 and 1.3. In Section 1.4 we prove the existence of a minimal sufficient statistic and give criteria for the minimality of a sufficient statistic. These concepts are illustrated by a number of examples in Sections 1.4 and 1.6.

To avoid technical problems without any practical relevance we restrict our considerations to families of p–measures which are dominated by a σ–finite measure. Even though several proofs are technically simpler if working with "sufficient sub–σ–algebras" rather than "sufficient statistics" we stick to the latter concept because it is meaningful from the operational point of view. Observe, however, that it is not the value a sufficient statistic attains, but the partition $S^{-1}(S\{x\})$, $x \in X$, it induces on X. Any 1–1 bimeasurable function of a sufficient statistic is sufficient. On the other hand, it appears difficult to interpret the sufficiency of a sub–σ–algebra which cannot be induced by a map.

Readers who want to get an idea of the technical problems arising in connection with non–dominated families of p–measures are invited to look into Bomze (1990) and into textbooks with detailed chapters on sufficiency, e.g., Witting (1985), Strasser (1985) and Torgersen (1991).

1.2 Sufficiency and factorization of densities

For dominated families, a statistic S is sufficient if and only if there are densities (with respect to an appropriate dominating measure) which depend on x through $S(x)$ only.

Throughout this section, $\mathfrak{P}$ is a dominated family of p–measures on $(X, \mathcal{A})$.

The following lemma exhibits the possibility of a special choice for the dominating measure which is technically convenient on certain occasions.

Lemma 1.2.1. (Halmos and Savage, 1949, p. 232, Lemma 7).
If $\mathfrak{P}|\mathcal{A} \ll \mu|\mathcal{A}$, there exists a countable subset $\{P_n : n \in \mathbb{N}\} \subset \mathfrak{P}$ such that $\mathfrak{P}|\mathcal{A} \ll P_*|\mathcal{A}$, with

$$P_* := \sum_1^\infty 2^{-n} P_n \,. \tag{1.2.2}$$

Proof. W.l.g. we assume that μ is finite. For every $P \in \mathfrak{P}$ we choose a fixed version $h_P \in dP/d\mu$ and define

$$S_P := \{x \in X : h_P(x) > 0\}.$$

Observe that $P(S_P^c) = 0$ for $P \in \mathfrak{P}$.
Since for all $P \in \mathfrak{P}$

$$P(A) = \int h_P(x) 1_A(x) \mu(dx) = \int h_P(x) 1_{A \cap S_P}(x) \mu(dx),$$

we have

$$P(A) = 0 \quad \text{iff} \quad \mu(A \cap S_P) = 0.$$

Let $\mathcal{T}_*$ denote the class of all countable unions of sets in $\{S_P : P \in \mathfrak{P}\}$. Since $\mathcal{T}_*$ is closed under countable unions, there exists $S_* \in \mathcal{T}_*$ such that

$$\mu(S_*) = \sup\{\mu(A) : A \in \mathcal{T}_*\} \tag{1.2.3}$$

Since $S_* \in \mathcal{T}_*$, there exist $P_n \in \mathfrak{P}$, $n \in \mathbb{N}$, such that $S_* = \bigcup_{n=1}^{\infty} S_{P_n}$. We shall show that $P_* := \sum_1^{\infty} 2^{-n} P_n$ has the asserted properties.

(i) Relation (1.2.3) implies $\mu(S_*^c \cap S_P) = 0$ for every $P \in \mathfrak{P}$, hence $P(S_*^c) = 0$ for every $P \in \mathfrak{P}$.

(ii) $P_*(A) = 0$ implies $P_n(A) = 0$ for $n \in \mathbb{N}$, hence $\mu(A \cap S_{P_n}) = 0$ for $n \in \mathbb{N}$ and therefore $\mu(A \cap S_*) = 0$. Since $P \ll \mu$, this implies $P(A \cap S_*) = 0$ for $P \in \mathfrak{P}$.

Since $P(A) = P(A \cap S_*) + P(A \cap S_*^c)$, the relation $P_*(A) = 0$ implies $P(A) = 0$ for $P \in \mathfrak{P}$, hence $\mathfrak{P} \ll P_*$. $\square$

Remark 1.2.4. If $\mathcal{A}$ is countably generated, the proof of Lemma 1.2.1 becomes much simpler. Since $\mathcal{L}_1(X, \mathcal{A}, \mu)$ is separable in this case, $\{h_P : P \in \mathfrak{P}\} \subset \mathcal{L}_1(X, \mathcal{A}, \mu)$ contains a countable subset, say $\{h_{P_n} : n \in \mathbb{N}\}$, which is dense in $\{h_P : P \in \mathfrak{P}\}$ (where h_P denotes a μ–density of P). The subset $\{P_n : n \in \mathbb{N}\}$ is dense in $\mathfrak{P}$ with respect to the sup–metric. Hence $P_n(A) = 0$ for all $n \in \mathbb{N}$ implies $P(A) = 0$ for $P \in \mathfrak{P}$, which implies $\mathfrak{P} \ll P_* := \sum_1^{\infty} 2^{-n} P_n$.

Throughout the following, S is a measurable map from $(X, \mathcal{A})$ to $(Y, \mathcal{B})$, and P^S denotes the conditional expectation operator, given S. Moreover, a function is called $\mathfrak{P}$–*integrable* if it is P-integrable for every $P \in \mathfrak{P}$.

Definition 1.1.4 of "sufficiency" requires that $\cap \{P^S 1_A : P \in \mathfrak{P}\} \neq \emptyset$ for every $A \in \mathcal{A}$. This is equivalent to the apparently stronger property

$$\cap \{P^S f : P \in \mathfrak{P}\} \neq \emptyset \text{ for every } \mathfrak{P}\text{-integrable function } f. \tag{1.2.5}$$

(Hint: Show that the class of all $\mathfrak{P}$–integrable functions f fulfilling (1.2.5) is closed under linear combinations and monotone limits; it contains 1_A by definition of sufficiency.)

Exercise 1.2.6. (i) If $\mathfrak{P}_0$ is dense in $\mathfrak{P}$ with respect to the sup–metric, sufficiency of S for $\mathfrak{P}_0$ implies sufficiency of S for $\mathfrak{P}$. (ii) This implication holds under the weaker condition that $\mathfrak{P}_0$ is dense in $\mathfrak{P}$ with respect to a topology for which $P \to P(A)$ is continuous for every $A \in \mathcal{A}$.

Definition 1.2.7. $\mathfrak{P}$ has densities *factorizing with respect to S* if there exists a function $g : (X, \mathcal{A}) \to (\mathbb{R}_+, \mathbb{B}_+)$, and for every $P \in \mathfrak{P}$ a function $h_P : (Y, \mathcal{B}) \to (\mathbb{R}_+, \mathbb{B}_+)$ such that

$$x \to g(x) h_P(S(x)) \in dP/d\mu. \tag{1.2.8}$$

The essential feature of this factorization is that the factor containing P depends on x through $S(x)$ only. The factor g depends on the dominating measure μ. If convenient one can achieve $g(x) \equiv 1$ by choosing $\mu = P_*$ from Lemma 1.2.1 (or a p–measure $P_0 \in \mathfrak{P}$ which dominates $\mathfrak{P}$, if any).

Proposition 1.2.9. *If $S : (X, \mathcal{A}) \to (Y, \mathcal{B})$ fulfills condition (1.2.8), then there exist densities in dP/dP_* which depend on x through $S(x)$ only.*

Proof. With $h_*(y) := \sum_1^\infty 2^{-n} h_{P_n}(y)$, we have $g(\cdot) h_*(S(\cdot)) \in dP_*/d\mu$. With $B_* := \{y \in Y : h_*(y) > 0\}$ let

$$\hat{h}_P(y) := \begin{cases} h_P(y)/h_*(y) & \text{for } y \in B_* \\ 0 & \text{for } y \in B_*^c. \end{cases}$$

Then $\hat{h}_P \circ S \in dP/dP_*$. We have to prove

$$\int \hat{h}_P(S(x)) 1_A(x) P_*(dx) = P(A) \quad \text{for } A \in \mathcal{A}.$$

This can be seen as follows:

$$\int \hat{h}_P(S(x)) 1_A(x) P_*(dx) = \int \hat{h}_P(S(x)) g(x) h_*(S(x)) 1_A(x) \mu(dx)$$

$$= \int h_P(S(x)) g(x) 1_{B_*}(S(x)) 1_A(x) \mu(dx) = P(A \cap S^{-1} B_*)$$

$$= P(A), \quad \text{since } P_*(S^{-1} B_*^c) = 0 \text{ implies } P(S^{-1} B_*^c) = 0.$$

$\square$

Some sort of "factorization theorem" occurs in Fisher (1922, pp. 330/1) and several succeeding papers. The present version is essentially due to Neyman (1935, p. 326, Teorema II).

Factorization Theorem 1.2.10. *S is sufficient for $\mathfrak{P}$ iff $\mathfrak{P}$ has densities factorizing with respect to S.*

Proof. (i) If S is sufficient for $\mathfrak{P}$, for any $A \in \mathcal{A}$ there exists $\varphi_A \in \cap \{P^S 1_A : P \in \mathfrak{P}\}$. This implies (see (1.10.2'))

$$\int \varphi_A(y) 1_B(\ddot{y}) P \circ S(dy) = \int 1_A(x) 1_B(S(x)) P(dx) \quad \text{for } B \in \mathcal{B}, \quad (1.2.11)$$

hence also (see Lemma 1.2.1)

$$\int \varphi_A(y) 1_B(y) P_* \circ S(dy) = \int 1_A(x) 1_B(S(x)) P_*(dx) \quad \text{for } B \in \mathcal{B},$$

and consequently (see Proposition 1.10.13)

$$\int \varphi_A(y) h(y) P_* \circ S(dy) = \int 1_A(x) h(S(x)) P_*(dx) \quad (1.2.12)$$

for every $P_* \circ S$–integrable function h.

Applied with $h = h_P \in dP \circ S/dP_* \circ S$, we obtain from (1.2.12) that

$$\int \varphi_A(y) P \circ S(dy) = \int 1_A(x) h_P(S(x)) P_*(dx).$$

Together with (1.2.11), applied with $B = Y$, this yields

$$P(A) = \int 1_A(x) h_P(S(x)) P_*(dx).$$

Since $A \in \mathcal{A}$ was arbitrary, this implies $h_P \circ S \in dP/dP_*$, hence (1.2.8) (with $\mu = P_*$).

(ii) If $\mathfrak{P}$ has densities factorizing with respect to S, there exist $h_P : (Y, \mathcal{B}) \to (\mathbb{R}_+, \mathbb{B}_+)$ such that $h_P \circ S \in dP/dP_*$, hence also $h_P \in dP \circ S/dP_* \circ S$.

For $A \in \mathcal{A}$ let $\varphi_A \in P_*^S 1_A$. This implies (see Proposition 1.10.13)

$$\int \varphi_A(y) h(y) P_* \circ S(dy) = \int 1_A(x) h(S(x)) P_*(dx)$$

for $P_* \circ S$–integrable functions $h|(Y, \mathcal{B})$. Applied with $h = 1_B h_P$ this yields

$$\int \varphi_A(y) 1_B(y) P \circ S(dy) = \int 1_A(x) 1_B(S(x)) P(dx) \quad \text{for } B \in \mathcal{B} \text{ and } P \in \mathfrak{P},$$

hence $\varphi_A \in \cap \{P^S 1_A : P \in \mathfrak{P}\}$, which implies that S is sufficient. $\qquad\square$

Remark 1.2.13. The Definition 1.1.4 of sufficiency refers to a σ–algebra $\mathcal{B}$ over Y. If the family $\mathfrak{P}$ is dominated, the reference to this more or less arbitrary σ–algebra over Y can be avoided by introducing the σ–algebra $\mathcal{B}_S := \{B \subset Y : S^{-1}B \in \mathcal{A}\}$. Measurability of $S : (X, \mathcal{A}) \to (Y, \mathcal{B})$ implies $\mathcal{B}_S \supset \mathcal{B}$.

If S is sufficient, then by Factorization Theorem 1.2.10 there exist P_*–densities of P which depend on x through $S(x)$ only, and which are, therefore, $\mathcal{B}_S$–measurable. Hence Factorization Theorem 1.2.10, applied with $(Y, \mathcal{B}_S)$ in place of $(Y, \mathcal{B})$, yields sufficiency of $S : (X, \mathcal{A}) \to (Y, \mathcal{B}_S)$.

Exercise 1.2.14. Let $\mathfrak{P}$ be a family of p–measures P on $(X, \mathcal{A})$ with sufficient statistic $S : (X, \mathcal{A}) \to (Y, \mathcal{B})$. For $P \in \mathfrak{P}$ let Q_P be a p–measure with P–density $x \to C(P) q(x)$. Show that S is sufficient for $\{Q_P : P \in \mathfrak{P}\}$. For $q = 1_{A_0}$ and $C(P) = 1/P(A_0)$ this yields that S remains sufficient for truncated distributions. (Hint: The factorization theorem yields a straightforward proof. An alternative proof which remains true for non–dominated families $\mathfrak{P}$ uses that $C(P) P^S q$ is a density of $Q_P \circ S$ with respect to $P \circ S$ to show that $P^S(fq)/P^S q$ (which can be chosen independent of P) is a conditional expectation of f, given S, with respect to Q_P. See Smith, 1957, p. 248, Theorem.)

1.3 Sufficiency and exhaustivity

Shows that sufficiency and exhaustivity are equivalent (under suitable regularity conditions), and adds another characterization by power functions of tests.

The reader interested in the characterization of sufficiency by means of decision theoretic concepts may consult Bahadur (1955), Sacksteder (1967), Heyer (1969, 1982) and Torgersen (1991).

Throughout this section, $\mathfrak{P}$ is a dominated family of p–measures on $(X, \mathcal{A})$.

Proposition 1.3.1. *If $S : (X, \mathcal{A}) \to (Y, \mathcal{B})$ is sufficient, then for every statistic T mapping $(X, \mathcal{A})$ into a Polish space $(Z, \mathcal{C})$ there exists a regular conditional distribution, given S, which does not depend on $P \in \mathfrak{P}$.*

Proof. Let P_* denote the dominating measure given by Lemma 1.2.1, and let $M|Y \times \mathcal{C}$ be a Markov kernel representing the conditional distribution of T, given S, with respect to P_* (see Theorem 1.10.17). By Proposition 1.10.13 this implies

$$\int M(y, C)h(y)P_* \circ S(dy) = \int 1_C(T(x))h(S(x))P_*(dx) \qquad (1.3.2)$$

for $C \in \mathcal{C}$, and every $P_* \circ S$–integrable function h.

By Factorization Theorem 1.2.10, for every $P \in \mathfrak{P}$ there exists $h_P \in dP \circ S/dP_* \circ S$ such that $h_P \circ S \in dP/dP_*$. Relation (1.3.2), applied with $h = 1_B h_P$, yields

$$\int M(y, C)1_B(y)P \circ S(dy) = \int 1_C(T(x))1_B(S(x))P(dx) \qquad (1.3.3)$$

for $C \in \mathcal{C}$, $B \in \mathcal{B}$ and $P \in \mathfrak{P}$.

Hence $M|Y \times \mathcal{C}$ is a conditional distribution of T, given S, not only with respect to P_*, but with respect to every $P \in \mathfrak{P}$. □

If $(X, \mathcal{A})$ itself is a Polish space, Proposition 1.3.1 implies the existence of a Markov kernel $M|Y \times \mathcal{A}$ which is a conditional distribution of "x", given S, with respect to $P \in \mathfrak{P}$, i.e.

$$\int M(y, A)1_B(y)P \circ S(dy) = P(A \cap S^{-1}B) \qquad (1.3.4)$$

for $A \in \mathcal{A}$, $B \in \mathcal{B}$ and $P \in \mathfrak{P}$. For the special case $B = Y$, this yields

$$\int M(y, A)P \circ S(dy) = P(A) \quad \text{for } A \in \mathcal{A}, \ P \in \mathfrak{P}. \qquad (1.3.5)$$

Leaving aside for the moment the technical condition that $(X, \mathcal{A})$ is a Polish space, we can say that any sufficient statistic is also exhaustive. Relation (1.3.4) asserts, however, a stronger property of M than just the possibility of regaining a random variable with the same distribution as x, if only $S(x)$ is known, as expressed by (1.3.5). Is there an intuitive interpretation for what (1.3.4) adds to (1.3.5)?

If $\mathcal{B}$ is countably generated and $\{y\} \in \mathcal{B}$ for every $y \in Y$, then (1.3.4) implies (see Proposition 1.10.25)

$$M\left(y, S^{-1}\{y\}\right) = 1 \quad \text{for } \mathfrak{P} \circ S\text{–a.a. } y \in Y. \tag{1.3.6}$$

If we determine a realization $\hat{x}$ from $M\left(S(x), \cdot\right)|\mathcal{A}$, we have $S(\hat{x}) = S(x)$ for almost all $(\hat{x}, x) \in X \times X$ (with respect to the p–measure on $\mathcal{A} \times \mathcal{A}$ defined by $A_1 \times A_2 \to \int M\left(S(x), A_2\right) 1_{A_1}(x) P_*(dx)$).

If we strengthen the definition of exhaustivity from (1.1.2) to

$$\int M\left(y, A \cap S^{-1}\{y\}\right) P \circ S(dy) = P(A) \quad \text{for } A \in \mathcal{A}, \ P \in \mathfrak{P},$$

then this implies sufficiency of S (by Proposition 1.10.25), and is equivalent to sufficiency if $(X, \mathcal{A})$ is a Polish space.

Relation (1.3.6) has yet another consequence of operational significance. If the randomization procedure described by the Markov kernel $M|Y \times \mathcal{A}$ is carried through twice (determining first a realization $\hat{x}$ from $M\left(S(x), \cdot\right)|\mathcal{A}$, then a realization $\hat{\hat{x}}$ from $M\left(S(\hat{x}), \cdot\right)|\mathcal{A}$), the pertaining Markov kernel is

$$(y, A) \to \int M\left(S(x), A\right) M(y, dx). \tag{1.3.7}$$

If M fulfills (1.3.6), the Markov kernel (1.3.7) is identical with M.

The definition (1.1.2) of exhaustivity requires nothing like this. But if an exhaustive Markov kernel (i.e. one fulfilling (1.3.5)) fulfills, in addition, (1.3.6), then it fulfills also (1.3.4) according to Proposition 1.10.25, i.e. it is a conditional distribution of x, given S.

Even if the Markov kernel M fulfills (1.3.5) only, the statistic S is nevertheless sufficient according to Theorem 1.3.9. An exhaustive Markov kernel may, therefore, be replaced by a Markov kernel on $Y \times \mathcal{A}$ fulfilling, in addition to (1.3.5), also (1.3.4) and (1.3.6), provided $(X, \mathcal{A})$ is a Polish space.

Remark 1.3.8. If S is a minimal sufficient statistic, then a Markov kernel fulfilling (1.1.2) is a conditional distribution, given S (i.e. it fulfills (1.3.4)), provided $\mathcal{A}$ is countably generated. (See Proposition 1.4.9.)

In Section 1.1 the concept of an exhaustive statistic was motivated by the fact that for every statistical procedure based on x there exists an equivalent (randomized) statistical procedure based on $S(x)$. If we now restrict our attention to the particular case of tests, it is enough to require that for every test based on x there exists a test based on $S(x)$ with the same power function. The following theorem (see Pfanzagl, 1974) shows that less is enough to establish that S is sufficient.

Theorem 1.3.9. *The statistic* $S : (X, \mathcal{A}) \to (Y, \mathcal{B})$ *is sufficient for* $\mathfrak{P}$ *iff for every pair* $P', P'' \in \mathfrak{P}$ *and every* $A \in \mathcal{A}$ *there exists* $\varphi_A : (Y, \mathcal{B}) \to ([0, 1], \mathbb{B}_0)$ *such that*

$$P' \circ S(\varphi_A) \leq P'(A) \quad \text{and} \quad P'' \circ S(\varphi_A) \geq P''(A). \tag{1.3.10}$$

Corollary 1.3.11. (i) *An exhaustive statistic is sufficient.* (ii) *A sufficient statistic is exhaustive if* $(X, \mathcal{A})$ *is a Polish spa ce.*

Proof of Theorem 1.3.9. (i) Let $P \in \mathfrak{P}$ and $A \in \mathcal{A}$ be fixed. Let P_* be defined as in Lemma 1.2.1. By assumption (1.3.10) for every $n \in \mathbb{N}$ there exists $\varphi_{A,n} : (Y, \mathcal{B}) \to ([0, 1], \mathbb{B}_0)$ such that

$$P \circ S(\varphi_{A,n}) \leq P(A) \quad \text{and} \quad P_n \circ S(\varphi_{A,n}) \geq P_n(A).$$

By Lemma 1.3.12 and Corollary 1.3.14, applied with $\varphi_{A,n} \circ S$ in place of $\hat{\varphi}_A$, there exists a density of P with respect to $P + P_n$, say p_n, which is $S^{-1}\mathcal{B}$-measurable. By Lemmas 1.3.18 and 1.10.6(i) this implies the existence of a density of P with respect to P_* which depends on x through $S(x)$ only. Hence S is sufficient for $\mathfrak{P}$.

(ii) If S is sufficient for $\mathfrak{P}$, there exists $\varphi_A \in \cap\{P^S 1_A : P \in \mathfrak{P}\}$, which implies

$$\int \varphi_A(y) P \circ S(dy) = P(A) \quad \text{for } A \in \mathcal{A}, \ P \in \mathfrak{P}.$$

$\square$

Lemma 1.3.12. *Let* P', P'' *be p-measures on* $\mathcal{A}$. *Assume that for every* $A \in \mathcal{A}$ *there exists* $\hat{\varphi}_A : (X, \mathcal{A}) \to ([0, 1], \mathbb{B}_0)$ *such that*

$$P'(\hat{\varphi}_A) \leq P'(A) \quad \text{and} \quad P''(\hat{\varphi}_A) \geq P''(A). \tag{1.3.13}$$

Then the densities of $P'|\mathcal{A}$ *and* $P''|\mathcal{A}$ *with respect to* $(P' + P'')|\mathcal{A}$ *are measurable with respect to the* σ-*algebra generated by* $\{A \in \mathcal{A} : \hat{\varphi}_A = 1_A \ (P' + P'')\text{-}a.e.\}$.

Corollary 1.3.14. *If* $\hat{\varphi}_A$ *is* $\mathcal{A}_0$-*measurable for every* $A \in \mathcal{A}$, *then there exist* $\mathcal{A}_0$-*measurable densities of* $P'|\mathcal{A}$ *and* $P''|\mathcal{A}$ *with respect to* $(P' + P'')|\mathcal{A}$.

Proof of Lemma 1.3.12. Since the assumption is symmetric with respect to P' and P'', it suffices to prove the assertion for P'. Let p' be a density of $P'|\mathcal{A}$

with respect to $(P' + P'')|\mathcal{A}$. For $r > 0$ let

$$A_r := \{x \in X : p'(x) < r(1 - p'(x))\}. \tag{1.3.15}$$

By (1.3.13) there exists $\hat{\varphi}_{A_r}$ such that

$$P'(\hat{\varphi}_{A_r}) \leq P'(A_r) \quad \text{and} \quad P''(\hat{\varphi}_{A_r}) \geq P''(A_r). \tag{1.3.16}$$

Since 1_{A_r} is of Neyman–Pearson type for $P':P''$, it is most powerful for $P':P''$, hence equality holds in (1.3.16) by Lemma 4.3.3(ii'). By Lemma 4.3.9 this implies

$$\hat{\varphi}_{A_r} = 1_{A_r} \quad (P' + P'')\text{-a.e.} \tag{1.3.17}$$

Let $\hat{\mathcal{A}}$ denote the σ–algebra generated by $\{A \in \mathcal{A} : \varphi_A = 1_A \ (P'+P'')\text{- a.e}\}$. Relation (1.3.17) implies $A_r \in \hat{\mathcal{A}}$. Since $r > 0$ was arbitrary, p' is $\hat{\mathcal{A}}$-measurable. $\square$

Proof of Corollary 1.3.14. $\varphi_A = 1_A \ (P' + P'')$–a.e. implies $A = \{x \in X : \hat{\varphi}_A(x) = 1\} \ (P' + P'')$. Since $\hat{\varphi}_A$ is $\mathcal{A}_0$-measurable, this implies $A \in \mathcal{A}_0 \ (P' + P'')$ (i.e. there exists $A_0 \in \mathcal{A}_0$ such that $(P' + P'')(A \triangle A_0) = 0$, where "$\triangle$" denotes the symmetric difference of sets). Hence $\hat{\mathcal{A}}$ is in the completion of $\mathcal{A}_0$ with respect to $(P' + P'')$. Since p' is $\hat{\mathcal{A}}$-measurable, there exists (see Lemma 1.10.3) an $\mathcal{A}_0$-measurable function which agrees $(P'+P'')$-a.e. with p'. This is the $\mathcal{A}_0$-measurable version of the density of $P'|\mathcal{A}$ with respect to $(P' + P'')|\mathcal{A}$. $\square$

The following lemma is due to Halmos and Savage (1949, p. 239, Lemma 12).

Lemma 1.3.18. *Let $\mathcal{A}_0 \subset \mathcal{A}$ be a sub-σ–algebra. Let $\mathfrak{P}$ be a dominated family of p-measures on $\mathcal{A}$. Assume that for every pair $P', P'' \in \mathfrak{P}$ there exists a density of $P'|\mathcal{A}$ with respect to $(P' + P'')|\mathcal{A}$ which is $\mathcal{A}_0$-measurable.*

Then for every $P \in \mathfrak{P}$ there exists a density with respect to $P_*|\mathcal{A}$ (defined in Lemma 1.2.1) which is $\mathcal{A}_0$-measurable.

Proof. Let $P_* = \sum_1^\infty 2^{-n} P_n$. For $n \in \mathbb{N}$ let $p_n \in dP/d(P + P_n)$ be $\mathcal{A}_0$-measurable. W.l.g. we assume $0 \leq p_n \leq 1$. From $P(A) = (P + P_n)(p_n 1_A)$ we obtain $P((1 - p_n)1_A) = P_n(p_n 1_A)$ for $A \in \mathcal{A}$, hence

$$P((1 - p_n)f) = P_n(p_n f) \quad \text{for every } \mathcal{A}\text{-measurable function } f \geq 0. \tag{1.3.19}$$

With $A_n := \{p_n = 1\}$ and $B_n := \{p_n = 0\}$ we obtain

$$P_n(A_n) = P_n(p_n 1_{A_n}) = P((1 - p_n)1_{A_n}) = 0$$

and

$$P(B_n) = P\big((1 - p_n)1_{B_n}\big) = P_n(p_n 1_{B_n}) = 0.$$

Let $A_* := \bigcap_1^\infty A_n$, $B_* := \bigcup_1^\infty B_n$. Then $P_n(A_*) = 0$ for $n \in \mathbb{N}$ implies $P_*(A_*) = 0$ and therefore $P(A_*) = 0$. Moreover, $P(B_n) = 0$ for $n \in \mathbb{N}$ implies $P(B_*) = 0$. Let

$$h_*(x) := \begin{cases} \left(\sum_1^\infty 2^{-n}\big(1 - p_n(x)\big)/p_n(x) \right)^{-1} & \text{for } x \underset{\notin}{\in} A_*^c \cap B_*^c. \\ 0 \end{cases}$$

From (1.3.19), applied with $f = 1_{A_*^c \cap B_*^c} 1_A h_*/p_n$, we obtain

$$P\big((1 - p_n)\frac{h_*}{p_n}1_{A_*^c \cap B_*^c}1_A\big) = P_n\big(h_* 1_{A_*^c \cap B_*^c}1_A\big).$$

Multiplication by 2^{-n} and summation over $n \in \mathbb{N}$ yields

$$P\big(1_{A_*^c \cap B_*^c}1_A\big) = P_*\big(h_* 1_{A_*^c \cap B_*^c}1_A\big). \tag{1.3.20}$$

Since $P(A_*^c \cap B_*^c) = 1$, we have $P\big(1_{A_*^c \cap B_*^c}1_A\big) = P(A)$. Since $h_*(x) = 0$ for $x \notin A_*^c \cap B_*^c$, we have $P\big(h_* 1_{A_*^c \cap B_*^c}1_A\big) = P_*(h_* 1_A)$. Hence (1.3.20) implies $P(A) = P_*(h_* 1_A)$, i.e. $h_* \in dP/dP_*$. Since h_* is $\mathcal{A}_0$–measurable, this proves the assertion. □

1.4 Minimal sufficiency

Introduces the concept of a "minimal" sufficient statistic, rendering a maximal reduction, and gives a criterion for "minimality".

Throughout this section, $\mathfrak{P}$ is a dominated family of p–measures on $(X, \mathcal{A})$.

To obtain a maximal reduction of the data x to $S(x)$ one should try to find a sufficient statistic S_* which is coarser than any other sufficient statistic.

Definition 1.4.1. The sufficient statistic $S_* : (X, \mathcal{A}) \to (Y_0, \mathcal{B}_0)$ is *minimal sufficient* if for any sufficient statistic $S : (X, \mathcal{A}) \to (Y, \mathcal{B})$ there exists a function $H : (Y, \mathcal{B}) \to (Y_0, \mathcal{B}_0)$ such that $S_* = H \circ S$ $\mathfrak{P}$–a.e.

Notice that some authors require $H : \big(S(X), \mathcal{B} \cap S(X)\big)$ only.

If S is sufficient, then with P_* defined by (1.2.2) there exist (by Factorization Theorem 1.2.10 and Proposition 1.2.9) P_*–densities of P which depend on x through $S(x)$ only. This suggests to start from densities $q_P \in dP/dP_*$ and to consider the partition consisting of the elements $\{\xi \in X : h_P(\xi) =$

$h_P(x)$ for $P \in \mathfrak{P}\}$, $x \in X$, with the intention to find a statistic $S_* | (X, \mathcal{A})$ inducing this partition (i.e., a statistic which is constant on each element, and attains different values on disjoint elements of this partition). Since densities q_P are unique up to P–null sets only, and the definition of a partition given above involves uncountably many $P \in \mathfrak{P}$, this intuitive idea has to be modified for technical reasons.

In the following we show first that a minimal sufficient statistic exists under mild conditions (see Theorem 1.4.2). More useful for practical purposes is Theorem 1.4.4 which can be applied to show that a given sufficient statistic is, in fact, minimal. Finally we show that a sufficient statistic S with $\mathfrak{P} \circ S$ boundedly complete is necessarily minimal (Proposition 1.4.8). These results are essentially due to Lehmann and Scheffé (1950, Section 6).

Theorem 1.4.2. *If $\mathcal{A}$ is countably generated, there exists a minimal sufficient statistic $S_* : (X, \mathcal{A}) \to (\mathbb{R}_+^{\mathbb{N}}, \mathbb{B}_+^{\mathbb{N}})$.*

Proof. Since $\mathcal{A}$ is countably generated, there exists a countable subset $\mathfrak{P}_0 := \{P_n : n \in \mathbb{N}\} \subset \mathfrak{P}$, which is dense in $\mathfrak{P}$ with respect to the sup–metric, such that $\mathfrak{P} \ll P_* := \sum_1^\infty 2^{-n} P_n$ (see Remark 1.2.4).

For $n \in \mathbb{N}$ let $p_n \in dP_n/dP_*$ be a fixed version. Let $S_* : X \to \mathbb{R}_+^{\mathbb{N}}$ be defined by

$$S_*(x) := \big(p_n(x)\big)_{n \in \mathbb{N}}, \qquad x \in X. \tag{1.4.3}$$

(i) S_* is sufficient for $\mathfrak{P}_0$, since $p_n(x) = \Pi_n\big(S_*(x)\big)$, $x \in X$ (with Π_n denoting the projection of $\mathbb{R}_+^{\mathbb{N}}$ onto its n–th component). Since $\mathfrak{P}_0$ is dense in $\mathfrak{P}$, S_* is sufficient for $\mathfrak{P}$ by Exercise 1.2.6(i).

(ii) If $S : (X, \mathcal{A}) \to (Y, \mathcal{B})$ is sufficient, there exist measurable functions $h_n : (Y, \mathcal{B}) \to (\mathbb{R}_+, \mathbb{B}_+)$ such that $h_n \circ S \in dP_n/dP_*$.

With the function $H : Y \to \mathbb{R}_+^{\mathbb{N}}$, defined by

$$H(y) = \big(h_n(y)\big)_{n \in \mathbb{N}}, \quad y \in Y,$$

we obtain

$$H\big(S(x)\big) = \big(h_n(S(x))\big)_{n \in \mathbb{N}}$$
$$= \big(p_n(x)\big)_{n \in \mathbb{N}} = S_*(x) \quad \text{for } P_*\text{–a.a. } x \in X,$$

hence for $\mathfrak{P}$–a.a. $x \in X$. $\qquad\square$

Since there exists a bimeasurable 1–1 map from $(\mathbb{R}_+^{\mathbb{N}}, \mathbb{B}_+^{\mathbb{N}})$ to $(\mathbb{R}, \mathbb{B})$ (see Parthasarathy, 1967, p. 14, Theorem 2.12), the minimal sufficient statistic given in Theorem 1.4.2 can always be assumed to be real valued. In spite of this, Theorem 1.4.2 is hardly useful for practical purposes. One will always prefer more natural versions of the minimal sufficient statistic, for instance continuous functions if X is a topological space (even if these statistics attain their values in $\mathbb{R}^k$, say, rather than in $\mathbb{R}$). Hence more useful than the general existence Theorem 1.4.2 are theorems which provide a tool for identifying a given sufficient statistic as a minimal one.

Theorem 1.4.4. *Let $\mathcal{A}$ be countably generated, and let $(Y, \mathcal{B})$ be a Polish space. Assume that $S : (X, \mathcal{A}) \to (Y, \mathcal{B})$ is sufficient for $\mathfrak{P}$, and that the functions h_P occurring in factorization (1.2.8) have the following property. There exists a countable subset $\mathfrak{P}_0 \subset \mathfrak{P}$ such that, for $y', y'' \in Y$,*

$$h_P(y') = h_P(y'') \quad \text{for all } P \in \mathfrak{P}_0, \text{ implies } y' = y''.$$

Then S is minimal sufficient.

Proof. W.l.g. we assume that $\mathfrak{P}_0 = \{P_n : n \in \mathbb{N}\}$ is dense in $\mathfrak{P}$ with respect to the sup–metric.

Let $H : Y \to \mathbb{R}_+^{\mathbb{N}}$ be defined by $H(y) = \big(h_{P_n}(y)\big)_{n \in \mathbb{N}}$. As can be seen from the proof of Theorem 1.4.2, $S_*(x) := H\big(S(x)\big)$ is minimal sufficient. By assumption, $y' \neq y''$ implies $H(y') \neq H(y'')$, hence H is a measurable 1–1 map of the Polish space Y into the Polish space $\mathbb{R}_+^{\mathbb{N}}$, endowed, for instance, with the metric

$$d\big((x_\nu)_{\nu \in \mathbb{N}}, \ (y_\nu)_{\nu \in \mathbb{N}}\big) = \sum_1^\infty 2^{-n} \frac{|x_\nu - y_\nu|}{1 + |x_\nu - y_\nu|} \ .$$

It is easy to see that the pertaining Borel algebra of $\mathbb{R}_+^{\mathbb{N}}$ is $\mathbb{B}_+^{\mathbb{N}}$. By Kuratowski's theorem (see, e.g., Parthasarathy, 1967, p. 21, Theorem 3.9), $H(Y) \in \mathbb{B}_+^{\mathbb{N}}$. Hence the function $\hat{H} : \mathbb{R}_+^{\mathbb{N}} \to Y$, defined by $\hat{H}(z) = H^{-1}(z)$ for $z \in H(Y)$, and $\hat{H}(z) = y_0$ (say) for $z \in H(Y)^c$, is measurable. Since $\hat{H}\big(H(y)\big) = y$ for $y \in Y$, we obtain

$$\hat{H}\big(S_*(x)\big) = \hat{H}\big(H(S(x))\big) = S(x) \quad \text{for all } x \in X.$$

Since S_* is minimal sufficient, S is minimal sufficient, too.   □

Example 1.4.5 (Logistic distribution). For $\vartheta \in \mathbb{R}$ let $P_\vartheta | \mathbb{B}$ denote the logistic distribution with location parameter ϑ. P_ϑ has Lebesgue density

$$x \to \frac{\exp[-(x - \vartheta)]}{(1 + \exp[-(x - \vartheta)])^2} \ , \qquad x \in \mathbb{R}.$$

We shall show that the order statistic is minimal sufficient for $\{P_\vartheta^n : \vartheta \in \mathbb{R}\}$.

$$h_\vartheta(x_1, \ldots, x_n) := \exp[n\vartheta] \prod_{\nu=1}^n \left(\frac{1 + \exp[-x_\nu]}{1 + \exp[-(x_\nu - \vartheta)]} \right)^2$$

is a P_0–density of P_ϑ. We shall show that

$$h_\vartheta(x_1', \ldots, x_n') = h_\vartheta(x_1'', \ldots, x_n'') \quad \text{for all } \vartheta \in \mathbb{Q} \tag{1.4.6}$$

implies that $(x_1', \ldots, x_n')$ equals $(x_1'', \ldots, x_n'')$ except for the order. Then minimal sufficiency of the order statistic follows from Theorem 1.4.4.

Since $\vartheta \to h_\vartheta(x_1, \ldots, x_n)$ is continuous, relation (1.4.6) implies

$$\prod_{\nu=1}^{n} \frac{1 - \xi \exp[-x_\nu']}{1 + \exp[-x_\nu']} = \prod_{\nu=1}^{n} \frac{1 - \xi \exp[-x_\nu'']}{1 + \exp[-x_\nu'']} \quad \text{for every } \xi < 0.$$

Either of these expressions is a polynomial in ξ with roots $\exp[x_\nu']$, $\nu = 1, \ldots, n$ and $\exp[x_\nu'']$, $\nu = 1, \ldots, n$, respectively. Since these polynomials agree for $\xi < 0$, the roots are identical up to the order.

Exercise 1.4.7 (Cauchy distribution). For $\vartheta \in \mathbb{R}$ let $P_\vartheta | \mathbb{B}$ denote the Cauchy distribution with location parameter ϑ. P_ϑ has Lebesgue density

$$x \to \frac{1}{\pi} \frac{1}{1 + (x - \vartheta)^2}, \quad x \in \mathbb{R}.$$

Show that the order statistic is minimal sufficient for $\{P_\vartheta^n : \vartheta \in \mathbb{R}\}$.

The following proposition can be applied to establish minimal sufficiency of certain statistics for exponential families (see Theorem 1.6.10) and minimal sufficiency of the order statistic in certain nonparametric families (see Theorem 1.5.10). It is based on the concept of bounded completeness given in Definition 1.5.1.

Proposition 1.4.8. *Let $S : (X, \mathcal{A}) \to (Y, \mathcal{B})$ be measurable, where $(Y, \mathcal{B})$ is a Polish space. If S is sufficient for $\mathfrak{P}$ and $\mathfrak{P} \circ S$ is boundedly complete, then S is minimal sufficient.*

Proof. Let $T : (X, \mathcal{A}) \to (Z, \mathcal{C})$ be a sufficient statistic. Hence for every $A \in \mathcal{A}$ there exists $\varphi_A : (Z, \mathcal{C}) \to ([0, 1], \mathbb{B}_0)$ such that $P(\varphi_A \circ T) = P(A)$ for every $P \in \mathfrak{P}$.

Since S is sufficient, there exists a conditional expectation of $\varphi_A \circ T$, given S, say ψ_A, which does not depend on $P \in \mathfrak{P}$. We have in particular $P(\psi_A \circ S) = P(\varphi_A \circ T)$ for $P \in \mathfrak{P}$, hence also $P(\psi_A \circ S) = P(A)$ for $P \in \mathfrak{P}$.

For $A = S^{-1}B$ (with $B \in \mathcal{B}$) we obtain

$$P(\psi_{S^{-1}B} \circ S) = P(1_{S^{-1}B}) = P(1_B \circ S) \quad \text{for } P \in \mathfrak{P}.$$

By bounded completeness of $\mathfrak{P} \circ S$ this implies $\psi_{S^{-1}B} = 1_B \quad P_* \circ S$–a.e., with P_* given by Lemma 1.2.1.

Hence 1_B is a conditional expectation of $\varphi_{S^{-1}B} \circ T$, given S, with respect to P_*.

If 1_B is a conditional expectation, given S, of some function $\varphi : (X, \mathcal{A}) \to ([0, 1], \mathbb{B}_0)$, we have

$$\int 1_B(y) 1_{B_0}(y) P_* \circ S(dy) = \int \varphi(x) 1_{B_0}(S(x)) P_*(dx) \quad \text{for } B_0 \in \mathcal{B}.$$

Applied for $B_0 = B$ and $B_0 = B^c$, this yields $\varphi = 1_B \circ S$ P_*-a.e.

This implies $\varphi_{S^{-1}B} \circ T = 1_B \circ S$ P_*-a.e., so that $S^{-1}\mathcal{B} \subset T^{-1}\mathcal{C}$ (P_*) (i.e., for $B \in \mathcal{B}$ there exists $C \in \mathcal{C}$ such that $P_*(S^{-1}B \triangle T^{-1}C) = 0$). By Lemmas 1.10.3 and 1.10.6(ii) this implies the existence of a function $H : (Z, \mathcal{C}) \to (Y, \mathcal{B})$ such that $S = H \circ T$ P_*-a.e. This proves that S is minimal sufficient. □

According to Corollary 1.3.11(i), any exhaustive statistic is sufficient. This does, however, not involve that the Markov kernel occurring in the definition of exhaustivity is a conditional distribution of x, given S. The situation is different if the sufficient statistic S is minimal.

Proposition 1.4.9. *Assume that $\mathcal{A}$ is countably generated, and $S : (X, \mathcal{A}) \to (Y, \mathcal{B})$ is a minimal sufficient statistic. Then any Markov kernel $M|Y \times \mathcal{A}$ fulfilling*

$$\int M(y, A) P \circ S(dy) = P(A) \quad \text{for } A \in \mathcal{A}, \ P \in \mathfrak{P}, \tag{1.4.10}$$

is a conditional distribution of x, given S, and therefore unique in the following sense: If $M_i|Y \times \mathcal{A}$, $i = 1, 2$, fulfill (1.4.10), then $M_1(y, \cdot) = M_2(y, \cdot)$ for $\mathfrak{P} \circ S$-a.a. $y \in Y$.

Proof. From Lemma 1.4.11, applied with $\varphi_A(x) = M(S(x), A)$ and P_* in place of P, we obtain

$$M(S(x), A) 1_{A_0}(x) = M(S(x), A \cap A_0) \quad \text{for } P_*\text{-a.a. } x \in X,$$

for $A \in \mathcal{A}$ and $A_0 \in \hat{\mathcal{A}} := \{A \in \mathcal{A} : M(S(\cdot), A) = 1_A \ P_*\text{-a.e.}\}$.

Since M is a Markov kernel, $\hat{\mathcal{A}}$ is a σ-algebra. By integration with respect to P we obtain

$$\int M(S(x), A) 1_{A_0}(x) P(dx) = P(A \cap A_0) \quad \text{for } A \in \mathcal{A}, \ A_0 \in \hat{\mathcal{A}}, \ P \in \mathfrak{P}.$$

That $M(\cdot, A)$ is a conditional expectation of 1_A, given S, follows if we prove that $S^{-1}\mathcal{B} \subset \hat{\mathcal{A}}$ (P_*). This follows from the minimality of S. From Lemmas 1.3.12 and 1.3.18, applied with $\hat{\varphi}_A(x) = M(S(x), A)$, $P' = P$ and $P'' = P_*$, we obtain that the density of $P|\mathcal{A}$ with respect to $P_*|\mathcal{A}$ is $\hat{\mathcal{A}}$-measurable. Since $\mathcal{A}$ is countably generated, the map $x \to (p_n(x))_{n \in \mathbb{N}}$ is sufficient (by Theorem

1.4.2). Since S is minimal sufficient, there exists $H : (\mathbb{R}_+^{\mathbb{N}}, \mathbb{B}_+^{\mathbb{N}}) \to (Y, \mathcal{B})$ such that $S(x) = H\big((p_n(x))_{n \in \mathbb{N}}\big)$ for P_*–a.a. $x \in X$. Since p_n is $\hat{\mathcal{A}}$–measurable, this implies $S^{-1}\mathcal{B} \subset \hat{\mathcal{A}}$ (P_*). □

Lemma 1.4.11. *Assume that for every $A \in \mathcal{A}$ there exists $\varphi_A : (X, \mathcal{A}) \to ([0,1], \mathbb{B}_0)$ with the following properties*
(i) $P(\varphi_A) = P(A)$ for $A \in \mathcal{A}$,
(ii) $\varphi_{A+B} = \varphi_A + \varphi_B$ P–a.e. if $A \cap B = \emptyset$.

   Let $\hat{\mathcal{A}} := \{A \in \mathcal{A} : \varphi_A = 1_A$ P–a.e. $\}$. Then

$$\varphi_{A \cap A_0} = \varphi_A 1_{A_0} \quad P\text{–a.e. for } A \in \mathcal{A}, \ A_0 \in \hat{\mathcal{A}}.$$

Proof. For every $A \in \mathcal{A}$, $A_0 \in \hat{\mathcal{A}}$

$$\varphi_{A \cap A_0} \underset{(P)}{\leq} \min\{\varphi_A, \varphi_{A_0}\} \underset{(P)}{=} \min\{\varphi_A, 1_{A_0}\} = \varphi_A 1_{A_0}.$$

Since $A_0 \in \hat{\mathcal{A}}$ implies $A_0^c \in \hat{\mathcal{A}}$, we have $\varphi_{A \cap A_0^c} \underset{(P)}{\leq} \varphi_A 1_{A_0^c}$, hence

$$\varphi_A \underset{(P)}{=} \varphi_{A \cap A_0} + \varphi_{A \cap A_0^c} \underset{(P)}{\leq} \varphi_A 1_{A_0} + \varphi_A 1_{A_0^c} = \varphi_A,$$

and therefore $\varphi_{A \cap A_0} = \varphi_A 1_{A_0}$ P–a.e. □

1.5 Completeness

Introduces the concept of [boundedly] complete families of p–measures and gives sufficient criteria for completeness and symmetric completeness.

Definition 1.5.1. The family of p–measures $\mathfrak{P} \mid \mathcal{A}$ is *[boundedly] complete* if for every [bounded] function $f : (X, \mathcal{A}) \to (\mathbb{R}, \mathbb{B})$, $P(f) = 0$ for all $P \in \mathfrak{P}$ implies $f = 0$ $\mathfrak{P}$–a.e.

If convenient, we follow the common abuse of language and speak of a [boundedly] complete statistic S if the family $\mathfrak{P} \circ S$ is [boundedly] complete.

The need for two different concepts, namely "completeness" and "bounded completeness", originates from statistical applications: Test theory uses "critical regions" or, more generally, "critical functions" which are bounded, whereas estimators are usually unbounded.

The concepts of completeness and bounded completeness were introduced by Lehmann and Scheffé (1947, 1950). They isolate the properties of families of p–

measures which are essential for the uniqueness of optimal unbiased estimators and optimal similar tests, respectively (which were implicitly used before by Scheffé, 1943 and Halmos, 1946).

Every complete family is boundedly complete, and there are examples of families which are boundedly complete without being complete. (A first example of this kind was given by Lehmann and Scheffé, 1950, p. 312, Example 3.1. For some recent results see Mattner, 1991.)

Lemma 1.5.2. *Let $\mathfrak{F}$ be a class of functions $f : (X, \mathcal{A}) \to (\mathbb{R}, \mathbb{B})$. The family $\mathfrak{P}$ of all p-measures $P|\mathcal{A}$ which are equivalent to some σ-finite measure μ, and which fulfill the condition $P(|f|) < \infty$ for $f \in \mathfrak{F}$, is complete.*

Proof. Assume there exists a function $f : (X, \mathcal{A}) \to (\mathbb{R}, \mathbb{B})$ with $P(|f|) < \infty$ for $P \in \mathfrak{P}$. Assume that $P_0\{f \neq 0\} > 0$ for some (and therefore all) $P_0 \in \mathfrak{P}$. We shall show that there exists $P_1 \in \mathfrak{P}$ such that $P_1(f) \neq 0$. If $P_0\{f > 0\} > 0$, there exists a p-measure P_1 with P_0-density $c'1_{\{f>0\}} + c''1_{\{f\leq 0\}}$ such that $P_1(f) > 0$. Since $P_1(|f|) \leq \max\{c', c''\}P_0(|f|) < \infty$, we have $P_1 \in \mathfrak{P}$.   □

Exercise 1.5.3. If $\mathfrak{P}_0 \subset \mathfrak{P}$ is [boundedly] complete and if $P(A) = 0$ for all $P \in \mathfrak{P}_0$ implies $P(A) = 0$ for all $P \in \mathfrak{P}$, then $\mathfrak{P}$ is [boundedly] complete

Regrettably, it is usually not an easy task to find out whether a given family of p-measures is complete or not. Moreover, the routine applications in statistical theory involve samples of arbitrary size. Of course, families of product measures with identical components, say $\{P^n : P \in \mathfrak{P}\}$, are usually not complete, even if $\mathfrak{P}$ is large. The relevant question is whether there is a sufficient statistic $S_n|X^n$ such that $\{P^n \circ S_n : P \in \mathfrak{P}\}$ is complete.

The most useful result for parametric theory concerns the completeness of various statistics in exponential families. It will be given in Theorem 1.6.10.

If $(x_1, \ldots, x_n)$ is a realization from a p-measure P^n, the order in which $x_1, \ldots, x_n$ occur is irrelevant: For any permutation $(i_1, \ldots, i_n)$ of $(1, \ldots, n)$, the realization $(x_{i_1}, \ldots, x_{i_n})$ has the same distribution as $(x_1, \ldots, x_n)$. A function $S_n|X^n$ which is maximal invariant under permutations (i.e. fulfills $S_n(x'_1, \ldots, x'_n) = S_n(x''_1, \ldots, x''_n)$ iff there is a permutation such that $x''_{i_j} = x'_j$ for $j=1, \ldots, n$) will be called *order statistic* (a name originating from the case $X \subset \mathbb{R}$ in which $S_n(x_1, \ldots, x_n)$ can be thought of as mapping $(x_1, \ldots, x_n)$ to $(x_{1:n}, \ldots \ldots, x_{n:n})$). Obviously, the order statistic is sufficient for every family $\{P^n : P \in \mathfrak{P}\}$. (This, by the way, is a case where the concept of a "sufficient sub-σ-algebra" — the sub-σ-algebra of all measurable sets in $\mathcal{A}^n$ which are invariant under permutations — is more natural than the concept of a "sufficient statistic".)

The functions on X^n depending on $(x_1, \ldots, x_n)$ through the order statistic are the functions of $(x_1, \ldots, x_n)$ which are invariant under all permutations of $(x_1, \ldots, x_n)$. "Completeness of the order statistic" for $\{P^n : P \in \mathfrak{P}\}$ is, therefore, the same as "symmetric completeness" of $\{P^n : P \in \mathfrak{P}\}$.

Definition 1.5.4. $\{P^n : P \in \mathfrak{P}\}$ is *symmetrically complete* if for any permutation invariant function $f_n | X^n$,

$$\int f_n(x_1, \ldots, x_n) P(dx_1) \ldots P(dx_n) = 0 \qquad \text{for all } P \in \mathfrak{P}$$

implies $f_n = 0$ P^n–a.e. for $P \in \mathfrak{P}$.

Exercise 1.5.5. If $\{P^n : P \in \mathfrak{P}\}$ is [boundedly] symmetrically complete for some sample size n, then $\mathfrak{P}$ is [boundedly] complete.

Symmetric completeness of $\{P^n : P \in \mathfrak{P}\}$ can be expected only if the family $\mathfrak{P}$ is very large. Completeness of $\mathfrak{P}$ is certainly a necessary condition for completeness of $\{P^n : P \in \mathfrak{P}\}$. If $\mathfrak{P}$ is complete and closed under convex combinations, this suffices for symmetric completeness for arbitrary sample sizes. This is the content of Theorem 1.5.10.

For smaller families $\mathfrak{P}$, $\{P^n : P \in \mathfrak{P}\}$ may fail to be symmetrically complete, even if the order statistic S_n is minimal sufficient. As an example consider a location parameter family $\mathfrak{P} = \{P_\vartheta : \vartheta \in \mathbb{R}\}$ with $P_\vartheta = P_0 \circ (x \to x + \vartheta)$. If P_0 is the Cauchy– or the logistic distribution, S_n is minimal sufficient for $\{P_\vartheta^n : \vartheta \in \mathbb{R}\}$ (see Example 1.4.5 and Exercise 1.4.7). Yet the order statistic fails to be complete for any location parameter family. (Let $\varphi | \mathbb{R}$ be a bounded function with $\varphi(x) + \varphi(-x) \neq 0$ and $\int \varphi(x_1 - x_2) P_0(dx_1) P_0(dx_2) = 0$. Then $\psi_2(x_1, x_2) := \varphi(x_1 - x_2) + \varphi(x_2 - x_1)$ is permutation invariant and $\int \psi_2(x_1, x_2) P_\vartheta(dx_1) P_\vartheta(dx_2) = 0$ for $\vartheta \in \mathbb{R}$.)

Some recent results on [bounded] completeness for location parameter families, sample size $n = 1$, can be found in Isenbeck and Rüschendorf (1992), and Mattner (1993).

The proof of Theorem 1.5.10 is prepared by two auxiliary results. (See Landers and Rogge, 1976.)

Proposition 1.5.6. *For $i \in \{1, \ldots, n\}$ let $(X_i, \mathcal{A}_i)$ be a measurable space and $\mathfrak{P}_i | \mathcal{A}_i$ a complete family of p–measures. Then the family*

$$\{ \underset{i=1}{\overset{n}{\times}} P_i : P_i \in \mathfrak{P}_i, \quad i = 1, \ldots, n \}$$

is complete.

Proof. Assume that for some measurable function $f_n : \left(\underset{1}{\overset{n}{\times}} X_i, \underset{1}{\overset{n}{\times}} \mathcal{A}_i \right) \to (\mathbb{R}, \mathbb{B})$

$$\left(\underset{i=1}{\overset{n}{\times}} P_i \right)(f_n) = 0 \qquad \text{for } P_i \in \mathfrak{P}_i, \; i = 1, \ldots, n. \tag{1.5.7}$$

We shall show that (1.5.7) implies

$$\int f_n(x_1, \ldots, x_n) \prod_1^n 1_{A_i}(x_i) P_1(dx_1) \ldots P_n(dx_n) = 0 \tag{1.5.8}$$

$$\text{for } A_i \in \mathcal{A}_i \text{ and } P_i \in \mathfrak{P}_i, \; i = 1, \ldots, n.$$

The system of all sets $A \in \underset{1}{\overset{n}{\times}} \mathcal{A}_i$ for which

$$\int f_n(x_1, \ldots, x_n) 1_A(x_1, \ldots, x_n) P_1(dx_1) \ldots P_n(dx_n) = 0$$

is a Dynkin system. By (1.5.8) it contains all sets $\underset{1}{\overset{n}{\times}} A_i$ with $A_i \in \mathcal{A}_i$ and is therefore equal to $\underset{1}{\overset{n}{\times}} \mathcal{A}_i$. This implies $f_n = 0 \; \underset{1}{\overset{n}{\times}} P_i$–a.e.

It remains to prove (1.5.8). Let

$$g(x_1) := \int f_n(x_1, x_2, \ldots, x_n) P_2(dx_2) \ldots P_n(dx_n),$$

with $P_i \in \mathfrak{P}_i$, $i = 2, \ldots, n$, arbitrary and fixed. Since $\int g(x_1) P_1(dx) = 0$ for all $P_1 \in \mathfrak{P}_1$, completeness of $\mathfrak{P}_1$ implies $g = 0$ $\mathfrak{P}_1$–a.e.

Therefore, $\int g(x_1) 1_{A_1}(x_1) P_1(dx_1) = 0$ for $A_1 \in \mathcal{A}_1$ and $P_1 \in \mathfrak{P}$. This implies

$$\int f_n(x_1, \ldots, x_n) 1_{A_1}(x_1) P_1(dx_1) P_2(dx_2) \ldots P_n(dx_n) = 0.$$

(1.5.8) follows by induction. $\qquad\qquad\qquad\qquad\qquad\qquad\qquad\qquad\qquad\quad$ $\square$

Lemma 1.5.9. *If $\mathfrak{P}$ is closed under convex combinations, the following holds true for every $n \in \mathbb{N}$: If a measurable function $f_n : X^n \to \mathbb{R}$ is permutation invariant, then $P^n(f_n) = 0$ for all $P \in \mathfrak{P}$ implies $\left(\underset{1}{\overset{n}{\times}} P_i \right)(f_n) = 0$ for all $P_i \in \mathfrak{P}$, $i = 1, \ldots, n$.*

Proof. $\alpha P + (1 - \alpha)Q \in \mathfrak{P}$ for $P, Q \in \mathfrak{P}$, $\alpha \in [0, 1]$ implies $\sum_1^n \alpha_i P_i \in \mathfrak{P}$ for $P_i \in \mathfrak{P}$, $i = 1, \ldots, n$ and $\alpha_i \geq 0$, $i = 1, \ldots, n$, with $\sum_1^n \alpha_i = 1$. Hence

$$0 = \left(\sum_1^n \alpha_i P_i \right)^n(f_n) = \sum_{i_1=1}^n \ldots \sum_{i_n=1}^n \alpha_{i_1} \ldots \alpha_{i_n} \left(\underset{j=1}{\overset{n}{\times}} P_{i_j} \right)(f_n).$$

The right hand side is homogeneous in $(\alpha_1, \ldots, \alpha_n)$. Hence this relation also holds true without the restriction $\sum_1^n \alpha_i = 1$. Being a polynomial in $(\alpha_1, \ldots, \alpha_n)$ which is identically 0 for all $\alpha_i \geq 0$, $i = 1, \ldots, n$, its coefficients are 0.

The coefficient of $\prod_1^n \alpha_i$ is

$$\sum_{(i_1,\ldots,i_n)} \left(\mathop{\times}_{j=1}^{n} P_{i_j} \right)(f_n) = 0$$

with the summation extending over all permutations $(i_1, \ldots, i_n)$ of $(1, \ldots, n)$. Since f_n is permutation invariant, all terms are equal and therefore equal to 0. □

Theorem 1.5.10. *Assume that $\mathfrak{P}$ is complete and closed under convex combinations. Then $\{P^n : P \in \mathfrak{P}\}$ is symmetrically complete for every $n \in \mathbb{N}$.*

Proof. Follows from Proposition 1.5.6 and Lemma 1.5.9. □

Without $\mathfrak{P}$ being closed under convex combinations, the assertion is not necessarily true. Example: $\{N_{(\mu,1)} : \mu \in \mathbb{R}\}$ is complete, yet $\{N_{(\mu,1)}^n : \mu \in \mathbb{R}\}$ is not symmetrically complete: The permutation invariant function

$$f_n(x_1, \ldots, x_n) = \sum_1^n x_\nu^2 - n\bar{x}_n^2 - (n-1)$$

fulfills $N_{(\mu,1)}^n(f_n) = 0$ for $\mu \in \mathbb{R}$.

Corollary 1.5.11. *Let $\mathfrak{F}$ be a class of functions $f : (X, \mathcal{A}) \to (\mathbb{R}, \mathbb{B})$. Let $\mathfrak{P}$ be the family of all p–measures $P|\mathcal{A}$ which are equivalent to some σ–finite measure μ and which fulfill $P(|f|) < \infty$ for $f \in \mathfrak{F}$. Then $\{P^n : P \in \mathfrak{P}\}$ is symmetrically complete for every $n \in \mathbb{N}$.*

Proof. Follows immediately from Lemma 1.5.2 and Theorem 1.5.10. □

If $\mathfrak{P}_0 = \{P \in \mathfrak{P} : P(u) = 0\}$, where u is a given function, the family $\{P^n : P \in \mathfrak{P}_0\}$ is not symmetrically complete: For any permutation invariant function f_{n-1},

$$\int \sum_{\nu=1}^n u(x_\nu) f_{n-1}(x_1, \ldots, x_{\nu-1}, x_{\nu+1}, \ldots, x_n) P(dx_1) \ldots P(dx_n) = 0 \quad (1.5.12)$$

$$\text{for } P \in \mathfrak{P}_0.$$

An interesting result of Hoeffding (1977, pp. 279/80, Theorems 1B, 2B; see also Fraser, 1954, p. 48, Theorem 2.1 for a more special result) implies that any permutation invariant function with expectation zero under P^n for every

$P \in \mathfrak{P}_0$ is of the type (1.5.12) if $\mathfrak{P}$ consists of all p-measures equivalent to a σ-finite measure μ. Moreover, if u is μ-unbounded (i.e. $\mu\{|u| > c\} > 0$ for every $c > 0$), then $\{P^n : P \in \mathfrak{P}_0\}$ is boundedly symmetrically complete.

1.6 Exponential families

Surveys basic properties of exponential families, including results on minimality and completeness of sufficient statistics.

Exponential families play an important role in nonasymptotic parametric theory for two reasons: (i) Many important families of distributions are exponential, (ii) the practically useful results of nonasymptotic statistical theory are more or less limited to exponential families.

The reader interested in a more complete presentation is referred to Barndorff–Nielsen (1978) or L.D. Brown (1986).

Definition 1.6.1. A family $\mathfrak{P}$ of p-measures is of *exponential type* if it has — with respect to some σ-finite measure — densities of the following type.

$$x \rightarrow C(P)g(x) \exp\left[\sum_1^m a_i(P)T_i(x)\right], \qquad (1.6.2)$$

with $T_i : (X, \mathcal{A}) \rightarrow (\mathbb{R}, \mathbb{B})$, $i = 1, \ldots, m$.

Observe that the factor g in (1.6.2) can always be eliminated by an appropriate choice of the dominating measure.

The p-measures in $\mathfrak{P}$ are mutually absolutely continuous. To simplify our notations we assume throughout the following that the dominating measure μ is equivalent to $\mathfrak{P}$. In the sequel we assume that (1.6.2) is a representation where the functions a_i, $i = 1, \ldots, m$, are affinely independent (i.e. $\sum_1^m c_i a_i(P) = c_0$ for all $P \in \mathfrak{P}$ implies $c_i = 0$ for $i = 0, 1, \ldots, m$) and the functions T_i, $i = 1, \ldots, m$, are affinely μ-independent (i.e. $\sum_1^m c_i T_i(x) = c_0$ for μ-a.a. $x \in X$ implies $c_i = 0$ for $i = 0, 1, \ldots, m$).

Exercise 1.6.3. Show that such a "minimal" representation always exists.

If $\mathfrak{P}$ is of exponential type, then so is $\{P^n : P \in \mathfrak{P}\}$:

$$\prod_{\nu=1}^n \left(C(P)g(x_\nu) \exp\left[\sum_{i=1}^m a_i(P)T_i(x_\nu)\right] \right)$$

$$= C_n(P) g_n(x_1, \ldots, x_n) \exp\left[\sum_{i=1}^{m} a_i(P) T_i^{(n)}(x_1, \ldots, x_n)\right]$$

with $C_n(P) = C(P)^n$, $g_n(x_1, \ldots, x_n) = \prod_1^n g(x_\nu)$ and $T_i^{(n)}(x_1, \ldots, x_n) = \sum_1^n T_i(x_\nu)$ for $i = 1, \ldots, m$.

The relevant properties of an exponential family (like minimal sufficiency and/or completeness of the statistic $(T_1^{(n)}, \ldots, T_m^{(n)})$) depend on the coefficients $(a_1(P), \ldots, a_m(P))$, $P \in \mathfrak{P}$, only. Hence it suffices to formulate the respective results without regard to the sample size.

Given the dominating measure μ and the functions g, $T_1, \ldots, T_m$, the set

$$A := \left\{ (a_1, \ldots, a_m) \in \mathbb{R}^m : \int g(x) \exp\left[\sum_1^m a_i T_i(x)\right] \mu(dx) < \infty \right\}$$

is the *natural parameter space*. If useful, we may embed any exponential family in the exponential family $\mathfrak{P}_0 := \{P_a : a \in A\}$, where P_a is defined by its μ–density

$$x \to C(a) g(x) \exp\left[\sum_1^m a_i T_i(x)\right], \tag{1.6.4}$$

with

$$C(a) = 1 / \int g(x) \exp\left[\sum_1^m a_i T_i(x)\right] \mu(dx).$$

Theorem 1.6.5. *The natural parameter space is convex.*

Proof. Let $(a_1, \ldots, a_m)$, $(b_1, \ldots, b_m) \in A$ and $\alpha \in (0, 1)$. By Hölder's inequality we have

$$\int g(x) \exp\left[\sum_{i=1}^m (\alpha a_i + (1 - \alpha) b_i) T_i(x)\right] \mu(dx)$$

$$= \int \left(g(x) \exp\left[\sum_{i=1}^m a_i T_i(x)\right]\right)^\alpha \left(g(x) \exp\left[\sum_{i=1}^m b_i T_i(x)\right]\right)^{1-\alpha} \mu(dx)$$

$$\leq \left(\int g(x) \exp\left[\sum_{i=1}^m a_i T_i(x)\right] \mu(dx)\right)^\alpha \left(\int g(x) \exp\left[\sum_{i=1}^m b_i T_i(x)\right] \mu(dx)\right)^{1-\alpha}.$$

Hence, $a, b \in A$ implies $\alpha a + (1 - \alpha) b \in A$.   □

Lemma 1.6.6. *Given functions* $f : (X, \mathcal{A}) \to (\mathbb{R}, \mathbb{B})$ *and* $T : (X, \mathcal{A}) \to (\mathbb{R}, \mathbb{B})$, *let*

$$A := \{a \in \mathbb{R} : \int |f(x)| \exp[aT(x)]\mu(dx) < \infty\}.$$

Then $\alpha \to \int f(x) \exp[\alpha T(x)]\mu(dx)$ *is a holomorphic function on* $\{a + ib : a \in A^\circ, \ b \in \mathbb{R}\}$, *with*

$$\frac{\partial}{\partial \alpha} \int f(x) \exp[\alpha T(x)]\mu(dx) = \int T(x) f(x) \exp[\alpha T(x)]\mu(dx). \qquad (1.6.7)$$

Proof. Let $\chi(\alpha) := \int f(x) \exp[\alpha T(x)]\mu(dx)$. For $a_0 \in A^\circ$ there exists $\varepsilon > 0$ such that $\chi(\alpha)$ exists and is finite for all $\alpha = a + ib$ with $|a - a_0| \leq \varepsilon$. Let $\alpha_0 := a_0 + ib$, and let $0 \neq c_n \in \mathbb{C}$ with $|c_n| < \varepsilon$ and $c_n \xrightarrow[n \in \mathbb{N}]{} 0$. Then we obtain by the Lebesgue dominated convergence theorem that

$$\frac{1}{c_n}(\chi(\alpha_0 + c_n) - \chi(\alpha_0))$$

$$= \int \frac{1}{c_n}(\exp[c_n T(x)] - 1) f(x) \exp[\alpha_0 T(x)]\mu(dx)$$

$$\xrightarrow[n \in \mathbb{N}]{} \int T(x) f(x) \exp[\alpha_0 T(x)]\mu(dx),$$

since

$$\left|\frac{1}{c_n}(\exp[c_n T(x)] - 1)\right| \leq \left|\sum_{j=1}^{\infty} \frac{c_n^{j-1} T(x)^j}{j!}\right| \leq \sum_{j=1}^{\infty} \frac{\varepsilon^{j-1} |T(x)|^j}{j!}$$

$$\leq \frac{1}{\varepsilon} \exp[\varepsilon |T(x)|] \leq \frac{1}{\varepsilon}(\exp[-\varepsilon T(x)] + \exp[\varepsilon T(x)]).$$

$\square$

Since $a \to \int g(x) \exp[\sum_1^m a_i T_i(x)]\mu(dx)$ is continuous in the interior of A, so is the density $a \to C(a) g(x) \exp[\sum_1^m a_i T_i(x)]$. Hence $a_n \to a_0$ implies $P_{a_n} \to P_{a_0}$ in the sup–metric. The following proposition, due to Barndorff–Nielsen (1969, p. 57, Proposition), provides a converse.

Proposition 1.6.8. *Let* $\{P_a : a \in A\}$ *be an exponential family with its natural parameter space. If* P_{a_n}, $n \in \mathbb{N}$, *converges weakly to* P_{a_0}, *then* $a_n \to a_0$.

For 1–dimensional exponential families this follows also from Propositions 1.7.9 and 1.7.15.

Proof. Let $(P_{a_n})_{n \in \mathbb{N}}$ converge weakly to P_{a_0}. It suffices to show that any subsequence of $(a_n)_{n \in \mathbb{N}}$ contains a subsequence converging to a_0. Let $\mathbb{N}_0 \subset \mathbb{N}$

be an arbitrary infinite subset. W.l.g. let $a_n \neq a_0$ for all $n \in \mathbb{N}_0$ and define $\varepsilon_n := (a_n - a_0)/\|a_n - a_0\|$ for $n \in \mathbb{N}_0$, where $\|\cdot\|$ denotes the Euclidean norm. As $\{\varepsilon_n : n \in \mathbb{N}_0\}$ is a bounded subset of $\mathbb{R}^m$, there exists ε_0 and $\mathbb{N}_1 \subset \mathbb{N}_0$ such that $(\varepsilon_n)_{n \in \mathbb{N}_1} \to \varepsilon_0$. We shall show that $(a_n)_{n \in \mathbb{N}_1} \to a_0$.

Since the representation (1.6.4) is assumed to be minimal, P_{a_0} is nondegenerate, i.e. its support is not contained in an $(m-1)$–dimensional flat. Hence, we have $P_{a_0}\{t \in \mathbb{R}^m : \varepsilon_0^\top t = r\} < 1$ for all $r \in \mathbb{R}$. It is easy to see that this implies the existence of $r', r'' \in \mathbb{R}$ with $r' < r''$ such that $P_{a_0}\{t \in \mathbb{R}^m : \varepsilon_0^\top t < r'\} > 0$ and $P_{a_0}\{t \in \mathbb{R}^m : \varepsilon_0^\top t > r''\} > 0$. Therefore, there exists a bounded open set $B' \subset \{t \in \mathbb{R}^m : \varepsilon_0^\top t < r'\}$ with $P_{a_0}(B') > 0$ and a bounded closed set $B'' \subset \{t \in \mathbb{R}^m : \varepsilon_0^\top t > r''\}$ with $P_{a_0}(B'') > 0$. Let $s', s'' \in \mathbb{R}$ be such that $r' < s' < s'' < r''$. As $|\varepsilon_n^\top t - \varepsilon_0^\top t| < \|\varepsilon_n - \varepsilon_0\| \cdot \|t\|$, the boundedness of B' implies the existence of $n' \in \mathbb{N}_1$ such that $n \geq n'$ for $n \in \mathbb{N}_1$ implies $|\varepsilon_n^\top t - \varepsilon_0^\top t| \leq s' - r'$, and therefore $\varepsilon_n^\top t < s'$ for all $t \in B'$. Hence $a_n^\top t - a_0^\top t < \|a_n - a_0\|s'$ for all $t \in B'$ and $n \in \mathbb{N}_1$ with $n \geq n'$. Since $(P_{a_n})_{n \in \mathbb{N}_1} \to P_{a_0}$ weakly, we obtain

$$P_{a_0}(B') \leq \lim_{n \in \mathbb{N}_1} P_{a_n}(B') = \lim_{n \in \mathbb{N}_1} P_{a_0}\Big(\frac{C(a_n)}{C(a_0)} \exp[a_n^\top t - a_0^\top t] 1_{B'}\Big)$$
$$\leq P_{a_0}(B')C(a_0)^{-1} \lim_{n \in \mathbb{N}_1} C(a_n) \exp[\|a_n - a_0\|s'].$$

As $P_{a_0}(B') > 0$, this implies

$$C(a_0) \leq \lim_{n \in \mathbb{N}_1} C(a_n) \exp[\|a_n - a_0\|\varepsilon'].$$

The dual argument yields

$$\overline{\lim_{n \in \mathbb{N}_1}}\, C(a_n) \exp[\|a_n - a_0\|s''] \leq C(a_0).$$

Hence

$$\overline{\lim_{n \in \mathbb{N}_1}}\, \exp[\|a_n - a_0\|(s'' - s')] \leq \frac{\overline{\lim_{n \in \mathbb{N}_1}}\, C(a_n) \exp[\|a_n - a_0\|s'']}{\lim_{n \in \mathbb{N}_1} C(a_n) \exp[\|a_n - a_0\|s']} \leq 1.$$

This implies $\lim_{n \in \mathbb{N}_1} \|a_n - a_0\| = 0$, since $s'' > s'$. $\square$

Theorem 1.6.9. *Let $\mathfrak{P}$ be an exponential family of p–measures P with densities (1.6.2). The statistic $S : (X, \mathcal{A}) \to (\mathbb{R}^m, \mathbb{B}^m)$, defined by $S(x) := (T_1(x), \ldots, T_m(x))$, is sufficient for $\mathfrak{P}$. S is minimal sufficient if the functions $P \to a_i(P)$, $i = 1, \ldots, m$, are affinely independent.*

Proof. With $P_0 \in \mathfrak{P}$ fixed, let

$$h_P(y_1, \ldots, y_m) := \frac{C(P)}{C(P_0)} \exp\Big[\sum_1^m (a_i(P) - a_i(P_0))y_i\Big].$$

Then P has P_0–density $h_P \circ S$.

(i) By Factorization Theorem 1.2.10 this implies that S is sufficient.

(ii) To prove minimal sufficiency, let $\mathfrak{P}_0 \subset \mathfrak{P}$ be a countable subset such that $\big\{(a_1(P), \ldots, a_m(P)) : P \in \mathfrak{P}_0\big\}$ is dense in $\big\{(a_1(P), \ldots, a_m(P)) : P \in \mathfrak{P}\big\}$. Then

$$h_P(y'_1, \ldots, y'_m) = h_P(y''_1, \ldots, y''_m) \quad \text{for all } P \in \mathfrak{P}_0$$

implies

$$\sum_1^m (a_i(P) - a_i(P_0))y'_i = \sum_1^m (a_i(P) - a_i(P_0))y''_i$$

for $P \in \mathfrak{P}_0$, and therefore for $P \in \mathfrak{P}$. Since $a_1, \ldots, a_m$ are affinely independent, this implies $y'_i = y''_i$ for $i = 1, \ldots, m$. Hence minimal sufficiency follows from Theorem 1.4.4. $\qquad\square$

The following theorem is due to Lehmann and Scheffé (1955, p. 223, Theorem 7.3). It occurs implicitly in Sverdrup (1953, p. 67, Theorem 1).

Theorem 1.6.10. *Assume that $\mathfrak{P}$ is an exponential family of p–measures P with densities (1.6.2).*

Then $\mathfrak{P} \circ (T_1, \ldots, T_m)$ is complete, provided $\big\{(a_1(P), \ldots, a_m(P)) : P \in \mathfrak{P}\big\}$ has a nonempty interior.

Proof. Let $P_0 \in \mathfrak{P}$ be such that $(a_1(P_0), \ldots, a_m(P_0))$ is in the interior of $A :=$ $\big\{(a_1(P), \ldots, a_m(P)) : P \in \mathfrak{P}\big\}$. W.l.g. we assume $g \equiv 1$ in (1.6.2) and $a_i(P_0) = 0$ for $i = 1, \ldots, m$. If

$$\int f\big(T_1(x), \ldots, T_m(x)\big)P(dx) = 0 \quad \text{for every } P \in \mathfrak{P},$$

this implies with $\nu = \mu \circ (T_1, \ldots, T_m)$

$$\int f(t_1, \ldots, t_m) \exp\Big[\sum_1^m a_j t_j\Big]\nu\big(d(t_1, \ldots, t_m)\big) = 0 \qquad (1.6.11)$$

for $(a_1, \ldots, a_m) \in A$. Since $(0, \ldots, 0)$ is in the interior of A, there exist open intervals $I_j \ni 0$, $j = 1, \ldots, m$, such that (1.6.11) holds true for $a_j \in I_j$, $j = 1, \ldots, m$.

By (1.6.11),

$$c := \int f^+(t_1,\ldots,t_m) \exp\left[\sum_1^m a_j t_j\right] \nu\big(d(t_1,\ldots,t_m)\big) \qquad (1.6.12)$$

$$= \int f^-(t_1,\ldots,t_m) \exp\left[\sum_1^m a_j t_j\right] \nu\big(d(t_1,\ldots,t_m)\big)$$

with $0 \le c < \infty$. We shall show that $c = 0$. If $c > 0$, let $Q^\pm|\mathbb{B}^m$ denote the p-measure with ν-density $(t_1,\ldots,t_m) \to f^\pm(t_1,\ldots,t_m)/c$. Relation (1.6.12) implies

$$\int \exp\left[\sum_1^m a_j t_j\right] Q^+\big(d(t_1,\ldots,t_m)\big) \qquad (1.6.13)$$

$$= \int \exp\left[\sum_1^m a_j t_j\right] Q^-\big(d(t_1,\ldots,t_m)\big)$$

for $(a_1,\ldots,a_m) \in \underset{1}{\overset{m}{\times}} I_j$. Let $(a_2,\ldots,a_m) \in \underset{2}{\overset{m}{\times}} I_j$ be fixed. Then, by Theorem 1.6.6, both sides of (1.6.13) are holomorphic functions of a_1 on $I_1 \times i\,\mathbb{R}$, hence they are identical on $I_1 \times i\,\mathbb{R}$, and therefore in particular on $\{0\} \times i\,\mathbb{R}$. By induction we obtain

$$\int \exp\left[i\sum_1^m s_j t_j\right] Q^+\big(d(t_1,\ldots,t_m)\big) = \int \exp\left[i\sum_1^m s_j t_j\right] Q^-\big(d(t_1,\ldots,t_m)\big)$$

for all $s_j \in \mathbb{R}$, $j = 1,\ldots,m$. This implies $Q^+ = Q^-$ by the uniqueness theorem for characteristic functions. Hence $f^+ = f^-$ ν-a.e., i.e. $f = 0$ ν-a.e.   □

Example 1.6.14 (Gamma distribution). For $a,b > 0$ let $\Gamma_{a,b}|\mathbb{B}_+$ denote the p-measure with Lebesgue density

$$x \to \frac{1}{a^b \Gamma(b)} x^{b-1} \exp[-x/a], \qquad x > 0.$$

$\{\Gamma_{a,b} : a,b > 0\}$ is an exponential family with

$$T_1(x) = x, \qquad\qquad T_2(x) = \log x,$$
$$a_1(\Gamma_{a,b}) = -1/a, \qquad a_2(\Gamma_{a,b}) = b.$$

Since $\{(a_1(\Gamma_{a,b}), a_2(\Gamma_{a,b})) : a,b > 0\} = (-\infty,0) \times (0,\infty)$, $S_n(x_1,\ldots,x_n) := (\sum_1^n x_\nu, \sum_1^n \log x_\nu)$ is a complete sufficient statistic. Equivalent statistics are $(\sum_1^n x_\nu, \prod_1^n x_\nu)$ or $(n^{-1}\sum_1^n x_\nu, (\prod_1^n x_\nu)^{1/n})$.

$\sum_1^n x_\nu$ is complete sufficient for every subfamily $\{\Gamma_{a,b}^n : a > 0\}$ with $b > 0$ fixed, and $\sum_1^n \log x_\nu$ is complete sufficient for every subfamily $\{\Gamma_{a,b}^n : b > 0\}$ with $a > 0$ fixed.

If the interior of $A := \{(a_1(P), \ldots, a_m(P)) : P \in \mathfrak{P}\}$ is empty, but A is not contained in a $(m-1)$–dimensional flat, $\mathfrak{P}$ is called a *curved exponential family*. In this case, the sufficient statistic $(T_1, \ldots, T_m)$ is minimal (according to Theorem 1.6.9.). It may be complete or not. Example 1.6.15 presents an exponential family, where this sufficient statistic is complete. Theorem 1.6.23 shows that $(T_1, \ldots, T_m)$ is not complete, if $(a_1(P), \ldots, a_m(P))$, $P \in \mathfrak{P}$, are polynomial dependent and $P \circ (T_1, \ldots, T_m) \ll \lambda^m$.

Example 1.6.15. A curved exponential family may be complete.

For $a > 0$ let Π_a denote the Poisson distribution with parameter a, and let

$$P_\vartheta := \Pi_{\vartheta^{-1}} \times \Pi_{\exp[-\vartheta]}, \qquad \vartheta > 0.$$

Since

$$P_\vartheta\{(k, \ell)\} = \frac{1}{k!\ell!} \exp[-\vartheta^{-1} - \exp[-\vartheta]]\vartheta^{-k} \exp[-\vartheta\ell],$$

this is a curved exponential family. We shall show that $\{P_\vartheta : \vartheta > 0\}$ is complete.

The existence of some function $f : (\{0\} \cup \mathbb{N})^2 \to \mathbb{R}$ fulfilling

$$\sum_{k=0}^{\infty} \sum_{\ell=0}^{\infty} f(k, \ell) P_\vartheta\{(k, \ell)\} = 0 \quad \text{for } \vartheta > 0$$

is equivalent to the existene of some function $g : (\{0\} \cup \mathbb{N})^2 \to \mathbb{R}$ fulfilling

$$\sum_{k=0}^{\infty} \sum_{\ell=0}^{\infty} g(k, \ell)\vartheta^{-k} \exp[-\vartheta\ell] = 0 \quad \text{for } \vartheta > 0. \tag{1.6.16}$$

In the following we shall show that $g \equiv 0$, hence $f \equiv 0$.

From (1.6.16), applied for $\vartheta = 1$, we obtain

$$\sum_{k=0}^{\infty} \sum_{\ell=0}^{\infty} |g(k, \ell)| \exp[-\ell] < \infty. \tag{1.6.17}$$

Since

$$|g(k, \ell)|\vartheta^m \exp[-\vartheta\ell] \le \vartheta^m \exp[1 - \vartheta]|g(k, \ell)| \exp[-\ell]$$

for $k \in \{0\} \cup \mathbb{N}$, $\ell \in \mathbb{N}$, $m \in \mathbb{Z}$ and $\vartheta \ge 1$, relation (1.6.17) implies

$$\lim_{\vartheta \to \infty} \sum_{k=0}^{\infty} \sum_{\ell=1}^{\infty} g(k, \ell)\vartheta^m \exp[-\vartheta\ell] = 0 \quad \text{for } m \in \mathbb{Z}. \tag{1.6.18}$$

Rewriting (1.6.16) as

$$\sum_{k=0}^{\infty} g(k, 0)\vartheta^{-k} + \sum_{k=0}^{\infty} \sum_{\ell=1}^{\infty} g(k, \ell)\vartheta^{-k} \exp[-\vartheta\ell] = 0 \quad \text{for } \vartheta > 0, \tag{1.6.19}$$

we obtain for $\vartheta \to \infty$ that $g(0,0) = 0$. Assume now that $g(k,0) = 0$ holds true for $k = 0, \ldots, K-1$. Multiplying both sides of (1.6.19) by ϑ^K and letting $\vartheta \to \infty$, we obtain from (1.6.18) and (1.6.19) that $g(K,0) = 0$. Hence $g(k,0) = 0$ for all $k \in \{0\} \cup \mathbb{N}$. Assume now that $g(k,\ell) = 0$ for all $k \in \{0\} \cup \mathbb{N}$ and $\ell = 0, \ldots, L-1$. Then (1.6.16) simplifies to

$$\sum_{k=0}^{\infty} \sum_{\ell=L}^{\infty} g(k,\ell)\vartheta^{-k} \exp[-\vartheta\ell] = 0 \quad \text{for all } \vartheta > 0. \tag{1.6.20}$$

Multiplying both sides by $\exp[L\vartheta]$ and replacing ℓ by $\ell + L$, we may rewrite relation (1.6.20) as

$$\sum_{k=0}^{\infty} \sum_{\ell=0}^{\infty} g(k, \ell+L)\vartheta^{-k} \exp[-\vartheta\ell] = 0 \quad \text{for all } \vartheta > 0.$$

This is equation (1.6.16) for the function $(k, \ell) \to g(k, \ell+L)$, so that $g(k, L) = 0$ for $k \in \{0\} \cup \mathbb{N}$. Hence $g(k,\ell) = 0$ for $k, \ell \in \{0\} \cup \mathbb{N}$.

Example 1.6.21. (Truncated exponential distribution). For $\vartheta > 0$ let $E_\vartheta | \mathbb{B}_+$ denote the exponential distribution with Lebesgue density

$$x \to \vartheta^{-1} \exp[-x/\vartheta], \qquad x > 0.$$

Given $c > 0$, let $P_\vartheta := E_\vartheta \circ (x \to x1_{(0,c)}(x) + c1_{[c,\infty)}(x))$. The p–measures $P_\vartheta | \mathbb{B} \cap (0,c]$ are mutually absolutely continuous. It is easy to check that P_ϑ has P_1–density

$$\vartheta^{-1} \exp[x] \exp[-\vartheta^{-1}x + (\log \vartheta)1_{\{c\}}(x)], \qquad x \in (0,c].$$

This is an exponential family with

$$T_1(x) = x, \qquad T_2(x) = 1_{\{c\}}(x),$$
$$a_1(P_\vartheta) = -\vartheta^{-1}, \qquad a_2(P_\vartheta) = \log \vartheta.$$

Since a_1, a_2 are affinely independent,

$$S_n(x_1, \ldots, x_n) := \left(\sum_{1}^{n} x_\nu, \sum_{1}^{n} 1_{\{c\}}(x_\nu)\right) \tag{1.6.22}$$

is minimal sufficient for $\{P_\vartheta^n : \vartheta > 0\}$ by Theorem 1.6.9, but not complete.

The following theorem, due to Wijsman (1958, p. 1031), explains why curved exponential families are usually not boundedly complete.

Theorem 1.6.23. *Let $\mathfrak{P}$ be an exponential family with coefficients*

$$(a_1(P), \ldots, a_m(P))$$

which are polynomial dependent, i.e.: There exists a polynomial $\Pi : \mathbb{R}^m \to$
$\mathbb{R}$ *such that* $\Pi(a_1(P), \ldots, a_m(P)) = 0$ *for every* $P \in \mathfrak{P}$. *Assume that* $P \circ$
$(T_1, \ldots, T_m) \ll \lambda^m$. *Then the statistic* $(T_1, \ldots, T_m)$ *is not complete.*

If the λ^m*–density of* $P \circ (T_1, \ldots, T_m)$ *is bounded away from* 0 *on some set of positive* λ^m*–measure, then* $(T_1, \ldots, T_m)$ *is not even boundedly complete.*

Since $P \circ (T_1, \ldots, T_m) \ll \lambda^m$ implies the same property for

$$P^n \circ \big((x_1, \ldots, x_n) \to (\sum_1^n T_1(x_\nu), \ldots, \sum_1^n T_m(x_\nu))\big),$$

the assertion of the theorem holds for arbitrary sample sizes n.

Proof. Any λ^m–density of $P \circ (T_1, \ldots, T_m)$ is of the type

$$C(P) g(t_1, \ldots, t_m) \exp\big[\sum_1^m a_i(P) t_i\big]. \tag{1.6.24}$$

Since (1.6.24) is the λ^m–density of a p–measure, the function g is positive on some rectangle, say $I = \overset{m}{\underset{1}{\times}}[t_i', t_i'']$.

Let k denote the degree of the polynomial Π. Let $b : I \to \mathbb{R}$ be defined by

$$b(t_1, \ldots, t_m) = \big(\prod_1^m (t_i - t_i')(t_i'' - t_i)\big)^k, \qquad t \in I.$$

Then b is k–times partially differentiable and fulfills

$$\frac{\partial^{k_1}}{\partial t_1^{k_1}} \cdots \frac{\partial^{k_m}}{\partial t_m^{k_m}} b(t_1, \ldots, t_m) = 0$$

on the boundary of I as long as $k_1 + \ldots + k_m < k$.

For any polynomial $\hat{\Pi}$ of degree k, we obtain by partial integration,

$$\int_I \big(\hat{\Pi}(\frac{\partial}{\partial t_1}, \ldots, \frac{\partial}{\partial t_m}) b(t_1, \ldots, t_m)\big) \exp\big[-\sum_1^m a_i t_i\big] d(t_1, \ldots, t_m)$$

$$= \hat{\Pi}(a_1, \ldots, a_m) \int_I b(t_1, \ldots, t_m) \exp\big[-\sum_1^m a_i t_i\big] d(t_1, \ldots, t_m).$$

Since $\Pi(a_1(P), \ldots, a_m(P)) = 0$ for every $P \in \mathfrak{P}$, this implies

$$\int_I \big(\Pi(-\frac{\partial}{\partial t_1}, \ldots, -\frac{\partial}{\partial t_m}) b(t_1, \ldots, t_m)\big) \exp\big[\sum_1^m a_i(P) t_i\big] d(t_1, \ldots, t_m) = 0$$

for $P \in \mathfrak{P}$. As g is positive on I and $\Pi\left(-\frac{\partial}{\partial t_1}, \ldots, -\frac{\partial}{\partial t_m}\right) b(t_1, \ldots, t_m) \not\equiv 0$, the statistic $(T_1, \ldots, T_m)$ is not complete.

If the λ^m–density of $P \circ (T_1, \ldots, T_m)$, and hence g, is bounded away from 0 on some set of positive λ^m–measure, then I can be chosen such that g is bounded away from 0 on I. $\hfill\square$

Example 1.6.25. The family of normal distributions $\{N_{(\mu,\sigma^2)} : \mu \in \mathbb{R}, \; \sigma^2 > 0\}$ is an exponential family with

$$T_1(x) = x, \qquad\qquad T_2(x) = x^2,$$
$$a_1\left(N_{(\mu,\sigma^2)}\right) = \mu/\sigma^2, \qquad a_2\left(N_{(\mu,\sigma^2)}\right) = -1/2\,\sigma^2.$$

Since

$$\left\{\left(a_1(N_{(\mu,\sigma^2)}), a_2(N_{(\mu,\sigma^2)})\right) : \mu \in \mathbb{R}, \sigma^2 > 0\right\} = \mathbb{R} \times (-\infty, 0),$$

$S_n(x_1, \ldots, x_n) = \left(\sum_1^n x_\nu, \sum_1^n x_\nu^2\right)$ is a complete sufficient statistic for $\{N_{(\mu,\sigma^2)}^n : \mu \in \mathbb{R}, \sigma^2 > 0\}$. An equivalent statistic is, for $n \geq 2$, $\left(\bar{x}_n, \sum_1^n (x_\nu - \bar{x}_n)^2\right)$.

$\sum_1^n x_\nu$ is complete sufficient for every subfamily $\{N_{(\mu,\sigma^2)}^n : \mu \in \mathbb{R}\}$ with $\sigma^2 > 0$ fixed, and $\sum_1^n (x_\nu - \mu)^2$ is complete sufficient for the subfamily $\{N_{(\mu,\sigma^2)}^n : \sigma^2 > 0\}$ with $\mu \in \mathbb{R}$ fixed.

For the subfamily with given coefficient of variation, $P_\mu := N_{(\mu,c^2\mu^2)}$, S_n remains minimal sufficient. $\{N_{(\mu,c^2\mu^2)} : \mu > 0\}$ is a curved exponential family with

$$T_1(x) = x, \qquad\qquad T_2(x) = x^2,$$
$$a_1(P_\mu) = 1/c^2\mu, \qquad a_2(P_\mu) = -1/2\,c^2\mu^2.$$

Since the functions a_1, a_2 are affinely independent, S_n is minimal sufficient for $\{N_{(\mu,c^2\mu^2)}^n : \mu > 0\}$. Since a_1, a_2 are polynomial dependent (we have $c^2 a_1(P_\mu)^2 + 2a_2(P_\mu) \equiv 0$), S_n is not complete. For instance

$$N_{(\mu,c^2\mu^2)}^n \left\{\sum_1^n x_\nu < 0\right\} = \Phi(-\sqrt{n}/c) \quad \text{for every } \mu > 0.$$

1.7 Auxiliary results on families with monotone likelihood ratios

Contains auxiliary results on families with monotone likelihood ratios, and establishes that exponential families are the only ones which have monotone likelihood ratios for arbitrary sample sizes.

Throughout this section let $\Theta = (\vartheta', \vartheta'')$ *be an interval. To avoid technicalities irrelevant for statistical applications, we assume that the p–measures P_ϑ, $\vartheta \in \Theta$, are mutually absolutely continuous.*

We start our considerations with some auxiliary results concerning p–measures $Q_\vartheta | \mathbb{B}$. Let $\mu | \mathbb{B}$ denote a dominating measure, $q(\cdot, \vartheta)$ a μ–density of Q_ϑ. If Q_ϑ, $\vartheta \in \Theta$, are mutually absolutely continuous, we may assume w.l.g. the existence of a set $B_0 \subset \mathbb{B}$ such that $q(t, \vartheta) > 0$ for $t \in B_0$, and $Q_\vartheta(B_0) = 1$ for $\vartheta \in \Theta$.

Definition 1.7.1. The family $\{Q_\vartheta : \vartheta \in \Theta\}$ has *isotone likelihood ratios* if the densities $q(\cdot, \vartheta)$ can be chosen such that

$$t \to q(t, \vartheta_2)/q(t, \vartheta_1) \text{ is isotone if } \vartheta_1 < \vartheta_2. \tag{1.7.2}$$

This condition is equivalent to

$$\vartheta \to q(t_2, \vartheta)/q(t_1, \vartheta) \text{ is isotone if } t_1 < t_2. \tag{1.7.3}$$

The following criterion is due to Karlin (1957, p. 283, Theorem 1).

Criterion 1.7.4. *If $\vartheta \to q(t, \vartheta)$ is differentiable for every $t \in B_0$, the family $\{Q_\vartheta : \vartheta \in \Theta\}$ has isotone likelihood ratios iff $t \to \frac{\partial}{\partial\vartheta} \log q(t, \vartheta)$ is isotone.*

Proof. (i) If $t \to q(t, \vartheta_2)/q(t, \vartheta_1)$ is isotone for all $\vartheta_1 < \vartheta_2$, the function

$$t \to \lim_{\tau \downarrow \vartheta} \frac{q(t, \tau) - q(t, \vartheta)}{(\tau - \vartheta)q(t, \vartheta)} = \left(\frac{\partial}{\partial\vartheta}q(t, \vartheta)\right)/q(t, \vartheta)$$

is isotone.

(ii) If $\left(\frac{\partial}{\partial\vartheta}q(t, \vartheta)\right)/q(t, \vartheta) \le \left(\frac{\partial}{\partial\vartheta}q(s, \vartheta)\right)/q(s, \vartheta)$ for $t < s$, we obtain

$$q(t, \vartheta)^2 \frac{\partial}{\partial\vartheta}\frac{q(s, \vartheta)}{q(t, \vartheta)} = q(t, \vartheta)\frac{\partial}{\partial\vartheta}q(s, \vartheta) - q(s, \vartheta)\frac{\partial}{\partial\vartheta}q(t, \vartheta) \ge 0.$$

Hence $\vartheta \to q(s, \vartheta)/q(t, \vartheta)$ is isotone, which is equivalent to (1.7.2). $\square$

Example 1.7.5. Let $p | \mathbb{R}$ be the positive Lebesgue density of a p–measure. Let $\{Q_\vartheta : \vartheta \in \mathbb{R}\}$ be the location parameter family generated by p, i.e. Q_ϑ has Lebesgue density $x \to p(x - \vartheta)$.

Then $\{Q_\vartheta : \vartheta \in \mathbb{R}\}$ has isotone likelihood ratios iff Q_0 is strongly unimodal (i.e. $\log p$ is concave). If p is differentiable, this follows immediately from Criterion 1.7.4. For the general case see Lehmann (1986, p. 509, Example 1). Examples of location parameter families having isotone likelihood ratios are normal, logistic and Laplace distributions.

Further examples of families with isotone likelihood ratios are the noncentral t-, χ^2- and F-distributions (see Lehmann, 1986, p. 295 and p. 428, Problem 4(i)), moreover the distributions of the correlation coefficients (see Example 1.7.12 and Exercise 1.7.13).

Monotonicity of likelihood ratios implies other order relations and relations of topological nature. (See Pfanzagl, 1969a, for more details.)

Definition 1.7.6. $\{Q_\vartheta : \vartheta \in \Theta\}$ is *stochastically isotone* if $\vartheta \to Q_\vartheta(-\infty, t]$ is antitone (equivalently: $\vartheta \to Q_\vartheta[t, \infty)$ isotone) for every $t \in \mathbb{R}$.

Proposition 1.7.7. *A family with isotone likelihood ratios is stochastically isotone.*
  More precisely: If $\{Q_\vartheta : \vartheta \in \Theta\}$ has isotone likelihood ratios, then
  (i) $t \to Q_{\vartheta_2}(-\infty, t]/Q_{\vartheta_1}(-\infty, t]$ *and* $t \to Q_{\vartheta_2}[t, \infty)/Q_{\vartheta_1}[t, \infty)$ *are isotone on the interior of the "convex support" if $\vartheta_1 < \vartheta_2$;*
  (ii) *either one of the relations under* (i) *implies that* $\vartheta \to Q_\vartheta(-\infty, t]$ *is antitone for every $t \in \mathbb{R}$.*

Proof. (i) $\vartheta_1 < \vartheta_2$ implies

$$q(s, \vartheta_2)/q(s, \vartheta_1) \underset{\leq}{\overset{\geq}{}} q(t, \vartheta_2)/q(t, \vartheta_1) \quad \text{for} \quad s \underset{<}{\overset{>}{}} t.$$

Hence integration over $s \in (t', t'')$ with respect to Q_{ϑ_1} yields for $t' < t < t''$

$$\frac{Q_{\vartheta_2}(t', t]}{Q_{\vartheta_1}(t', t]} \leq \frac{q(t, \vartheta_2)}{q(t, \vartheta_1)} \leq \frac{Q_{\vartheta_2}[t, t'')}{Q_{\vartheta_1}[t, t'')} . \tag{1.7.8}$$

From this the relations under (i) follow easily.
  (ii) By (i), $\vartheta_1 < \vartheta_2$ implies

$$Q_{\vartheta_2}(-\infty, t]/Q_{\vartheta_1}(-\infty, t] \leq Q_{\vartheta_2}(-\infty, s]/Q_{\vartheta_1}(-\infty, s]$$

for $s > t$. Hence (ii) follows for $s \to \infty$. $\qquad\qquad\square$

Proposition 1.7.9. *For families with isotone likelihood ratios, weak convergence implies pointwise convergence of densities (hence also convergence in the sup–metric by Scheffé's lemma).*

Proof. W.l.g. we assume $\vartheta_n \geq \vartheta_0$. Analogous to (1.7.8) we obtain for every $t \in \mathbb{R}$

$$\frac{Q_{\vartheta_n}(-\infty, t)}{Q_{\vartheta_0}(-\infty, t)} \leq \frac{q(t, \vartheta_n)}{q(t, \vartheta_0)} \leq \frac{Q_n[t, \infty)}{Q_0[t, \infty)} .$$

Weak convergence implies

$$\lim_{n \to \infty} Q_{\vartheta_n}(-\infty, t) \geq Q_{\vartheta_0}(-\infty, t) \quad \text{and} \quad \overline{\lim_{n \to \infty}} Q_{\vartheta_n}[t, \infty) \leq Q_{\vartheta_0}[t, \infty),$$

whence

$$\lim_{n \to \infty} \frac{q(t, \vartheta_n)}{q(t, \vartheta_0)} = 1.$$

□

We now turn to families of mutually absolutely continuous p–measures on an arbitrary measurable space $(X, \mathcal{A})$. Let μ be a σ–finite dominating measure.

Definition 1.7.10. The family $\{P_\vartheta : \vartheta \in \Theta\}$ admits a *sufficient statistic* $S : (X, \mathcal{A}) \to (\mathbb{R}, \mathbb{B})$ *with isotone likelihood ratios* if for every $\vartheta \in \Theta$ there exists a μ–density

$$x \to g(x) h_\vartheta(S(x)), \qquad x \in X, \tag{1.7.11}$$

with $g : (X, \mathcal{A}) \to (\mathbb{R}_+, \mathbb{B}_+)$ and $h_\vartheta : (Y, \mathbb{B} \cap Y) \to (\mathbb{R}_+, \mathbb{B}_+)$, where Y denotes the convex hull of $S(X)$, such that

$$y \to h_{\vartheta_2}(y)/h_{\vartheta_1}(y) \quad \text{is isotone on } Y \text{ if } \vartheta_1 < \vartheta_2.$$

Recall that S is sufficient for $\{P_\vartheta : \vartheta \in \Theta\}$ by Factorization Theorem 1.2.10.

An alternative definition starts from the weaker assumption that for arbitrary $\vartheta_i \in \Theta$ with $\vartheta_1 < \vartheta_2$, there exists an isotone function $H_{\vartheta_1, \vartheta_2}$ such that $p(x, \vartheta_2)/p(x, \vartheta_1) = H_{\vartheta_1, \vartheta_2}(T(x))$ for μ–a.a. $x \in X$, with the exceptional μ–null set depending on ϑ_1, ϑ_2. Then the densities can always be chosen such that (1.7.11) holds true (see Pfanzagl, 1967).

If $\{P_\vartheta : \vartheta \in \Theta\}$ admits a sufficient statistic S with isotone likelihood ratios, then the family $\{P_\vartheta \circ S : \vartheta \in \Theta\}$ of p–measures on $\mathbb{B}$ has isotone likelihood ratios in the sense of Definition 1.7.1. Observe that Definition 1.7.10 requires much more: that S is, in addition, sufficient for $\{P_\vartheta : \vartheta \in \Theta\}$.

Example 1.7.12. Let $\mathfrak{P} = \{N^n_{(\mu_1, \mu_2, \sigma_1^2, \sigma_2^2, \varrho)} : \mu_i \in \mathbb{R}, \ \sigma_i^2 > 0, \ \varrho \in (-1, 1)\}$ and

$$r_n\big((x_1, y_1), \ldots, (x_n, y_n)\big) := \frac{\displaystyle\sum_{\nu=1}^{n} (x_\nu - \bar{x}_n)(y_\nu - \bar{y}_n)}{\Big(\displaystyle\sum_{\nu=1}^{n} (x_\nu - \bar{x}_n)^2 \sum_{\nu=1}^{n} (y_\nu - \bar{y}_n)^2\Big)^{1/2}}.$$

According to Fisher (1915) the distribution of r_n under $N^n_{(\mu_1,\mu_2,\sigma_1^2,\sigma_2^2,\varrho)}$ has Lebesgue density

$$p_n(r,\varrho) := \frac{2^{n-3}}{\pi(n-3)!}(1-\varrho^2)^{\frac{n-1}{2}}(1-r^2)^{\frac{n-4}{2}}G_n(\varrho r), \qquad r \in (-1,1),$$

with

$$G_n(u) := \sum_{k=0}^{\infty} \Gamma\left(\frac{n+k-1}{2}\right)^2 \frac{2^k}{k!} u^k.$$

Since $\frac{\partial}{\partial r}\left(p_n(r,\varrho_2)/p_n(r,\varrho_1)\right) > 0$ for $\varrho_1 < \varrho_2$, the family $\{N^n_{(\mu_1,\mu_2,\sigma_1^2,\sigma_2^2,\varrho)} \circ r_n :$ $\mu_i \in \mathbb{R}, \sigma_i^2 > 0, \varrho \in (-1,1)\}$ has isotone likelihood ratios. The statistic r_n is, however, not sufficient for $\mathfrak{P}$. Therefore, Theorem 4.5.2 on the existence of uniform most powerful tests does not apply. Tests for the hypothesis $\varrho \leq \varrho_0$ using a critical region $\{r_n > c_n(\varrho_0)\}$ are certainly useful (thanks to their asymptotic properties for large n). They are, however, not uniformly most powerful (with n fixed).

Exercise 1.7.13. Let $\mathfrak{P} = \{N^n_{(\mu_1,\mu_2,\sigma^2,\sigma^2,\varrho)} : \mu_i \in \mathbb{R}, \sigma^2 > 0, \varrho \in (-1,1)\}$, and

$$R_n\big((x_1,y_1),\ldots,(x_n,y_n)\big) :=$$
$$2\sum_{\nu=1}^{n}(x_\nu - \bar{x}_n)(y_\nu - \bar{y}_n) \Big/ \left(\sum_{\nu=1}^{n}(x_\nu - \bar{x}_n)^2 + \sum_{\nu=1}^{n}(y_\nu - \bar{y}_n)^2\right).$$

According to Hsu (1940, p. 418, relation (36)) the distribution of R_n under $N^n_{(\mu_1,\mu_2,\sigma^2,\sigma^2,\varrho)}$ has Lebesgue density

$$p_n(r,\varrho) := B\left(\frac{n-1}{2},\frac{1}{2}\right)^{-1}(1-\varrho^2)(1-r^2)^{\frac{n-3}{2}}(1-\varrho r)^{-(n-1)}, \qquad r \in (-1,1).$$

Since $\frac{\partial^2}{\partial\varrho\partial r}\log p_n(r,\varrho) = (n-1)(1-\varrho r)^{-2} > 0$, the family

$$\{N^n_{(\mu_1,\mu_2,\sigma^2,\sigma^2,\varrho)} \circ R_n : \mu_i \in \mathbb{R}, \sigma^2 > 0, \varrho \in (-1,1)\}$$

has isotone likelihood ratios by Criterion 1.7.4 (but R_n is not sufficient for $\mathfrak{P}$).

Example 1.7.14 (see Ghurye and Wallace, 1959). Let $\{P_\vartheta | \mathbb{B} : \vartheta \in \mathbb{R}\}$ be a location parameter family with positive Lebesgue densities and isotone likelihood ratios. Then, for every $n \in \mathbb{N}$, the family of convolution products (i.e. $\{P_\vartheta^n \circ ((x_1,\ldots,x_n) \to \sum_1^n x_\nu) : \vartheta \in \mathbb{R}\})$ has isotone likelihood ratios, but $\sum_1^n x_\nu$ is sufficient for the location parameter family $\{P_\vartheta^n : \vartheta \in \mathbb{R}\}$, $n \geq 2$, only if P_0 is a normal distribution (see Dynkin, 1951, and Ferguson, 1962).

Proof. The location parameter family generated by P_0 has isotone likelihood ratios iff P_0 is strongly unimodal (see Example 1.7.5). Since strong unimodality of P_0 implies strong unimodality of the convolution product P^{*n} (by Corollary 2.3.24), the location parameter family generated by P^{*n} has isotone likelihood ratios by Example 1.7.5. □

Proposition 1.7.15. *Let* $\{P_\vartheta : \vartheta \in \Theta\}$ *be an exponential family, with* $\mu-$*densities*

$$x \to C(\vartheta)g(x)\exp[a(\vartheta)S(x)], \qquad \vartheta \in \Theta.$$

If a *is isotone, the family* $\{P_\vartheta : \vartheta \in \Theta\}$ *has isotone likelihood ratios in* S, *with*

$$h_\vartheta(y) = C(\vartheta)\exp[a(\vartheta)y], \qquad y \in \mathbb{R}.$$

Recall that this proposition covers, in particular, the following special cases: Binomial, Poisson, $\{N_{(\mu,\sigma^2)} : \mu \in \mathbb{R}, \ \sigma^2 > 0\}$ and $\{\Gamma_{a,b} : a, b > 0\}$ with either parameter fixed.

Since families of p–measures admitting a sufficient statistic with isotone likelihood ratios have various favorable statistical properties (see Theorem 4.5.2 and Theorem 5.4.3), it is of interest to know which families have a sufficient statistic with isotone likelihood ratios for arbitrary sample sizes. Regrettably, one–parameter exponential families are the only ones.

For the following theorem see Borges and Pfanzagl (1963, p. 112, Theorem 1) or Heyer (1982, p. 98, Theorem 14.2). It characterizes 1–dimensional exponential families by the existence of isotone likelihood ratios for infinitely many sample sizes. A characterization without monotony of likelihood ratios requires severe regularity conditions on the densities and/or the sufficient statistic. The first results of this kind go back to Darmois (1935), Koopman (1936) and Pitman (1936). The best result available so far is due to Hipp (1974) who requires for some sample size $n \geq 2$ the existence of a sufficient statistic fulfilling a local Lipschitz condition, and p–measures on $(\mathbb{R}, \mathbb{B})$ equivalent to the Lebesgue measure. That something beyond continuity of the sufficient statistic is needed follows from the existence of a continuous function from $\mathbb{R}^n$ into $\mathbb{R}$ which is $1-1$ λ^n–a.e. (see Denny, 1964).

Theorem 1.7.16. *Let* $\mathfrak{P}$ *be a family of mutually absolutely continuous* $p-$*measures on a measurable space* $(X, \mathcal{A})$. *Let* $P_0|\mathcal{A}$ *be a* $p-$*measure equivalent to* $\mathfrak{P}$.

Assume there exists an infinite subset $\mathbb{N}_0 \subset \mathbb{N}$, *containing 1, with the following property: For every* $n \in \mathbb{N}_0$ *there exists a function* $T_n : (X^n, \mathcal{A}^n) \to (\mathbb{R}, \mathbb{B})$ *and isotone functions* $H_P^{(n)}$, $P \in \mathfrak{P}$, *such that* $H_P^{(n)} \circ T_n$ *is a* P_0^n–*density of* P^n.

Then there exists a function $S : (X, \mathcal{A}) \to (\mathbb{R}, \mathbb{B})$ and functions $C : \mathfrak{P} \to (0, \infty)$ and $a : \mathfrak{P} \to \mathbb{R}$ such that

$$x \to C(P) \exp\big[a(P)S(x)\big],$$

is a P_0-density of P.

Proof. (i) Writing T and H_P for T_1 and $H_P^{(1)}$ we obtain for all $n \in \mathbb{N}_0$

$$H_P^{(n)}\big(T_n(x_1, \ldots, x_n)\big) = \prod_{i=1}^{n} H\big(T(x_i)\big) \qquad P_0^n\text{-a.e.} \tag{1.7.17}$$

We shall replace (1.7.17) by a similar set of equations which hold everywhere. These equations will then yield exponentiality by Lemma 1.7.41.

(ii) Let $\mathcal{C}$ denote the class of all sets $\{x \in X : H_P\big(T(x)\big) \leq c\}$ with $P \in \mathfrak{P}$, $c \geq 0$. Since the elements of $\mathcal{C}$ are of the form $\{x \in X : T(x) < k\}$ or $\{x \in X : T(x) \leq k\}$, the class $\mathcal{C}$ is totally ordered by inclusion.

By Lemma 1.7.27 there exists a function $S_0 : X \to [0, 1]$ fulfilling

$$S_0(y) = P_0\{x \in X : S_0(x) \leq S_0(y)\} \quad \text{for all } y \in X \tag{1.7.18}$$

such that for every $P \in \mathfrak{P}$, $c \geq 0$,

$$\{x \in X : H_P\big(T(x)\big) \leq c\} = \{x \in X : S_0(x) \leq F_P(c)\} \quad (P_0) \tag{1.7.19}$$

with $F_P(c) := P_0\{x \in X : H_P\big(T(x)\big) \leq c\}$.

By Lemma 1.7.40 for every $P \in \mathfrak{P}$ there exists an isotone left continuous function G_P such that $r \leq F_P(c)$ iff $G_P(r) \leq c$. Hence (1.7.19) implies

$$\{x \in X : H_P\big(T(x)\big) \leq c\} = \{x \in X : G_P\big(S_0(x)\big) \leq c\} \quad (P_0) \tag{1.7.20}$$

for every $c \geq 0$.

By Lemma 1.7.48 this implies

$$H_P\big(T(x)\big) = G_P\big(S_0(x)\big) \quad P_0\text{-a.e.} \tag{1.7.21}$$

Inserted in (1.7.17) this implies for all $n \in \mathbb{N}_0$

$$H_P^{(n)}\big(T_n(x_1, \ldots, x_n)\big) = \prod_{\nu=1}^{n} G_P\big(S_0(x_\nu)\big) \quad P_0^n\text{-a.e.} \tag{1.7.22}$$

(iii) Let now $\mathcal{C}$ denote the class of all sets

$$A_P(c) := \Big\{(x_1, \ldots, x_n) \in X^n : \prod_{\nu=1}^{n} G_P\big(S_0(x_\nu)\big) \leq c\Big\}, \quad P \in \mathfrak{P}, \; c \geq 0.$$

By (1.7.22) these sets are — up to P_0^n-null sets — of the form $\{(x_1, \ldots, x_n) \in X^n : T_n(x_1, \ldots, x_n) < k\}$ or $\{(x_1, \ldots, x_n) \in X^n : T_n(x_1, \ldots, x_n) \leq k\}$. Hence $\mathcal{C}$ is ordered by inclusion (P_0^n). Since we wish to apply Lemma 1.7.27 in the version of the Addendum, we have to verify condition (1.7.30). Assume that $P_0^n\left(\bigcap_1^\infty A_{P_i}(c_i)\right) \leq P_0^n(A_P(c))$. Since $\mathcal{C}$ is ordered (P_0^n), this implies

$$\bigcap_{i=1}^\infty A_{P_i}(c_i) \subset A_P(c) \qquad (P_0^n). \qquad (1.7.23)$$

We shall show that the inclusion holds without the restriction "(P_0^n)". For this purpose, let $(y_1, \ldots, y_n) \in \bigcap_1^\infty A_{P_i}(c_i)$ be arbitrary. This implies $\prod_{\nu=1}^n G_{P_i}(S_0(y_\nu)) \leq c_i$ for all $i \in \mathbb{N}$. Hence

$$\{(x_1, \ldots, x_n) \in X^n : S_0(x_\nu) \leq S_0(y_\nu) \quad \text{for } \nu = 1, \ldots, n\} \subset \bigcap_{i=1}^\infty A_{P_i}(c_i).$$

Together with (1.7.23) this implies

$$\{(x_1, \ldots, x_n) \in X^n : S_0(x_\nu) \leq S_0(y_\nu) \quad \text{for } \nu = 1, \ldots, n\} \qquad (1.7.24)$$

$$\subset \{(x_1, \ldots, x_n) \in X^n : \prod_{\nu=1}^n G_P(S_0(x_\nu)) \leq c\} \quad (P_0^n).$$

Furthermore, for arbitrary $r_\nu < S_0(y_\nu)$, $\nu = 1, \ldots, n$,

$$P_0^n\{(x_1, \ldots, x_n) \in X^n : r_\nu < S_0(x_\nu) \leq S_0(y_\nu) \quad \text{for } \nu = 1, \ldots, n\} > 0,$$

since by (1.7.18)

$$P_0\{x \in X : r < S_0(x) \leq S_0(y)\} = S_0(y) - \sup\{S_0(z) : S_0(z) \leq r\} \geq S_0(y) - r > 0.$$

Hence (1.7.24) implies that $(S_0(y_1), \ldots, S_0(y_n))$ is a left accumulation point of points $(S_0(x_1), \ldots, S_0(x_n))$ for which $\prod_1^n G_P(S_0(x_\nu)) \leq c$.

Since G_P is left continuous, this implies $\prod_1^n G_P(S_0(y_\nu)) \leq c$. Hence $(y_1, \ldots, y_n) \in A_P(c)$. Since $(y_1, \ldots, y_n) \in \bigcap_1^\infty A_{P_i}(c_i)$ was arbitrary, this implies $\bigcap_1^\infty A_{P_i}(c_i) \subset A_P(c)$. Similarly, $P_0^n\left(\bigcap_1^\infty A_{P_i}(c_i)\right) \geq P_0^n(A_P(c))$ implies $\bigcap_1^\infty A_{P_i}(c_i) \supset A_P(c)$. This verifies condition (1.7.30).

Hence Lemma 1.7.27 implies the existence of a measurable function $S_n : X^n \to [0, 1]$ such that for every $P \in \mathfrak{P}$, $c \geq 0$,

$$\{(x_1, \ldots, x_n) \in X^n : \prod_{\nu=1}^n G_P(S_0(x_\nu)) \leq c\} \qquad (1.7.25)$$

$$= \{(x_1, \ldots, x_n) \in X^n : S_n(x_1, \ldots, x_n) \leq F_P^{(n)}(c)\}$$

with $F_P^{(n)}(c) := P_0^n\{(x_1, \ldots, x_n) \in X^n : \prod_1^n G_P(S_0(x_\nu)) \leq c\}$.

By Lemma 1.7.40 there exists an isotone function $G_P^{(n)}$ such that

$$r \leq F_P^{(n)}(c) \quad \text{iff} \quad G_P^{(n)}(r) \leq c.$$

Hence (1.7.25) implies for $n \in \mathbb{N}_0$, $c \geq 0$

$$\{(x_1, \ldots, x_n) \in X^n : \prod_{\nu=1}^{n} G_P(S_0(x_\nu)) \leq c\}$$
$$= \{(x_1, \ldots, x_n) \in X^n : G_P^{(n)}(S_n(x_1, \ldots, x_n)) \leq c\},$$

i.e.

$$\prod_{\nu=1}^{n} G_P(S_0(x_\nu)) = G_P^{(n)}(S_n(x_1, \ldots, x_n)) \qquad (1.7.26)$$

for all $n \in \mathbb{N}_0$, $(x_1, \ldots, x_n) \in X^n$.

(iv) By (1.7.26), for all $n \in \mathbb{N}_0$, the inequality

$$\prod_{\nu=1}^{n} G_{P_1}(S_0(x_\nu)) < \prod_{\nu=1}^{n} G_{P_1}(S_0(y_\nu)) \qquad \text{for some } P_1 \in \mathfrak{P}$$

implies $S_n(x_1, \ldots, x_n) < S_n(y_1, \ldots, y_n)$ which, in turn, implies by (1.7.26)

$$\prod_{\nu=1}^{n} G_P(S_0(x_\nu)) \leq \prod_{\nu=1}^{n} G_P(S_0(y_\nu)) \qquad \text{for all } P \in \mathfrak{P}.$$

By Lemma 1.7.41, applied for $f_1 = \log G_{P_1} \circ S_0$, $f_2 = \log G_P \circ S_0$ and $X = \{x : G_{P_1}(S_0(x)), G_P(S_0(x)) \in (0, \infty)\}$, there exist functions $a : \mathfrak{P} \to [0, \infty)$ and $C : \mathfrak{P} \to (0, \infty)$ and a measurable function $S : X \to \mathbb{R}$ such that

$$G_P(S_0(x)) = C(P) \exp[a(P)S(x)] \qquad \text{for } P_0\text{–a.a. } x \in X.$$

Together with (1.7.21) this implies the assertion. $\square$

The following Lemma 1.7.27 combines Hilfssatz 1 in Pfanzagl (1960, p. 171) and Lemma 1 in Borges and Pfanzagl (1963, p. 112).

Lemma 1.7.27. *Let $P|\mathcal{A}$ be a p–measure and $\mathcal{C} \subset \mathcal{A}$ a system of sets which is totally ordered by inclusion (P), (i.e., if $A, B \in \mathcal{A}$, then $A \subset B$ (P) or $B \subset A$ (P), where $A \subset B$ (P) means $P(A - B) = 0$).*
Then there exists an $\mathcal{A}$–measurable function $S : X \to [0, 1]$ such that

$$A = \{x \in X : S(x) \leq P(A)\} \quad (P) \quad \text{for each } A \in \mathcal{C}, \qquad (1.7.28)$$

and

$$S(y) = P\{x \in X : S(x) \leq S(y)\} \quad \text{for each } y \in X. \qquad (1.7.29)$$

Addendum. *If for all $A_i \in C$, $i \in \{0\} \cup \mathbb{N}$*

$$P(A_0) = P(\bigcap_{i=1}^{\infty} A_i) \quad \text{implies} \quad A_0 = \bigcap_{i=1}^{\infty} A_i, \tag{1.7.30}$$

then (1.7.28) holds without the restriction "(P)".

Proof. (i) Let first $\mathcal{D}$ be a subsystem of $\mathcal{A}$ which is totally ordered by inclusion, and for which $D_1, D_2 \in \mathcal{D}$, $D_1 \neq D_2$ implies $P(D_1) \neq P(D_2)$. W.l.g. we may assume that $P(D_0) = 1$ for some $D_0 \in \mathcal{D}$ (for otherwise we may add X to the system $\mathcal{D}$).

Let

$$S(x) := \begin{cases} \inf\{P(A) : A \in \mathcal{D}, \ x \in A\} & x \in D_0. \\ 1 & x \notin D_0. \end{cases} \tag{1.7.31}$$

As an immediate consequence of (1.7.31), we obtain for all $r \in [0,1]$

$$\{x \in X : S(x) < r\} = \cup\{A : A \in \mathcal{D}, \ P(A) < r\}. \tag{1.7.32}$$

Let

$$s_0 := \sup\{P(A) : A \in \mathcal{D}, \ P(A) < r\}.$$

If there exists $A_0 \in \mathcal{D}$, $P(A_0) < r$ with $P(A_0) = s_0$, then

$$\cup\{A : A \in \mathcal{D}, \ P(A) < r\} = A_0. \tag{1.7.33}$$

If not, let $A_n \in \mathcal{D}$, $P(A_n) < r$ be such that $P(A_n) \uparrow s_0$. Then

$$\cup\{A : A \in \mathcal{D}, \ P(A) < r\} = \bigcup_{n=1}^{\infty} A_n. \tag{1.7.34}$$

In either case, the set on the right side of (1.7.32) is measurable, which implies, in particular, that S is measurable.

Let $D \in \mathcal{D}$ be arbitrary. (1.7.32) implies for all $r > P(D)$

$$D \subset \{x \in X : S(x) < r\},$$

hence

$$D \subset \{x \in X : S(x) \leq P(D)\} \qquad \text{for each } D \in \mathcal{D}. \tag{1.7.35}$$

Furthermore, from (1.7.32), (1.7.33) and (1.7.34),

$$P\{x \in X : S(x) < r\} \leq r \qquad \text{for all } r \in [0,1],$$

whence

$$P\{x \in X : S(x) \leq r\} \leq r \qquad \text{for all } r \in [0,1]. \tag{1.7.36}$$

Applied for $r = P(D)$ this implies, together with (1.7.35),

$$D = \{x \in X : S(x) \leq P(D)\} \quad (P) \qquad \text{for each } D \in \mathcal{D}. \tag{1.7.37}$$

(ii) If $\mathcal{C}$ fulfills assumption (1.7.30), then it is totally ordered by inclusion and for $C_1, C_2 \in \mathcal{C}$, $C_1 \neq C_2$ implies $P(C_1) \neq P(C_2)$. Hence we may apply the results obtained under (i) for $\mathcal{D} = \mathcal{C}$. To prove that (1.7.35) is, in this case, an equality, we proceed as follows: If there exists $r_0 > P(D)$ such that $\{A \in \mathcal{C} : P(D) < P(A) < r_0\} = \emptyset$, we have for all $r \in (P(D), r_0]$

$$D = \cup\{A : A \in \mathcal{C}, \ P(A) < r\} = \{x \in X : S(x) < r\},$$

hence

$$D = \{x \in X : S(x) \leq P(D)\}. \tag{1.7.38}$$

If $P(D)$ is a right accumulation point of $\{P(A) : A \in \mathcal{C}, \ P(D) < P(A)\}$, we choose a decreasing sequence A_n such that $P(A_n) \downarrow P(D)$, $P(A_n) > P(D)$. Then by (1.7.35) and (1.7.32)

$$D \subset \{x \in X : S(x) \leq P(D)\} \subset \{x \in X : S(x) < P(A_n)\}$$
$$= \cup\{A : A \in \mathcal{C}, \ P(A) < P(A_n)\} \subset A_n,$$

and therefore

$$D \subset \{x \in X : S(x) \leq P(D)\} \subset \bigcap_{n=1}^{\infty} A_n. \tag{1.7.39}$$

Since $P(\bigcap_{n=1}^{\infty} A_n) = \lim_{n \to \infty} P(A_n) = P(D)$, we obtain from (1.7.30) that $D = \bigcap_{n=1}^{\infty} A_n$, i.e. (1.7.38) holds true also in this case.

(iii) If $\mathcal{C}$ is totally ordered by inclusion (P), we choose a countable subclass $\mathcal{C}_0 \subset \mathcal{C}$ such that $\{P(A) : A \in \mathcal{C}_0\}$ has the following properties: It is dense in $M := \{P(A) : A \in \mathcal{C}\}$, contains all points of M which are not right accumulation points of M, and $A, B \in \mathcal{C}_0$, $A \neq B$ implies $P(A) \neq P(B)$.

Since $\mathcal{C}_0$ is countable, there exists a P–null set N such that $\mathcal{D} := \{C \cup N : C \in \mathcal{C}_0\}$ is totally ordered by inclusion. If we apply the result of (i) for $\mathcal{D}$ we obtain that (1.7.37) holds true for all $D \in \mathcal{C}_0$. To prove this relation for all elements of $\mathcal{C}$, we remark that by definition of $\mathcal{C}_0$ for every $C \in \mathcal{C}$ and every $r > P(C)$ there exists $D \in \mathcal{C}_0$ such that $P(C) \leq P(D) < r$. Hence $C \subset D$ (P). Together with (1.7.37)

$$C \subset \{x \in X : S(x) \leq P(D)\} \subset \{x \in X : S(x) < r\} \quad (P).$$

Since $r > P(C)$ was arbitrary, this implies

$$C \subset \{x \in X : S(x) \leq P(C)\} \quad (P).$$

Together with (1.7.36) this proves (1.7.28).

(iv) Now we shall prove (1.7.29). Let $y \in X$ be arbitrary. By (1.7.31) (applied for $\mathcal{D} = \{C \cup N : C \in \mathcal{C}_0\}$) for every $r > S(y)$ there exists $A \in \mathcal{C}_0$ such that

$$S(y) \leq P(A) < r.$$

By (1.7.28), $A = \{x \in X : S(x) \leq P(A)\} \subset \{x \in X : S(x) < r\}$ (P), whence

$$S(y) \leq P(A) \leq P\{x \in X : S(x) < r\}.$$

Since $r > S(y)$ was arbitrary, this implies

$$S(y) \leq P\{x \in X : S(x) \leq S(y)\}.$$

Together with (1.7.36) this implies (1.7.29).   □

Lemma 1.7.40. *If the function $F : \mathbb{R} \to \mathbb{R}$ is isotone and right continuous, then there exists an isotone left continuous function $G : F(\mathbb{R}) \to [-\infty, \infty)$ such that for all $r \in F(\mathbb{R})$, $c \in \mathbb{R}$,*

$$r \leq F(c) \quad \text{iff} \quad G(r) \leq c.$$

Proof. Let $a := \inf F(\mathbb{R})$, $b := \sup F(\mathbb{R})$, and assume w.l.g. that $a < b$. Define G by

$$G(r) = \begin{cases} \sup\{t \in \mathbb{R} : F(t) < r\} & r \in (a, b) \cap F(\mathbb{R}), \\ -\infty & r = a \in F(\mathbb{R}), \\ \inf\{t \in \mathbb{R} : F(t) = b\} & r = b \in F(\mathbb{R}). \end{cases}$$

Then $G : F(\mathbb{R}) \to [-\infty, \infty)$, and $G(r) = -\infty$ iff $r = a \in F(\mathbb{R})$.

Let $r \in F(\mathbb{R})$ and $c \in \mathbb{R}$ with $r \leq F(c)$. If $r = a$, then $G(r) = -\infty \leq c$. If $r = b$, then $F(c) = b$ and $G(b) \leq c$ by the definition of $G(b)$. If $r \in (a, b) \cap F(\mathbb{R})$, then for all $t \in \mathbb{R}$ with $F(t) < r$ we have $t < c$ which implies that $G(r) = \sup\{t \in \mathbb{R} : F(t) < r\} \leq c$. Conversely, let $r \in F(\mathbb{R})$ and $c \in \mathbb{R}$ with $G(r) \leq c$. Since $a \leq F(c)$ is true for all $c \in \mathbb{R}$ we can assume that $r \in (a, b] \cap F(\mathbb{R})$. For $r = b$ we have $F(c) = b$. If $r \in (a, b) \cap F(\mathbb{R})$ then $G(r) \leq c$ implies $\{t \in \mathbb{R} : F(t) < r\} \subset (-\infty, c]$ or equivalently $(c, \infty) \subset \{t \in \mathbb{R} : F(t) \geq r\}$. This, together with right continuity of F, implies $F(c) \geq r$. Hence for all $r \in F(\mathbb{R})$ and $c \in \mathbb{R}$ we have $r \leq F(c)$ iff $G(r) \leq c$.

To prove left continuity of G let $r_n \in F(\mathbb{R})$, $n \in \mathbb{N}$, with $\lim_{n \to \infty} r_n = r \in F(\mathbb{R})$, and $r_n < r$ for all $n \in \mathbb{N}$. If $r < b$, then $a \leq r_n < b$ and $G(r_n) \leq G(r)$ for all $n \in \mathbb{N}$. Furthermore, for every $\varepsilon > 0$ there exists $t \in \mathbb{R}$ such that $G(r) \leq t + \varepsilon$ and $F(t) < r$. Then $F(t) < r_n$ for all but finitely many $n \in \mathbb{N}$, and therefore $G(r_n) \geq t \geq G(r) - \varepsilon$ for these $n \in \mathbb{N}$. As $\varepsilon > 0$ was arbitrary, this proves $\lim_{n \to \infty} G(r_n) = G(r)$. If $r = b$, then $t < G(b)$ implies $F(t) < b$, and therefore $F(t) < r_n$ for all but finitely many $n \in \mathbb{N}$. This implies $G(r_n) \geq t$ for these $n \in \mathbb{N}$. Since $t < G(b)$ was arbitrary, we obtain $\underline{\lim}_{n \to \infty} G(r_n) \geq G(b)$.

This together with $G(r_n) \leq G(b)$ for all $n \in \mathbb{N}$ implies $\lim_{n\to\infty} G(r_n) = G(b)$.

$\square$

The following lemma is due to Borges and Pfanzagl (1963, p. 113, Lemma 2).

Lemma 1.7.41. *Let $f_k : X \to \mathbb{R}$, $k = 1, 2$, be two functions with the following property:*

There exists an infinite subset $\mathbb{N}_0 \subset \mathbb{N}$ such that for all $n \in \mathbb{N}_0$ and all $x_\nu, y_\nu \in X$, $\nu = 1, \ldots, n$,

$$\sum_{\nu=1}^{n} f_1(x_\nu) < \sum_{\nu=1}^{n} f_1(y_\nu) \tag{1.7.42}$$

implies

$$\sum_{\nu=1}^{n} f_2(x_\nu) \leq \sum_{\nu=1}^{n} f_2(y_\nu). \tag{1.7.43}$$

Then there exists a function $f_0 : X \to \mathbb{R}$ and constants $a_k \geq 0$ and b_k, $k = 1, 2$, such that

$$f_k(x) = a_k f_0(x) + b_k, \qquad k = 1, 2. \tag{1.7.44}$$

Proof. If f_1 or f_2 is constant, the assertion is trivial. Hence assume that there exists $x_1, x_2 \in X$ such that $f_1(x_1) < f_1(x_2)$ and $f_2(x_1) \neq f_2(x_2)$. For every $\nu \in \mathbb{N}$ and $x \in X$ there exists an integer $m_\nu(x)$ such that

$$\frac{m_\nu(x)}{\nu}\left(f_1(x_2) - f_1(x_1)\right) < f_1(x) - f_1(x_1) \tag{1.7.45}$$

$$< \frac{m_\nu(x) + 2}{\nu}\left(f_1(x_2) - f_1(x_1)\right).$$

Both (strict) inequalities are easily transformed into (strict) inequalities of the type (1.7.42) (with $n = \nu + |m_\nu(x)| + 2$; if these n do not belong to $\mathbb{N}_0$, add identical terms on both sides). By assumption, these (strict) inequalities imply corresponding (weak) inequalities with f_1 replaced by f_2. From these (weak) inequalities we obtain

$$\frac{m_\nu(x)}{\nu}\left(f_2(x_2) - f_2(x_1)\right) \leq f_2(x) - f_2(x_1) \tag{1.7.46}$$

$$\leq \frac{m_\nu(x) + 2}{\nu}\left(f_2(x_2) - f_2(x_1)\right).$$

Hence $f_2(x_2) - f_2(x_1) > 0$. Let

$$f_0(x) := \frac{f_1(x) - f_1(x_1)}{f_1(x_2) - f_1(x_1)}. \qquad (1.7.47)$$

Relations (1.7.45) and (1.7.46) together imply for all $\nu \in \mathbb{N}_0$:

$$f_0(x) - \frac{2}{\nu} < \frac{m_\nu(x)}{\nu} \leq \frac{f_2(x) - f_2(x_1)}{f_2(x_2) - f_2(x_1)} \leq \frac{m_\nu(x) + 2}{\nu} < f_0(x) + \frac{2}{\nu},$$

hence

$$\frac{f_2(x) - f_2(x_1)}{f_2(x_2) - f_2(x_1)} = f_0(x).$$

Together with (1.7.47) this implies (1.7.44) with $a_k = f_k(x_2) - f_k(x_1)$, $b_k = f_k(x_1)$, $k = 1, 2$. We remark that the assumption of the Lemma implies $a_k \geq 0$, $k = 1, 2$. $\qquad\square$

Lemma 1.7.48. *Let $P|A$ be a p–measure and let f and g be real valued functions fulfilling*

$$\{x \in X : f(x) \leq c\} = \{x \in X : g(x) \leq c\} \quad (P) \quad \text{for } c \in \mathbb{R}. \qquad (1.7.49)$$

Then $f = g$ P-a.e.

Proof. By assumption (1.7.49), we have $P(A_{k,n}) = 0$ for $k \in \mathbb{Z}$, $n \in \mathbb{N}$, where

$$A_{k,n} = \{x \in X : f(x) \in (\frac{k}{n}, \frac{k+1}{n}] \quad \text{and} \quad g(x) \notin (\frac{k}{n}, \frac{k+1}{n}]\}.$$

Since

$$\{x \in X : f(x) \neq g(x)\} = \bigcup_{n=1}^{\infty} \{x \in X : |f(x) - g(x)| > \frac{1}{n}\} = \bigcup_{n \in \mathbb{N}, k \in \mathbb{Z}} A_{k,n},$$

this implies the assertion. $\qquad\square$

1.8 Ancillary statistics

Presents some results for ancillary statistics, including Basu's theorem. These results are illustrated by several examples.

Definition 1.8.1. A function $T : (X, \mathcal{A}) \rightarrow (Z, \mathcal{C})$ is *ancillary* for $\mathfrak{P}$ if the induced p–measure $P \circ T$ is the same for every $P \in \mathfrak{P}$.

Hence knowing $T(x)$ only (instead of the observation x) gives no information about the underlying p–measure P. This must not be interpreted in the sense that $T(x)$ "carries no information about P". Given a statistic $S|X$, the combined statistic $x \to (S(x), T(x))$ will, in general, enable more accurate assertions about P than $S(x)$ alone (excepting, of course, the case of S being sufficient). For the purpose of illustration, consider Example 9.1.1 referring to the family $\{N^n_{(\vartheta, a^2 \vartheta^2)} : \vartheta > 0\}$, with $a > 0$ known. Consider the statistic $S(x_1, \ldots, x_n) = \bar{x}_n$, and the ancillary statistic $T(x_1, \ldots, x_n) = \bar{x}_n/s_n$. The statistic (S, T) is sufficient and renders estimator sequences for ϑ with asymptotic variance $a^2\vartheta^2/(1 + 2a^2)$, the asymptotically optimal estimator sequence for ϑ, based on $S(x)$, has asymptotic variance $a^2\vartheta^2$.

Theorem 1.8.2. (D.Basu, 1955a and 1958). *Let $\mathfrak{P}|\mathcal{A}$ be dominated. Assume that $S : (X, \mathcal{A}) \to (Y, \mathcal{B})$ is sufficient.*

(i) *If S, T are P_0–independent for some $P_0 \in \mathfrak{P}$ with $\mathfrak{P} \ll P_0$, then T is ancillary and S, T are P–independent for every $P \in \mathfrak{P}$.*

(ii) *If T is ancillary and $\mathfrak{P} \circ S$ boundedly complete, then S, T are P–independent for every $P \in \mathfrak{P}$.*

Proof. Let $C \in \mathcal{C}$ be arbitrary. Since S is sufficient, there exists $\varphi_{T^{-1}C} : (X, \mathcal{A}) \to ([0,1], \mathbb{B}_0)$ which is a conditional expectation of $1_{T^{-1}C}$, given S, with respect to every $P \in \mathfrak{P}$, i.e.

$$P(\varphi_{T^{-1}C} \circ S 1_{S^{-1}B}) = P(T^{-1}C \cap S^{-1}B) \quad \text{for } B \in \mathcal{B}, \ P \in \mathfrak{P}. \quad (1.8.3)$$

(i) P_0–independence of S, T implies (see Criterion 1.8.13)

$$\varphi_{T^{-1}C} \circ S = P_0(T^{-1}C) \quad P_0\text{–a.e.}, \quad (1.8.4)$$

and therefore P–a.e. for every $P \in \mathfrak{P}$. Together with (1.8.3), applied with $B = Y$, this implies

$$P(T^{-1}C) = P_0(T^{-1}C) \quad \text{for } P \in \mathfrak{P}. \quad (1.8.5)$$

Since $C \in \mathcal{C}$ was arbitrary, $P \circ T|\mathcal{C} = P_0 \circ T|\mathcal{C}$ for $P \in \mathfrak{P}$. Applying (1.8.3) again, now with arbitrary $B \in \mathcal{B}$, we obtain from (1.8.4) and (1.8.5) that

$$P(\varphi_{T^{-1}C} \circ S)P(S^{-1}B) = P(T^{-1}C \cap S^{-1}B).$$

Since $P(\varphi_{T^{-1}C} \circ S) = P(T^{-1}C)$, the functions S, T are P–independent.

(ii) Let $P_0 \in \mathfrak{P}$ be fixed. Since $P \circ T|\mathcal{C} = P_0 \circ T|\mathcal{C}$, relation (1.8.3), applied with $B = Y$, yields

$$P(\varphi_{T^{-1}C} \circ S - P_0(T^{-1}C)) = 0 \quad \text{for } P \in \mathfrak{P}.$$

Since $\mathfrak{P} \circ S$ is boundedly complete, this implies $\varphi_{T^{-1}C} \circ S = P_0(T^{-1}C)$ P–a.e. From this, P–independence of S, T follows as in (i). $\qquad \square$

Observe that bounded completeness in 1.8.2(ii) cannot be dispensed with. The sample correlation coefficient r_n is ancillary for $\{N^n_{(\mu_1,\mu_2,\sigma_1^2,\sigma_2^2,\varrho_0)} : \mu_i \in \mathbb{R}, \ \sigma_i^2 > 0\}$ (with $\varrho_0 \neq 0$ fixed) and (see (4.7.16)) $S_*((x_1,y_1),\ldots,(x_n,y_n)) = (\bar{x}_n, \bar{y}_n, s_n^2(\underline{x}), s_n^2(\underline{y}), r_n(\underline{x},\underline{y}))$ is equivariant and minimal sufficient for this family, but r_n and S_* are obviously not stochastically independent.

Remark 1.8.6. A number of applications of Theorem 1.8.2(ii) is of the following type: $\{P_{\vartheta,\tau} : \vartheta \in \Theta, \ \tau \in T\}$ is such that the conditions of 1.8.2(ii) are fulfilled for every subfamily $\{P_{\vartheta,\tau} : \vartheta \in \Theta\}$, with $\tau \in T$ fixed, i.e. there exists a statistic $S|X$ which is sufficient for $\{P_{\vartheta,\tau} : \vartheta \in \Theta\}$, for every $\tau \in T$, and $\{P_{\vartheta,\tau} \circ S : \vartheta \in \Theta\}$ is boundedly complete. Then S and T are stochastically independent under $P_{\vartheta,\tau}$ for every $\vartheta \in \Theta$, $\tau \in T$, provided $P_{\vartheta,\tau} \circ T$ depends on τ only.

Example 1.8.7. For $a, b > 0$ let $\Gamma_{a,b}|\mathbb{B}_+$ denote the gamma distribution with Lebesgue density

$$x \to \frac{1}{a^b \Gamma(b)} x^{b-1} \exp[-x/a], \qquad x > 0.$$

Then $\bar{x}_n$ and $\bar{x}_n / \left(\prod_1^n x_\nu\right)^{1/n}$ are stochastically independent under $\Gamma^n_{a,b}$ for all $a, b > 0$. The statistic $\bar{x}_n$ is sufficient for $\{\Gamma^n_{a,b} : a > 0\}$ for every $b > 0$, and $\{\Gamma^n_{a,b} \circ \bar{x}_n : a > 0\}$ is complete.

Moreover, the distribution of $\bar{x}_n / \left(\prod_1^n x_\nu\right)^{1/n}$ under $\Gamma^n_{a,b}$ does not depend on a:

$$\Gamma^n_{a,b} \circ \left(\bar{x}_n / \left(\prod_1^n x_\nu\right)^{1/n}\right) = \Gamma^n_{1,b} \circ \left(n^{-1} \sum_1^n a x_\nu / \left(\prod_1^n a x_\nu\right)^{1/n}\right)$$

$$= \Gamma^n_{1,b} \circ \bar{x}_n / \left(\prod_1^n x_\nu\right)^{1/n}.$$

Hence the independence of $\bar{x}_n$ and $\bar{x}_n / \left(\prod_1^n x_\nu\right)^{1/n}$ follows according to Remark 1.8.6.

Example 1.8.7 infers properties of statistics for a given family of p–measures. For a converse result, characterizing the gamma distributions by properties of statistics, see Laha (1954) and Lukacs (1956, Section 7).

Example 1.8.8. For the family $\{N_{(\mu,\sigma^2)} : \mu \in \mathbb{R}, \ \sigma^2 > 0\}$, the statistics $\bar{x}_n$ and $T_n(x_1,\ldots,x_n)$ are stochastically independent under $N^n_{(\mu,\sigma^2)}$, for every $\mu \in \mathbb{R}$, $\sigma^2 > 0$, provided $T_n(x_1+\mu,\ldots,x_n+\mu) = T_n(x_1,\ldots,x_n)$. Since $\bar{x}_n$ is sufficient for $\{N^n_{(\mu,\sigma^2)} : \mu \in \mathbb{R}\}$ for every fixed $\sigma^2 > 0$, and $\{N^n_{(\mu,\sigma^2)} \circ \bar{x}_n : \mu \in \mathbb{R}\}$ is complete, this follows according to Remark 1.8.6. As a particular consequence we obtain that $\bar{x}_n$ and $\left(n^{-1}\sum_1^n(x_\nu - \bar{x}_n)^2\right)^{1/2}$ are stochastically independent under $N_{(\mu,\sigma^2)}$ for all $\mu \in \mathbb{R}$, $\sigma^2 > 0$.

Exercise 1.8.9. The stochastic independence of $\bar{x}_n$ and $\left(n^{-1}\sum_1^n(x_\nu - \bar{x}_n)^2\right)^{1/2}$ is not implied by the equivariance and invariance, respectively, under the transformation $x \to x + \mu$. Show that $\tilde{x}_n$ (the sample median) and $n^{-1}\sum_1^n |x_\nu - \tilde{x}_n|$ are not stochastically independent. (Hint:

$$\frac{N_{(0,1)}^3\left\{\tilde{x}_3 \in (m, m+1), \sum_1^3 |x_\nu - \tilde{x}_3| \in (0,1)\right\}}{N_{(0,1)}^3\left\{\tilde{x}_3 \in (m, m+1)\right\}N_{(0,1)}^3\left\{\sum_1^3 |x_\nu - \tilde{x}_3| \in (0,1)\right\}}$$

tends to 0 as m tends to infinity.)

Exercise 1.8.10. Show that $(\bar{x}_n, \bar{y}_n)$ and $\left(s_n^2(\underline{x}), s_n^2(\underline{y}), r_n(\underline{x}, \underline{y})\right)$ are stochastically independent under $N_{(\mu_1,\mu_2,\sigma_1^2,\sigma_2^2,\varrho)}^n$. If $\varrho = 0$, then $r_n(\underline{x}, \underline{y})$ is stochastically independent of $\left(\bar{x}_n, \bar{y}_n, s_n^2(\underline{x}), s_n^2(\underline{y})\right)$.

Proposition 1.8.11. *Let* $T : (X, \mathcal{A}) \to (Z, \mathcal{C})$ *be ancillary. Assume that* $S : (X, \mathcal{A}) \to (Y, \mathcal{B})$ *and* T *are stochastically independent under every* $P \in \mathfrak{P}$. *Then the following holds true.*

For $f : (Y \times Z, \mathcal{B} \times \mathcal{C}) \to (W, \mathcal{D})$, *the conditional distribution of* $x \to f(S(x), T(x))$, *given* S, *under* P *can be described by the Markov kernel (not depending on* P*)*

$$M(y, D) := \int 1_D(f(y, z)) P_0 \circ T(dz), \tag{1.8.12}$$

where $P_0 \in \mathfrak{P}$ *is arbitrarily fixed.*

Hence S *is sufficient for* $\mathfrak{P}$, *restricted to the sub-σ-algebra generated by* S *and* T.

Addendum. *If there exists* $f_0 : (Y \times Z, \mathcal{B} \times \mathcal{C}) \to (X, \mathcal{A})$ *such that*

$$f_0(S(x), T(x)) = x \qquad for \ x \in X,$$

then S *is sufficient for* $\mathfrak{P}$.

This result has the following intuitive interpretation: Let x be a realization governed by P. If instead of x only $S(x)$ is known, compute $f_0(S(x), T(u))$, where u is a realization governed by P_0. The random variable generated by this combined experiment (which can be carried through without the knowledge of P) is distributed according to P.

Proof. Since $M(\cdot, D)$ is measurable for every $D \in \mathcal{D}$, and since $M(y, \cdot)|\mathcal{D}$ is a p–measure for $y \in Y$, it remains to be shown that $y \to M(y, D)$ is a conditional expectation of $1_D(f \circ (S, T))$, given S.

Since S, T are P–independent, and since $P \circ T = P_0 \circ T$, this follows from

$$\int 1_D\big(f(S(x), T(x))\big)1_B(S(x))P(dx)$$

$$= \int 1_D\big(f(y, z)\big)1_B(y)P \circ S(dy)P \circ T(dz)$$

$$= \int 1_D\big(f(y, z)\big)1_B(y)P \circ S(dy)P_0 \circ T(dz)$$

$$= \int M(y, D)1_B(y)P \circ S(dy) \qquad \text{for } B \in \mathcal{B}. \qquad \square$$

Criterion 1.8.13. *Let $S : (X, \mathcal{A}) \to (Y, \mathcal{B})$ and $T : (X, \mathcal{A}) \to (Z, \mathcal{C})$ be measurable functions, $P|\mathcal{A}$ a p–measure.*

(i) If there exists a constant version of $P^S 1_{T^{-1}C}$ for every $C \in \mathcal{C}$ (which necessarily is $P(T^{-1}C)$), then S and T are P–independent.

(ii) If S and T are P–independent, then $P(T) \in P^S T$ (if T is P–integrable). Since P–independence of S and T implies P–independence of S and $1_{T^{-1}C}$ for every $C \in \mathcal{C}$, this implies $P(T^{-1}C) \in P^S 1_{T^{-1}C}$.

Proof. (i) If there exists a constant $K_C \in P^S 1_{T^{-1}C}$, we have for all $B \in \mathcal{B}$

$$P(T^{-1}C \cap S^{-1}B) = \int K_C 1_B(y)P \circ S(dy) = K_C P(S^{-1}B).$$

Applied for $B = Y$ this yields $K_C = P(T^{-1}C)$, hence $P(T^{-1}C \cap S^{-1}B) = P(T^{-1}C)P(S^{-1}B)$.

(ii) P–independence of S and T implies P–independence of $1_{S^{-1}B}$ and T for every $B \in \mathcal{B}$, hence

$$\int T(x)1_{S^{-1}B}(x)P(dx) = P(T)P(S^{-1}B) = \int P(T)1_{S^{-1}B}(x)P(dx).$$

Therefore, $P(T) \in P^S T$. $\qquad \square$

The reader interested in more details about the concept of ancillarity may consult the survey paper by Lehmann and Scholz (1992).

1.9 Equivariance and invariance

Introduces transformation groups on X and equivariant statistics. A family of mutually absolutely continuous p–measures generated by a transformation group is dominated by the Haar measure. Every minimal sufficient statistic is almost equivariant. The conditional expectation of an almost invariant function with respect to an equivariant sufficient statistic is almost invariant.

Let $(X, \mathcal{A})$ be a measurable space and G a group of $1-1$ transformations $a : X \to X$, endowed with a σ–algebra $\mathcal{D}$, such that

$$(a, b) \to ab \text{ is } \mathcal{D} \times \mathcal{D}, \mathcal{D}\text{–measurable}$$
$$a \to a^{-1} \text{ is } \mathcal{D}, \mathcal{D}\text{–measurable}, \tag{1.9.1'}$$

and

$$(a, x) \to ax \text{ is } \mathcal{D} \times \mathcal{A}, \mathcal{A}\text{–measurable}. \tag{1.9.1''}$$

Throughout the following we assume that $a\mathcal{A} = \mathcal{A}$ and $a\mathcal{D} = \mathcal{D}$ for $a \in G$. $Gx := \{ax : a \in G\}$ is the *orbit* of x, generated by G. Two orbits Gx', Gx'' are either identical or disjoint.

Definition 1.9.2. A function $S : X \to Y$ is *equivariant* if

$$S(x') = S(x'') \text{ implies } S(ax') = S(ax'') \quad \text{for } x', x'' \in X, \ a \in G. \tag{1.9.3}$$

The function S is *invariant* if

$$S(ax) = S(x) \quad \text{for } x \in X, \ a \in G. \tag{1.9.4}$$

An invariant function is constant on each orbit. A function is *maximal invariant* if it attains different values on different orbits.

Given a measure $\mu | \mathcal{A}$, a measurable function S is *μ–almost equivariant* [*μ–almost invariant*] if relation (1.9.3) [respectively (1.9.4)] holds for all x', x'' outside a μ–null set which may depend on $a \in G$.

If $S : X \to Y$ is equivariant, the transformation group G, acting on X, induces a transformation group, say $\overline{G}$, acting on $\overline{Y} := S(X)$, with elements $\overline{a}$ defined by

$$\overline{a}y := S(aS^{-1}\{y\}), \qquad y \in \overline{Y}.$$

(Notice that $S(aS^{-1}\{y\})$ contains a unique element of $\overline{Y}$.) The induced transformation group $\overline{G}$ fulfills (1.9.1) with $\overline{\mathcal{B}} = \mathcal{B} \cap \overline{Y}$ in place of $\mathcal{A}$ and $\overline{\mathcal{D}}$, the set of all subsets $\overline{D}$ of $\overline{G}$ such that $\{a \in G : \overline{a} \in \overline{D}\} \in \mathcal{D}$, in place of $\mathcal{D}$.

The map $a \to \overline{a}$ is a homomorphism, i.e. $\overline{ab} = \overline{a}\overline{b}$.

A measure $\nu|\mathcal{D}$ is *left invariant* if

$$\nu(aD) = \nu(D) \qquad \text{for } D \in \mathcal{D}, \ a \in G.$$

The definition of right invariance is an obvious modification.

The group G, endowed with a Hausdorff topology $\mathcal{U}$, is a *topological group* if the operations $(a, b) \to ab$ and $a \to a^{-1}$ are continuous.

Theorem 1.9.5. *If G is a locally compact topological group, there exists a left (as well as a right) invariant regular measure, the so-called left invariant Haar measure. Either of these measures is unique up to multiplication by a constant and finite [σ-finite] if G is compact [σ-compact].*

Proof. Nachbin (1965), p. 65, Theorem 1, and p. 75, Proposition 4.   □

As a typical example of a transformation group we mention the linear transformations on $\mathbb{R}^p$, assigning to each pair $(a, c) \in \mathbb{R} \times (0, \infty)$ the transformation $x \to a + cx$, $x \in \mathbb{R}^p$. The group operation is $(a'', c'')(a', c') = (a'' + c''a', c''c')$; the pertaining left invariant Haar measure on $(\mathbb{R} \times (0, \infty), \ \mathbb{B} \times \mathbb{B}_+)$ has λ^2-density $(u, v) \to v^{-1}$, $(u, v) \in \mathbb{R} \times (0, \infty)$.

Starting from a p-measure $P|\mathcal{A}$, let $P_a|\mathcal{A}$ be the p-measure defined for $a \in G$ by

$$P_a(A) := P(a^{-1}A), \qquad A \in \mathcal{A}.$$

In other words, P_a is the p-measure induced by P and the map $x \to ax$. The relationship $(P_a)_b = P_{ba}$ is straightforward.

Lemma 1.9.6. *Let $\nu|\mathcal{D}$ be left invariant and σ-finite, and $\mu|\mathcal{D}$ σ-finite. Then $\nu(D) = 0$ implies $\mu(aD) = 0$ for ν-a.a. $a \in G$.*

Proof. For $D \in \mathcal{D}$ with $\nu(D) = 0$ we have

$$\int \mu(bD)\nu(db) = \mu \times \nu\{(a, b) \in G^2 : b^{-1}a \in D\}$$
$$= \int \nu(aD^{-1})\mu(da) = \int \nu(D^{-1})\mu(da) = 0,$$

since $\nu(D) = 0$ implies $\nu(D^{-1}) = 0$.   □

Proposition 1.9.7. *Given a p-measure $Q|\mathcal{D}$, let $Q_a(D) := Q(a^{-1}D)$ for $D \in \mathcal{D}$, $a \in G$.*

If $\{Q_a : a \in G\}$ is dominated by some σ-finite measure, then this family is dominated by the Haar measure.

Proof. Let $\mu|\mathcal{D}$ denote the dominating σ–finite measure. If $\nu(D) = 0$, there exists by Lemma 1.9.6 an element $a_0 \in G$ such that $\mu(a_0 D) = 0$. Since $Q_{a_0 a} \ll \mu$ for every $a \in G$, this implies $Q_a(D) = Q_{a_0 a}(a_0 D) = 0$. $\square$

The following theorem, giving conditions under which any almost invariant statistic is equivalent to an invariant statistic, is due to Stein. The same result under conditions of a different nature occurs in Berk and Bickel (1968, p. 1573, Theorem).

Theorem 1.9.8. *Assume there exists a p–measure $\Pi|\mathcal{D}$ such that $\Pi(D) = 0$ implies $\Pi(Da) = 0$ for $a \in G$. Assume, moreover, that $T : (X, \mathcal{A}) \to (Z, \mathcal{C})$ is almost invariant with respect to $P|\mathcal{A}$, where $\mathcal{C}$ is countably generated, and $\{z\} \in \mathcal{C}$ for $z \in Z$. Then there exists an invariant function $\overline{T} : (X, \mathcal{A}) \to (Z, \mathcal{C})$ such that $T = \overline{T}$ P–a.e.*

Proof. Let $\mathcal{A}_0 := \{A \in \mathcal{A} : P(A \,\Delta\, aA) = 0$ for $a \in G\}$ and

$$\mathcal{A}_I := \{A \in \mathcal{A} : A = aA \quad \text{for } a \in G\}.$$

We have $\mathcal{A}_I \subset \mathcal{A}_0$. We shall show that $\mathcal{A}_0 \subset \mathcal{A}_I$ (P). By Lemmas 1.10.3, 1.10.4 and 1.10.6(ii), this implies that for any $\mathcal{A}_0$–measurable function T there exists an $\mathcal{A}_I$–measurable function $\overline{T}$ such that $P\{T \neq \overline{T}\} = 0$. We have

$$(T^{-1}C \,\Delta\, aT^{-1}C) \subset \{x \in X : T(a^{-1}x) \neq T(x)\}.$$

Hence T is $\mathcal{A}_0$–measurable if it is almost invariant. Since any $\mathcal{A}_I, \mathcal{C}$–measurable function is invariant, this implies the assertion.

It remains to prove $\mathcal{A}_0 \subset \mathcal{A}_I$ (P). For $A \in \mathcal{A}_0$ and $a \in G$ we have $1_A(x) = 1_A(ax)$ for P–a.a. $x \in X$. This suggests to define the invariant set equivalent to A by $A_I := \{x \in X : \int 1_A(bx)\Pi(db) = 1\}$. We shall show that $A_I \in \mathcal{A}_I$. $x \in A_I$ implies that $D_x := \{b \in G : bx \notin A\}$ is a Π–null set. Hence $D_x a = \{b \in G : ba^{-1}x \notin A\}$ is a Π–null set too, so that $\int 1_A(ba^{-1}x)\Pi(db) = 1$, i.e. $a^{-1}x \in A_I$, or $x \in aA_I$. Therefore, $A_I \subset aA_I$ for $a \in G$, which implies $A_I = aA_I$ for $a \in G$.

It remains to be shown that $P(A \,\Delta\, A_I) = 0$ for $A \in \mathcal{A}_0$. If $A \in \mathcal{A}_0$, we have

$$\int |1_A(x) - 1_A(ax)| P(dx) = 0 \quad \text{for } a \in G,$$

hence, by Fubini's theorem,

$$\int |1_A(x) - 1_A(ax)|\Pi(da) = 0 \quad \text{for } P\text{–a.a. } x \in X.$$

This implies $1_A(x) = \int 1_A(ax)\Pi(da)$ for P–a.a. $x \in X$, hence $P(A \,\Delta\, A_I) = 0$. $\square$

Lemma 1.9.9. *If G is σ–locally compact, then there exists a p–measure $\Pi|D$ such that $\Pi(D) = 0$ implies $\Pi(Da) = 0$ for $a \in G$.*

Proof. By Theorem 1.9.5 there exists a right invariant Haar measure which is σ–finite, say $\nu|\mathcal{D}$. Let $G = \sum_1^\infty D_n$ with $D_n \in \mathcal{D}$, $0 < \nu(D_n) < \infty$. The p–measure $\Pi|\mathcal{D}$, defined by

$$\Pi(D) := \sum_1^\infty 2^{-n} \nu(D \cap D_n)/\nu(D_n), \qquad D \in \mathcal{D},$$

is mutually absolutely continuous with respect to ν. Hence Π has the asserted property if ν does. This is the case, since ν is right invariant.   □

Theorem 1.9.10. *Let $\mathfrak{P}$ be a family of mutually absolutely continuous p–measures $P|\mathcal{A}$ which is closed under G, i.e. $P \in \mathfrak{P}$ implies $P_a \in \mathfrak{P}$ for $a \in G$.*
If $S : (X, \mathcal{A}) \to (Y, \mathcal{B})$ is minimal sufficient for $\mathfrak{P}$, then S is $\mathfrak{P}$–almost equivariant.

Proof. Let $A \in \mathcal{A}$ be arbitrary. Since S is sufficient, there exists a conditional expectation of 1_{aA}, given S, say $\varphi_a : (Y, \mathcal{B}) \to ([0, 1], \mathbb{B}_0)$, which does not depend on P, i.e.

$$\int \varphi_a(S(x)) 1_B(S(x)) P(dx) = \int 1_{aA}(x) 1_B(S(x)) P(dx)$$

$$\text{for } B \in \mathcal{B}, \ P \in \mathfrak{P}.$$

Applied with P replaced by P_a, we obtain from $P_a = P \circ (x \to ax)$ that

$$\int \varphi_a(S(ax)) 1_B(S(ax)) P(dx) = \int 1_{aA}(ax) 1_B(S(ax)) P(dx).$$

Since $1_{aA}(ax) = 1_A(x)$, the function φ_a is a conditional expectation of 1_A, given $S \circ a$, which does not depend on $P \in \mathfrak{P}$. Hence $S \circ a$ is sufficient for $\mathfrak{P}$.

If S is minimal sufficient, there exists a function $H_a : (Y, \mathcal{B}) \to (Y, \mathcal{B})$ such that $S(x) = H_a(S(ax))$ for $x \notin N_a$, with $P(N_a) = 0$ for $P \in \mathfrak{P}$.

Since $P_c \ll P$, $P(N) = 0$ implies $P(c^{-1}N) = P_c(N) = 0$. Hence $P(aN_a) = 0$. Moreover, $x', x'' \notin aN_a$ implies $a^{-1}x', a^{-1}x'' \notin N_a$, so that $S(a^{-1}x') = H_a(S(x'))$ and $S(a^{-1}x'') = H_a(S(x''))$. Therefore, for x', x'' outside the P–null set aN_a, $S(x') = S(x'')$ implies $S(a^{-1}x') = S(a^{-1}x'')$, i.e. S is $\mathfrak{P}$–almost equivariant.   □

Proposition 1.9.11. *Let $\mathfrak{P}$ be a family of mutually absolutely continuous p–measures which is closed under G. Assume that $S : (X, \mathcal{A}) \to (Y, \mathcal{B})$ is equivariant and sufficient for $\mathfrak{P}$.*

If $h : (X, \mathcal{A}) \to (\mathbb{R}, \mathbb{B})$ *is* $\mathfrak{P}$*-almost invariant and* $h_0 : (Y, \mathcal{B}) \to (\mathbb{R}, \mathbb{B})$ *is a conditional expectation of* h*, given* S*, under* $\mathfrak{P}$*, then* $x \to h_0(S(x))$ *is* $\mathfrak{P}$*-almost invariant.*

Proof. Let $\mathcal{A}_S := \{A \in \mathcal{A} : S^{-1}SA = A\}$. We shall show that $a\mathcal{A}_S = \mathcal{A}_S$ for $a \in G$.

$x \in S^{-1}S(aA)$ is equivalent to $S(x) \in S(aA)$, i.e. to the existence of $x' \in A$ such that $S(x) = S(ax')$. Since S is equivariant, this implies $S(a^{-1}x) = S(x')$, i.e. $a^{-1}x \in S^{-1}SA = A$ for $A \in \mathcal{A}_S$. Hence $S^{-1}S(aA) \subset aA$. Since $S^{-1}S(aA) \supset aA$ in general, this implies $S^{-1}S(aA) = aA$ for $A \in \mathcal{A}_S$, i.e. $aA \in \mathcal{A}_S$.

Let $\mathcal{B}_S := \{B \subset Y : S^{-1}B \in \mathcal{A}\}$. Since $S^{-1}SS^{-1}B = S^{-1}B$ for $B \subset Y$, we have $S^{-1}\mathcal{B}_S = \mathcal{A}_S$. Hence $aS^{-1}\mathcal{B}_S = S^{-1}\mathcal{B}_S$.

Sufficiency of $S : (X, \mathcal{A}) \to (Y, \mathcal{B})$ implies sufficiency of $S : (X, \mathcal{A}) \to (Y, \mathcal{B}_S)$ by Remark 1.2.13.

Let $h_0 : (Y, \mathcal{B}_S) \to (\mathbb{R}, \mathbb{B})$ be a conditional expectation of h, given S, which does not depend on $P \in \mathfrak{P}$, i.e.

$$\int h_0(S(x)) 1_B(S(x)) P(dx) = \int h(x) 1_B(S(x)) P(dx) \qquad (1.9.12)$$

$$\text{for } B \in \mathcal{B}_S, \ P \in \mathfrak{P}.$$

If h is $\mathfrak{P}$-almost invariant, relation (1.9.12), applied with P_a in place of P, implies

$$\int h_0(S(ax)) 1_B(S(ax)) P(dx) = \int h(ax) 1_B(S(ax)) P(dx)$$

$$= \int h(x) 1_B(S(ax)) P(dx) \qquad \text{for } B \in \mathcal{B}_S.$$

Since $aS^{-1}\mathcal{B}_S = S^{-1}\mathcal{B}_S$, this implies

$$\int h_0(S(ax)) 1_B(S(x)) P(dx) \qquad (1.9.13)$$

$$= \int h(x) 1_B(S(x)) P(dx) \qquad \text{for } B \in \mathcal{B}_S.$$

From (1.9.12) and (1.9.13),

$$\int \Big(h_0(S(ax)) - h_0(S(x)) \Big) 1_B(S(x)) P(dx) = 0 \qquad \text{for } B \in \mathcal{B}_S.$$

Since S and $x \to S(ax)$ are $\mathcal{B}_S$-measurable, this implies that

$$h_0(S(ax)) = h_0(S(x)) \qquad \text{for } P\text{-a.a. } x \in X. \qquad \square$$

Corollary 1.9.14. *Let* T *be a* $\mathfrak{P}$*-almost invariant statistic which maps* $(X, \mathcal{A})$ *into a Polish space, say* $(Z, \mathcal{C})$*. If* $M|Y \times \mathcal{C}$ *is a Markov kernel representing the* *conditional distribution of* T, *given* S, *with respect to* $\mathfrak{P}$, *then the* p*-measure* $x \to M\big(S(x),\cdot\big)|\mathcal{C}$ *is almost invariant in the sense that, for every* $a \in G$, *there exists a* $\mathfrak{P}$*-null set* N_a *such that*

$$M\big(S(ax),C\big) = M\big(S(x),C\big) \qquad for\ C \in \mathcal{C},\ if\ x \notin N_a.$$

Proof. Apply Proposition 1.9.11 with $h = 1_C \circ T$ and $h_0 = M(\cdot,C)$, for $C \in \mathcal{C}$, and use that $\mathcal{C}$ is countably generated. $\qquad\qquad\qquad\qquad\qquad\qquad\qquad$ $\Box$

Corollary 1.9.15. *If G is a σ-locally compact topological group, and if $\{S(ax) : a \in G\} = S(X)$ for every $x \in X$, then any $\mathfrak{P}$-almost invariant statistic is stochastically P-independent of S, for every $P \in \mathfrak{P}$.*

Proof. If $T : (X,\mathcal{A}) \to (Z,\mathcal{C})$ is $\mathfrak{P}$-almost invariant, and if h_C is any conditional expectation of $1_{T^{-1}C}$, given S, then $h_C \circ S$ is $\mathfrak{P}$-almost invariant by Proposition 1.9.11. According to Theorem 1.9.8, Lemma 1.9.9 and Lemma 1.10.6(i), there exists a version $\hat{h}_C$ of the conditional expectation such that $\hat{h}_C \circ S$ is invariant. Since $\{S(ax) : a \in G\} = S(X)$, this version is constant on $S^{-1}Y$ and therefore equal to $P(T^{-1}C)$. Hence

$$\int 1_C\big(T(x)\big)1_B\big(S(x)\big)P(dx) = \int P(T^{-1}C)1_B\big(S(x)\big)P(dx)$$

$$= P(T^{-1}C)P(S^{-1}B) \qquad for\ B \in \mathcal{B},\ C \in \mathcal{C}.\ \Box$$

Since any invariant statistic is ancillary for $\{P_a : a \in G\}$, this result comes close to Basu's Theorem 1.8.2 (which requires the sufficient statistic to be boundedly complete). It seems doubtful whether there are natural applications which are not covered by Basu's theorem.

A result which comes close to Corollary 1.9.15 for the special case of a location parameter family was obtained by Ghurye (1958, p. 160, Lemma 2). Corollary 1.9.15 was stated explicitly (without regularity conditions) by Fraser (1966, p. 148, Theorem 2). For a more detailed study of this and related questions see Hall, Wijsman and Ghosh (1965), Berk (1972), and Landers and Rogge (1973).

To avoid pitfalls, the reader interested in results relating "invariance" and "sufficiency" should be aware of the fact that some results of this type might be close to trivial: If the equivariant sufficient statistic is real valued and continuous (for some sample size greater than one), then the family of p-measures can be represented as a location parameter family of normal distributions, or a scale parameter family of gamma distributions. This theorem was proved by Dynkin (1951) for 1-dimensional exponential families. For a proof under mild regularity conditions (and more details on the history of this subject) see Pfanzagl (1972). In this paper, the transformation group is assumed to be commutative (a condition which is necessary anyway). Hipp (1975) avoids assuming commutativity from the beginning by strengthening the topological conditions on the transformation group (assuming local compactness).

1.10 Appendix: Conditional expectations, conditional distributions

Contains definitions and results on conditional expectations and conditional distributions which are needed in the text. Proofs can be found in Ash (1972), Witting (1985), and Bauer (1990b).

Throughout this section, P is a p–measure on a measurable space $(X, \mathcal{A})$, and $\mathcal{A}_0 \subset \mathcal{A}$ is a sub–σ–algebra.

Definition 1.10.1. For a sub–σ–algebra $\mathcal{A}_0 \subset \mathcal{A}$, $f_0 : (X, \mathcal{A}_0) \to (\mathbb{R}, \mathbb{B})$ is a *conditional expectation* of $f \in \mathcal{L}_1(X, \mathcal{A}, P)$, given $\mathcal{A}_0$, if

$$P(f_0 1_{A_0}) = P(f 1_{A_0}) \quad \text{for } A_0 \in \mathcal{A}_0. \tag{1.10.2}$$

A function f_0 fulfilling (1.10.2) always exists. It is unique up to a P–null set. The symbol $P^{\mathcal{A}_0} f$ denotes the equivalence class of all functions fulfilling (1.10.2).

In many cases the following variant of Definition 1.10.1 is more natural.

Definition 1.10.1'. For a function $S : (X, \mathcal{A}) \to (Y, \mathcal{B})$, $f_0 : (Y, \mathcal{B}) \to (\mathbb{R}, \mathbb{B})$ is a *conditional expectation* of $f \in \mathcal{L}_1(X, \mathcal{A}, P)$, given S, if

$$P \circ S(f_0 1_B) = P(f 1_{S^{-1}B}) \quad \text{for every } B \in \mathcal{B}. \tag{1.10.2'}$$

The symbol $P^S f$ denotes the equivalence class of all functions fulfilling (1.10.2').

Abusing this notation we occasionally write $P^S f$ and $P^{\mathcal{A}_0} f$ for a particular element.

If f_0 is a conditional expectation of f, given S, then $f_0 \circ S$ is a conditional expectation of f, given $S^{-1}\mathcal{B}$. Conversely, any conditional expectation f_0 of f, given $S^{-1}\mathcal{B}$, can be represented as a contraction of S by the Factorization Lemma 1.10.6.

Lemma 1.10.3. *Let $(X, \mathcal{A})$ be a measurable space, and $(Y, \mathcal{B})$ a Polish space. If $f : X \to Y$ fulfills $f^{-1}\mathcal{B} \subset \mathcal{A}$ (P) for a p-measure $P|\mathcal{A}$, then there exists $f' : (X, \mathcal{A}) \to (Y, \mathcal{B})$ such that $f = f'$ P-a.e.*

Proof. There exists (apply Dudley (1989), p. 97, Proposition 4.2.6) a sequence of $f^{-1}\mathcal{B}$–measurable simple functions f_n, $n \in \mathbb{N}$, converging to f. for every $n \in \mathbb{N}$ there exists an $\mathcal{A}$–measurable function $f'_n = f_n$ P-a.e. Hence f is P-a.e. the limit of a sequence of $\mathcal{A}$–measurable functions. Since the domain of convergence of f'_n, $n \in \mathbb{N}$, is measurable, this implies the assertion. □

Lemma 1.10.4. *If $\mathcal{A}$ is countably generated, then it can be induced by a real function, i.e. there exists $f : X \to \mathbb{R}$ such that $\mathcal{A} = f^{-1}\mathbb{B}$.*

Proof. Let $\mathcal{A}$ be generated by A_n, $n \in \mathbb{N}$, and let

$$f := \sum_1^\infty 3^{-n} 1_{A_n}. \tag{1.10.5}$$

As f is $\mathcal{A}$–measurable, we have $f^{-1}\mathbb{B} \subset \mathcal{A}$.

It remains to be shown that $A_n \in f^{-1}\mathbb{B}$ for $n \in \mathbb{N}$. Let

$$B_n := \bigcup_{\substack{0 \leq i \leq n-1 \\ 1 \leq j_1 < \ldots < j_i \leq n-1}} \left[\sum_1^i 3^{-j_\nu} + 3^{-n}, \sum_1^i 3^{-j_\nu} + 3^{-(n-1)}/2 \right].$$

We have $B_n \in \mathbb{B}$ and, by (1.10.5), $x \in A_n$ iff $f(x) \in B_n$. Hence, $A_n = f^{-1}B_n \in f^{-1}\mathbb{B}$. □

Lemma 1.10.6. *Let $S : (X, \mathcal{A}) \to (Y, \mathcal{B})$ and $T : (X, \mathcal{A}) \to (Z, \mathcal{C})$ be measurable maps such that $S^{-1}\mathcal{B} \subset T^{-1}\mathcal{C}$.*

(i) If $\{y\} \in \mathcal{B}$ for $y \in Y$, then there exists a map $H : (T(X), \mathcal{C} \cap T(X)) \to (Y, \mathcal{B})$ with $S = H \circ T$.

(ii) If $(Y, \mathcal{B})$ is a Polish space, then there exists a map $H : (Z, \mathcal{C}) \to (Y, \mathcal{B})$ with $S = H \circ T$.

Proof. (i) Witting (1985), p. 407, Hilfssatz 3.90.

(ii) The map obtained in (i) can be extended to a $\mathcal{C}$–measurable map on Z. (See Dudley (1989), p. 97, Corollary 4.2.7.) □

The following properties of conditional expectations are formulated for Definition 1.10.1. The same properties hold true for Definition 1.10.1′.

Theorem 1.10.7. *Let $\mathcal{A}_0 \subset \mathcal{A}$ be a sub-σ-algebra and let $f, g \in \mathcal{L}_1(X, \mathcal{A}, P)$. The map $f \to P^{\mathcal{A}_0} f$ is linear and monotone, i.e.*

$$P^{\mathcal{A}_0}(af + bg) = aP^{\mathcal{A}_0} f + bP^{\mathcal{A}_0} g \qquad \text{for } a, b \in \mathbb{R}, \tag{1.10.8}$$

and

$$f \leq g \quad P\text{-a.e.} \quad \text{implies} \quad P^{\mathcal{A}_0} f \leq P^{\mathcal{A}_0} g. \tag{1.10.9}$$

In particular, $f \to P^{\mathcal{A}_0} f$ is positive, i.e.

$$P^{\mathcal{A}_0} f \geq 0 \quad \text{for nonnegative functions } f. \tag{1.10.10}$$

Proof. Straightforward (Ash, 1972, Theorems 6.5.2 and 6.5.1(b)). □

Theorem 1.10.11 (Jensen's inequality). *Let $f : (X, \mathcal{A}) \to (\mathbb{R}^p, \mathbb{B}^p)$ be P-integrable, $B \in \mathbb{B}^p$ a convex set containing $f(X)$, and $C : B \to \mathbb{R}$ a convex function. Then $P^{\mathcal{A}_0} f \in B$ P-a.e., and*

$$C(P^{\mathcal{A}_0} f) \leq P^{\mathcal{A}_0}(C \circ f) \qquad P\text{-a.e.} \tag{1.10.12}$$

If C is strictly convex, this inequality is strict unless $f = $ const P-a.e.

In particular, $P\big(C(P^{\mathcal{A}_0} f)\big) \leq P(C \circ f)$. Applied with $C(y) := \|y\|$, this implies $P(\|P^{\mathcal{A}_0} f\|) \leq P(\|f\|)$, i.e. the map $f \to P^{\mathcal{A}_0} f$ is contractive.

Proof. Witting (1985), pp. 125/6, or Dudley (1989), p. 274. □

The following proposition generalizes (1.10.2) from $g_0 = 1_{\mathcal{A}_0}$ to arbitrary $\mathcal{A}_0$-measurable functions g_0.

Proposition 1.10.13. *Let $f \in \mathcal{L}_1(X, \mathcal{A}, P)$ and let $g_0 : (X, \mathcal{A}_0) \to (\mathbb{R}, \mathbb{B})$ be such that $f g_0$ is P-integrable. Then*

$$(P^{\mathcal{A}_0} f) g_0 = P^{\mathcal{A}_0}(f g_0). $$

Proof. Straightforward (Ash, 1972, p. 261, Theorem 6.5.11(a)). □

Proposition 1.10.14. *If $f \in \mathcal{L}_2(X, \mathcal{A}, P)$, then $P^{\mathcal{A}_0} f \in L_2(X, \mathcal{A}_0, P)$, and $f - P^{\mathcal{A}_0} f$ is orthogonal to $L_2(X, \mathcal{A}_0, P)$. In other words: $P^{\mathcal{A}_0} f$ is the projection of f on $L_2(X, \mathcal{A}_0, P)$.*

Proof. Follows immediately from Proposition 1.10.13. □

Proposition 1.10.15. *If* $\mathcal{A}_0 \subset \mathcal{A}_1$ *are sub–σ–algebras, then*

$$P^{\mathcal{A}_0}(P^{\mathcal{A}_1} f) = P^{\mathcal{A}_0} f \qquad \text{for } f \in \mathcal{L}_1(X, \mathcal{A}, P).$$

Proof. Straightforward (Ash, 1972, p. 260, Theorem 6.5.10(a)).   □

Conditional distributions

Throughout this section let $S : (X, \mathcal{A}) \rightarrow (Y, \mathcal{B})$ and $T : (X, \mathcal{A}) \rightarrow (Z, \mathcal{C})$ be measurable maps.

In view of the applications in Section 1.3, it appears convenient to write the results on conditional distributions for "conditional distributions, given S". Obvious modifications lead to the corresponding versions for "conditional distributions, given $\mathcal{A}_0$".

For every $C \in \mathcal{C}$ there exists a conditional expectation of $1_{T^{-1}C}$, given S, i.e. a function $\varphi_C : (Y, \mathcal{B}) \rightarrow (\mathbb{R}, \mathbb{B})$ such that

$$\int \varphi_C(y) 1_B(y) P \circ S(dy) = P(T^{-1}C \cap S^{-1}B) \qquad \text{for } B \in \mathcal{B}.$$

W.l.g. we may assume that $\varphi_C \in [0, 1]$ for every $C \in \mathcal{C}$. From the uniqueness P–a.e. of conditional expectations we obtain
(i) $\varphi_{C_1} + \varphi_{C_2} = \varphi_{C_1 \cup C_2}$ P–a.e. if C_1, C_2 are disjoint,
(ii) $C_n \uparrow C$ implies $\varphi_{C_n} \uparrow \varphi_C$ P–a.e.

Moreover, we have $\varphi_\emptyset = 0$ P–a.e. and $\varphi_Z = 1$ P–a.e.

This suggests that $C \rightarrow \varphi_C(y)$, with y fixed, may be considered as a p–measure. This is not quite so since relations (i) and (ii) hold up to a P–null set only, and there are uncountably many relations involved. Hence it is not possible, in general, to coordinate the (uncountably many) choices of φ_C for $C \in \mathcal{C}$ in such a way that $C \rightarrow \varphi_C(y)$ is a p–measure on $\mathcal{C}$ for every $y \in Y$. (For a counterexample due to Dieudonné, see Doob, 1953, p. 624.) Of importance is the positive result formulated as Theorem 1.10.17.

Definition 1.10.16. Let $(Y, \mathcal{B})$ and $(Z, \mathcal{C})$ be measurable spaces. A function $M : Y \times \mathcal{C} \rightarrow [0, 1]$ is a *Markov kernel* if
(i) $y \rightarrow M(y, C)$ is $\mathcal{B}$–measurable for every $C \in \mathcal{C}$,
(ii) $C \rightarrow M(y, C)$ is a p–measure for every $y \in Y$.

Theorem 1.10.17. *Assume that $(Z, \mathcal{C})$ is a Polish space. Then for arbitrary maps $S : (X, \mathcal{A}) \to (Y, \mathcal{B})$ and $T : (X, \mathcal{A}) \to (Z, \mathcal{C})$ there exists a Markov kernel $M | Y \times \mathcal{C}$ such that $M(\cdot, C)$ is a conditional expectation of $1_{T^{-1}C}$, given S, i.e.*

$$\int M(y, C) 1_B(y) P \circ S(dy) = P(T^{-1}C \cap S^{-1}B) \quad \text{for } B \in \mathcal{B}, \ C \in \mathcal{C}. \quad (1.10.18)$$

Proof. Ash (1972), p. 266, or Dudley (1989), p. 270, Theorem 10.2.2.   □

A Markov kernel fulfilling relation (1.10.18) is called *conditional distribution* of T, given S, with respect to P. (We omit "with respect to P", if P is understood.) Observe that (1.10.18) implies

$$\int M(y, C) f(y) P \circ S(dy) = \int 1_C(T(x)) f(S(x)) P(dx) \quad (1.10.19)$$

for any $P \circ S$–integrable function f.

If $\mathcal{C}$ is countably generated, a Markov kernel fulfilling (1.10.18) is unique in the following sense: If M_i, $i = 1, 2$, are two Markov kernels fulfilling (1.10.18), then there exists a $P \circ S$–null set $N \in \mathcal{B}$ such that the measures $M_i(y, \cdot) | \mathcal{C}$ are identical for all $y \in N^c$.

Remark 1.10.20. If $(X, \mathcal{A})$ itself is a Polish space, then Theorem 1.10.17 applies for $T(x) \equiv x$, and we obtain a Markov kernel $M | Y \times \mathcal{A}$ which is a conditional distribution of "x", given S, i.e.

$$\int M(y, A) 1_B(y) P \circ S(dy) = P(A \cap S^{-1}B) \quad \text{for } A \in \mathcal{A}, \ B \in \mathcal{B}. \quad (1.10.21)$$

By definition, $M(y, \cdot)$ is a p–measure on the measurable space $(X, \mathcal{A})$, but the intuitive interpretation is that of the distribution of x within the set $S^{-1}\{y\}$. This interpretation requires that $M(y, A)$ depends, in fact, only on that part of A which is in $S^{-1}\{y\}$. It requires that

$$M(y, A) = M(y, A \cap S^{-1}\{y\}) \quad \text{for } A \in \mathcal{A}. \quad (1.10.22)$$

Example 1.10.23. Let $P | \mathbb{B}^2$ be a p–measure with λ^2–density $(u, v) \to p(u^2 + v^2)$, $u, v \in \mathbb{R}$. The conditional distribution of (u, v), given $(u^2 + v^2)^{1/2} = y$, is the uniform distribution over the circle with radius y and center 0.

Proposition 1.10.25 below shows that any conditional distribution of x, given S, fulfills relation (1.10.22) for $P \circ S$–a.a. $y \in Y$. Even if $(X, \mathcal{A})$ is a Polish space, it is not possible in general to obtain a conditional distribution fulfilling relation (1.10.22) for all $y \in Y$ (see Blackwell and Ryll–Nardzewski, 1963).

Lemma 1.10.24. *Let $M|Y \times A$ be a Markov kernel. Let $y \in Y$ with $\{y\} \in B$ be fixed. Then the following conditions are equivalent:*
(i) $M(y, S^{-1}\{y\}) = 1$.
(ii) $M(y, A) = M(y, A \cap S^{-1}\{y\})$ *for $A \in A$.*
(iii) $M(y, A \cap S^{-1}B) = M(y, A)1_B(y)$ *for $A \in A$, $B \in B$.*

Proof. The implications (i) $\Rightarrow$ (ii) and (iii) $\Rightarrow$ (i) are straightforward.
(ii) $\Rightarrow$ (iii): By (ii), we have

$$M(y, A \cap S^{-1}B) = M(y, A \cap S^{-1}B \cap S^{-1}\{y\}).$$

For $y \in B$, this implies by (ii)

$$M(y, A \cap S^{-1}B) = M(y, A \cap S^{-1}\{y\}) = M(y, A).$$

On the other hand, $y \in B^c$ implies $M(y, A \cap S^{-1}B) = 0$.   □

Proposition 1.10.25. *Let $M|Y \times A$ be a Markov kernel. Let $S : (X, A) \to (Y, B)$ be measurable, where B is countably generated and $\{y\} \in B$ for $y \in Y$. Then the following conditions (i), (ii), (iii) are equivalent:*
(i) $\int M(y, A)1_B(y)P \circ S(dy) = P(A \cap S^{-1}B)$ *for $A \in A$, $B \in B$;*
(ii') $\int M(y, A)P \circ S(dy) = P(A)$ *for $A \in A$,*
(ii") $M(y, S^{-1}\{y\}) = 1$ *for $P \circ S$-a.a. $y \in Y$;*
(iii) $\int M(y, A \cap S^{-1}\{y\})P \circ S(dy) = P(A)$ *for $A \in A$.*

Proof. (i) $\Rightarrow$ (ii'): Relation (ii') follows from (i), applied with $B = Y$.
(i) $\Rightarrow$ (ii"): Let $B_0 \subset B$ be a countable algebra generating B. From (i) we obtain with $B_0 \in B_0$

$$\int 1_{B_0}(y)1_B(y)P \circ S(dy) = \int M(y, S^{-1}B_0)1_B(y)P \circ S(dy)   \text{ for } B \in B,$$

hence $M(y, S^{-1}B_0) = 1_{B_0}(y)$ for $P \circ S$-a.a. $y \in Y$. Denoting the exceptional null set by N_{B_0} and letting $N := \cup\{N_{B_0} : B_0 \in B_0\}$, we have $P \circ S(N) = 0$ and $M(y, S^{-1}B_0) = 1_{B_0}(y)$ for $y \in N^c$ and $B_0 \in B_0$. Since, for $y \in N^c$, the p-measures $B \to M(y, S^{-1}B)$ and $B \to 1_B(y)$ are identical on B_0, they are identical on B, too. Hence, $M(y, S^{-1}\{y\}) = 1$ for $y \in N^c$.
(ii) $\Rightarrow$ (iii): Relation (ii") implies by Lemma 1.10.24

$$M(y, A \cap S^{-1}\{y\}) = M(y, A)   \text{ for } P \circ S\text{-a.a. } y \in Y.$$

Thus (iii) follows from (ii').
(iii) $\Rightarrow$ (i): Relation (iii) implies $M(y, S^{-1}\{y\}) = 1$ for $P \circ S$-a.a. $y \in Y$, and hence by Lemma 1.10.24

$$M(y, A)1_B(y) = M(y, A \cap S^{-1}B) = M(y, A \cap S^{-1}B \cap S^{-1}\{y\})$$

for $P \circ S$–a.a. $y \in Y$. This implies (i) by (iii). □

Proposition 1.10.26. *Let $M|Y \times C$ be a conditional distribution of T, given S, with respect to P. Then for any $f : (Z \times Y, C \times B) \to (\mathbb{R}, \mathbb{B})$, the function*

$$y \to \int f(t,y) M(y, dt)$$

is a conditional expectation of $x \to f(T(x), S(x))$, given S, provided this function is P–integrable.

Corollary 1.10.27. *Let $M|Y \times A$ be a conditional distribution of x, given S, with respect to P.*
Then for any $f \in \mathcal{L}_1(X, A, P)$, the function $y \to \int f(x) M(y, dx)$ is a conditional expectation of f, given S.

Proof of Proposition 1.10.26. We have to show that $\int f(t, y) M(y, dt)$ exists for $P \circ S$–a.a. $y \in Y$, that

$$y \to \int f(t,y) M(y, dt) \quad \text{is } B\text{–measurable}, \tag{1.10.28}$$

and that

$$\int \int f(t, y) M(y, dt) 1_B(y) P \circ S(dy) \tag{1.10.29}$$

$$= \int f(T(x), S(x)) 1_{S^{-1}B}(x) P(dx) \qquad \text{for } B \in B.$$

Since f is approximable by $C \times B$–measurable elementary functions, it suffices to prove relations (1.10.28) and (1.10.29) for $f = 1_D$, with $D \in C \times B$.
Let S denote the class of all sets $D \in C \times B$ such that (1.10.28) and (1.10.29) hold true for $f = 1_D$. The class S contains $C \times B$ for $C \in C$, $B \in B$, since $M|Y \times C$ is a conditional distribution of T, given S. As S is a Dynkin system, this implies $S = C \times B$. □

Proposition 1.10.30. *Let (X, A) and (Y, B) be measurable spaces, and let $\mu|A$, $\nu|B$ be σ–finite measures. Let $P|A \times B$ be a p–measure with $\mu \times \nu$–density p.*
Then the following holds true:
(i) $p_2(y) := \int p(x, y) \mu(dx)$ *is a ν–density of $P \circ \Pi_2$, the 2nd marginal of P.*
(ii) $p(x|y) := p(x, y) / p_2(y)$ *is a μ–density of a Markov kernel $M|Y \times A$ representing the conditional distribution of x, given Π_2, i.e.*

$$M(y, A) := \int p(x|y) 1_A(x) \mu(dx) \quad \text{for } y \in Y, \ A \in A.$$

(For $p_2(y) = 0$, the definition of $p(x|y)$ is arbitrary, subject to the condition $\int p(x|y)\mu(dx) = 1$, e.g., $p(x|y) = p_1(x)$.)

Proof. Straightforward (Witting, 1985, p. 128, Satz 1.126a).   □

Example 1.10.31. Let $P = N_{(\mu_1,\mu_2,\sigma_1^2,\sigma_2^2,\varrho)}|\,\mathbb{B}^2$ be a bivariate normal distribution. Then the 2nd marginal is $P_2 = N_{(\mu_2,\sigma_2^2)}$, and the conditional distribution of x, given Π_2, is

$$M(y,\cdot) = N_{(\mu_1+\varrho\sigma_1\sigma_2^{-1}(y-\mu_2),\sigma_1^2(1-\varrho^2))}.$$

Lemma 1.10.32. *Let P and Q be two mutually absolutely continuous p-measures on $(X,\mathcal{A})$, with $x \to p(T(x),S(x)) \in dP/dQ$. Assume there exists a Markov kernel $\hat{Q}|Y \times C$ representing the conditional distribution of T, given S, with respect to Q.*

Then the following holds true.

(i) $\overline{p}(y) := \int p(t,y)\hat{Q}(y,dt)$ is a $Q \circ S$–density of $P \circ S$, and $0 < \overline{p}(y) < \infty$ for $P \circ S$-a.a. $y \in Y$.

(ii) There exists a conditional distribution of T, given S, with respect to P, say $\hat{P}|Y \times C$, with the following property: For every $y \in Y$, $t \to p(t,y)/\overline{p}(y)$ is a density of $\hat{P}(y,\cdot)$ with respect to $\hat{Q}(y,\cdot)$. (Observe that $\overline{p}(y) = 0$ implies $p(t,y) = 0$ for $\hat{Q}(y,\cdot)$-a.a. $t \in Z$.)

Proof. (i) We have to show that

$$\int \overline{p}(y)1_B(y)Q \circ S(dy) = P \circ S(B) \quad \text{for } B \in \mathcal{B}.$$

This follows immediately from

$$\int\int p(t,y)\hat{Q}(y,dt)1_B(y)Q \circ S(dy)$$
$$= \int p(T(x),S(x))1_B(S(x))Q(dx).$$

Since $\overline{p}$ is a $Q \circ S$–density of $P \circ S$, we have $\int \overline{p}(y)1_B(y)Q \circ S(dy) = P \circ S(B)$ for $B \in \mathcal{B}$. Applied with $B = \{y \in Y : 0 < \overline{p}(y) < \infty\}$, this implies $P \circ S\{y \in Y : 0 < \overline{p}(y) < \infty\} = 1$.

(ii) We have to show that

$$\hat{P}(y,C) := \int \frac{p(t,y)}{\overline{p}(y)}1_C(t)\hat{Q}(y,dt), \qquad y \in Y,\ C \in \mathcal{C},$$

defines a Markov kernel fulfilling

$$\int \hat{P}(y,C)1_B(y)P \circ S(dy) = P(T^{-1}C \cap S^{-1}B) \quad \text{for } B \in \mathcal{B}.$$

This follows immediately from

$$\int \int \frac{p(t,y)}{\bar{p}(y)} 1_C(t)\hat{Q}(y,dt)1_B(y)P \circ S(dy)$$

$$= \int \int p(t,y)1_C(t)\hat{Q}(y,dt)1_B(y)Q \circ S(dy)$$

$$= \int p(T(x),S(x))1_C(T(x))1_B(S(x))Q(dx)$$

$$= P(T^{-1}C \cap S^{-1}B). \qquad \square$$

Theorem 1.10.33. *Let $(Y,\mathcal{B})$ be a measurable space, $(Z,\mathcal{C})$ a Polish space, and $M : Y \times \mathcal{C} \to [0,1]$ a Markov kernel.*

Then there exists a $\mathcal{B} \times \mathbb{B}_0, \mathcal{C}$–measurable map $m : Y \times (0,1) \to Z$ such that, for every $y \in Y$, the p–measure $M(y,\cdot)|\mathcal{C}$ is induced by U and $u \to m(y,u)$, i.e.

$$M(y,C) = U\{u \in (0,1) : m(y,u) \in C\} \qquad \text{for } C \in \mathcal{C}, \ y \in Y,$$

where $U|\mathbb{B}_0$ denotes the uniform distribution.

Proof. (i) Let $\varphi : Z \to (0,1)$ be a $1-1$ $\mathcal{C}, \mathbb{B}_0$–measurable map such that $\varphi(C) \in \mathbb{B}_0$ for $C \in \mathcal{C}$. (See Parthasarathy, 1967, p. 12, Theorem 2.8, and p. 14, Theorem 2.12 for existence.) Let $\psi : (0,1) \to Z$ be defined by $\psi(u) = z$ if $u = \varphi(z)$, and $\psi(u) = z_0$ (an arbitrary element in Z) if $u \notin \varphi(Z)$. Since $\varphi(Z) \in \mathbb{B}_0$, ψ is $\mathbb{B}_0, \mathcal{C}$–measurable. Moreover, $\psi(\varphi(z)) = z$ for $z \in Z$.

(ii) Any measure $\mu|\mathcal{C}$ can be induced by the measure $\mu \circ \varphi| \mathbb{B}_0$ and the map $\psi : (0,1) \to Z$. Let $\nu = \mu \circ \varphi$. We shall show that $\mu = \nu \circ \psi$.

$\nu(B) = \mu\{z \in Z : \varphi(z) \in B\}$ for $B \in \mathbb{B}_0$ implies

$$\nu(\psi^{-1}C) = \mu\{z \in Z : \varphi(z) \in \psi^{-1}C\}$$
$$= \mu\{z \in Z : \psi(\varphi(z)) \in C\} = \mu(C) \qquad \text{for } C \in \mathcal{C},$$

i.e. $\mu = \nu \circ \psi$.

(iii) Given a Markov kernel $N|Y \times \mathbb{B}_0$, let

$$F(y,t) := N(y,(0,t]) \qquad \text{for } y \in Y, \ t \in (0,1).$$

It is straightforward to show that $y \to F(y,t)$ is measurable for $t \in (0,1)$, and that $t \to F(y,t)$ is isotone and right continuous for $y \in Y$. Hence $(y,t) \to F(y,t)$ is $\mathcal{B} \times \mathbb{B}_0$–measurable by Lemma 6.7.3(ii). Let $G : Y \times (0,1) \to (0,1)$ be defined by

$$G(y,u) \doteq \inf\{t \in (0,1) : F(y,t) \geq u\}.$$

Since $F(y, \cdot)$ is isotone and right continuous, we have

$$G(y, u) \leq t \quad \text{iff} \quad F(y, t) \geq u.$$

Hence $(y, u) \to G(y, u)$ is $\mathbb{B} \times \mathbb{B}_0$–measurable, and

$$U\{u \in (0, 1) : G(y, u) \leq t\} = U\{u \in (0, 1) : F(y, t) \geq u\} = F(y, t).$$

Hence the Markov kernel $N|Y \times \mathbb{B}_0$ can be induced by U and the map $u \to G(y, u)$, i.e.

$$N(y, B) = U\{u \in (0, 1) : G(y, u) \in B\} \quad \text{for } y \in Y, \ B \in \mathbb{B}_0.$$

(iv) Let now $N|Y \times \mathbb{B}_0$ be the Markov kernel defined by

$$N(y, B) = M(y, \varphi^{-1}B), \qquad B \in \mathbb{B}_0.$$

By (ii) and (iii) we have

$$\begin{aligned}
M(y, C) &= N(y, \psi^{-1}C) = U\{u \in (0, 1) : G(y, u) \in \psi^{-1}C\} \\
&= U\{u \in (0, 1) : \psi(G(y, u)) \in C\} \qquad \text{for } C \in \mathcal{C}.
\end{aligned}$$

Hence the Markov kernel $M|Y \times \mathcal{C}$ is induced by U and the map $m : Y \times (0, 1) \to Z$, defined by $m(y, u) = \psi(G(y, u))$. $\qquad\qquad \square$

Chapter 2
The evaluation of estimators

2.1 Introduction

Introduces basic concepts like functional, (randomized) estimator, and discusses basic
questions concerning the comparability of estimators.

Let $\mathfrak{P}$ be a family of p–measures on a measurable space $(X, \mathcal{A})$, and let
$\kappa : \mathfrak{P} \to (Y, \mathcal{B})$ be a *functional* mapping $\mathfrak{P}$ into a measurable space $(Y, \mathcal{B})$. The
following considerations are mostly restricted to the case $(Y, \mathcal{B}) = (\mathbb{R}^p, \mathbb{B}^p)$.

An *estimator* of κ is a map $\hat{\kappa} : (X, \mathcal{A}) \to (Y, \mathcal{B})$. The value $\hat{\kappa}(x)$ is called
"estimate". The estimator is *proper* if $\hat{\kappa}(X) \subset \kappa(\mathfrak{P})$, i.e. if it attains only such
values in Y which are possible values of the functional.

In most applications, $(X, \mathcal{A})$ is replaced by $(X^n, \mathcal{A}^n)$, and $\mathfrak{P}_n$ is a family of
product measures, usually the product of identical components, i.e. $\mathfrak{P}_n = \{P^n : P \in \mathfrak{P}\}$. In this case, it is convenient to consider the functional $\kappa | \mathfrak{P}_n$ as defined on
$\mathfrak{P}$. The estimator of κ is then a map $\kappa^{(n)} : (X^n, \mathcal{A}^n) \to (Y, \mathcal{B})$. This is, however,
of no relevance for the concepts used for the evaluation of estimators as long as n is
fixed (i.e., as long as the considerations are nonasymptotic).

For theoretical purposes, at least, we need the more general concept of
a *randomized estimator*. By this we mean a Markov kernel $K | X \times \mathcal{B}$ with
the following interpretation: If x is observed, the estimate is obtained as a
realization from $K(x, \cdot) | \mathcal{B}$. The distribution of the estimator under $P | \mathcal{A}$ is then
given by the p–measure $P \circ K | \mathcal{B}$, defined by

$$P \circ K(B) = \int K(x, B) P(dx), \qquad B \in \mathcal{B}.$$

Observe that any nonrandomized estimator $\hat{\kappa}$ can be represented by the
Markov kernel $(x, B) \to 1_B(\hat{\kappa}(x))$, $x \in X$, $B \in \mathcal{B}$.

Remark 2.1.1. According to Theorem 1.10.33, for Polish spaces $(Y, \mathcal{B})$, for any
Markov kernel $K | X \times \mathcal{B}$ there exists a measurable function $\hat{\kappa} : X \times (0, 1) \to Y$

such that the distribution of $u \to \hat{\kappa}(x, u)$ under U (the uniform distribution over $(0,1)$) equals $M(x, \cdot)$, for every $x \in X$. Conversely, any randomization procedure based on such an auxiliary random variable u can be expressed by a Markov kernel, defined by

$$K(x, B) := U\{u \in (0,1) : \hat{\kappa}(x, u) \in B\}, \qquad x \in X, \ B \in \mathcal{B}.$$

The representation of a Markov kernel by an auxiliary random variable makes it possible to consider any randomized estimator as a nonrandomized one (namely $(x, u) \to \hat{\kappa}(x, u)$), if we replace P by $P \times U$. Moreover, it offers the possibility of an alternative randomization procedure: Instead of obtaining the estimate as a realization from $M(x, \cdot)$, after x has been observed, one may select a (nonrandomized) estimator from the class $\{\hat{\kappa}(\cdot, u) : u \in (0,1)\}$ by determining u as a realization from U.

The reader interested in the exchangeability between these two types of randomization in the framework of decision theory is referred to von Weizsäcker (1974). In the following, we state all definitions for nonrandomized estimators.

Our basic principle is to base the evaluation of estimators on their distribution. Hence concepts for the comparison of estimators are, in fact, concepts for the comparison of p–measures.

It is surprising to observe how long it took to develop the idea that the quality of an estimator has to be judged by its distribution. Throughout the 19th century the authors were hypnotized by the idea that $x_1, \ldots, x_n$ are n observations of one and the same unknown quantity, and that these observations are to be averaged in order to obtain the true value of the unknown quantity. It was common to discuss the properties of such an average in relation to the given observations which it was meant to summarize, rather than by its distribution. (Some statements illustrating this spirit: "The position of the average should be somewhere in the center of the sample". "Should the extreme observations be given smaller weight, since they are obviously less accurate than the observations in the middle of the sample?" etc.) Even now there are a few authors who try to justify estimation procedures on other grounds than by the good performance of the resulting estimators. One instance is the so-called "maximum likelihood principle" which is rather plausible and, in fact, a sound heuristic procedure for the construction of good estimators. But it misses the magic power, attributed to it by some scholars, which makes it superfluous to study the properties of the resulting estimators.

The definition of an estimator given so far is almost empty. It is not yet carrying what the intuitive concept of an estimator requires, namely: that the distribution of the estimator under P should be concentrated as closely as possible about $\kappa(P)$, simultaneously for all $P \in \mathfrak{P}$. There arise some difficulties if we try to make this idea precise.

Comparing the quality of estimators $\hat{\kappa}_i$, $i = 1, 2$, means to compare the concentration of $P \circ \hat{\kappa}_1$ and $P \circ \hat{\kappa}_2$ about $\kappa(P)$. This demands, first of all, a suitable concept for the "concentration" of p-measures. Of particular interest for statistical theory are estimators which are optimal in the sense that they are more concentrated about $\kappa(P)$ than any other estimator. For this purpose, it is not necessary that any two estimators are comparable with respect to concentration; it is enough that one particular estimator is comparable with all other estimators, and comes out as the best one in all comparisons. This problem occurs, in particular, in connection with multidimensional functionals. Suitable concepts will be discussed in Sections 2.3 and 2.4.

Since other qualities of an estimator (like amount of computing, robustness etc.) are practically relevant, an ordering of estimators according to their concentration may not be enough; it might be desirable to *measure differences in concentration*. This includes, in particular, the problem of evaluating the difference in concentration between the distributions $P^n \circ \kappa^{(n)}$ and $P^m \circ \kappa^{(m)}$ of one and the same estimator for different sample sizes n and m. This problem will be discussed in Section 2.6.

What has been said so far is not the whole truth. What we have to evaluate is the difference in concentration about $\kappa(P)$ between $P \circ \hat{\kappa}_1$ and $P \circ \hat{\kappa}_2$ *for every* $P \in \mathfrak{P}$. But this is a point, where statistical theory has not much to contribute (unless one is willing to assign prior probabilities to the different members of $\mathfrak{P}$, and to "average" the difference between $P \circ \hat{\kappa}_1$ and $P \circ \hat{\kappa}_2$ over $P \in \mathfrak{P}$). Fortunately, the simultaneous comparison does not offer serious problems for the applied statistician. For all natural estimators — at least for parametric families — the order between $P^n \circ \kappa_1^{(n)}$ and $P^n \circ \kappa_2^{(n)}$ is the same for all $P \in \mathfrak{P}$ (and all $n \in \mathbb{N}$), i.e. if $P^n \circ \kappa_1^{(n)}$ is more concentrated than $P^n \circ \kappa_2^{(n)}$ for *one* $P \in \mathfrak{P}$, it is usually so for *all* $P \in \mathfrak{P}$.

Depending on the problem, it might be desirable to impose on the estimators an additional condition that they are properly centered. If $P \circ \hat{\kappa}$ were symmetric, one would naturally require that $P \circ \hat{\kappa}$ should be symmetric about $\kappa(P)$. In the absence of symmetry it is not so clear what "properly centered" means. In Section 2.2 we discuss several "unbiasedness"–concepts suggested in literature for this purpose.

Now we consider the interplay between "unbiasedness" and "concentration". Our problem has the following formal structure: For every $P \in \mathfrak{P}$ let $\mathcal{U}(P)$ denote the class of all estimators fulfilling for P a certain condition (like mean- or median unbiasedness) and let $<_P$ denote a partial order between estimators (like concentration of $P \circ \hat{\kappa}$). Let $\mathcal{U}(\mathfrak{P}) = \cap\{\mathcal{U}(P) : P \in \mathfrak{P}\}$.

We say that $\hat{\kappa}_0$ is optimal on $\mathfrak{P}$ if $\hat{\kappa}_0 \in \mathcal{U}(\mathfrak{P})$, and if $\hat{\kappa}_0 <_P \hat{\kappa}$ for every $P \in \mathfrak{P}$ and every $\hat{\kappa} \in \mathcal{U}(\mathfrak{P})$. This somewhat complex interplay is responsible for the fact that "optimality on $\mathfrak{P}$" is usually neither passed on to subfamilies of $\mathfrak{P}$, nor to families containing $\mathfrak{P}$. Hence the paradoxical situation may occur that an estimator is optimal on $\mathfrak{P}_2$ and optimal on $\mathfrak{P}_0 \subset \mathfrak{P}_2$, but not optimal on $\mathfrak{P}_1$ inbetween, i.e. $\mathfrak{P}_0 \subset \mathfrak{P}_1 \subset \mathfrak{P}_2$ (see Example 3.2.9). However: If $\hat{\kappa}$ is optimal on $\mathfrak{P}_0$, and if it happens that $\hat{\kappa} \in \mathcal{U}(\mathfrak{P}_1)$, then $\hat{\kappa}$ is optimal on $\mathfrak{P}_1$, provided there is an optimal estimator on $\mathfrak{P}_1$.

Assume that $\mathfrak{P}$ is partitioned into subfamilies $\mathfrak{P}_i$, $i \in I$, where I is an arbitrary index set. (As an example think of $\mathfrak{P} = \{P_{\vartheta,\tau} : \vartheta \in \Theta, \ \tau \in T\}$ and $\mathfrak{P}_\tau := \{P_{\vartheta,\tau} : \vartheta \in \Theta\}$, with the index set T.) An estimator in $\mathcal{U}(\mathfrak{P})$ (say $\hat{\kappa}_0$) which is optimal on each $\mathfrak{P}_i$, $i \in I$, is optimal on $\mathfrak{P}$: Let $P \in \mathfrak{P}$ and $\hat{\kappa} \in \mathcal{U}(\mathfrak{P})$ be arbitrary. We have to show that $\hat{\kappa}_0 <_P \hat{\kappa}$. Since $\mathfrak{P} = \bigcup_{i \in I} \mathfrak{P}_i$, we have $P \in \mathfrak{P}_i$ for some $i \in I$. Since $\hat{\kappa}_0$ is optimal on $\mathfrak{P}_i$, $\hat{\kappa} \in \mathcal{U}(\mathfrak{P}) \subset \mathcal{U}(\mathfrak{P}_i)$ implies $\hat{\kappa}_0 <_P \hat{\kappa}$. See Remark 3.2.8 and Example 3.2.9 for some more details.

For proving the nonexistence of an estimator which is optimal on $\mathfrak{P}$, one may proceed as follows: Choose two mutually absolutely continuous p–measures $P_i \in \mathfrak{P}$ and determine the estimator $\hat{\kappa}_i$ which is in $\mathcal{U}(\mathfrak{P})$ optimal with respect to P_i. If $\hat{\kappa}_i$ is unique and $P_i\{\hat{\kappa}_1 \neq \hat{\kappa}_2\} > 0$, then there is no optimal estimator on $\mathfrak{P}$. (Assume that, to the contrary, $\hat{\kappa}_0$ is optimal on $\mathfrak{P}$. Then $\hat{\kappa}_0 <_P \hat{\kappa}$ for every $P \in \mathfrak{P}$ and every $\hat{\kappa} \in \mathcal{U}(\mathfrak{P})$. In particular: $\hat{\kappa}_0 <_{P_i} \hat{\kappa}_i$. Since $\hat{\kappa}_i$ is unique, this implies $P_i\{\hat{\kappa}_0 = \hat{\kappa}_i\} = 1$, for $i = 1, 2$, a relation which is in contradiction to $P_i\{\hat{\kappa}_1 \neq \hat{\kappa}_2\} > 0$.) See Example 3.1.6.

2.2 Unbiasedness of estimators

Discusses the motivation for "equivariance", "mean unbiasedness" and "median unbiasedness". Illustrates the meaning of these concepts by examples and counterexamples.

In the absence of any prior knowledge about P, it appears natural to require that the estimator be "unbiased" in the sense that it treats all $P \in \mathfrak{P}$ equal, i.e. that it has no preference for particular p–measures in $\mathfrak{P}$.

To begin with we consider the special case where $\mathfrak{P}$ is generated by a transformation group: Let G be a group of transformations a, which transform $x \in X$ to $ax \in X$. For any $P \in \mathfrak{P}$ let $P_a := P \circ (x \to ax)$. We assume that the functional $\kappa : \mathfrak{P} \to (Y, \mathcal{B})$ is equivariant, i.e.: If $\kappa(P') = \kappa(P'')$, then $\kappa(P'_a) = \kappa(P''_a)$ for every $a \in G$. This induces a transformation group $\overline{G}$ on $\kappa(\mathfrak{P})$, with $\overline{a} \in \overline{G}$ defined by $\kappa(P_a) = \overline{a}\kappa(P)$.

In this case, a natural requirement on the estimator is equivariance. Recall that the estimator $\hat{\kappa}$ is equivariant if

$$\hat{\kappa}(ax) = \overline{a}\hat{\kappa}(x) \qquad \text{for } x \in X, \ a \in G. \tag{2.2.1}$$

In some situations the requirement of equivariance is cogent: Assume that $(X, \mathcal{A}) = (\mathbb{R}_+, \mathbb{B}_+)$, $G = (0, \infty)$, with ax denoting multiplication, and $P_a = P \circ (x \to ax)$. Any estimator of the functional $\kappa(P_a) = a$ has to fulfill the condition

$$\kappa^{(n)}(ax_1, \ldots, ax_n) = a\kappa^{(n)}(x_1, \ldots, x_n).$$

Otherwise, a change of the scale from inches to cm would result in a different type of estimator.

Less convincing is the requirement of equivariance in the following example. This example also illustrates a case where equivariant estimators do not exist for all sample sizes.

Example 2.2.2. Let $\mathfrak{P}$ be the family of all nondegenerate p–dimensional distributions $N_{(0,\Sigma)}$. The matrix Σ is the unknown parameter, the set of all symmetric and positive definite $p \times p$–matrices is the parameter space. Every nonsingular $p \times p$–matrix A defines a transformation of $\mathbb{R}^p$ into itself by $x \to Ax$. This induces in the parameter space the transformation $\Sigma \to A\Sigma A^\top$ (see Lemma 2.4.14). The functional $\kappa(N_{(0,\Sigma)}) = \det \Sigma$ is equivariant:

$$\kappa(N_{(0,\Sigma)} \circ (x \to Ax)) = (\det A)^2 \kappa(N_{(0,\Sigma)}).$$

An estimator for κ, based on a sample of size n, say $\kappa^{(n)} : (\mathbb{R}^p)^n \to (0, \infty)$ is equivariant if

$$\kappa^{(n)}(Ax_1, \ldots, Ax_n) = (\det A)^2 \kappa^{(n)}(x_1, \ldots, x_n) \qquad (2.2.3)$$

for every nonsingular $p \times p$–matrix A and all $x_\nu \in \mathbb{R}^p$, $\nu = 1, \ldots, n$.

It seems doubtful whether the requirement of equivariance is cogent in this case. Moreover, equivariant estimators do not exist if $n < p$.

Proof. Let A be a matrix with elements a_{ij}, $i, j = 1, \ldots, p$, subject to the condition $a_{11} = 2$ and $a_{ij} = \delta_{ij}$ for $i = 2, \ldots, p$, $j = 1, \ldots, p$. Relation (2.2.3) implies

$$\kappa^{(n)}(Ax_1, \ldots, Ax_n) = 4\kappa^{(n)}(x_1, \ldots, x_n) \quad \text{for } x_\nu \in \mathbb{R}^p, \ \nu = 1, \ldots, n. \quad (2.2.4)$$

Let $\xi = (x_{\nu j})_{\substack{\nu=1,\ldots,n \\ j=2,\ldots,p}}$. If $n < p$, for $\lambda^{n(p-1)}$–a.a. $\xi \in \mathbb{R}^{n(p-1)}$ there exist $a_{12}, \ldots, a_{1p}$ such that $\sum_{j=2}^p a_{1j}x_{\nu j} = -x_{\nu 1}$ for $\nu = 1, \ldots, n$, whence $Ax_\nu = x_\nu$ for $\nu = 1, \ldots, n$. Hence (2.2.4) implies $\kappa^{(n)} = 0$ λ^{np}–a.e. $\qquad \square$

The idea to use equivariance to guarantee that the estimator treats all parameter values in a homogeneous way is of limited reach for several reasons. The obvious one: In most cases the underlying family $\mathfrak{P}$ is not generated by a transformation group. This concerns, in particular, the usual discrete distributions. Moreover, the restriction to equivariant estimators may exclude some useful estimators. If equivariance is not a necessity if judged from the substantial meaning of the problem (like equivariance under changes of scales), it is hard to justify from a purely formal point.

That abstention from equivariance may, indeed, yield better estimators, is illustrated by the so–called Stein–estimators (see Example 2.7.1 which, by the way, concerns a model, where the requirement of equivariance appears quite natural).

In the absence of a transformation group, the usual vehicle for achieving equal treatment of all p–measures in the estimation of a real valued functional is mean unbiasedness.

Definition 2.2.5. The estimator $\hat{\kappa}$ is *mean unbiased* if $\kappa(P)$ is the mean of $P \circ \hat{\kappa}$ for $P \in \mathfrak{P}$, i.e. if

$$\int \hat{\kappa}(x)P(dx) = \kappa(P) \qquad \text{for } P \in \mathfrak{P}.$$

Since we give about the same weight to median unbiasedness (as defined in Definition 2.2.13) we speak of "mean unbiasedness" rather than just "unbiasedness" for the sake of clarity.

The following example illustrates a situation where mean unbiasedness is highly desirable.

Example 2.2.6. Assume that $\vartheta \in (0, \infty)$ is an unknown weight, and $P_\vartheta | \mathbb{B}_+$ the distribution of measures obtained for the weight ϑ by a certain measurement technique (think, e.g., of the determination of the dry weight or sugar content). Based on n measurements $x_1, \ldots, x_n$, an estimate $\vartheta^{(n)}(x_1, \ldots, x_n)$ for the weight ϑ is obtained. If the value of a lot is computed from this estimate, and if this is done recurrently, then the vendor will be privileged if the estimator has an upward mean bias, i.e. if $P_\vartheta^n(\vartheta^{(n)}) > \vartheta$ for every $\vartheta > 0$.

There are other situations where conditions inherently related to mean unbiasedness, yet slightly different, seem appropriate.

Example 2.2.7. Let $\mathfrak{P}$ be a family of p–measures $P | \mathbb{B}$ with positive Lebesgue density. For given $\beta \in (0, 1)$ let $q(P)$ denote the upper β–quantile of P, defined by $P(q(P), \infty) = \beta$. The problem is to estimate $q(P)$ from a sample $x_1, \ldots, x_n$. Let $q^{(n)}(x_1, \ldots, x_n)$ be such an estimate. Would it not be disturbing if, in the long run, the fraction of P falling above the estimated β–quantile were larger than β? To prevent this, one has to require that

$$\int P(q^{(n)}(x_1,\ldots,x_n),\infty)P(dx_1)\ldots P(dx_n) = \beta \quad \text{for } P \in \mathfrak{P}. \qquad (2.2.8)$$

If $q^{(n)}$ fulfills (2.2.8), then $(q^{(n)},\infty)$ is a β-*expectation tolerance interval*. Relation (2.2.8) is some sort of unbiasedness condition, yet something else than mean unbiasedness which would require that

$$\int q^{(n)}(x_1,\ldots,x_n)P(dx_1)\ldots P(dx_n) = q(P) \quad \text{for } P \in \mathfrak{P}.$$

The situations considered in these examples have a special feature in common, namely: The estimator is applied recurrently, and the sum or the average of the estimates is operationally meaningful. The requirement of mean unbiasedness is hard to justify if the estimation procedure is applied only once.

In many situations, the statistician is unable to foresee the possible uses of an estimator. Why should an estimator for the variance be mean unbiased? Considered as a measure of "spread" the variance has the wrong dimension (say cm^2 rather than cm). Hence the possibility cannot be excluded that the square root of this estimator will be used as an estimator for the standard deviation, which makes mean unbiasedness of the original estimator useless.

If mean unbiasedness is not required for the uses of the estimator, does it guarantee, at least, that the estimator treats all p-measures equal? The following example (modelled after an example by D. Basu, 1955b) demonstrates that this is not necessarily so.

Example 2.2.9. Let $\mathfrak{P} = \{N^n_{(\vartheta,1)} : \vartheta \in \mathbb{R}\}$. With $B_n := \{(x_1,\ldots,x_n) \in \mathbb{R}^n : x_1 \leq x_n\}$ let $\vartheta^{(n)}(x_1,\ldots,x_n) := 2\bar{x}_n 1_{B_n}(x_1,\ldots,x_n)$. Despite the strong preference for $\vartheta = 0$ (we have $N^n_{(\vartheta,1)}\{\underline{x} \in \mathbb{R}^n : \vartheta^{(n)}(\underline{x}) = 0\} = \frac{1}{2}$ for all $\vartheta \in \mathbb{R}$), this estimator is mean unbiased.

There are situations in which mean unbiasedness is highly desirable if not cogent. Yet mean unbiased estimators may not exist, or they may have undesirable properties. Whether useful mean unbiased estimators exist or not depends on certain mathematical features of the basic family and has nothing to do with whether they are desirable for a particular application. Since statisticians are prepared to settle with biased estimators if mean unbiased estimators do not exist, why should one insist on mean unbiased estimators if such ones exist, but are inferior?

That mean unbiasedness is of greater concern to theoretically minded statisticians is illustrated by the fact that the mean unbiased estimator for the correlation coefficient in the bivariate normal distribution, discovered by Olkin and Pratt (1958), is hardly ever used, for no other reason than that of beeing difficult to compute.

The following examples show that one has to live with situations where mean unbiased estimators may be desirable, but do not exist or are absurd.

Example 2.2.10. For the family of binomial distributions $\{B_{n,p} : p \in (0,1)\}$, the functional $\kappa : (0,1) \to \mathbb{R}$ admits a mean unbiased estimator iff κ is a polynomial of degree less than or equal to n.

Proof. (i) If $\hat{\kappa}$ is mean unbiased for κ, we have

$$\sum_{k=0}^{n} \hat{\kappa}(k) \binom{n}{k} p^k (1-p)^{n-k} = \kappa(p) \qquad \text{for } p \in (0,1).$$

Hence κ is a polynomial of degree $\leq n$.

(ii) For every $m \in \{0, 1, \ldots, n\}$,

$$\sum_{k=m}^{n} k(k-1) \ldots (k-m+1) B_{n,p}\{k\} = n(n-1) \ldots (n-m+1) p^m.$$

Hence $\hat{\kappa}_m$, defined by

$$\hat{\kappa}_m(k) := \frac{k(k-1) \ldots (k-m+1)}{n(n-1) \ldots (n-m+1)}$$

is mean unbiased for p^m, and $\sum_0^n a_m \hat{\kappa}_m$ is mean unbiased for $\kappa(p) = \sum_0^n a_m p^m$.

$\square$

The result of Example 2.2.10 is due to Kolmogorov (1950, p. 156). In spite of its simplicity, it is, perhaps, not generally recognized.

An approach to mean unbiased estimators by Washio, Morimoto and Ikeda (1956), based on operator theoretic methods, yields $\hat{\kappa}(k) := k/(n-k+1)$ as a mean unbiased estimator for $p/(1-p)$ (see p. 92, Example 5).

A similar criterion for the existence of mean unbiased estimators holds for convex families $\mathfrak{P}$: If there exists a mean unbiased estimator of κ for the sample size n, then $\kappa\big((1-\alpha)P_0 + \alpha P_1\big)$ is — for all $P_0, P_1 \in \mathfrak{P}$ — a polynomial in α of degree at most n, since

$$\kappa\big((1-\alpha)P_0 + \alpha P_1\big) = \int \kappa^{(n)}(x_1, \ldots, x_n) \prod_{\nu=1}^{n} \big((1-\alpha)P_0(dx_\nu) + \alpha P_1(dx_\nu)\big).$$

If $\mathfrak{P}$ is the family of *all* p–measures dominated by some σ–finite measure, then this necessary condition becomes sufficient, provided κ is bounded on $\mathfrak{P}$ (see Bickel and Lehmann, 1969, p. 1531, Theorem 4.1).

Exercise 2.2.11. Let $\mathfrak{P}$ be the family of all p–measures with positive Lebesgue density. Show that there is no mean unbiased estimator for the functional

$$\kappa(P) := \left(\int (x_1 - x_2)^2 P(dx_1) P(dx_2) \right)^{1/2}.$$

Exercise 2.2.12. Let $X = \mathbb{N}$, and let P_a, $a > 0$, be the Poisson distribution truncated at 0, defined by

$$P_a\{k\} = \frac{1}{e^a - 1} \frac{a^k}{k!} , \qquad k \in \mathbb{N}.$$

Show that
(i) the family $\{P_a : a > 0\}$ is complete, hence mean unbiased estimators are unique (provided they exist at all).
(ii) $\hat{\kappa}(1) = 0$ and $\hat{\kappa}(k) = k$ for $k = 2, 3, \ldots$ is mean unbiased for a. This estimator is not proper.
(iii) $\hat{\kappa}(k) = (-1)^{k+1}$ is mean unbiased for e^{-a}, the probability of 0 for the non–truncated Poisson distribution.

As an alternative to mean unbiasedness we consider median unbiasedness.

Definition 2.2.13. The estimator $\hat{\kappa}$ is *median unbiased* if $\kappa(P)$ is a median of $P \circ \hat{\kappa}$ for $P \in \mathfrak{P}$, i.e. if for $P \in \mathfrak{P}$

$$P\{x \in X : \hat{\kappa}(x) \geq \kappa(P)\} \geq \frac{1}{2} \quad \text{and} \quad P\{x \in X : \hat{\kappa}(x) \leq \kappa(P)\} \geq \frac{1}{2}.$$

That the true parameter value $\kappa(P)$ is overestimated and underestimated with about the same probability is of immediate intuitive appeal. Furthermore, for any monotone function h, the function $h \circ \hat{\kappa}$ is a median unbiased estimator of $h(\kappa(P))$ if $\hat{\kappa}$ is a median unbiased estimator of $\kappa(P)$ — a property not shared by mean unbiasedness.

Example 2.2.14. For radio–active material, the probability that an atom which exists at time 0 will still exist at time t is $\exp[-\vartheta t]$, ϑ being the decay constant, typical for the material. The decay constant is closely related to another physically meaningful quantity, the half–life, which equals $(\log 2)/\vartheta$. It will not always be clear whether the quantity to be estimated is the decay constant or the half–life. Working with a median unbiased estimator, this creates no problems: If $\hat{\vartheta}$ is a median unbiased estimator for the decay constant, then $(\log 2)/\hat{\vartheta}$ is a median unbiased estimator for the half–life.

In spite of its greater intuitive appeal, median unbiasedness has only been used sporadically in theoretical investigations. This is, perhaps, due to the fact that it was considered as mathematically difficult to handle. Fraser (1957, p. 49) says so explicitly: "Median unbiasedness has found little application in estimation theory primarily because it does not seem to lend itself to the mathematical analysis needed to find minimum risk estimates."

Historical Remark 2.2.15. Laplace (1774, pp. 636–644) shows that an estimator T minimizing the loss function $|T - \vartheta|$ is median unbiased. Gauß (1821) who replaced this loss function by the loss function $(T - \vartheta)^2$ ends up with the method of least squares and mean unbiasedness. From then on, the concept of mean unbiasedness is prevalent.

G.W. Brown (1947, p. 583) seems to have been the first to revive the concept of median unbiasedness. In connection with control charts, the usefulness of median unbiased estimators was early recognized (see Eisenhart and Martin, 1948, Eisenhart, 1949).

General discussions of the concept of bias are contained in Lehmann (1951), Birnbaum (1961, 1964), van der Vaart (1961).

The median unbiasedness of special estimators is investigated also by Hodges and Lehmann (1963, Section 9) and by Vogt (1968). The first general result (corresponding to our Theorem 5.4.3) is the existence of maximally concentrated median unbiased estimators for families with monotone likelihood ratios noticed by Lehmann (1959, p. 83; see also 1986, p. 95).

In connection with the existence of median unbiased estimators there is one problem which does not occur with mean unbiased estimators: If P is a discrete distribution, then so is P^n and $P^n \circ \kappa^{(n)}$ for any estimator $\kappa^{(n)}$. Usually, there are gaps in the support of $P^n \circ \kappa^{(n)}$. Then nonrandomized estimators cannot be median unbiased, in general.

As an example, consider the family of Poisson distributions, $\{\Pi_a : a > 0\}$. Assume that $\hat{\kappa} : \{0\} \cup \mathbb{N} \to (0, \infty)$ is increasing. If (a', a'') is a gap in $\hat{\kappa}(\{0\} \cup \mathbb{N})$, then $\{k \in \{0\} \cup \mathbb{N} : \hat{\kappa}(k) < a\} = \{k \in \{0\} \cup \mathbb{N} : \hat{\kappa}(k) \le a\} = \{0, 1, \ldots, k_0\}$ for every $a \in (a', a'')$. Hence median unbiasedness of $\hat{\kappa}$ requires that $\Pi_a\{0, 1, \ldots, k_0\} = \frac{1}{2}$ for every $a \in (a', a'')$, which is impossible (since $a \to \Pi_a\{0, 1, \ldots, k_0\}$ is decreasing). To achieve median unbiasedness, one has to use randomized estimators bridging such gaps. This will be discussed in Sections 5.1 and 5.3.

2.3 The concentration of real valued estimators

Introduces "concentration on intervals around $\kappa(P)$" and investigates its relation to "spread".

Comparing the concentration of two estimators $\hat{\kappa}_i : X \to \mathbb{R}$ about $\kappa(P)$ can be reformulated as the problem of comparing the concentration of the p–measures $P \circ (\hat{\kappa}_i - \kappa(P))$ about 0. Hence we may simplify our notations by speaking of the concentration of p–measures $Q_i | \mathbb{B}$ about 0.

Definition 2.3.1. $Q_1 | \mathbb{B}$ is *more concentrated about* 0 *than* $Q_2 | \mathbb{B}$ if

$$Q_1(-t', t'') \geq Q_2(-t', t'') \qquad \text{for } t', t'' \geq 0.$$

Observe that Q_1 is more concentrated on all open intervals about 0 than Q_2 iff it is more concentrated on all closed intervals about 0.

Lemma 2.3.2. *For $i = 1, 2$ let $Q_i | \mathbb{B}$ be a p–measure with Lebesgue density q_i and median 0 such that $\{t \in \mathbb{R} : q_1(t) \geq q_2(t)\}$ is an interval containing 0. Then Q_1 is more concentrated about 0 than Q_2. The inequality in Definition 2.3.1 is strict if $\{t \in \mathbb{R} : q_1(t) > q_2(t)\}$ is an interval.*

Proof. If $Q_1(0, t_0) < Q_2(0, t_0)$ for some $t_0 > 0$, then $q_1(t) < q_2(t)$ for $t \geq t_0$. This implies $Q_1(0, \infty) < Q_2(0, \infty)$, in contradiction to the assumption that 0 is a median of Q_1 and Q_2. Hence, $Q_1(0, t) \geq Q_2(0, t)$ for $t > 0$.

Together with the corresponding inequality for intervals $(-t, 0)$ this proves the assertion. $\qquad\square$

Comparability in the strong sense of Definition 2.3.1 implies $Q_1[0, \infty) \geq Q_2[0, \infty)$ and $Q_1(-\infty, 0] \geq Q_2(-\infty, 0]$. If $Q_1\{0\} = 0$, it follows that $Q_2\{0\} = 0$, and therefore $Q_1[0, \infty) = Q_2[0, \infty)$ and $Q_1(-\infty, 0] = Q_2(-\infty, 0]$. Hence estimators with nonatomic distributions cannot be comparable in the strong sense of Definition 2.3.1, unless $Q_1(-\infty, 0] = Q_2(-\infty, 0]$. In particular: *A median unbiased estimator with nonatomic distribution can be comparable with median unbiased estimators only.*

Restricting Definition 2.3.1 to intervals symmetric about 0 is not always opportune. If the problem is to evaluate an equivariant estimator of a scale parameter c, natural intervals are (ct^{-1}, ct) with $t > 1$ (rather than $(c-t, c+t)$ with $t > 0$).

There is another concept of concentration which does not refer to a particular "center".

Definition 2.3.3. $Q_1 | \mathbb{B}$ is *less spread out* than $Q_2 | \mathbb{B}$ (for short $Q_1 <_s Q_2$) if

$$F_1^{-1}(\beta) - F_1^{-1}(\alpha) \leq F_2^{-1}(\beta) - F_2^{-1}(\alpha) \qquad \text{for } 0 < \alpha < \beta < 1. \qquad (2.3.4)$$

An equivalent condition is

$$F_2^{-1} - F_1^{-1} \text{ is isotone on } (0, 1). \qquad (2.3.5)$$

Here the left continuous quantile function F^{-1} is defined by

$$F^{-1}(\alpha) := \inf\{x \in \mathbb{R} : F(x) \geq \alpha\}, \qquad 0 < \alpha < 1,$$

where F denotes the (right continuous) distribution function of $Q|\mathbb{B}$.

Historical Remark 2.3.6. The spread order was introduced independently by Saunders and Moran (1978) and by Bickel and Lehmann (1979). In a form essentially equivalent to (2.3.5) it was earlier used by Doksum (1969).

Example 2.3.7. For $a \in \mathbb{R}$, $c > 0$ let $Q_{a,c} := Q \circ (x \to a + cx)$. Then $Q_{a_1,c_1} <_s Q_{a_2,c_2}$ for all $a_1, a_2 \in \mathbb{R}$ and $0 < c_1 \leq c_2$. In particular: $Q_a := Q \circ (x \to x + a)$ is equivalent to Q in the spread order. Using (2.3.5) this follows immediately from $F_{a,c}^{-1}(\alpha) = a + cF^{-1}(\alpha)$ for $\alpha \in (0,1)$ (where F is the distribution function of Q).

Example 2.3.8. For $\Gamma_{a,b}|\mathbb{B}_+$, the gamma distribution with shape parameter b and scale parameter a, we have $\Gamma_{a,b_1} <_s \Gamma_{a,b_2}$ for $a > 0$ and $0 < b_1 \leq b_2$. (For $b_1 \geq 1$ the assertion follows from Proposition 2.3.21(i), applied with $Q = \Gamma_{a,b_1}$ and $R_2 = \Gamma_{a,b_2-b_1}$. The proof for arbitrary $b_1 > 0$ is given in Saunders and Moran, 1978.) Since $a_1 \leq a_2$ implies $\Gamma_{a_1,b} <_s \Gamma_{a_2,b}$ by Example 2.3.7, we obtain $\Gamma_{a_1,b_1} <_s \Gamma_{a_2,b_2}$ if $0 < a_1 \leq a_2$ and $0 < b_1 \leq b_2$.

Exercise 2.3.9. The support of $Q|\mathbb{B}$ is the smallest interval I with $Q(I) = 1$. Two p-measures with the same bounded support are comparable in the spread order only if they are identical. (Hint: Use relation (2.3.4). Supplement the definition of $F^{-1}(\alpha)$ for $\alpha = 0$ and $\alpha = 1$ appropriately.)

Proposition 2.3.10. *For $i = 1,2$ let $Q_i|\mathbb{B}$ be a p-measure with Lebesgue density q_i. Assume that $\{x \in \mathbb{R} : q_i(x) > 0\}$ is an interval. Then $Q_1 <_s Q_2$ is equivalent to*

$$q_1\big(F_1^{-1}(\alpha)\big) \geq q_2\big(F_2^{-1}(\alpha)\big) \qquad \text{for } \lambda\text{-a.a. } \alpha \in (0,1). \tag{2.3.11}$$

Proof. If Q admits a Lebesgue density q, we have $(F^{-1})' = 1/q \circ F^{-1}$ λ-a.e. on $(0,1)$. Therefore, (2.3.5) is equivalent to

$$\frac{1}{q_2 \circ F_2^{-1}} - \frac{1}{q_1 \circ F_1^{-1}} \geq 0 \qquad \lambda\text{-a.e. on } (0,1). \tag{2.3.12}$$

This, in turn, is equivalent to (2.3.11).

(That relation (2.3.12) implies (2.3.5) follows from the equality

$$\int\limits_{\alpha}^{\beta} \frac{1}{q \circ F^{-1}} d\lambda = \int\limits_{F^{-1}(\alpha)}^{F^{-1}(\beta)} \frac{F'}{q} d\lambda = F^{-1}(\beta) - F^{-1}(\alpha)$$

for $0 < \alpha < \beta < 1$, which holds true with $F = F_i$ and $q = q_i$ since F is increasing and absolutely continuous on $[F^{-1}(\alpha), F^{-1}(\beta)]$ with inverse F^{-1}.)

$\square$

The intuitive interpretation of the spread order is clear from (2.3.4) and/or (2.3.11): $Q_1 <_s Q_2$ means that Q_1 is more concentrated than Q_2 about each quantile.

Spread order is defined without reference to a particular "center" and there-fore not in an immediate relation to "concentration about 0" as defined in Definition 2.3.1. Recall that nonatomic measures can be comparable with respect to their concentration about 0 only if $Q_1(-\infty, 0] = Q_2(-\infty, 0]$. If this necessary condition is fulfilled, $Q_1 <_s Q_2$ implies that Q_1 is more concentrated about 0 than Q_2 (see Proposition 2.3.17). In fact, comparability in "spread order" is a much stronger concept than comparability in "concentration about a given point".

Lemma 2.3.13. $Q_1 <_s Q_2$ *implies*

$$F_2\big(F_2^{-1}(\alpha) + t\big) \le F_1\big(F_1^{-1}(\alpha) + t\big) \quad \text{for } t > 0, \ \alpha \in (0,1) \quad (2.3.14')$$

and

$$F_2\big(F_2^{-1}(\alpha) + t\big) \ge F_1\big(F_1^{-1}(\alpha) + t\big) \quad \text{for } t < 0, \ \alpha \in (0,1). \quad (2.3.14'')$$

Either of these inequalities implies $Q_1 <_s Q_2$.

Proof. The following conditions are equivalent for $0 < \alpha \le \beta < 1$:

$$F_1^{-1}(\beta) - F_1^{-1}(\alpha) \le F_2^{-1}(\beta) - F_2^{-1}(\alpha),$$
$$\beta \le F_1\big(F_1^{-1}(\alpha) + F_2^{-1}(\beta) - F_2^{-1}(\alpha)\big).$$

(i) If $Q_1 <_s Q_2$ and $t > 0$, let $\beta := F_2\big(F_2^{-1}(\alpha) + t\big)$. We assume w.l.g. that $\beta > \alpha$. Since $F_2^{-1}(\beta) \le F_2^{-1}(\alpha) + t$, we obtain $\beta \le F_1\big(F_1^{-1}(\alpha) + t\big)$.

(ii) If (2.3.14') holds true, let $t := F_2^{-1}(\beta) - F_2^{-1}(\alpha)$. Since $\beta \le F_2\big(F_2^{-1}(\alpha) + t\big) \le F_1\big(F_1^{-1}(\alpha) + F_2^{-1}(\beta) - F_2^{-1}(\alpha)\big)$, $Q_1 <_s Q_2$ follows. (Observe that (2.3.14') holds true for $t = 0$ since distribution functions are right continuous.)

This proves the assertion for $t > 0$. The case $t < 0$ follows similarly. $\square$

Definition 2.3.15. A p-measure $Q|\mathbb{B}$ is *unimodal* about a *mode* x if its distribution function is convex on $(-\infty, x)$ and concave on (x, ∞).

Notice that unimodal p–measures have no atoms, except, perhaps, for the mode. If Q is absolutely continuous with respect to Lebesgue measure, then unimodality of Q implies the existence of a density q which is superconvex, i.e. the set $\{y \in \mathbb{R} : q(y) \geq r\}$ is convex for each $r > 0$.

Lemma 2.3.16. *If $Q|\mathbb{B}$ is symmetric about 0 and unimodal, then*

$$Q(u - t, u + t) \leq Q(-t, t) \qquad \text{for } u \in \mathbb{R}, \ t > 0.$$

Proof. For $0 < u \leq t$ we have

$$Q(u - t, u + t) = Q(-t, t) + Q(t, u + t) - Q(-t, u - t],$$

with

$$Q(t, u + t) = F(u + t) - F(t) \leq F(t) - F(t - u) = Q(t - u, t) \leq Q(-t, u - t]$$

since $\big(F(u + t) + F(t - u)\big)/2 \leq F(t)$. For $u > t$ we show that

$$Q(u - t, u + t) \leq Q(0, 2t), \quad \text{i.e. } F(u + t) + F(0) \leq F(2t) + F(u - t).$$

This follows from

$$F(2t) \geq \frac{2t}{u + t} F(u + t) + \frac{u - t}{u + t} F(0)$$

and

$$F(u - t) \geq \frac{u - t}{u + t} F(u + t) + \frac{2t}{u + t} F(0).$$

$\square$

Proposition 2.3.17. *Assume that $Q_1 <_s Q_2$.*

(i) If 0 is a common quantile of Q_1 and Q_2 in the sense that $F_1^{-1}(\alpha) = 0 = F_2^{-1}(\alpha)$ for some $\alpha \in (0, 1)$, then

$$Q_1(-t', t'') \geq Q_2(-t', t'') \qquad \text{for } t', t'' > 0.$$

(ii) If Q_1 is symmetric about 0 and unimodal, then

$$Q_1(-t, t) \geq Q_2(-t, t) \qquad \text{for } t > 0.$$

Proof. (i) By assumption, there exists $\alpha \in (0, 1)$ with $F_1^{-1}(\alpha) = 0 = F_2^{-1}(\alpha)$. Therefore the assertion follows from (2.3.14).

(ii) We assume w.l.g. that Q_1 is nondegenerate and $\alpha := F_2(0) \in (0, 1)$. By (2.3.14') we have

$$F_2(t) \leq F_2\big(F_2^{-1}(\alpha + \tfrac{1}{n}) + t\big) \leq F_1\big(F_1^{-1}(\alpha + \tfrac{1}{n}) + t\big) \qquad \text{for } n \in \mathbb{N}.$$

Since F_1^{-1} is continuous (as F_1 is strictly increasing), this implies $F_2(t) \leq F_1\big(F_1^{-1}(\alpha) + t\big)$. Moreover, (2.3.14″) yields

$$F_2(-t) \geq F_2\big(F_2^{-1}(\alpha) - t\big) \geq F_1\big(F_1^{-1}(\alpha) - t\big).$$

Hence

$$Q_1(-t,t] = F_1(t) - F_1(-t) \geq F_1\big(F_1^{-1}(\alpha) + t\big) - F_1\big(F_1^{-1}(\alpha) - t\big)$$
$$\geq F_2(t) - F_2(-t) = Q_2(-t,t],$$

where the first inequality follows from Lemma 2.3.16.

From $Q_1(-t,t] \geq Q_2(-t,t]$ for $t > 0$, the assertion follows easily (let $t_n \uparrow t$). $\square$

Even the weaker concept of "concentration about 0" for $P \circ \big(\hat{\kappa} - \kappa(P)\big)$, i.e. concentration about $\kappa(P)$ of $P \circ \hat{\kappa}$, applies only now and then for finite sample size comparisons of estimators. It is the comparison of limit distributions, where these concepts display their full force. The following results will be applied in Sections 8.5.

Throughout the rest of this section $Q|\mathbb{B}$ is a p-measure with Lebesgue density q, such that $I := \{x \in \mathbb{R} : q(x) > 0\}$ is some open interval (including $I = \mathbb{R}$ and $I = (0,\infty)$). Furthermore, F denotes the distribution function of Q.

Definition 2.3.18. A p-measure Q on $\mathbb{B} \cap I$ is *strongly unimodal* if $\log q$ is concave on I.

"Strongly unimodal" implies "unimodal" (since "$\log p$ concave" implies "$\log p$ superconvex" implies "p superconvex", see Lemma 2.4.5(i),(ii)).

The following distributions are strongly unimodal: normal, logistic, Laplace.

Lemma 2.3.19. *A real valued function on an open interval is concave iff it is continuous everywhere and differentiable with antitone derivative, except for countably many points.*

Proof. Bourbaki (1958), p. 49, Proposition 8. $\square$

Lemma 2.3.20. *Let $Q|\mathbb{B} \cap I$ be a p-measure with Lebesgue density q. Then $q \circ F^{-1}$ is concave on $(0,1)$ iff Q is strongly unimodal.*

Proof. By Lemma 2.3.19, the following relations are equivalent (e.c. means "except for countably many points").

$q \circ F^{-1}$ concave on $(0,1)$;
$q \circ F^{-1}$ continuous on $(0,1)$ and $(q \circ F^{-1})'$ is antitone e.c.;
$q \circ F^{-1}$ continuous on $(0,1)$ and $q' \circ F^{-1}/q \circ F^{-1}$ is antitone e.c.;
q continuous on I and q'/q is antitone e.c.;
$\log q$ continuous on I and $(\log q)'$ is antitone e.c.;
$\log q$ concave on I. □

Part (ii) of the following proposition is due to Droste and Wefelmeyer (1985). The proof follows Klaassen (1985). Part (i) is due to Lewis and Thompson (1981).

Proposition 2.3.21. *For any p-measure $Q|\mathbb{B} \cap I$ with Lebesgue density the following holds true.*
(i) *If Q is strongly unimodal, then $R_1 <_s R_2$ implies $Q * R_1 <_s Q * R_2$.*
(ii) *If $Q <_s Q * R$ for every $R|\mathbb{B}$, then Q is strongly unimodal.*

Proof. (i) Let h_i and H_i be the density and the distribution function of $Q * R_i$, respectively. We shall show first that there exists $u_0 \in [-\infty, \infty]$ such that

$$G_1\big(H_1^{-1}(\alpha) - u\big) \gtrless G_2\big(H_2^{-1}(\alpha) - u\big) \qquad \text{if } u \lessgtr u_0, \qquad (2.3.22)$$

where G_i denotes the distribution function of R_i.
    For $y = H_1^{-1}(\alpha)$ and $z = H_2^{-1}(\alpha)$ let

$$u_0 := \inf\{u : G_1(y - u) < G_2(z - u)\}.$$

Then we have

$$G_1(y - u) \geq G_2(z - u) \qquad \text{if } u \leq u_0.$$

On the other hand, $G_1(y - v) < \alpha < G_2(y - v)$ implies $y - v \leq G_1^{-1}(\alpha)$ and $z - v \geq G_2^{-1}(\alpha)$, and hence by Lemma 2.3.13

$$G_1\big(y - (v + c)\big) \leq G_1\big(G_1^{-1}(\alpha) - c\big) \leq G_2\big(G_2^{-1}(\alpha) - c\big) \leq G_2\big(z - (v + c)\big)$$

for $c > 0$. Thus, if $G_1(y - v) < G_2(z - v)$, then $G_1(y - u) \leq G_2(z - u)$ for $u > v$. Therefore,

$$G_1(y - u) \leq G_2(z - u) \qquad \text{if } u > u_0.$$

Since $H_1\big(H_1^{-1}(\alpha)\big) = \alpha = H_2\big(H_2^{-1}(\alpha)\big)$, we have

$$\int q(u)\big[G_1\big(H_1^{-1}(\alpha) - u\big) - G_2\big(H_2^{-1}(\alpha) - u\big)\big]\lambda(du) = 0. \qquad (2.3.23)$$

Furthermore

$$h_1\big(H_1^{-1}(\alpha)\big) - h_2\big(H_2^{-1}(\alpha)\big)$$
$$= \int q'(u)\big[G_1\big(H_1^{-1}(\alpha) - u\big) - G_2\big(H_2^{-1}(\alpha) - u\big)\big]\lambda(du).$$

By Proposition 2.3.10 it suffices to show that the right hand side is nonnegative. In view of (2.3.22) and (2.3.23) we assume w.l.g. that $u_0 \in \mathbb{R}$. Then

$$h_1\big(H_1^{-1}(\alpha)\big) - h_2\big(H_2^{-1}(\alpha)\big)$$
$$= \int q(u)\big[G_1\big(H_1^{-1}(\alpha) - u\big) - G_2\big(H_2^{-1}(\alpha) - u\big)\big]\Big(\frac{q'(u)}{q(u)} - \frac{q'(u_0)}{q(u_0)}\Big)\lambda(du)$$

by (2.3.23), and the right hand side is nonnegative since the integrand is non-negative by (2.3.22) and Lemma 2.3.19.

(ii) We first show the assertion for a density q which is continuous on I.

Let $R|\mathbb{B}$ be an arbitrary p–measure, and denote the density and the distribution function of $Q * R$ by h and H, respectively. Since q is continuous, Proposition 2.3.10 implies $h\big(H^{-1}(\alpha)\big) \le q\big(F^{-1}(\alpha)\big)$ for $0 < \alpha < 1$, hence $h(x) \le q\big(F^{-1}(H(x))\big)$ for $x \in H^{-1}(0,1)$. This can be rewritten as

$$\int q\big(F^{-1}(F(x-y))\big)R(dy) \le q\big(F^{-1}(\int F(x-y)R(dy))\big) \quad \text{for } x \in H^{-1}(0,1),$$

where $q\big(F^{-1}(0)\big) := 0$ and $q\big(F^{-1}(1)\big) := 0$.

Since $R|\mathbb{B}$ was arbitrary, this implies

$$\int q \circ F^{-1} dP \le q \circ F^{-1}\big(\int x P(dx)\big) \text{ for each } p\text{–measure } P|\mathbb{B} \cap (0,1).$$

Therefore $q \circ F^{-1}$ is concave on $(0,1)$, hence Q is strongly unimodal by Lemma 2.3.20.

We now infer the general case. As $N_{(0,1/n)}$ is strongly unimodal, $Q <_s Q * R$ implies $N_{(0,1/n)} * Q <_s N_{(0,1/n)} * Q * R$ by (i). Since the density, p_n say, of $N_{(0,1/n)} * Q$ is positive and continuous on $\mathbb{R}$, $N_{(0,1/n)} * Q$ is strongly unimodal. Since $\lim_{n\to\infty} p_n = q$ λ–a.e., $\log q$ is concave. $\square$

Corollary 2.3.24. *Strong unimodality of Q_i, $i = 1,2$, implies strong unimodality of $Q_1 * Q_2$.*

Proof. For any R, $Q_2 <_s Q_2 * R$ by 2.3.21(i). This implies $Q_1 * Q_2 <_s Q_1 * (Q_2 * R) = (Q_1 * Q_2) * R$ by 2.3.21(i). Hence $Q_1 * Q_2$ is strongly unimodal by 2.3.21(ii). $\square$

A similar characterization of "strong unimodality" is due to Ibragimov (1956): Q is strongly unimodal if and only if $Q * R$ is unimodal for any unimodal R.

Corollary 2.3.25. *Let $R|\,\mathbb{B}$ be an arbitrary p–measure. If $Q|\,\mathbb{B}\cap I$ is strongly unimodal and $Q(-\infty, 0] = Q * R(-\infty, 0]$, then*

$$Q(-t', t'') \geq Q * R(-t', t'') \qquad \text{for } t', t'' > 0.$$

Proof. Follows from Propositions 2.3.21(i) and 2.3.17(i).   □

Proposition 2.3.26. *Let $R|\,\mathbb{B}$ be an arbitrary p–measure. If $Q|\,\mathbb{B}$ is symmetric about 0 and unimodal, then*

$$Q(-t, t) \geq Q * R(-t, t) \qquad \text{for } t > 0.$$

Proof. By Lemma 2.3.16, we have

$$Q(-t-y, t-y) \leq Q(-t, t) \qquad \text{for } y \in \mathbb{R}.$$

From this the assertion follows by integration with respect to R.   □

What we shall need for the comparison of 1–dimensional limit distributions is Corollary 2.3.25. Regrettably, this result does not generalize to multidimensional limit distributions. The multidimensional result, Corollary 2.4.13(ii), specialized for $p = 1$, becomes Proposition 2.3.26. It has been included here because the 1–dimensional version has a comparatively simple proof (which does not depend on Anderson's theorem).

2.4 Concentration of multivariate estimators

"Concentration on intervals about $\kappa(P)$" is generalized to "concentration on convex sets symmetric about $\kappa(P)$". The main result is Anderson's theorem. The section is concluded with results on the concentration of normal distributions.

As in Section 2.3, we reformulate the problem of comparing the concentration of two estimators $\hat{\kappa}_i : X \to \mathbb{R}^p$ about $\kappa(P)$ as the problem of comparing the concentration of the p–measures $P \circ (\hat{\kappa}_i - \kappa(P))|\,\mathbb{B}^p$ about 0.

To generalize Definition 2.3.1 to the multivariate case, we have, first of all, to generalize the concept of an "interval containing 0". Natural generalizations are "rectangles containing 0" or "convex sets containing 0".

In defining the concept of "multivariate concentration" we have to have an eye not only on its intuitive interpretation, but also on its fruitfulness from the mathematical point of view. Already in the univariate case the most useful applications are to limit distributions. This holds all the more for a multivariate concept of concentration. A concept of concentration based on the probability $P\{\hat{\kappa} \in C\}$ for arbitrary convex sets containing $\kappa(P)$ is certainly meaningful from the intuitive point of view. Yet it is hardly useful. Generally, there are no estimators comparable in this strong sense. The following definition is coined with an eye on the application to limit distributions.

Definition 2.4.1. $Q_1 | \mathbb{B}^p$ is *more concentrated about* 0 than $Q_2 | \mathbb{B}^p$ if $Q_1(C) \geq Q_2(C)$ for every convex set $C \in \mathbb{B}^p$ which is symmetric about 0.

The essential distinction between the univariate Definition 2.3.1 and the multivariate Definition 2.4.1 is the restriction to *symmetric* sets in the latter. Whereas univariate normal distributions with mean 0 are comparable in the sense of Definition 2.3.1, multivariate normal distributions with mean vector 0 are comparable in the sense of Definition 2.4.1 iff the difference in their covariance matrices is nonnegative definite (i.e. if their covariance matrices are comparable in the Löwner order (which is defined by $\Sigma_1 \leq_L \Sigma_2$ iff $\Sigma_2 - \Sigma_1$ is positive semidefinite; see Corollary 2.4.17). They are comparable on arbitrary convex sets containing the origin iff their covariance matrices are proportional. See Proposition 2.4.18.

That Definition 2.4.1 is based on convex sets has the advantage that the comparability of p-measures is retained under linear transformations. Restricted to linear transformations, this seems to be of but limited value. It is, however, all we need for the comparison of limit distributions (see Proposition 7.2.1).

Proposition 2.4.2. *If $Q_1 | \mathbb{B}^p$ is more concentrated than Q_2 on all convex sets in $\mathbb{B}^p$ [containing 0/symmetric about 0], then $Q_1 \circ (u \to Au)$ is more concentrated than $Q_2 \circ (u \to Au)$ on all convex sets in $\mathbb{B}^q$ [containing 0/symmetric about 0] for any $q \times p$–matrix A.*

Proof. $Q_i \circ (u \to Au)(C) = Q_i\{u \in \mathbb{R}^p : Au \in C\}$. If C is a measurable convex set [containing 0/symmetric about 0], then $\{u \in \mathbb{R}^p : Au \in C\}$ has the same properties. $\square$

Proposition 2.4.2 applies, in particular, for $q = 1$. If $Q_1 | \mathbb{B}^p$ is more concentrated than $Q_2 | \mathbb{B}^p$ on all measurable convex sets symmetric about 0, then this carries over to all 1–dimensional "marginal" distributions, $Q_i \circ (u \to a^\top u)$, with $a \in \mathbb{R}^p$. It would be technically useful (in dealing with limit distributions) if this could be inverted, i.e. if $Q_1\{u \in \mathbb{R}^p : a^\top u \in I\} \geq Q_2\{u \in \mathbb{R}^p : a^\top u \in I\}$ for all $a \in \mathbb{R}^p$ and all I symmetric about 0 would imply that $Q_1(C) \geq Q_2(C)$

for all measurable convex sets $C \subset \mathbb{R}^p$ which are symmetric about 0. The following example demonstrates that this is not generally true. It is, however, true if Q_1 and Q_2 are normal.

Example 2.4.3. Let $Q_0 = N_{(0,I)}$ and let $Q_1 | \mathbb{B}^2$ be the p–measure with λ^2–density $u \rightarrow \varphi_I(u)(1 + q(u))$, where $\varphi_I(u) = \frac{1}{2\pi} \exp[-u^T u/2]$ and $q(u) = \frac{4}{17}(2 - 5u^T u + (u^T u)^2)$. Then we have, for all $0 \neq a \in \mathbb{R}^2$ and $t > 0$,

$$Q_0\{u \in \mathbb{R}^2 : -t < a^T u < t\} > Q_1\{u \in \mathbb{R}^2 : -t < a^T u < t\},$$

whereas

$$Q_0\{u \in \mathbb{R}^2 : u^T u \leq r\} < Q_1\{u \in \mathbb{R}^2 : u^T u \leq r\} \quad \text{for } 0 < r < 1.$$

To prove the first inequality, let $0 \neq a = (a_1, a_2) \in \mathbb{R}^2$. W.l.g. we assume $a_1 = 1$ and (by rotation invariance) $a_2 = 0$. Then the derivative with respect to t of $\int \varphi_I(u)q(u)1_{(-t,t) \times \mathbb{R}}(u)\lambda^2(du)$ equals

$$\frac{8}{17}\varphi_I(t)t^2(t^2 - 3) \overset{<}{\underset{>}{=}} 0 \quad \text{for } t \overset{<}{\underset{>}{=}} \sqrt{3}.$$

This implies $\int \varphi_I(u)q(u)1_{(-t,t) \times \mathbb{R}}(u)\lambda^2(du) < 0$, thereby proving the first relation. The second inequality is verified by direct computation which yields

$$\int \varphi_I(u)q(u)1_{(0,r]}(u^T u)\lambda^2(du) = \frac{4}{17}r(1 - r)e^{-r/2} \quad \text{for } r > 0.$$

The main motivation for Definition 2.4.1 is Anderson's theorem (see Corollary 2.4.13), according to which $N_{(0,\Sigma)}$ is more concentrated (in the sense of Definition 2.4.1) than $N_{(0,\Sigma)} * R$, for any p–measure $R | \mathbb{B}^p$. Even if $N_{(0,\Sigma)} * R$ is symmetric about 0, $N_{(0,\Sigma)}(C) \leq N_{(0,\Sigma)} * R(C)$ is not necessarily true for *all* measurable convex sets C containing 0. (Hint: Choose $R = N_{(0,\hat{\Sigma} - \Sigma)}$ with $\hat{\Sigma}$ not proportional to Σ, and apply Proposition 2.4.18.)

For the main result of this section, Anderson's theorem, we need some preparatory lemmas. Following Anderson (1955) the proof is usually based on the Brunn–Minkowski theorem (see, e.g., Strasser, 1985, pp. 193/4 or Ibragimov and Has'minskii, 1981, p. 155, Lemma 10.1). The proof given below uses elementary properties of convex sets only.

Definition 2.4.4. A function $f : \mathbb{R}^p \rightarrow (-\infty, \infty]$ is *superconvex* if the set $\{x \in \mathbb{R}^p : f(x) \geq r\}$ is convex for each $r \in \mathbb{R}$.

The following lemma collects some properties of superconvex functions.

Lemma 2.4.5. (i) *Every concave function is superconvex.*

(ii) *If $f : \mathbb{R}^p \to \mathbb{R}$ is superconvex and $m : \mathbb{R} \to \mathbb{R}$ isotone, then $m \circ f$ is superconvex.*

(iii) *$f : \mathbb{R}^p \to \mathbb{R}$ is superconvex on $\mathbb{R}^p$ iff the function $\alpha \to f((1 - \alpha)x_0 + \alpha x_1)$ is superconvex on $[0, 1]$ for all $x_i \in \mathbb{R}^p$, $i = 0, 1$.*

(iv) *If $f : \mathbb{R}^p \to \mathbb{R}$ is superconvex with mode 0, then the function $r \to f(rx_0)$ is antitone on $[0, \infty)$ for every $x_0 \in \mathbb{R}^p$.*

Proof. (i) Let f be concave. $f(x_i) \geq c$ for $i = 0, 1$ implies for $\alpha \in [0, 1]$

$$f((1 - \alpha)x_0 + \alpha x_1) \geq (1 - \alpha)f(x_0) + \alpha f(x_1) \geq c.$$

Hence $\{x \in \mathbb{R}^p : f(x) \geq c\}$ is convex.

(ii) Let $m(f(x_i)) \geq c$ for $i = 0, 1$. If f is superconvex, then $\{x \in \mathbb{R}^p : f(x) \geq \min\{f(x_0), f(x_1)\}\}$ is convex, hence

$$f((1 - \alpha)x_0 + \alpha x_1) \geq \min\{f(x_0), f(x_1)\} \quad \text{for } \alpha \in [0, 1].$$

Since m is isotone, $m(f((1 - \alpha)x_0 + \alpha x_1)) \geq c$ follows.

(iii) Given $x_i \in \mathbb{R}^p$, $i = 0, 1$, let $h(\alpha) := f((1 - \alpha)x_0 + \alpha x_1)$, $\alpha \in [0, 1]$.

(iii') $h(\alpha_i) \geq c$ for $i = 0, 1$ is equivalent to $f(y_i) \geq c$ for $y_i = (1 - \alpha_i)x_0 + \alpha_i x_1$. Superconvexity of f implies $f((1 - \delta)y_0 + \delta y_1) \geq c$ for $\delta \in [0, 1]$. Since

$$(1 - \delta)y_0 + \delta y_1 = (1 - [(1 - \delta)\alpha_0 + \delta\alpha_1])x_0 + [(1 - \delta)\alpha_0 + \delta\alpha_1]x_1,$$

we obtain $h((1 - \delta)\alpha_0 + \delta\alpha_1) \geq c$. Hence $\{\alpha \in [0, 1] : h(\alpha) \geq c\}$ is convex.

(iii'') $f(x_i) \geq c$ for $i = 0, 1$ is equivalent to $h(\alpha) \geq c$ for $\alpha = 0$ and $\alpha = 1$. If h is superconvex, this implies $h(\alpha) \geq c$ for all $\alpha \in [0, 1]$, i.e. $f((1 - \alpha)x_0 + \alpha x_1) \geq c$. Hence $\{x \in \mathbb{R}^p : f(x) \geq c\}$ is convex.

(iv) It suffices to prove $f(rx_0) \geq f(x_0)$ for $r \in [0, 1]$. Since f is superconvex, so is the function $r \to f(rx_0)$ ($r \in [0, 1]$). Therefore, we obtain

$$f(rx_0) \geq \min\{f(0), f(x_0)\} = f(x_0)$$

from $rx_0 = (1 - r)0 + rx_0$.   □

Lemma 2.4.6. *Let $q \geq 2$. If $C \in \mathbb{B}^q$ is convex, then the function $y \to \lambda(C_y)$, $y \in \mathbb{R}^{q-1}$, is concave.*

Proof. (i) For $y \in \mathbb{R}^{q-1}$, C_y is an interval. This implies for $\alpha \in [0, 1]$

$$\lambda((1 - \alpha)C_{y_0} + \alpha C_{y_1}) = (1 - \alpha)\lambda(C_{y_0}) + \alpha\lambda(C_{y_1}). \tag{2.4.7}$$

This can be seen as follows. If $(a_i, b_i) \subset I_i \subset [a_i, b_i]$, $i = 0, 1$, let $a_\alpha = (1 - \alpha)a_0 + \alpha a_1$, $b_\alpha = (1 - \alpha)b_0 + \alpha b_1$. Then we have

$$(a_\alpha, b_\alpha) \subset (1 - \alpha)I_0 + \alpha I_1 \subset [a_\alpha, b_\alpha],$$

since

$$(1 - u)a_\alpha + ub_\alpha = (1 - \alpha)[(1 - u)a_0 + ub_0] + \alpha[(1 - u)a_1 + ub_1]$$
$$\in (1 - \alpha)I_0 + \alpha I_1,$$

and $u \in (1 - \alpha)I_0 + \alpha I_1$ implies

$$u \le (1 - \alpha)b_0 + \alpha b_1 = b_\alpha$$

and

$$u \ge (1 - \alpha)a_0 + \alpha a_1 = a_\alpha.$$

Since C is convex,

$$(1 - \alpha)C_{y_0} + \alpha C_{y_1} \subset C_{(1-\alpha)y_0 + \alpha y_1}. \tag{2.4.8}$$

($x_i \in C_{y_i}$ is equivalent to $(x_i, y_i) \in C$. Since C is convex, $((1 - \alpha)x_0 + \alpha x_1, (1 - \alpha)y_0 + \alpha y_1) \in C$ follows, i.e. $(1 - \alpha)x_0 + \alpha x_1 \in C_{(1-\alpha)y_0 + \alpha y_1}.$)
From (2.4.7) and (2.4.8),

$$(1 - \alpha)\lambda(C_{y_0}) + \alpha\lambda(C_{y_1}) \le \lambda(C_{(1-\alpha)y_0 + \alpha y_1}), \tag{2.4.9}$$

i.e. $y \to \lambda(C_y)$ is concave. $\square$

Corollary 2.4.10. *Let $q \ge 2$. If a measurable function $f : \mathbb{R}^q \to [0, \infty]$ is superconvex, then the function $y \to \int f(x, y)\lambda^p(dx)$, $y \in \mathbb{R}^{q-p}$, is concave for every $p \in \{1, \ldots, q - 1\}$.*

Proof. We prove the assertion for $p = 1$. Since any concave function is superconvex, the assertion for arbitrary $p < q$ follows by induction. We define

$$f_n := 2^{-n} \sum_{i=1}^{n2^n} 1_{C_{i,n}} \text{ with } C_{i,n} := \{(x, y) \in \mathbb{R}^1 \times \mathbb{R}^{q-1} : f(x, y) \ge i2^{-n}\}.$$

Since the sets $C_{i,n}$ are convex, the functions $y \to \int f_n(x, y)\lambda(dx)$ are concave by Lemma 2.4.6. Since $f_n \uparrow f$ implies

$$\int f_n(x, y)\lambda(dx) \uparrow \int f(x, y)\lambda(dx),$$

the function $y \to \int f(x, y)\lambda(dx)$ is concave. $\square$

Lemma 2.4.11. *Let $f : \mathbb{R}^p \to [0, \infty)$ be measurable and superconvex, and $C \in \mathbb{B}^p$ convex. Then the function*

$$y \to \int f(x)1_C(x - y)\lambda^p(dx)$$

is concave.

Proof. Let the function $g : \mathbb{R}^{2p} \to [0, \infty)$ be defined by

$$g(x, y) := f(x) 1_C(x - y).$$

Since $g(x_i, y_i) \geq r > 0$ implies $(x_i - y_i) \in C$ and $f(x_i) \geq r$ for $i = 0, 1$, we have

$$((1 - \alpha)x_0 + \alpha x_1) - ((1 - \alpha)y_0 + \alpha y_1) \in C,$$

and therefore

$$g((1 - \alpha)x_0 + \alpha x_1, (1 - \alpha)y_0 + \alpha y_1) = f((1 - \alpha)x_0 + \alpha x_1) \geq r.$$

Hence g is superconvex, and the assertion follows from Corollary 2.4.10.   □

Theorem 2.4.12 (Anderson, 1955). *If a measurable function $f : \mathbb{R}^p \to [0, \infty)$ is symmetric about 0 λ^p-a.e. and superconvex, and if $C \in \mathbb{B}^p$ is convex and symmetric about 0, then*

$$r \to \int_{C + r x_0} f(x) \lambda^p(dx)$$

is antitone on $[0, \infty)$ for every $x_0 \in \mathbb{R}^p$.

Proof. By Lemma 2.4.11, the function $y \to \int f(x) 1_C(x - y) \lambda^p(dx)$ is concave and therefore (see Lemma 2.4.5(i)) superconvex. Being symmetric about 0, this function has mode 0. Hence the assertion follows from Lemma 2.4.5(iv).   □

Corollary 2.4.13. *Let $Q | \mathbb{B}^p$ be a p-measure with a Lebesgue density which is symmetric about 0 λ^p-a.e. and superconvex. Let $C \in \mathbb{B}^p$ be convex and symmetric about 0. Then we have*
(i) *$r \to Q(C + r x_0)$ is antitone on $[0, \infty)$ for every $x_0 \in \mathbb{R}^p$;*
(ii) *$Q(C) \geq Q * R(C)$ for all p-measures $R | \mathbb{B}^p$.*

Proof. (i) Follows immediately from Theorem 2.4.12.
(ii) By (i), we have

$$Q * R(C) = Q \times R\{(x, y) \in \mathbb{R}^{2p} : x + y \in C\}$$
$$= \int Q(C - y) R(dy) \leq \int Q(C) R(dy) = P(C).$$ □

We conclude this section with a few remarks on the order between multidimensional normal distributions.

Lemma 2.4.14. *For any nondegenerate normal distribution $N_{(\mu, \Sigma)} | \mathbb{B}^p$ and any $q \times p$-matrix A of rank q,*

$$N_{(\mu, \Sigma)} \circ (u \to Au) = N_{(A\mu, A\Sigma A^\top)} \quad \text{(which is nondegenerate).} \tag{2.4.15}$$

Proof. Anderson (1984), p. 31, Theorem 2.4.4.   □

Exercise 2.4.16. Show that the density of a p–dimensional normal distribution with mean 0 is symmetric about 0 and superconvex.

Corollary 2.4.17. *Let* Σ_i, $i = 1, 2$, *be symmetric, positive definite* $p \times p$–*matrices. Then* $N_{(0,\Sigma_1)}(C) \geq N_{(0,\Sigma_2)}(C)$ *for every set* $C \in \mathbb{B}^p$ *which is symmetric about 0 and convex iff* $\Sigma_1 \leq_L \Sigma_2$.

Proof. Follows from Lemma 2.4.14 and Corollary 2.4.13(ii), applied with $Q = N_{(0,\Sigma_1)}$ and $R = N_{(0,\Sigma_2 - \Sigma_1)}$.   □

Proposition 2.4.18. (i) *If* $N_{(0,\Sigma_1)}(C) \geq N_{(0,\Sigma_2)}(C)$ *for arbitrary rectangles* C *in* $\mathbb{B}^p$ *containing the origin, then* $\Sigma_1 = \alpha\Sigma_2$ *with* $\alpha \leq 1$.
   (ii) *If* $\Sigma_1 = \alpha\Sigma_2$ *with* $\alpha \leq 1$, *then* $N_{(0,\Sigma_1)}(C) \geq N_{(0,\Sigma_2)}(C)$ *for all convex sets* $C \in \mathbb{B}^p$ *containing the origin.*

Proof. (i) The case $p = 1$ is obvious. It suffices to consider the case $p = 2$ (from this, the general case follows by induction).

Since any similarity transformation transforms the class of all rectangles containing the origin onto itself, we may assume w.l.g. that Σ_1 is a diagonal matrix, and that

$$\Sigma_2(\varrho) = \begin{pmatrix} 1 & \varrho \\ \varrho & 1 \end{pmatrix} \quad \text{for some } \varrho \in (-1, 1).$$

For each quadrant C with vertex 0, we have

$$N_{(0,\Sigma_1)}(C) = N_{(0,\Sigma_2(\varrho))}(C).$$

Applied for the quadrant $C_0 := \{(x, y) \in \mathbb{R}^2 : x, y \geq 0\}$ we obtain

$$\frac{1}{4} = N_{(0,\Sigma_1)}(C_0) = N_{(0,\Sigma_2(\varrho))}(C_0).$$

Let $p(x, y, \varrho) = \frac{1}{2\pi\sqrt{1-\varrho^2}} \exp\left[-\frac{1}{2(1-\varrho^2)}(x^2 - 2\varrho xy + y^2)\right]$ denote the λ^2–density of $N_{(0,\Sigma_2(\varrho))}$. Since

$$\frac{\partial}{\partial\varrho} p(x, y, \varrho) = \frac{\partial^2}{\partial y \partial x} p(x, y, \varrho) \quad \text{(direct computation)}$$

and

$$\frac{\partial}{\partial x} p(x, y, \varrho) = \left(\frac{\varrho}{1 - \varrho^2} y - \frac{1}{1 - \varrho^2} x\right) p(x, y, \varrho),$$

we have

$$\frac{\partial}{\partial \varrho} N_{(0,\Sigma_2(\varrho))}(C_0) = -\int_0^\infty \frac{\partial}{\partial x} p(x,0,\varrho)\lambda(dx) = p(0,0,\varrho) > 0 \quad \text{for } \varrho \in (-1,1).$$

Therefore, $N_{(0,\Sigma_2(\varrho))}(C_0) = 1/4$ implies $\varrho = 0$.

Let $C_1 := \{(x,y) \in \mathbb{R}^2 : |y| \le x\}$ and $\Sigma_1 = \begin{pmatrix} \sigma_1^2 & 0 \\ 0 & \sigma_2^2 \end{pmatrix}$. Since

$$N_{(0,\Sigma_1)}(C_1) = \int_0^\infty (\Phi(x/\sigma_2) - \Phi(-x/\sigma_2)) N_{(0,\sigma_1^2)}(dx)$$

$$= \int_0^\infty \left(2\Phi\left(\frac{\sigma_1}{\sigma_2}x\right) - 1\right) N_{(0,1)}(dx),$$

$N_{(0,\Sigma_1)}(C_1)$ is increasing as a function of σ_1/σ_2. Therefore, we obtain (since C_1 is a quadrant with vertex 0) $\sigma_1 = \sigma_2$, i.e. $\Sigma_1 = \alpha\Sigma_2$. It remains to be shown that $\alpha \le 1$. For $C := \{(x,y) \in \mathbb{R}^2 : |x|,|y| \le 1\}$, we have (by assumption)

$$N_{(0,\Sigma_1)}(C) = N_{(0,\Sigma_2)}(\alpha^{-1/2}C) \le N_{(0,\Sigma_1)}(\alpha^{-1/2}C).$$

Hence $\alpha \le 1$ follows.

(ii) Let $C \in \mathbb{B}^p$ be a convex set containing the origin. Then we have

$$N_{(0,\Sigma_1)}(C) = N_{(0,\alpha\Sigma_2)}(C) = N_{(0,\Sigma_2)}(\alpha^{-1/2}C) \ge N_{(0,\Sigma_2)}(C),$$

since $\alpha^{-1/2}C \supset C$ for $\alpha \le 1$. □

Lemma 2.4.19. *Let Σ be a symmetric, positive definite $p \times p$-matrix and $R|\,\mathbb{B}^p$ a p-measure. Then $N_{(0,\Sigma)}\{u \in \mathbb{R}^p : |u_i| \le t\} = N_{(0,\Sigma)} * R\{u \in \mathbb{R}^p : |u_i| \le t\}$ for $t > 0$ and $i = 1,\dots,p$ implies $R\{0\} = 1$.*

Proof. Let Q_i and R_i denote the i-th marginals of $N_{(0,\Sigma)}$ and R, respectively. By assumption,

$$\int Q_i(-t+v,t+v)R_i(dv) = Q_i(-t,t) \quad \text{for } t > 0. \tag{2.4.20}$$

Since $Q_i(-t+v,t+v) < Q_i(-t,t)$ for $v \ne 0$, relation (2.4.20) implies $R_i\{0\} = 1$. Since $\{u \in \mathbb{R}^p : u \ne 0\} \subset \bigcup_1^p \{u \in \mathbb{R}^p : u_i \ne 0\}$, we obtain $R\{u \in \mathbb{R}^p : u \ne 0\} = 0$. □

2.5 Evaluating estimators by loss functions

Introduces "loss functions" and "risk" and discusses the connection with concentration on convex sets.

Let $\mathfrak{P}$ be a family of p–measures and $\kappa : \mathfrak{P} \to \mathbb{R}^p$ a functional.

The following definition is a mathematical expression for the assumption that the loss which occurs if the estimate attains the value u, whereas the true value of the functional is $\kappa(P)$, can be quantified. For the applications in Chapter 3 it is not necessary to assume that the loss depends on $u - \kappa(P)$ only (and not otherwise on P).

Definition 2.5.1. A function $L : \mathbb{R}^p \times \mathfrak{P} \to \mathbb{R}_+$, with $u \to L(u, P)$ measurable for $P \in \mathfrak{P}$, is a *loss function for* κ if

$$L\big(\kappa(P), P\big) = 0 \qquad \text{for } P \in \mathfrak{P}.$$

The loss function is *subconvex* if $\{u \in \mathbb{R}^p : L(u, P) \leq r\}$ is convex for $r \geq 0$ and $P \in \mathfrak{P}$.
The loss function is *convex* if $u \to L(u, P)$ is convex for $P \in \mathfrak{P}$.

For $p = 1$ the condition "subconvex" reduces to the requirement that $u \to L(u, P)$ is isotone as u moves away from $\kappa(P)$ in either direction.
Obviously, any convex loss function is subconvex (since $L(u_i, P) \leq r$ for $i = 0, 1$ implies $L\big((1 - \alpha)u_0 + \alpha u_1, P\big) \leq (1 - \alpha)L(u_0, P) + \alpha L(u_1, P) \leq r$ for every $\alpha \in (0, 1)$). There are important examples of subconvex loss functions which are not convex, for instance $L(\cdot, P) = 1 - 1_{C_P}$, where C_P is a measurable convex set containing $\kappa(P)$.
A natural convex loss function is $L(u, P) := |u - \kappa(P)|$, a mathematically more convenient one is $L(u, P) := \big(u - \kappa(P)\big)^2$.
More generally, $L(u, P) := C\big(u - \kappa(P)\big)$ is a convex loss function if C is a convex function attaining its minimum 0 at 0.
The comparison of estimators is based on the "expected loss", usually named "risk", which is defined as

$$L_*(\hat{\kappa}, P) := \int L\big(\hat{\kappa}(x), P\big) P(dx). \tag{2.5.2}$$

Given the loss function, any two estimators are comparable with regard to their risk. In other words: With a single loss function we obtain a total order between the estimators. Thinking of real applications, loss functions appear as a construct with only a loose connection to reality. Examples of "real" loss functions are still wanted. Therefore the comparison between two estimators is practically relevant only if the outcome is the same for a class of loss functions large enough to contain the unknown "true" loss function.

There is a beautiful result — the so-called Theorem of Lehmann and Scheffé (see Theorem 3.2.5) — which guarantees for certain families of p-measures the existence of a mean unbiased estimator which is of minimal risk, simultaneously for all convex loss functions. Even though the true loss function is unknown, we know for sure that it will never be convex: Losses increase with the distance from the true value $\kappa(P)$, but the rate of increase will certainly fall off, eventually. This somewhat diminishes the practical relevance of the Lehmann–Scheffé theorem.

A more natural class of loss functions which is more likely to include the "true" loss function (if any) is the class of all subconvex loss functions. However: If two estimators are comparable, simultaneously for all [symmetric] subconvex loss functions, they are comparable with respect to their concentration on all convex sets containing $\kappa(P)$ [symmetric about $\kappa(P)$]. This is made precise in the following

Proposition 2.5.3. *The following assertions are equivalent.*

(i) $P\{\hat{\kappa}_1 \in C\} \geq P\{\hat{\kappa}_2 \in C\}$ for every measurable convex set C containing $\kappa(P)$ [symmetric about $\kappa(P)$].

(ii) For every subconvex [and symmetric] loss function $L(\cdot, P)$, the distribution of losses $L(\hat{\kappa}_i, P)$ under P is more concentrated about 0 for $i = 1$ than for $i = 2$.

(iii) For every subconvex [and symmetric] loss function $L(\cdot, P)$, the risk of $\hat{\kappa}_1$ is smaller than the risk of $\hat{\kappa}_2$.

Proof. (i) implies (ii): Since $L(\cdot, P)$ is a [symmetric] subconvex loss function, $C_r := \{u \in \mathbb{R}^p : L(u, P) \leq r\}$ is a measurable convex set containing $\kappa(P)$ [symmetric about $\kappa(P)$]. Since $P\{x \in X : L(\hat{\kappa}(x), P) \leq r\} = P\{x \in X : \hat{\kappa}(x) \in C_r\}$, relation (ii) follows.

(ii) implies (iii) since $L_*(\hat{\kappa}, P) := \int L(\hat{\kappa}(x), P) P(dx) = \int L(u, P) P \circ \hat{\kappa}(du) = \int_0^\infty P \circ \hat{\kappa}\{L(\cdot, P) > r\}\lambda(dr)$.

(iii) implies (i) since the loss function $L(\cdot, P) = 1 - 1_C$ is [symmetric] and subconvex if C is a measurable convex set containing $\kappa(P)$ [symmetric about $\kappa(P)$]. □

If the loss function depends on P through $\kappa(P)$ only, one might establish an inherent relationship to "unbiasedness" (see Lehmann, 1951). Writing $L(\cdot, \kappa(P))$ instead of $L(\cdot, P)$ we arrive at the following

Definition 2.5.4. The estimator $\hat{\kappa}$ for $\kappa(P)$ is *L-unbiased* if the function $t \to \int L(\hat{\kappa}(x), t) P(dx)$ attains its minimum for $t = \kappa(P)$.

Proposition 2.5.5. *With the loss function $L(u, t) = |u - t|^k$, we obtain median unbiasedness for $k = 1$, and mean unbiasedness for $k = 2$.*

Proof. Let $Q := P \circ \hat{\kappa}$.

(i) For $k = 2$, we assume that $\mu := \int y Q(dy)$ is finite. Then the assertion follows from

$$\int (y - t)^2 Q(dy) = \int (y - \mu)^2 Q(dy) + (\mu - t)^2.$$

(ii) For $k = 1$ we assume w.l.g. that the median of Q is 0, and that $t > 0$. We have

$$\begin{aligned} |y| &\leq |y - t| + t \\ |y| &= |y - t| - t \end{aligned} \quad \text{for} \quad y \overset{>}{\underset{\leq}{}} 0.$$

Hence

$$\int |y| Q(dy) \leq \int |y - t| Q(dy) + t(Q(0, \infty) - Q(-\infty, 0]) \leq \int |y - t| Q(dy). \quad \square$$

In spite of the inherent relationship between special loss functions and special types of unbiasedness, there seems to be no mathematical advantage in using together with each loss function the pertaining type of unbiasedness. From the mathematical point of view, mean unbiasedness goes well with convex loss functions, and median unbiasedness with subconvex loss functions. From the applied point of view, the appropriate notion of unbiasedness should be chosen according to the necessities of the problem, not by formal considerations.

Historical Remark 2.5.6. The concept of a loss function was introduced by Laplace (1774, p. 636 and 1812, Sections 20ff.) who used the loss function $L(u, t) = |u - t|$. Gauß (1821, Section 6) replaced this loss function by the quadratic loss function $L(u, t) = (u - t)^2$, exclusively for technical reasons. Gauß (in his letter to Bessel, dated February 28, 1839) says "... ohne die außerordentlichen Vorteile, die die Wahl des Quadrates gewährt, könnte man jede andere jenen Bedingungen [positive and subconvex] entsprechende Funktion wählen ...". Arbitrary loss functions are also used by Edgeworth (1883, p. 362), Czuber (1891, p. 289), Pizzetti (1892, p. 179), furthermore by Neyman and Pearson (1933b, p. 195) in connection with hypothesis testing. The interest in this approach was revived by Wald in connection with his decision theory, starting with (1939) and culminating in (1950).

2.6 The relative efficiency of estimators

The preceding sections define a (partial) order between estimators with respect to their concentration. In the present section we try to quantify these differences.

As in the preceding sections, $\mathfrak{P}$ is a family of p–measures and $\kappa : \mathfrak{P} \to \mathbb{R}$ the functional to be estimated. Even if the estimator $\hat{\kappa}_1$ is — for every $P \in \mathfrak{P}$ — more concentrated about $\kappa(P)$ than $\hat{\kappa}_2$, this does not necessarily mean that $\hat{\kappa}_1$ is preferable in practice. Other properties (such as amount of computing, robustness, etc.) will influence the choice of the estimator. For such an examination it is not enough to know that $\hat{\kappa}_1$ — judged by its concentration — is better than $\hat{\kappa}_2$. We need to know: how much better?

If one takes the concept of a loss function serious, the answer is straightforward. One has just to compare the risks $L_*(\hat{\kappa}_1, P)$ and $L_*(\hat{\kappa}_2, P)$, for all $P \in \mathfrak{P}$. With the true loss function unknown, statistical theory approaches the problem from a different angle.

The basic idea is to start from estimator sequences, i.e. to compare $\hat{\kappa}_1^{(n)}$, $n \in \mathbb{N}$, with $\hat{\kappa}_2^{(n)}$, $n \in \mathbb{N}$. For a given sample size n we determine a sample size m_n such that $\hat{\kappa}_2^{(m_n)}$ is as good as $\hat{\kappa}_1^{(n)}$. The ratio n/m_n is the *relative efficiency* of $\hat{\kappa}_2^{(n)}$, $n \in \mathbb{N}$, with respect to $\hat{\kappa}_1^{(n)}$.

Example 2.6.1. Let $\mathfrak{P} := \{N_{(0,\sigma^2)} : \sigma^2 > 0\}$. The estimator $\hat{\sigma}_n^2(\underline{x}) := \frac{1}{n}\sum_1^n x_\nu^2$ is of minimal convex risk among all mean unbiased estimators for σ^2. The function $n\hat{\sigma}_n^2/\sigma^2$ is distributed as χ_n^2. As an alternative one might consider the estimator $s_n^2(\underline{x}) := \frac{1}{n-1}\sum_1^n(x_\nu - \bar{x}_n)^2$ which is mean unbiased for σ^2 even if the true distribution is $N_{(\mu,\sigma^2)}$ with $\mu \neq 0$. Since $(n-1)s_n^2/\sigma^2$ is distributed as χ_{n-1}^2, we have $m_n = n+1$: The distribution of s_{n+1}^2 is the same as that of $\hat{\sigma}_n^2$, for every p–measure in the family $\mathfrak{P} = \{N_{(0,\sigma^2)} : \sigma^2 > 0\}$.

A clear–cut answer, as in Example 2.6.1, is rare. Usually we have to face the following problems.

a) We have to make precise the meaning of "$\kappa_2^{(m_n)}$ is as good as $\kappa_1^{(n)}$". A reasonable approach is the following:
(i) Specify a covering probability $\beta \in (0,1)$ and determine u_n such that

$$|P^n\{|\kappa_1^{(n)} - \kappa(P)| \leq u\} - \beta| = \min.$$

(ii) Determine m_n such that

$$|P^m\{|\kappa_2^{(m)} - \kappa(P)| \leq u_n\} - \beta| = \min.$$

Then both, $\kappa_1^{(n)}$ and $\kappa_2^{(m_n)}$, are in the interval $\big(\kappa(P) - u_n, \kappa(P) + u_n\big)$ with approximately the same probability β.

The relative efficiency depends, in general, on the arbitrarily chosen covering probability β. Moreover, the relative efficiency of $\kappa_1^{(n)}$ with respect to $\kappa_2^{(n)}$ is not necessarily the reciprocal of the relative efficiency of $\kappa_2^{(n)}$ with respect to $\kappa_1^{(n)}$.

In asymptotic considerations, the sequence of relative efficiencies, n/m_n, $n \in \mathbb{N}$, converges usually to a limit. If this is the case, $\lim_{n \to \infty} n/m_n$ is called *asymptotic relative efficiency*. Usually, the asymptotic relative efficiency is independent of β, a fact which is essential for the operational significance of this concept. Moreover, the asymptotic relative efficiency of $\kappa_1^{(n)}$, $n \in \mathbb{N}$, with respect to $\kappa_2^{(n)}$, $n \in \mathbb{N}$, is the reciprocal of the asymptotic relative efficiency of $\kappa_2^{(n)}$, $n \in \mathbb{N}$, with respect to $\kappa_1^{(n)}$, $n \in \mathbb{N}$.

b) In the consideration indicated under a), the p–measure P was considered fixed. In fact, u_n as well as m_n will depend on P. We therefore obtain, in general, an [asymptotic] relative efficiency depending on P.

c) Replacing estimators by estimator sequences is not without problems. If we wish to compare the efficiency of sample mean and sample median as estimators for μ in the family $\{N_{(\mu,\sigma^2)}^n : \mu \in \mathbb{R}, \ \sigma^2 > 0\}$, this goes off well. On the other hand if we are given 25 observations and consider the possibility of using a trimmed mean, excluding the two smallest and the two largest observations, i.e. $\mu^{(25)}(x_1, \ldots, x_{25}) = \frac{1}{21} \sum_3^{23} x_{i:25}$, what is the sequence $\mu^{(n)}$, $n \in \mathbb{N}$, in which we embed the estimator $\mu^{(25)}$? An experimenter who wants to know the loss which results from using $\mu^{(25)}$ instead of the sample mean might be surprised to learn that he has first to define $\mu^{(n)}$ for every $n \in \mathbb{N}$.

So far our endeavours to define "relative efficiency" have been restricted to real valued estimators. It is, in fact, hard to find a meaningful concept of "joint relative efficiency" for multidimensional estimators. Think of two 2–dimensional estimators where one component has relative efficiency 2, the other relative efficiency $\frac{1}{2}$: How could one find a meaningful "average" relative efficiency without knowing the possible uses of these estimators? Among the many ill motivated proposals is the one by Wilks (1932, Section 3) to use the "generalized variance", $\det \Sigma_i$, for defining the asymptotic relative efficiency for two asymptotically normal estimator sequences.

2.7 Examples on the evaluation of estimators

As indicated in Sections 2.3 and 2.4, the main application of the concepts of concentration is to the comparison of limit distributions (see Section 8.5). In the present section we give some examples illustrating the usefulness and the limitations of these concepts for finite sample sizes.

Omitting the condition of equivariance may lead to estimators with smaller quadratic risk.

Example 2.7.1. For $\mathfrak{P} = \{N_{(\mu,\sigma^2)} : \mu \in \mathbb{R},\ \sigma^2 > 0\}$, the functional $\kappa(N_{(\mu,\sigma^2)}) = \sigma^2$ is equivariant under the transformations $x \to a + cx$, $a \in \mathbb{R}$, $c > 0$: Since $N_{(\mu,\sigma^2)} \circ (x \to a + cx) = N_{(a+c\mu,\ c^2\sigma^2)}$, we have

$$\kappa\big(N_{(\mu,\sigma^2)} \circ (x \to a + cx)\big) = c^2 \kappa(N_{(\mu,\sigma^2)}).$$

The estimator $\hat{s}_n^2(x_1, \ldots, x_n) := (n+1)^{-1} \sum_1^n (x_\nu - \bar{x})^2$ minimizes the quadratic risk in the class of all equivariant estimators (see Example 2.7.2). Omitting the condition of equivariance, one can find estimators of smaller quadratic risk, for instance $\hat{\hat{s}}_n^2(x_1, \ldots, x_n) := \min\{\hat{s}_n^2(x_1, \ldots, x_n), \sum_1^n x_\nu^2/(n+2)\}$:

$$\int (\hat{\hat{s}}_n^2 - \sigma^2)^2 dN_{(\mu,\sigma^2)}^n < \int (\hat{s}_n^2 - \sigma^2)^2 dN_{(\mu,\sigma^2)}^n \quad \text{for } \mu \in \mathbb{R},\ \sigma^2 > 0$$

(see Stein, 1964, p. 155; see also Zacks, 1971, pp. 396/7). Observe that $\hat{\hat{s}}_n^2(x_1 + a, \ldots, x_n + a) \neq \hat{\hat{s}}_n^2(x_1, \ldots, x_n)$ with positive probability under $N_{(\mu,\sigma^2)}^n$ for $a \neq 0$.

Omitting the condition of mean unbiasedness may lead to estimators with smaller quadratic risk. This is demonstrated by the straightforward Example 2.7.2. More subtle is Example 2.7.3, demonstrating that this effect is not necessarily negligible as the sample size tends to infinity.

Example 2.7.2. The problem is to estimate σ^2 in the family $\{N_{(\mu,\sigma^2)}^n : \mu \in \mathbb{R},\ \sigma^2 > 0\}$. The estimator $s_n^2(x_1, \ldots, x_n) := (n-1)^{-1} \sum_1^n (x_\nu - \bar{x}_n)^2$ is mean unbiased with minimal convex risk (see Theorem 3.2.5). Evaluated by its quadratic risk, $\frac{n-1}{n+1} s_n^2$ is better: We have

$$\int (s_n^2 - \sigma^2)^2 dN_{(\mu,\sigma^2)}^n = \frac{2}{n-1} \sigma^4,$$

but

$$\int \Big(\frac{n-1}{n+1} s_n^2 - \sigma^2\Big)^2 dN_{(\mu,\sigma^2)}^n = \frac{2}{n+1} \sigma^4.$$

In fact, $\frac{n-1}{n+1}s_n^2$ has minimal quadratic risk in the class of all equivariant estimators (see Lehmann, 1983, p. 179, Example 3.6).

Example 2.7.3. (Pfanzagl, 1993). A sequence of mean unbiased estimators with minimal convex risk may be asymptotically inefficient.

For $a > 0$ let Π_a denote the Poisson distribution with parameter a, and let

$$P_\vartheta := \Pi_{\vartheta^{-1}} \times \Pi_{\exp[-\vartheta]}, \qquad \vartheta > 0,$$

i.e.

$$P_\vartheta\{(k, \ell)\} = \frac{1}{k!\ell!} \exp[-\vartheta^{-1} - \exp[-\vartheta]]\vartheta^{-k} \exp[-\vartheta\ell], \qquad k, \ell \in \{0\} \cup \mathbb{N}.$$

The statistic

$$S_n\big((k_\nu, \ell_\nu)_{\nu=1,\ldots,n}\big) = \Big(\sum_1^n k_\nu, \sum_1^n \ell_\nu\Big)$$

is sufficient for $\{P_\vartheta^n : \vartheta > 0\}$, and $P_\vartheta^n \circ S_n = \Pi_{n\vartheta^{-1}} \times \Pi_{n\exp[-\vartheta]}$. As in Example 1.6.15, it can be shown that for every $n \in \mathbb{N}$, $\{P_\vartheta^n \circ S_n : \vartheta > 0\}$ is complete. Hence $n^{-1}\sum_1^n k_\nu$ is the mean unbiased estimator with minimal convex risk for ϑ^{-1} (see Theorem 3.2.5). Yet the sequence of these estimators is asymptotically inefficient. The (asymptotic) variance of $n^{1/2}(n^{-1}\sum_1^n k_\nu - \vartheta^{-1})$ is ϑ^{-1}, the minimal asymptotic variance is $\vartheta^{-1}\big(1 + \vartheta^3 \exp[-\vartheta]\big)^{-1}$. An asymptotically efficient estimator sequence is

$$\bar{k}_n + \frac{\bar{\ell}_n \exp[1/\bar{k}_n] - 1}{\bar{k}_n \exp[1/\bar{k}_n] + 1/\bar{k}_n^2}, \qquad n \in \mathbb{N},$$

which can be obtained by the improvement procedure (7.5.10).

If an estimator, say $\kappa^{(n)} : X^n \to \mathbb{R}$, is defined for every sample size n, one expects that $\kappa^{(n+1)}$ is more concentrated about $\kappa(P)$ than $\kappa^{(n)}$. In the following examples we discuss in which sense and to which extent this is true.

Since comparability on *all* intervals containing $\kappa(P)$ is generally impossible unless $P\{\hat{\kappa}_1 \geq \kappa(P)\} = P\{\hat{\kappa}_2 \geq \kappa(P)\}$ and $P\{\hat{\kappa}_1 \leq \kappa(P)\} = P\{\hat{\kappa}_2 \leq \kappa(P)\}$, one cannot expect comparability in this strong sense between mean unbiased estimators since such ones will usually fail to be median unbiased (unless $P \circ \hat{\kappa}_i$ is symmetric about $\kappa(P)$).

Even in the case of median unbiased estimators, $\kappa^{(n+1)}$ is not necessarily more concentrated about $\kappa(P)$ than $\kappa^{(n)}$. The trivial example: Let P be the p-measure with characteristic function $\exp[-|t|^{1/2}]$. P is symmetric about 0 and unimodal with bounded density, the distribution of $n^{-1}\sum_1^n x_\nu$, under P^n, being $P \circ (x \to nx)$, drifts apart as n increases. (See Lukacs, 1970, pp. 136, 138 and 158.) Hence the sample mean cannot be used as an estimator for ϑ in the family $\{P_\vartheta : \vartheta \in \mathbb{R}\}$ with $P_\vartheta := P \circ (x \to x + \vartheta)$.

Example 2.7.4. Let $\mathfrak{P} \,|\, \mathbb{B}$ denote the family of all p–measures with positive Lebesgue density. The functional to be estimated is the median. Then the following holds true.

(i) For every odd sample size, the sample median $\tilde{x}_{2m+1}$ is a median unbiased estimator.

(ii) $\tilde{x}_{2m+1}$ is more concentrated than $\tilde{x}_{2m-1}$.

Proof. Let F denote the distribution function of P, and F_n (for odd n) the distribution function of the sample median. We have (see, e.g., Reiß, 1989, p. 21, Theorem 1.3.2)

$$F_{2m+1}(t) = G_m\big(F(t)\big),$$

with

$$G_m(\alpha) := \frac{(2m+1)!}{(m!)^2} \int\limits_0^\alpha u^m (1-u)^m du \qquad \text{for } \alpha \in [0,1].$$

The assertion follows from $G_m\big(\tfrac{1}{2}\big) = \tfrac{1}{2}$ and

$$G_m(\alpha) \mathrel{\substack{>\\<}} G_{m-1}(\alpha) \qquad \text{for} \qquad \alpha \mathrel{\substack{>\\<}} \frac{1}{2}.$$

$\square$

If we define the sample median for even sample sizes $n = 2m$ by

$$\tilde{x}_{2m} := \frac{1}{2}(x_{m:2m} + x_{m+1:2m}),$$

the following holds true for *symmetric* distributions.

(iii) $\tilde{x}_{2m}$ is a median unbiased estimator.

(iv) $\tilde{x}_{2m+1}$ is not necessarily more concentrated than $\tilde{x}_{2m}$.

To see this, consider the following example. Let P be a p–measure with continuous Lebesgue density p, symmetric about 0, such that $p(0) = 0$ and $p(x) > 0$ for $x \neq 0$. Since the distribution of $\tilde{x}_n$ has a continuous Lebesgue density p_n with $p_{2m}(0) > 0 = p_{2m+1}(0)$, we have $p_{2m} > p_{2m+1}$ on some interval containing 0. Hence there are intervals containing 0 on which the distribution of $\tilde{x}_{2m}$ is more concentrated than the distribution of $\tilde{x}_{2m+1}$.

A nicer example of this kind occurs in Asrabadi (1985, p. 717, Table I), who shows by numerical computation that the variance of $\tilde{x}_{2m}$ is smaller than the variance of $\tilde{x}_{2m+1}$ (for $m = 3, 5, 7, 9, 11$) if P is the Laplace distribution. This is, however, not true (Asrabadi's values for the variance of $\tilde{x}_{2m}$ are wrong.)

In case of the Laplace distribution, $\tilde{x}_{2m+1}$ is more concentrated than $\tilde{x}_{2m}$ on arbitrary intervals containing the median.

It suffices to prove the assertion for the standard Laplace distribution. In this case the distributions of $\tilde{x}_{2m+1}$ and $\tilde{x}_{2m}$ have Lebesgue densities

$$p_{2m+1}(x) = (m+1) \binom{2m+1}{m+1} \left(\tfrac{1}{2}\right)^{m+1} e^{-(m+1)|x|} \left(1 - \tfrac{1}{2} e^{-|x|}\right)^m$$

and

$$p_{2m}(x) = m \binom{2m}{m} \left(\tfrac{1}{2}\right)^m e^{-2m|x|} \left[\left(\tfrac{1}{2}\right)^m + m \int_0^{|x|} \left(e^y - \tfrac{1}{2}\right)^{m-1} dy\right],$$

respectively.

Since $p_{2m+1}(x) \geq p_{2m}(x)$ is a nondegenerate interval containing 0, the assertion follows from Lemma 2.3.2.

Example 2.7.5. Let $\mathfrak{P} = \{N_{(\mu,\sigma^2)} : \mu \in \mathbb{R}, \sigma^2 > 0\}$. The problem is to estimate μ. Since the sample mean $\bar{x}_n$ is distributed as $N_{(\mu,\sigma^2/n)}$, the concentration of $N_{(\mu,\sigma^2)}^n \circ \bar{x}_n$ on arbitrary intervals containing μ increases with n. In fact, more is true: $N_{(\mu,\sigma^2/(n+1))}$ is more concentrated than $N_{(\mu,\sigma^2/n)}$ in the spread order.

Example 2.7.6. Let $\mathfrak{P} = \{N_{(\mu,\sigma^2)} : \mu \in \mathbb{R}, \sigma^2 > 0\}$. The problem is to estimate σ^2. The estimator $s_n^2(\underline{x}) := (n-1)^{-1} \sum_1^n (x_\nu - \bar{x}_n)^2$ minimizes the convex risk among all mean unbiased estimators (see Theorem 3.2.5). If μ is known, a better estimator is $\hat{\sigma}_n^2(\underline{x}) := n^{-1} \sum_1^n (x_\nu - \mu)^2$. Since

$$N_{(\mu,\sigma^2)}^{n+1} \circ s_{n+1}^2 = N_{(\mu,\sigma^2)}^n \circ \hat{\sigma}_n^2 \qquad (= \Gamma_{\frac{2\sigma^2}{n},\frac{n}{2}}),$$

not knowing μ can be compensated by 1 additional observation. But is s_n^2 really less concentrated than $\hat{\sigma}_n^2$? Mathematically the same question with a different interpretation: Is s_n^2 really less concentrated than s_{n+1}^2? Thanks to the optimum property of s_{n+1}^2 among all mean unbiased estimators of σ^2 based on $(n+1)$ observations, the risk of s_{n+1}^2 is smaller than the risk of s_n^2 for every convex loss function.

Since $N_{(0,\sigma^2)}^{n+1}\{s_{n+1}^2 \leq \sigma^2\} \neq N_{(0,\sigma^2)}^n\{s_n^2 \leq \sigma^2\}$, s_{n+1}^2 cannot be more concentrated than s_n^2 on all intervals containing σ^2. Hence the risk of s_{n+1}^2 is smaller than the risk of s_n^2 for all convex, but not for all subconvex loss functions.

The situation is different with median unbiased estimators: With c_m denoting the median of $\Gamma_{\frac{2}{m},\frac{m}{2}}$, the estimator s_n^2/c_{n-1} is median unbiased for σ^2. It is not straightforward to see that the distribution of this estimator, $\Gamma_{\frac{2\sigma^2}{(n-1)c_{n-1}},\frac{n-1}{2}}$, becomes more concentrated about σ^2 if n increases. This follows, however, from Corollary 5.5.15, according to which s_{n+1}^2/c_n is maximally concentrated among all median unbiased estimators for σ^2, based on a sample of size $n + 1$. Since s_n^2/c_{n-1} is another such estimator, we have

$$N_{(\mu,\sigma^2)}^{n+1}\{s_{n+1}^2/c_n \in (t',t'')\} \geq N_{(\mu,\sigma^2)}^n\{s_n^2/c_{n-1} \in (t',t'')\}$$

for $0 < t' \leq \sigma^2 \leq t'' < \infty$.

Example 2.7.7. Let $\mathfrak{P} = \{N_{(\mu_1,\mu_2,\sigma^2,\sigma^2,\varrho)} : \mu_1, \mu_2 \in \mathbb{R}, \sigma^2 > 0, \varrho \in (0,1)\}$. The problem is to estimate ϱ. Let

$$R_n((x_1,y_1),\ldots,(x_n,y_n)) :=$$

$$2\sum_1^n (x_\nu - \bar{x}_n)(y_\nu - \bar{y}_n) \Big/ \left(\sum_1^n (x_\nu - \bar{x}_n)^2 + \sum_1^n (y_\nu - \bar{y}_n)^2 \right).$$

The distribution of $(1+\varrho)(1-R_n)/2(1-\varrho R_n)$ under $N_{(\mu_1,\mu_2,\sigma^2,\sigma^2,\varrho)}^n$ is the beta distribution $B_{\frac{n-1}{2},\frac{n-1}{2}}$ (see De Lury, 1938).

Using that $B_{\frac{n-1}{2},\frac{n-1}{2}}$ is symmetric about $\frac{1}{2}$, one can easily see that R_n is median unbiased for ϱ. According to Lemma 2.3.2, $B_{\alpha,\alpha}$ is more concentrated on intervals about $1/2$ than $B_{\beta,\beta}$ if $\alpha > \beta$. Since $R \to (1+\varrho)(1-R)/2(1-\varrho R)$ is monotone (antitone, to be more precise), R_n is more concentrated than R_{n-1} on arbitrary intervals containing ϱ. However: R_n is not more concentrated than R_{n-1} in the spread order, since $B_{\frac{n-1}{2},\frac{n-1}{2}}$ and $B_{\frac{n-2}{2},\frac{n-2}{2}}$ have the same bounded support $(0,1)$ (see Exercise 2.3.9).

The results of Section 5.5 are not suitable to prove that R_n is maximally concentrated among *all* median unbiased estimators for ϱ, even though this is probably the case.

Chapter 3
Mean unbiased estimators and convex loss functions

3.1 Introduction

States the problem of finding a mean unbiased estimator of minimal risk, and stresses the limitations of this approach.

Given the family $\mathfrak{P}$, the functional $\kappa : \mathfrak{P} \to \mathbb{R}^p$ and the loss function $L : \mathbb{R}^p \times \mathfrak{P} \to \mathbb{R}_+$, the problem is to find an estimator $\hat{\kappa} : X \to \mathbb{R}^p$ which minimizes the risk, $L_*(\hat{\kappa}, P)$, simultaneously for all $P \in \mathfrak{P}$. Formulated in this way, the problem is not meaningful: The estimator $\hat{\kappa}(x) \equiv \kappa(P_0)$ minimizes the risk at P_0, but we obtain a different minimizing estimator for each $P \in \mathfrak{P}$. The situation is totally other if we restrict the class of estimators. Then it may happen that in the restricted class of estimators there is one which minimizes the risk simultaneously for all $P \in \mathfrak{P}$. What we can formulate as a meaningful problem is the following:

Find an estimator $\hat{\kappa} : X \to \mathbb{R}$ such that

$$\int \hat{\kappa}(x) P(dx) = \kappa(P) \quad \text{for every } P \in \mathfrak{P} \tag{3.1.1$'$}$$

and

$$\int L\big(\hat{\kappa}(x), P_0\big) P_0(dx) = \min . \tag{3.1.1$''$}$$

What we would like to obtain is an estimator $\hat{\kappa}$ fulfilling (3.1.1$''$) simultaneously for all $P_0 \in \mathfrak{P}$. What we can do is to find an estimator fulfilling (3.1.1$''$) for a given P_0, and to hope that this estimator is the same for every $P_0 \in \mathfrak{P}$. The suggestion, popular in statistical theory, to replace (3.1.1$''$) by $\sup\{\int L(\hat{\kappa}(x), P) P(dx) : P \in \mathfrak{P}\} = \min$ is hardly justified from the practical point of view.

The following proposition is essentially due to Lehmann and Scheffé (1950, p. 324, Lemma 5.1).

Proposition 3.1.2. *If the loss function is strictly convex, then for each $P_0 \in \mathfrak{P}$ there exists at most one mean unbiased estimator which minimizes the risk for P_0.*

Proof. Assume that $\hat{\kappa}_i$, $i = 1, 2$, are mean unbiased. Then $\hat{\kappa}_0 := \frac{1}{2}\hat{\kappa}_1 + \frac{1}{2}\hat{\kappa}_2$ is mean unbiased too, and

$$L\big(\hat{\kappa}_0(x), P_0\big) \le \frac{1}{2}L\big(\hat{\kappa}_1(x), P_0\big) + \frac{1}{2}L\big(\hat{\kappa}_2(x), P_0\big) \quad \text{for all } x \in X. \tag{3.1.3}$$

Integrating over x with respect to P_0 we obtain (see (2.5.2))

$$L_*(\hat{\kappa}_0, P_0) \le \frac{1}{2}L_*(\hat{\kappa}_1, P_0) + \frac{1}{2}L_*(\hat{\kappa}_2, P_0). \tag{3.1.4}$$

If both, $\hat{\kappa}_1$ and $\hat{\kappa}_2$, minimize the risk for P_0, equality holds in (3.1.4) which implies equality in (3.1.3) for P_0-a.a. $x \in X$. Since $L(\cdot, P_0)$ is strictly convex, this implies $\hat{\kappa}_1(x) = \hat{\kappa}_2(x)$ for P_0-a.a. $x \in X$.   □

For $L(u, P_0) = |u - \kappa(P_0)|^s$, $s > 1$, Barankin (1949, p. 483, Theorem 2(iii)) proved the existence of an estimator $\hat{\kappa}$ fulfilling (3.1.1') and (3.1.1'') for P_0 under the assumption that $\int \big(p(x)/p_0(x)\big)^{s/(s-1)} P_0(dx) < \infty$ (with p denoting a density of P) for every $P \in \mathfrak{P}$. The theorem was extended to convex loss functions by Kozek (1977, p. 188, Theorem 4.4), after an attempt by M.M. Rao (1965). (Using Orlicz space theory, M.M. Rao (1965) attacks the problem $L(u, P) = C(u - \kappa(P))$, C being a symmetric convex function with $C(0) = 0$. But his attack fails, because on p. 135 he "recasts" the problem by exchanging the risk function against a related, but different function which lacks any statistical meaning.)

The results mentioned here are existence theorems. It appears difficult to obtain workable procedures for the construction of such optimal estimators. Moreover, the estimators fulfilling (3.1.1) depend in general on the loss function L and on P_0. If they are different for different P_0, the problem of finding an "optimal mean unbiased estimator" for κ has no solution.

To illustrate that the solution to (3.1.1) depends, in general, on P_0 we choose an example where an explicit solution is available.

Exercise 3.1.5. For $\vartheta \in (0, 1)$ let

$$P_\vartheta\{0\} = \vartheta \quad \text{and} \quad P_\vartheta\{k\} = (1 - \vartheta)^2 \vartheta^{k-1} \quad \text{for } k \in \mathbb{N}.$$

(i) Show that an estimator $\hat{\vartheta} \mid \{0\} \cup \mathbb{N}$ for ϑ is mean unbiased if and only if $\hat{\vartheta}(k) = (1 - \hat{\vartheta}(0))(k - 1)$ for $k \in \mathbb{N}$. Hence the class of all mean unbiased estimators for ϑ can be represented as

$$\hat{\vartheta}(k, a) := \begin{cases} a \\ (1 - a)(k - 1) \end{cases} \quad \text{for} \quad \begin{matrix} k = 0 \\ k \in \mathbb{N} \end{matrix} \quad \text{with } a \in \mathbb{R}.$$

(ii) Show that $\hat{\vartheta}(\cdot, \vartheta_0(1 + \vartheta_0)/(1 + \vartheta_0^2))$ is the unique mean unbiased estimator for ϑ which minimizes the quadratic risk for $\vartheta = \vartheta_0$. Since $\hat{\vartheta}(\cdot, a_1) \neq \hat{\vartheta}(\cdot, a_2)$ for $a_1 \neq a_2$, there is no mean unbiased estimator which mimimizes the quadratic risk simultaneously for all $\vartheta \in (0, 1)$.

Example 3.1.6. For $\mu \in \mathbb{R}$, $\sigma > 0$ let $P_{\mu,\sigma} := N_{(\mu,1)} \times N_{(\mu,\sigma^2)}$. The problem is to estimate μ in the presence of the unknown nuisance parameter σ, based on a sample of size n. We shall show that there is no mean unbiased estimator which minimizes the risk for some strictly convex loss function.

Proof. $P_{\mu,\sigma}$ has λ^2–density

$$\frac{1}{2\pi\sigma} \exp\left[-\frac{1}{2\sigma^2}(\sigma^2 x^2 + y^2 - 2\mu(\sigma^2 x + y) + \mu^2(\sigma^2 + 1))\right].$$

Hence $\sum_1^n (\sigma_0^2 x_\nu + y_\nu)$ is complete sufficient for $\{P_{\mu,\sigma_0}^n : \mu \in \mathbb{R}\}$. The estimator

$$\mu^{(n)}\big((x_1, y_1), \ldots, (x_n, y_n); \sigma_0\big) := \sum_1^n (\sigma_0^2 x_\nu + y_\nu)/n(\sigma_0^2 + 1)$$

is mean unbiased for μ on $\{P_{\mu,\sigma}^n : \mu \in \mathbb{R}, \ \sigma > 0\}$ and has minimal convex risk under P_{μ,σ_0} for every $\mu \in \mathbb{R}$ by Theorem 3.2.5. According to Proposition 3.1.2 such estimators are unique λ^{2n}–a.e. if the loss function is strictly convex. Since $\mu^{(n)}\big((x_1, y_1), \ldots, (x_n, y_n); \sigma_2\big) \neq \mu^{(n)}\big((x_1, y_1), \ldots, (x_n, y_n); \sigma_1\big)$ for λ^{2n}–a.a. $(x_\nu, y_\nu)_{\nu=1,\ldots,n}$ if $\sigma_1 \neq \sigma_2$, this establishes the non–existence of a mean unbiased estimator for μ which minimizes the risk simultaneously for all $\sigma^2 > 0$ for some strictly convex loss function. $\square$

The situation is more favorable for one–parametric families generated by a transformation group, in particular location– or scale parameter families. Under these circumstances, the theory of Pitman estimators can be applied to obtain, for arbitrary sample sizes, estimators for the location (respectively scale–) parameter which minimize the quadratic risk in the class of all equivariant estimators. See Witting (1985), Section 3.5.6. Without the requirement of equivariance, mean unbiased estimators minimizing the quadratic risk for all $\vartheta \in \mathbb{R}$ usually do not exist (see Bondesson, 1975).

Let $\overline{G}$ denote the transformation group induced on $\mathbb{R}^p$ by an equivariant functional $\kappa : \mathfrak{P} \rightarrow \mathbb{R}^p$. A loss function $L : \mathbb{R}^p \times \mathfrak{P} \rightarrow [0, \infty)$ is equivariant if $L(\bar{a}u, P_a) = \Delta(a)L(u, P)$ for $a \in G$ (with $\Delta(a) : G \rightarrow \mathbb{R}$ not depending on P). Notice that $\Delta(ab) = \Delta(a) \cdot \Delta(b)$. An example of such a loss function is $(u, P) \rightarrow \sum_1^p (u_i - \kappa_i(P))^2$.

Proposition 3.1.7. *Let $\mathfrak{P}$ be a family of mutually absolutely continuous p-measures which is closed under a transformation group G. Let $\kappa : \mathfrak{P} \rightarrow \mathbb{R}^p$ be an equivariant functional, and L a strictly convex equivariant loss function.*

An estimator, which minimizes on $\mathfrak{P}$ the risk under L in the class of all mean unbiased estimators, is $\mathfrak{P}$–almost equivariant.

Proof. By assumption, the following relation holds for arbitrary $a, b \in G$.

$$\int \hat{\kappa}(a^{-1}x)P_b(dx) = \int \hat{\kappa}(x)P_{ba^{-1}}(dx) = \kappa(P_{ba^{-1}})$$

$$= \bar{a}^{-1}\kappa(P_b) = \bar{a}^{-1}\int \hat{\kappa}(x)P_b(dx).$$

Hence $x \rightarrow \bar{a}\hat{\kappa}(a^{-1}x)$ is mean unbiased, for any $a \in G$. Since $\hat{\kappa}$ minimizes on $\mathfrak{P}$ the risk under L in the class of all mean unbiased estimators, we obtain

$$\int L(\hat{\kappa}(x), P_{ba})P_{ba}(dx) \leq \int L(\bar{a}\hat{\kappa}(a^{-1}x), P_{ba})P_{ba}(dx),$$

hence

$$\int L(\hat{\kappa}(ax), P_{ba})P_b(dx) \leq \int L(\bar{a}\hat{\kappa}(x), P_{ba})P_b(dx).$$

Using the equivariance of L, this implies

$$\int L(\bar{a}^{-1}\hat{\kappa}(ax), P_b)P_b(dx) \leq \int L(\hat{\kappa}(x), P_b)P_b(dx).$$

Since $L(\cdot, P)$ is strictly convex, Proposition 3.1.2 yields

$$\bar{a}^{-1}\hat{\kappa}(ax) = \hat{\kappa}(x) \qquad \text{for } \mathfrak{P}\text{–a.a. } x \in X.$$

Hence $\hat{\kappa}$ is $\mathfrak{P}$–almost equivariant.   □

3.2 The Rao–Blackwell–Lehmann–Scheffé–Theorem

A sufficient statistic can be used to improve mean unbiased estimators with respect to convex loss functions. If the sufficient statistic is complete, the improved estimators have minimal convex risk.

Section 3.1 poses in (3.1.1) the problem of finding a mean unbiased estimator which minimizes the risk for a given convex loss function. The solution depends, in general, on the loss function as well as on the p–measure with respect to which the risk is minimized. In the present section we turn to a more special situation in which mean unbiased estimators exist which fulfill (3.1.1″) simultaneously for all convex loss functions, and all $P_0 \in \mathfrak{P}$.

Throughout this section we assume the existence of a mean unbiased estimator $\hat{\kappa} : X \to \mathbb{R}^p$, and of a sufficient statistic S, mapping $(X, \mathcal{A})$ into some measurable space $(Y, \mathcal{B})$.

For one–dimensional parameters and quadratic loss functions the following theorem was obtained independently by C.R. Rao (1945, p. 83) and Blackwell (1947, p. 106, Theorem 2). It was extended to power loss functions by Barankin (1950, p. 284, Corollary 2), and to multidimensional parameters and convex loss functions by Hodges and Lehmann (1950, p. 186, Theorem 3.3).

Theorem 3.2.1 (Rao–Blackwell). *Let* $\hat{\kappa} = (\hat{\kappa}_1, \ldots, \hat{\kappa}_p)$ *be an estimator of* $\kappa(P) \in \mathbb{R}^p$ *and* $S|X$ *a statistic sufficient for* $\mathfrak{P}$. *Let* $k = (k_1, \ldots, k_p)$ *be a conditional expectation of* $\hat{\kappa}$, *given* S *(i.e.* $k_i \in P^S \hat{\kappa}_i$*), which is independent of* $P \in \mathfrak{P}$. *(For the existence see Proposition 1.3.1.) Then the following holds true for* $\hat{\kappa}^0 := k \circ S$.

(i) *For any convex loss function* L,

$$L_*(\hat{\kappa}^0, P) \le L_*(\hat{\kappa}, P) \qquad \text{for all } P \in \mathfrak{P}. \tag{3.2.2}$$

(ii) *If the loss function is strictly convex, then the inequality in (3.2.2) is strict unless* $\hat{\kappa}^0 = \hat{\kappa}$ *P–a.e.*

(iii) *If* $\hat{\kappa}$ *is mean unbiased for* κ, *then so is* $\hat{\kappa}^0$.

Proof. (i) By Jensen's inequality (see Theorem 1.10.11),

$$L(k, P) \le P^S L(\hat{\kappa}, P) \qquad P \circ S\text{–a.e.} \tag{3.2.3}$$

(3.2.2) follows from (3.2.3) by integration with respect to $P \circ S$.

(ii) Equality in (3.2.2) implies equality in (3.2.3) $P \circ S$–a.e. If $L(\cdot, P)$ is strictly convex, this implies $\hat{\kappa}^0 = \hat{\kappa}$ P–a.e.

(iii) Follows immediately from $P(\hat{\kappa}^0) = P(\hat{\kappa})$ for all $P \in \mathfrak{P}$ (see (1.10.2′)).
□

A more general version of this theorem, for randomized estimators, can be proved in the same way.

For families of p–measures with a sufficient statistic S and convex loss functions, the class of all nonrandomized estimators which are contractions of S is complete in the following sense: For any randomized estimator there exists a nonrandomized one, depending on x through $S(x)$ only, the convex risk of which is not greater. The same assertion holds true within the subclass of all mean unbiased estimators.

Notice that the "improved" estimator $\hat{\kappa}^0$ obtained by taking conditional expectations, may have a serious defect compared with the original estimator: It may not be proper any more (see Examples 3.3.5 and 3.4.16–3.4.18).

In spite of the fact that the Rao–Blackwell theorem gives a precise advice how an estimator can be improved, this may be impracticable for technical reasons.

Example 3.2.4. For $b > 0$ let $\Gamma_{1,b}$ be the gamma distribution, given by the Lebesgue density

$$x \to \frac{1}{\Gamma(b)} x^{b-1} \exp[-x], \qquad x > 0.$$

Then the function $(x_1, \ldots, x_n) \to x_1$ is mean unbiased for b.

Since $(x_1, \ldots, x_n) \to \prod_1^n x_i$ is complete sufficient, the conditional expectation of x_1, given $\prod_1^n x_i$, minimizes the convex risk among all mean unbiased estimators for b (see Theorem 3.2.5). It is, however, impossible to express this conditional expectation as a function of $(x_1, \ldots, x_n)$ in closed form.

The Rao–Blackwell theorem presumes that some mean unbiased estimator is already known. Usually, such an estimator can be found ad hoc. This is the case for all functionals of the type $\kappa(P) = \int f(x) P(dx)$. In this case, $f(x_1)$ (or $n^{-1} \sum_1^n f(x_\nu)$) is a mean unbiased estimator.

Systematic methods for constructing mean unbiased estimators of functionals defined on location and/or scale parameter families are given by Tate (1959). For exponential families see Kitigawa (1956).

The improved estimator obtained according to the Rao–Blackwell theorem is optimal if the family $\mathfrak{P} \circ S$ is complete. This is the content of the following theorem, due to Lehmann and Scheffé (1950, p. 321, Theorem 5.1).

The idea that mean unbiased estimators depending on the data through a [minimal?] sufficient statistic minimize the quadratic risk, occurs in a vague form in Geary (1942, pp. 216/7) and C.R. Rao (1947, pp. 280/1, Theorem 1). In Blackwell (1947, p. 106, Theorem 2) it occurs under the explicit assumption that there is only one unbiased estimator which is a contraction of S.

Theorem 3.2.5 (Lehmann–Scheffé). *Assume that S is complete sufficient for $\mathfrak{P}$. If a contraction of S is mean unbiased for κ, then it minimizes the convex risk among all mean unbiased estimators.*

Proof. Let $\hat{\kappa}_0$ be a mean unbiased estimator of κ, depending on x through $S(x)$, say $\hat{\kappa}_0 = k_0 \circ S$. If $\hat{\kappa}$ is an arbitrary mean unbiased estimator of κ, there exists by Theorem 3.2.1 an estimator $\hat{\kappa}_1 = k_1 \circ S$ such that $L_*(\hat{\kappa}_1, P) \leq L_*(\hat{\kappa}, P)$ for every $P \in \mathfrak{P}$. Since $\mathfrak{P} \circ S$ is complete, there exists at most one mean unbiased estimator of κ depending on x through $S(x)$. Hence $\hat{\kappa}_0 = \hat{\kappa}_1$ P–a.e. Therefore, $L_*(\hat{\kappa}_0, P) = L_*(\hat{\kappa}_1, P) \leq L_*(\hat{\kappa}, P)$, for every $P \in \mathfrak{P}$. $\qquad\square$

If $\mathfrak{P} \circ S$ is complete, there is only one mean unbiased estimator depending on x through $S(x)$. Hence taking the conditional expectation of *some* initial mean unbiased estimator leads always to the *same* optimal estimator, however poor the initial estimator might be.

The most fruitful application of Theorem 3.2.5 is to exponential families. The conditions for this application are specified in the following Theorem 3.2.6. Recall that for *curved* exponential families the minimal sufficient statistic is not complete in most cases (see Theorem 1.6.23).

Theorem 3.2.6. *Let $\mathfrak{P} | \mathcal{A}$ be an exponential family having densities*

$$x \to C(P)g(x)\exp\Big[\sum_1^k a_j(P)T_j(x)\Big].$$

Assume that

$$\{(a_1(P), \ldots, a_k(P)) : P \in \mathfrak{P}\}$$

has a nonempty interior.

If for some functional $\kappa : \mathfrak{P} \to \mathbb{R}^p$ there exists a mean unbiased estimator for some sample size m, then there exists for every sample size $n \geq m$ a mean unbiased estimator which depends on $(x_1, \ldots, x_n)$ through

$$\Big(\sum_1^n T_1(x_\nu), \ldots, \sum_1^n T_k(x_\nu)\Big)$$

only, and this estimator minimizes the convex risk among all mean unbiased estimators.

Proof. Follows immediately from Theorem 3.2.5, since the statistic

$$(x_1, \ldots, x_n) \to \Big(\sum_1^n T_1(x_\nu), \ldots, \sum_1^n T_k(x_\nu)\Big)$$

is sufficient and complete by Theorem 1.6.10. $\qquad\square$

Another typical — nonparametric — application of Theorem 3.2.5 is the following theorem (going back to Halmos, 1946).

Theorem 3.2.7. *Let $\mathfrak{P} \,|\, \mathcal{A}$ be a family of p–measures which is complete and closed under convex combinations. If for some functional $\kappa : \mathfrak{P} \to \mathbb{R}^p$ there exists a mean unbiased estimator $\kappa^{(m)} : X^m \to \mathbb{R}^p$, then the estimator*

$$\kappa^{(n)}(x_1, \ldots, x_n) := \frac{1}{\binom{n}{m}} \sum \kappa^{(m)}(x_{i_1, \ldots, i_m})$$

(with $n \geq m$ and the summation extending over all m–tuples $(i_1, \ldots, i_m)$ from $(1, \ldots, n)$) minimizes the convex risk among all mean unbiased estimators for κ.

Proof. Since $\{P^n : P \in \mathfrak{P}\}$ is symmetrically complete by Theorem 1.5.10, and $\kappa^{(n)}$ is symmetric, the assertion follows from Theorem 3.2.5.   □

Remark 3.2.8. Some scholars (see, e.g., Zacks, 1971, p. 114) are of the opinion that a mean unbiased estimator of minimal variance cannot exist unless there is a complete sufficient statistic. This is obviously wrong.

Without a complete sufficient statistic, the following procedure works in some cases: (i) prove that $\hat{\kappa}$ is mean unbiased for κ in the whole family, and (ii) represent $\mathfrak{P}$ as a union of sub–families such that $\hat{\kappa}$ is optimal for κ within each subfamily. To illustrate this idea, let $\mathfrak{P} = \{P_{\vartheta,\eta} : \vartheta \in \Theta, \ \eta \in H\}$. Assume there exists a statistic $S : (X, \mathcal{A}) \to (Y, \mathcal{B})$ which is sufficient for $\{P_{\vartheta,\eta} : \vartheta \in \Theta\}$ and $\{P_{\vartheta,\eta} \circ S : \vartheta \in \Theta\}$ is complete, for every $\eta \in H$. If a mean unbiased estimator for $\kappa(P_{\vartheta,\eta}) = K(\vartheta)$ depends on x through $S(x)$ only, then this estimator is of minimal convex risk. This follows from Theorem 3.2.5, applied to each of the subfamilies $\{P_{\vartheta,\eta} : \vartheta \in \Theta\}$ with η fixed.

The situation is different if there is a complete sufficient statistic $S(x, \eta)$ for $\{P_{\vartheta,\eta} : \vartheta \in \Theta\}$ which depends on η. In this case, an estimator of minimal convex risk for the family $\{P_{\vartheta,\eta} : \vartheta \in \Theta, \ \eta \in H\}$ may not exist. (Example 3.1.6 was of this type.)

Example 3.2.9. It may occur that an estimator $\hat{\kappa}$ which is mean unbiased on $\mathfrak{P}_2$ and of minimal convex risk on $\mathfrak{P}_2$ and on $\mathfrak{P}_0 \subset \mathfrak{P}_2$ is not of minimal convex risk for a family $\mathfrak{P}_1$ inbetween, i.e. $\mathfrak{P}_0 \subset \mathfrak{P}_1 \subset \mathfrak{P}_2$. A special example of this paradoxical phenomenon, due to Lehmann (1983, p. 102, Example 4.2) is: $\mathfrak{P}_0 = \{N_{(\mu,1)} : \mu \in \mathbb{R}\}$, $\mathfrak{P}_2$ = family of all p–measures on $\mathbb{B}$ with positive Lebesgue density, fulfilling $\int |x| P(dx) < \infty$, and $\mathfrak{P}_1$ the subfamily of all symmetric p–measures in $\mathfrak{P}_2$. For every $n \in \mathbb{N}$, $\overline{x}_n$ is mean unbiased on $\mathfrak{P}_2$ for $\kappa(P) := \int x P(dx)$, and of minimal convex risk on $\mathfrak{P}_2$ (by Theorem 3.2.7) and on $\mathfrak{P}_0$ (by Theorem 3.2.6). Against that, the sample median $\tilde{x}_n$ is mean unbiased on $\mathfrak{P}_1$, and has smaller variance than $\overline{x}_n$ for the Laplace distribution (provided $n \geq 3$).

Remark 3.2.10. Assume that $S : (X, \mathcal{A}) \to (Y, \mathcal{B})$ is complete sufficient, and $T : (X, \mathcal{A}) \to (Z, \mathcal{C})$ is ancillary. If a mean unbiased estimator $\hat{\kappa}$ depends on x through $\big(S(x), T(x)\big)$ only, say $\hat{\kappa}(x) = k\big(S(x), T(x)\big)$, then

$$\hat{\kappa}_*\big(S(x)\big) := \int k(S(x), T(u)) P_0(du) \qquad (3.2.10)$$

(with $P_0 \in \mathfrak{P}$ arbitrarily fixed) minimizes the convex risk among all mean unbiased estimators. This is a theorem of Eaton and Morris (1970, p. 1709, Theorem 2.1). It follows immediately from Theorem 3.2.5. Since S and T are stochastically independent by Basu's Theorem 1.8.2, $\hat{\kappa}_*$ is a conditional expectation of $\hat{\kappa}$, given S, with respect to P (apply Proposition 1.10.26 with the Markov kernel $M(y, \cdot) \equiv P_0 \circ T$).

The theorem of Lehmann–Scheffé asserts that completeness of the minimal sufficient statistic guarantees that every functional admitting a mean unbiased estimator admits a (unique) mean unbiased estimator of minimal convex risk. This is what is of interest for applications. Theoretically minded statisticians want to know whether there is a converse to the Lehmann–Scheffé theorem. The answer is to the affirmative: If *every* functional admitting a mean unbiased estimator admits a mean unbiased estimator of minimal variance, then the minimal sufficient sub-σ-algebra is necessarily complete. (See Schmetterer and Strasser, 1974, following the pioneering paper by Bahadur, 1957. For a more abstract treatment see Torgersen, 1981.)

3.3 Examples of mean unbiased estimators with minimal convex risk

Illustrates how mean unbiased estimators with minimal convex risk can be obtained by means of the Rao–Blackwell–Lehmann–Scheffé theorem.

Example 3.3.1. For $a > 0$ let Π_a denote the Poisson distribution, and $P_a := \Pi_a \times \Pi_{a^{-1}}$, $a > 0$. The functional to be estimated is $\kappa(P_a) = a$. We have

$$P_a\{(k, \ell)\} = \frac{\exp[-(a + a^{-1})]}{k!\ell!} a^{k-\ell}, \qquad k, \ell \in \{0\} \cup \mathbb{N}.$$

$S(k, \ell) = k - \ell$ is sufficient, and $\{P_a \circ S : a > 0\}$ is complete. Moreover, k is mean unbiased. Hence the conditional expectation of k, given $k - \ell$, under P_a is a mean unbiased estimator of minimal convex risk.

The joint distribution of $T(k, \ell) = k$ and $S(k, \ell) = k - \ell$ under P_a is

$$Q_a\{(k, s)\} = \frac{\exp[-(a + a^{-1})]}{k!(k - s)!} a^s, \qquad k \in \{0\} \cup \mathbb{N}, \; s \in \mathbb{Z}, \; s \le k.$$

For $s \in \mathbb{Z}$, the conditional distribution of k, given $S = s$, under P_a is

$$M(s, k) = c(s)/k!(k - s)!, \qquad k \in \{0\} \cup \mathbb{N}, \; k \ge s,$$

with $c(s) = \left(\sum (k!(k - s)!)^{-1}\right)^{-1}$, where the summation extends over all $k \in \{0\} \cup \mathbb{N}$, $k \ge s$. (Observe that, thanks to sufficiency of S for $\{P_a : a > 0\}$, this conditional distribution does not depend on a.)

The conditional expectation of k, given $S = s$, is

$$\sum_{\substack{k \in \{0\} \cup \mathbb{N} \\ k \ge s}} k M(s, k) = c(s)/c(s - 1).$$

Hence $c(k - \ell)/c(k - \ell - 1)$ is the mean unbiased estimator for a with minimal convex risk.

Example 3.3.2 (see Tweedie, 1957, Sections 3 and 4). For $\mu > 0$, $\lambda > 0$ let $P_{\mu,\lambda} | \mathbb{B}_+$ denote the inverse normal distribution, with Lebesgue density

$$x \to \sqrt{\frac{\lambda}{2\pi}} x^{-3/2} \exp\left[-\frac{\lambda(x - \mu)^2}{2\mu^2 x}\right], \qquad x > 0.$$

According to Tweedie, $\bar{x}_n$ and $\lambda \sum_1^n (x_\nu^{-1} - \bar{x}_n^{-1})$ are independently distributed as $P_{\mu,n\lambda}$ and χ_{n-1}^2, respectively. We will show that

(i) the statistic $(x_1, \ldots, x_n) \to (\sum_1^n x_\nu, \sum_1^n x_\nu^{-1})$ is complete sufficient for $\{P_{\mu,\lambda}^n : \mu, \lambda > 0\}$;

(ii) the estimator $(x_1, \ldots, x_n) \to (\bar{x}_n, (n - 3)/\sum_1^n (x_\nu^{-1} - \bar{x}_n^{-1}))$ is mean unbiased for $\kappa(P_{\mu,\lambda}) = (\mu, \lambda)$ with minimal convex risk among all mean unbiased estimators.

ad (i): $\{P_{\mu,\lambda} : \mu, \lambda > 0\}$ is an exponential family with

$$T_1(x) = x, \qquad\qquad T_2(x) = x^{-1}$$
$$a_1(P_{\mu,\lambda}) = -\lambda/2\mu^2, \qquad a_2(P_{\mu,\lambda}) = -\lambda/2.$$

Completeness follows from Theorem 1.6.10 since

$$\{(a_1(P_{\mu,\lambda}), a_2(P_{\mu,\lambda})) : \mu, \lambda > 0\} = (-\infty, 0)^2.$$

ad (ii): Since $\int x P_{\mu,x}(dx) = \mu$, and

$$\int (n - 3)/\sum_1^n (x_\nu^{-1} - \bar{x}_n^{-1}) P_{\mu,\lambda}^n(d\underline{x}) = \lambda(n - 3) \int x^{-1} \chi_{n-1}^2(dx) = \lambda,$$

the assertion follows.

Example 3.3.3. For $a > 0$, $b > 0$ let $\Gamma_{a,b}| \mathbb{B}_+$ denote the gamma distribution with Lebesgue density

$$x \rightarrow \frac{1}{a^b \Gamma(b)} x^{b-1} \exp[-x/a], \qquad x > 0.$$

By Example 1.6.14 the statistic $(x_1, \ldots, x_n) \rightarrow (\sum_1^n x_\nu, \sum_1^n \log x_\nu)$ is complete sufficient for $\{\Gamma_{a,b}^n : a, b > 0\}$. Therefore, $\bar{x}_n$ is mean unbiased with minimal convex risk for $\kappa(\Gamma_{a,b}) = ab$, the expectation of $\Gamma_{a,b}$, among all mean unbiased estimators.

Example 3.3.4. Let $\mathfrak{P} = \{N_{(\mu,\sigma^2)} : \mu \in \mathbb{R}, \sigma^2 > 0\}$ and $\kappa(N_{(\mu,\sigma^2)}) := \mu/\sigma$. $\bar{x}_n$ is mean unbiased for μ, and

$$\frac{\sqrt{2}\Gamma\left(\frac{n-1}{2}\right)}{\Gamma\left(\frac{n-2}{2}\right)} \left(\sum_1^n (x_\nu - \bar{x}_n)^2\right)^{-1/2}$$

is mean unbiased for $1/\sigma$ (if $n \geq 2$). Since $\bar{x}_n$ and $\sum_1^n (x_\nu - \bar{x}_n)^2$ are stochastically independent (see Example 1.8.8),

$$\frac{\sqrt{2}\Gamma\left(\frac{n-1}{2}\right)}{\Gamma\left(\frac{n-2}{2}\right)} \bar{x}_n \left(\sum_1^n (x_\nu - \bar{x}_n)^2\right)^{-1/2}$$

is mean unbiased for μ/σ. As a contraction of the complete sufficient statistic $(\sum_1^n x_\nu, \sum_1^n x_\nu^2)$, this estimator minimizes the convex risk among all mean unbiased estimators.

Example 3.3.5. Let $\mathfrak{P} = \{N_{(\mu,\sigma^2)} : \mu > 0, \sigma^2 > 0\}$ and $\kappa(N_{(\mu,\sigma^2)}) := \sigma/\mu$, the coefficient of variation. A reasonable estimator for σ/μ is $s_n/\bar{x}_n$ (with $s_n^2 := n^{-1}\sum_1^n (x_\nu - \bar{x}_n)^2$). This estimator is, however, not mean unbiased. Below we shall show that $g_n(s_n/\bar{x}_n)$ is mean unbiased for σ/μ, with

$$g_n(u) := \sqrt{\frac{n}{2\pi}} u \int\limits_0^1 (1-t)^{(n-3)/2} t^{-1/2}(1+u^2 t)^{-1} dt.$$

Since $(\bar{x}_n, s_n)$ is a complete sufficient statistic for $\mathfrak{P}$, $g_n(s_n/\bar{x}_n)$ is of minimal convex risk among all mean unbiased estimators for σ/μ. Notice that this estimator shares a defect with the "naive" estimator $s_n/\bar{x}_n$: It may attain negative values (but with a probability converging quickly to 0).

The following proof uses ideas of Voinov (1985) who obtained a mean unbiased estimator for $1/\mu$. (It is not a straightforward application, since Voinov's estimator for $1/\mu$ is not stochastically independent of s_n.)

Proof. For $c > 0$, we have

$$\int_0^\infty g(\sqrt{y}/x) y^{\frac{n-3}{2}} \exp[-cy] dy$$

$$= \sqrt{\frac{n}{2\pi}} \int_0^\infty \int_0^y \frac{(y-t)^{\frac{n-3}{2}} x t^{-1/2}}{t + x^2} dt \exp[-cy] dy,$$

$$\int_0^\infty y^{\frac{n-3}{2}} \exp[-cy] dy = \Gamma(\frac{n-1}{2}) c^{-\frac{n-1}{2}}$$

and (by Erdélyi et al., 1954, p. 266, relation 5.12.12)

$$\int_0^\infty \frac{x t^{-1/2}}{t + x^2} \exp[-ct] dt = 2\pi \big(1 - \Phi(\sqrt{2cx})\big) \exp[cx^2].$$

By the convolution theorem for Laplace transforms this implies

$$\int_0^\infty g(\sqrt{y}/x) y^{\frac{n-3}{2}} \exp[-cy] dy \tag{3.3.6}$$

$$= \sqrt{2\pi n} \Gamma(\frac{n-1}{2}) c^{-\frac{n-1}{2}} \big(1 - \Phi(\sqrt{2cx})\big) e^{cx^2}.$$

Using that $\bar{x}_n$ and s_n^2 are independently distributed as $N_{(\mu,\sigma^2/n)}$ and $\Gamma_{2\sigma^2/n,(n-1)/2}$, respectively, and that (see Voinov, 1985, p. 357, relation (3.4))

$$\sqrt{2\pi n} \int \big(1 - \Phi(\sqrt{n}\,x/\sigma)\big) \exp\Big[\frac{nx^2}{2\sigma^2}\Big] N_{(\mu,\sigma^2/n)}(dx) = \sigma/\mu,$$

the assertion follows from (3.3.6) applied with $c = n/2\sigma^2$.   □

Examples on mean unbiased estimators with minimal convex risk for log-normal distributions can be found in Shimizu (1988, pp. 29ff.).

3.4 Mean unbiased estimation of probabilities

Illustrates the general results of Section 3.2 on the existence of mean unbiased estimators with minimal convex risk by applications to the estimation of probabilities.

The problem is to estimate the functional $\kappa(P) := P(A)$ for a given set $A \in \mathcal{A}$. For any family $\mathfrak{P}$ of p–measures, 1_A is a mean unbiased estimator of $P(A)$.

Hence $n^{-1}\sum_1^n 1_A(x_\nu)$ is a mean unbiased estimator of κ in the family $\{P^n : P \in \mathfrak{P}\}$. If $\mathfrak{P}$ is complete and closed under convex combinations, $\{P^n : P \in \mathfrak{P}\}$ is symmetrically complete (Theorem 1.5.10). In this case, $n^{-1}\sum_1^n 1_A(x_\nu)$ minimizes the convex risk among all mean unbiased estimators (see Theorem 3.2.7). More interesting are cases in which $\{P^n : P \in \mathfrak{P}\}$ admits a sufficient statistic coarser than the order statistic.

Example 3.4.1. Let $\mathfrak{P} \,|\, \mathbb{B}^2$ be the family of all p–measures with λ^2–density. The problem is to estimate $P(A)$ (for a given set $A \in \mathbb{B}^2$) based on a realization $((x_1, y_1), \ldots, (x_n, y_n))$ from P^n.

a) By Theorem 3.2.7, the estimator

$$\kappa^{(n)}\big((x_1, y_1), \ldots, (x_n, y_n)\big) := n^{-1}\sum_1^n 1_A(x_\nu, y_\nu) \qquad (3.4.2)$$

minimizes the convex risk among all mean unbiased estimators for $P(A)$ in the family $\{P^n : P \in \mathfrak{P}\}$.

b) Let $\mathfrak{P}_0 \subset \mathfrak{P}$ be the subfamily of all p–measures P with rotation invariant λ^2–density p. In this case p depends on (x, y) through $S(x, y) := \sqrt{x^2 + y^2}$ only. $\mathfrak{P}_0 \circ S$ is complete, since the subfamily $\{N^2_{(0,\sigma^2)} \circ S : \sigma^2 > 0\}$ is complete and $\mathfrak{P}_0 \circ S \ll \lambda^2 \circ S \ll N^2_{(0,1)} \circ S$ (see Exercise 1.5.3). Moreover, $\mathfrak{P}_0 \circ S$ is closed under convex combinations. Hence $\{(P \circ S)^n : P \in \mathfrak{P}_0\}$ is symmetrically complete by Theorem 1.5.10.

Let φ_A be a conditional expectation of 1_A, given S, with respect to $P \in \mathfrak{P}_0$. We shall show below that

$$\varphi_A(r) = \frac{1}{2\pi}\int_0^{2\pi} 1_A(r\cos\alpha, r\sin\alpha)d\alpha, \qquad r > 0. \qquad (3.4.3)$$

The estimator

$$\kappa_0^{(n)}\big((x_1, y_1), \ldots, (x_n, y_n)\big) := n^{-1}\sum_1^n \varphi_A\big(\sqrt{x_\nu^2 + y_\nu^2}\big) \qquad (3.4.4)$$

is mean unbiased for 1_A in the family $\{P^n : P \in \mathfrak{P}_0\}$, and it is a permutation invariant function of $\big(\sqrt{x_1^2 + y_1^2}, \ldots, \sqrt{x_n^2 + y_n^2}\big)$. Hence it minimizes the convex risk among all mean unbiased estimators for $P(A)$ in the family $\mathfrak{P}_0$ (by Theorem 3.2.7).

To prove (3.4.3), let $f : (\mathbb{R}^2, \mathbb{B}^2) \to (\mathbb{R}, \mathbb{B})$ be integrable. We have for every $B \in \mathbb{B}_+$

$$\int f(x, y)1_{S^{-1}B}(x, y)p(S(x, y))\lambda^2(d(x, y))$$

$$= \int_0^\infty \int_0^{2\pi} f(r\cos\alpha, r\sin\alpha) 1_B(r) p(r) r \, dr \, d\alpha$$

$$= \int_0^\infty \frac{1}{2\pi} \int_0^{2\pi} f(r\cos\alpha, r\sin\alpha) d\alpha 1_B(r) P \circ S(dr).$$

(Since p is rotation invariant, $r \to 2\pi r p(r)$ is a Lebesgue density of $P \circ S$.)
Applied for $f = 1_A$ this proves (3.4.3).

c) Let $\mathfrak{P}_1 := \{N^2_{(0,\sigma^2)} : \sigma^2 > 0\}$. The function $S_n((x_1,y_1), \ldots, (x_n,y_n)) := \sum_1^n (x_\nu^2 + y_\nu^2)$ is sufficient for $\{P^n : P \in \mathfrak{P}_1\}$, and $\{P^n \circ S_n : P \in \mathfrak{P}_1\}$ is complete. Below we shall show that

$$\psi_A(w) := \frac{n-1}{\pi w} \int 1_A(x,y) 1_{(0,1)}\left(\frac{x^2+y^2}{w}\right)\left(1 - \frac{x^2+y^2}{w}\right)^{n-2} \lambda^2(d(x,y)),$$

$$w > 0, \ A \in \mathbb{B}^2, \tag{3.4.5}$$

is a conditional expectation of 1_A, given S_n (independent of $P \in \mathfrak{P}_1$). Since $\{P^n \circ S : P \in \mathfrak{P}_1\}$ is complete,

$$\kappa_1^{(n)}((x_1,y_1), \ldots, (x_n,y_n)) := \psi_A\left(\sum_1^n (x_\nu^2 + y_\nu^2)\right) \tag{3.4.6}$$

minimizes the convex risk among all mean unbiased estimators for $P(A)$ in the family $\mathfrak{P}_1$ (by Theorem 3.2.5).

It remains to prove (3.4.5). We have

$$\left(N^2_{(0,\sigma^2)}\right)^n \circ \left(((x_1,y_1), \ldots, (x_n,y_n)) \to ((x_1,y_1), \sum_2^n (x_\nu^2 + y_\nu^2))\right)$$

$$= N^2_{(0,\sigma^2)} \times \Gamma_{2\sigma^2, n-1}$$

with λ^3–density

$$(x,y,z) \to \frac{1}{\pi\Gamma(n-1)} (2\sigma^2)^{-n} z^{n-2} \exp\left[-\frac{x^2+y^2+z}{2\sigma^2}\right], \quad x, y \in \mathbb{R}, \ z > 0.$$

Hence $\left(N^2_{(0,\sigma^2)}\right)^n \circ \left(((x_1,y_1), \ldots, (x_n,y_n)) \to ((x_1,y_1), \sum_1^n (x_\nu^2 + y_\nu^2))\right)$ has λ^3–density

$$(u,v,w) \to \frac{1}{\pi\Gamma(n-1)} (w - (u^2 + v^2))^{n-2} \exp\left[-\frac{w}{2\sigma^2}\right], \quad 0 < u^2 + v^2 < w.$$

This implies (3.4.5) since

$$\left(\int (w - (u^2 + v^2))^{n-2} 1_{(0,w)} (u^2 + v^2) \lambda^2(d(u,v))\right)^{-1} = \frac{n-1}{\pi w^{n-1}}.$$

Specializing these results for $A = \{(x, y) : x + y > 1\}$ we obtain from (3.4.3)

$$\varphi_A(r) = \frac{1}{\pi} \arccos \frac{1}{\sqrt{2r}} \quad \text{if} \quad r > \frac{1}{\sqrt{2}}, \text{ and } 0 \text{ elsewhere.}$$

For $A = \{(x, y) : x^2 + y^2 > 1\}$ we obtain

$$\varphi_A\left(\sqrt{x^2 + y^2}\right) = 1_A(x, y).$$

Hence the restriction from $\mathfrak{P}$ to $\mathfrak{P}_0$ leads in this case to no better estimators of $P(A)$. The further restriction to $\mathfrak{P}_1$ leads to (see (3.4.5))

$$\psi_A(w) = \left(1 - \frac{1}{w}\right)^{n-1} \quad \text{if} \quad w > 1, \text{ and } 0 \text{ elsewhere.}$$

So far, $A \in \mathcal{A}$ was fixed. If we have an estimator for $P(A)$, say $\underline{x} \to \kappa^{(n)}(\underline{x}, A)$ for every $A \in \mathcal{A}$, it is natural to change the interpretation and to consider $A \to \kappa^{(n)}(\underline{x}, A)$ as an estimate for the p–measure $P|\mathcal{A}$. This requires that $A \to \kappa^{(n)}(\underline{x}, A)$ is, in fact, a p–measure for every $\underline{x} \in X^n$, in other words that $\kappa^{(n)} \mid X^n \times \mathcal{A}$ is a Markov kernel.

$\kappa^{(n)}$ is a *mean unbiased* estimator for P in $\mathfrak{P}$ if

$$\int \kappa^{(n)}(\underline{x}, A) P^n(d\underline{x}) = P(A) \quad \text{for every } A \in \mathcal{A}, \ P \in \mathfrak{P}.$$

The natural loss function for estimates Q of a p–measure P is

$$L(Q, P) = \sup_{A \in \mathcal{A}} |Q(A) - P(A)|. \tag{3.4.7}$$

Observe that this loss function is convex, i.e.

$$L\big((1 - \alpha)Q_0 + \alpha Q_1, P\big) \leq (1 - \alpha)L(Q_0, P) + \alpha L(Q_1, P).$$

If $\mathfrak{P}$ is dominated, the problem of estimating p–measures in $\mathfrak{P}$ is inherently related to the problem of estimating densities. Let p denote a μ–density of P. The estimating density based on the sample $\underline{x}$ will be denoted by $p_n(\underline{x}, \cdot)$. Throughout the following we assume that $(\underline{x}, \xi) \to p_n(\underline{x}, \xi)$ is $\mathcal{A}^n \times \mathcal{A}$-measurable.

Proposition 3.4.8. *The following assertions are equivalent:*
(i) *For μ-a.a. $\xi \in X$, $\underline{x} \to p_n(\underline{x}, \xi)$ is mean unbiased for $p(\xi)$.*
(ii) *For every $A \in \mathcal{A}$, $\underline{x} \to \int 1_A(\underline{x}) p_n(\underline{x}, \xi) \mu(d\xi)$ is mean unbiased for $P(A)$.*

Proof. (ii) is equivalent to: for all $A \in \mathcal{A}$

$$\int \left(\int p_n(\underline{x}, \xi) 1_A(\xi) \mu(d\xi) \right) P^n(d\underline{x}) = P(A) = \int p(\xi) 1_A(\xi) \mu(d\xi).$$

By Fubini's theorem, this is equivalent to

$$\int \left(\int p_n(\underline{x}, \xi) P^n(d\underline{x}) \right) 1_A(\xi) \mu(d\xi) = \int p(\xi) 1_A(\xi) \mu(d\xi) \quad \text{for all } A \in \mathcal{A},$$

which, in turn, is equivalent to

$$\int p_n(\underline{x}, \xi) P^n(d\underline{x}) = p(\xi) \quad \text{for } \mu\text{–a.a. } \xi \in X.$$

$\square$

The natural loss function for estimates q of a density $p \in dP/d\mu$ is

$$L(q, p) := \int |q(\xi) - p(\xi)| \mu(d\xi). \tag{3.4.9}$$

Since

$$\sup_{A \in \mathcal{A}} |Q(A) - P(A)| = \frac{1}{2} \int |q(\xi) - p(\xi)| \mu(d\xi),$$

optimality of a mean unbiased density estimator with respect to the loss function (3.4.9) coincides with optimality of the pertaining mean unbiased estimator of the p–measure with respect to the loss function (3.4.7).

However: Mean unbiased estimators for a dominated family of p–measures cannot be represented by a mean unbiased density estimator if this family is large.

Proposition 3.4.10. *Let $\mathcal{A}$ be countably generated, with $\{x\} \in \mathcal{A}$ for $x \in X$. Let $\mathfrak{P} | \mathcal{A}$ be a complete and convex family of p–measures, dominated by some nonatomic measure $\mu | \mathcal{A}$. Then an estimator for the μ–density of P which is mean unbiased on $\mathfrak{P}$ does not exist (in spite of the fact that there is an estimator of P which is mean unbiased on $\mathfrak{P}$).*

Proof. Let $p^{(n)}(\underline{x}, \xi)$ be mean unbiased for $p(\xi)$ for μ–a.a. ξ. W.l.g. we may assume that $p^{(n)}(\underline{x}, \xi)$ is invariant under permutations.

$$\int p^{(n)}(\underline{x}, \xi) P^n(d\underline{x}) = p(\xi) \quad \text{for } \mu\text{–a.a. } \xi \in X$$

implies by Fubini's theorem

$$\int \left(\int p^{(n)}(\underline{x}, \xi) 1_A(\xi) \mu(d\xi) \right) P^n(d\underline{x}) = P(A) \quad \text{for all } A \in \mathcal{A}.$$

The estimator $\underline{x} \to \int p^{(n)}(\underline{x}, \xi) 1_A(\xi) \mu(d\xi)$ is permutation invariant, and unbiased for $P(A)$. Since $\{P^n : P \in \mathfrak{P}\}$ is symmetrically complete by Theorem 1.5.10, there exists a P^n–null set N_A such that for $\underline{x} \notin N_A$

$$\int p^{(n)}(\underline{x}, \xi) 1_A(\xi) \mu(d\xi) = n^{-1} \sum_1^n 1_A(x_\nu). \tag{3.4.11}$$

Let $\mathcal{A}_0$ be a countable algebra generating $\mathcal{A}$. For $\underline{x} \notin N_0 := \cup\{N_A : A \in \mathcal{A}_0\}$ relation (3.4.11) holds for all $A \in \mathcal{A}_0$, and therefore for all $A \in \mathcal{A}$. Applied for $A = \{x_1, \ldots, x_n\}$ this leads to a contradiction. $\qquad\square$

Theorem 3.4.12. *Assume that $\mathcal{A}$ is countably generated. Assume that $S_n : (X^n, \mathcal{A}^n) \to (Y, \mathcal{B})$ is sufficient for $\{P^n : P \in \mathfrak{P}\}$ and that $\{P^n \circ S_n : P \in \mathfrak{P}\}$ is complete.*

If a Markov kernel on $X^n \times \mathcal{A}$ — which depends on $\underline{x} \in X^n$ through $S_n(\underline{x})$ only — is mean unbiased for $P|\mathcal{A}$ on $\mathfrak{P}$, it minimizes the risk for the loss function (3.4.7) among all estimators for $P|\mathcal{A}$ which are mean unbiased on $\mathfrak{P}$ (and it is the only estimator with this property).

Proof. Let $K^{(n)}|X^n \times \mathcal{A}$ be an arbitrary Markov kernel which is mean unbiased for $P|\mathcal{A}$ on $\mathfrak{P}$. Let $\mathcal{A}_0$ be a countable algebra generating $\mathcal{A}$. For $A \in \mathcal{A}$ let $\varphi_A^{(n)}$ be a conditional expectation of $K^{(n)}(\cdot, A)$, given S_n. Since $t \to |t - P(A)|$ is convex, we have by Jensen's inequality

$$|\varphi_A^{(n)} - P(A)| \le (P^n)^{S_n}(|K^{(n)}(\cdot, A) - P(A)|)$$
$$\le (P^n)^{S_n}\left(\sup_{B \in \mathcal{A}_0} |K^{(n)}(\cdot, B) - P(B)| \right) \quad P^n\text{–a.e.}$$

Since $\mathcal{A}_0$ is countable, and

$$\sup_{A \in \mathcal{A}_0} |Q(A) - P(A)| = \sup_{A \in \mathcal{A}} |Q(A) - P(A)|, \tag{3.4.13}$$

this implies

$$P^n\left(\sup_{A \in \mathcal{A}} |\varphi_A^{(n)} \circ S_n - P(A)| \right) \le P^n\left(\sup_{A \in \mathcal{A}} |K^{(n)}(\cdot, A) - P(A)| \right). \tag{3.4.14}$$

Let $K_*^{(n)}|X^n \times \mathcal{A}$ denote a mean unbiased Markov kernel for $P|\mathcal{A}$ depending on $\underline{x}$ through $S_n(\underline{x})$ only. Since

$$P^n(\varphi_A^{(n)} \circ S_n) = P^n(K^{(n)}(\cdot, A)) = P(A) = P^n(K_*^{(n)}(\cdot, A)),$$

completeness of $\{P^n \circ S_n : P \in \mathfrak{P}\}$ implies for every $A \in \mathcal{A}$

$$\varphi_A^{(n)} \circ S_n = K_*^{(n)}(\cdot, A) \quad P^n\text{–a.e.,}$$

hence (use (3.4.13) and (3.4.14))

$$P^n\big(\sup_{A\in\mathcal{A}}|K_*^{(n)}(\cdot,A)-P(A)|\big)\le P^n\big(\sup_{A\in\mathcal{A}}|K^{(n)}(\cdot,A)-P(A)|\big). \qquad \square$$

There is a weak point in this theorem: For the optimal Markov kernel $K^{(n)}|X^n\times\mathcal{A}$, the p-measure $K^{(n)}(\underline{x},\cdot)|\mathcal{A}$ is usually not an element of $\mathfrak{P}$. (Asymptotically this effect disappears, since $K^{(n)}(\underline{x},\cdot)|\mathcal{A}$ converges to $P|\mathcal{A}$ for $P^{\mathbf{N}}$-a.a. $\underline{x}\in X^{\mathbf{N}}$.)

We conclude this section by a few examples illustrating the application of Theorem 3.4.12. In all these examples, X is a subset of $\mathbb{R}$, and the sufficient statistic is of the special type $S_n(x_1,\dots,x_n)=\sum_1^n x_\nu$.

Assume that $P^n\circ S_n$ has density $q_n(\cdot,P)$ with respect to a σ-finite measure μ (which is, in our examples, either the Lebesgue- or the counting measure). Then $P^n\circ\big((x_1,\dots,x_n)\to(x_1,\sum_2^n x_\nu)\big)$ has μ^2-density $(x,y)\to q_1(x,P)q_{n-1}(y,P)$, and $P^n\circ\big((x_1,\dots,x_n)\to(x_1,\sum_1^n x_\nu)\big)$ has μ^2-density $(x,y)\to q_1(x,P)q_{n-1}(y-x,P)$. From this we obtain the μ-density of the conditional distribution of x_1, given $\sum_1^n x_\nu=y$, as

$$x\to\tilde q_n(y,x):=q_1(x,P)q_{n-1}(y-x,P)/q_n(y,P). \tag{3.4.15}$$

Since $\sum_1^n x_\nu$ is sufficient for $\mathfrak{P}$, the density $\tilde q_n$ does not depend on $P\in\mathfrak{P}$. Hence $p_n(\underline{x},\cdot):=\tilde q_n(\sum_1^n x_\nu,\cdot)$ is mean unbiased for the μ-density of P (which is equal to $q_1(\cdot,P)$). Starting from $\tilde q_n(y,\cdot)$ we define, for $Y\subset\mathbb{R}$, a Markov kernel $M^{(n)}|Y\times\mathcal{A}$ by

$$M^{(n)}(y,A):=\int 1_A(x)\tilde q_n(y,x)\mu(dx) \qquad y\in Y,\ A\in\mathcal{A}.$$

As a consequence of (3.4.15),

$$\int M^{(n)}(\sum_1^n x_\nu,A)P(dx_1)\dots P(dx_n)=P(A) \qquad\text{for } A\in\mathcal{A}\text{ and }P\in\mathfrak{P},$$

i.e. $(x_1,\dots,x_n)\to M^{(n)}(\sum_1^n x_\nu,\cdot)|\mathcal{A}$ is mean unbiased for $P|\mathcal{A}$.

If the sufficient statistic is complete, these estimators are, according to Theorem 3.4.12, of minimal convex risk with respect to the loss functions (3.4.9) and (3.4.7), respectively.

Example 3.4.16. For $a>0$ let Π_a denote the Poisson distribution with mean a. The statistic $S_n(x_1,\dots,x_n)=\sum_1^n x_\nu$ is sufficient for $\{\Pi_a^n:a>0\}$, and $\{\Pi_a^n\circ S_n:a>0\}$ is complete by Theorem 1.6.10. Since $k\to\Pi_{na}\{k\}$ is a density of $\Pi_a^n\circ S_n$ with respect to counting measure, we obtain from (3.4.15) the conditional probability of $x=k$, given $\sum_1^n x_\nu=y$ as

$$\Pi_a\{k\}\Pi_{(n-1)a}\{y-k\}/\Pi_{na}\{y\}=B_{y,\frac1n}\{k\}$$

for $k \in \{0, 1, \ldots, y\}$ and $y \in \{0\} \cup \mathbb{N}$, if we define

$$B_{0,\frac{1}{n}}\{k\} = \begin{cases} 1 & = \\ 0 & > \end{cases} \quad \text{for} \quad k \quad 0.$$

Observe that $B_{y,\frac{1}{n}}$ is independent of a, thanks to sufficiency of $\sum_1^n x_\nu$.

According to Theorem 3.4.12, $(x_1, \ldots, x_n) \to B_{\sum_1^n x_\nu, \frac{1}{n}}$ is the unique mean unbiased estimator for Π_a which minimizes the risk for the loss function (3.4.7).

Example 3.4.17. For $\vartheta > 0$ let $E_\vartheta | \mathbb{B}_+$ denote the exponential distribution. $S_n(x_1, \ldots, x_n) = \sum_1^n x_\nu$ is sufficient for $\{E_\vartheta^n : \vartheta > 0\}$, and $\{E_\vartheta^n \circ S_n : \vartheta > 0\}$ is complete by Theorem 1.6.10.

$E_\vartheta^n \circ S_n$ has Lebesgue density

$$y \to \frac{1}{\vartheta^n (n-1)!} y^{n-1} \exp[-y/\vartheta], \quad y > 0.$$

According to (3.4.15), the conditional distribution of x, given $\sum_1^n x_\nu = y$, is

$$\tilde{q}_n(y, x) = \frac{n-1}{y} \left(1 - \frac{x}{y}\right)^{n-2}, \quad 0 < x < y.$$

Observe that $\tilde{q}_n(y, \cdot)$ is independent of ϑ, thanks to sufficiency of $\sum_1^n x_\nu$. According to Theorem 3.4.12,

$$(x_1, \ldots, x_n) \to \frac{n-1}{\sum_1^n x_\nu} \left(1 - \frac{x}{\sum_1^n x_\nu}\right)^{n-2}, \quad 0 < x < \sum_1^n x_\nu,$$

is the unique mean unbiased estimator for $\vartheta^{-1} \exp[-x/\vartheta]$ which minimizes the risk for the loss function (3.4.9).

Exercise 3.4.18. For $\vartheta > 0$ let $P_\vartheta | \mathbb{B}_+$ denote the exponential distribution truncated at T, with Lebesgue density

$$x \to \frac{1}{\vartheta} \exp[-x/\vartheta] (1 - \exp[-T/\vartheta])^{-1}, \quad 0 < x < T. \tag{3.4.19}$$

By rescaling we may assume w.l.g. that $T = 1$.

Let $c_n(x) := \sum_{k=0}^{[x]} (-1)^k \binom{n}{k} (x - k)^{n-1}$, with $[x]$ denoting the largest integer $\leq x$.

Starting from

$$q_n(x, \vartheta) = \frac{\exp[-x/\vartheta]}{(n-1)! \vartheta^n (1 - \exp[-1/\vartheta])^n} c_n(x)$$

(see Bain and Weeks, 1964) use relation (3.4.15) to show that

$$\tilde{q}_n(y, x) = (n-1)c_{n-1}(y-x)/c_n(y), \qquad 0 < x < 1, \ 0 < y - x < n - 1.$$

Hence $(x_1, \ldots, x_n) \to \tilde{q}_n(\sum_1^n x_\nu, \cdot)$ is a mean unbiased estimator for (3.4.19) which minimizes the risk for the loss function (3.4.9).

Comparable results can be found in Holla (1967), and Sathe and Varde (1969).

Basically the same idea was applied by a number of authors in connection with various examples, so for instance: *Normal distribution with μ, σ^2 unknown:* Kolmogorov (1950, pp. 166/7) [observe that on p. 164 the factor $1/\sqrt{2}$ is missing in formula (14)], Lieberman and Resnikoff (1955, p. 469), Schmetterer (1960, p. 657, Theorem 1), Barton (1961, p. 228), Patil and Wani (1966, p. 43, Section 2.3). *Poisson distribution:* Barton (1961, p. 228), Glasser (1962, p. 410). *Gamma distribution* with known shape parameter: Tate (1959, p. 354), Pugh (1963, p. 60), A.P. Basu (1964, p. 216), Patil and Wani (1966, p. 43). A straightforward modification of the results on the exponential distribution leads to a result for the scale parameter family of Laplace distributions (Asrabadi, 1985, p. 732).

The reader interested in equivariant density estimators is referred to Wertz (1978).

3.5 A result on bounded mean unbiased estimators

A bounded estimator is of minimal convex risk if it is of minimal quadratic risk.

The following Theorem 3.5.1 goes back to C.R. Rao (1952, p. 30, Theorem 2) and Bahadur (1957, Theorem 6, p. 218). The version presented below is essentially due to Padmanabhan (1970, p. 109, Theorem 3.1). For mean unbiased estimators the quadratic loss function $L(t, P) := (t - \kappa(P))^2$ occurring in the assumption of Theorem 3.5.1 can be replaced by the loss function $L(t, P) := t^2$. In this form the theorem was generalized by Schmetterer and Strasser (1974, p. 61, Satz 2) who prove that the bounded mean unbiased estimator minimizes the risk for every convex loss function among all mean unbiased estimators if it minimizes the risk for one convex loss function (fulfilling certain regularity conditions) which does not depend on P.

Theorem 3.5.1. *Let $\kappa : \mathfrak{P} \to \mathbb{R}$ be an arbitrary functional. If a bounded mean unbiased estimator for κ minimizes the quadratic risk in the class of all mean unbiased estimators, then it minimizes the convex risk in this class.*

For unbounded estimators this is not true any more (see Bahadur, 1957, p. 222).

Proof. (i) Let $\mathfrak{H} = \{h \in \mathcal{L}_2(X, \mathcal{A}, \mathfrak{P}) : P(h) = 0 \text{ for } P \in \mathfrak{P}\}$, and

$$\mathfrak{F}_0 = \{f \in \mathcal{L}_\infty(X, \mathcal{A}) : P(hf) = 0 \quad \text{for } h \in \mathfrak{H}, \ P \in \mathfrak{P}\}.$$

$\mathfrak{F}_0$ is closed under multiplication, linear combinations and (uniform) limits in $\mathcal{L}_\infty(X, \mathcal{A})$. Hence $\mathcal{A}_0 := \{A \in \mathcal{A} : 1_A \in \mathfrak{F}_0\}$ is a $\cap$-closed Dynkin system and therefore a σ-algebra.

We shall show that every $f \in \mathfrak{F}_0$ is $\mathcal{A}_0$-measurable. Let $f \in \mathfrak{F}_0$ and $P \in \mathfrak{P}$ be fixed. We have to show that $P(h1_{f^{-1}B}) = 0$ for every $B \in \mathbb{B}$ and $h \in \mathfrak{H}$. To see this, we define for fixed $h \in \mathfrak{H}$ the signed measure $\mu | \mathbb{B}$ by $\mu(B) = P(h1_{f^{-1}B})$, $B \in \mathbb{B}$. For any μ-integrable function $g : (\mathbb{R}, \mathbb{B}) \to (\mathbb{R}, \mathbb{B})$, we have $\mu(g) = P(hg \circ f)$. For $g(u) = u^k$, we obtain $\int u^k \mu(du) = P(hf^k) = 0$ for $k \in \mathbb{N}$. Since f is bounded, $\mu | \mathbb{B}$ has a bounded support and is, therefore, uniquely determined by its moments (see Chow and Teicher, 1988, p. 283, Proposition 4, and p. 285, Proposition 6). This implies $\mu \equiv 0$, i.e. $P(h1_{f^{-1}B}) = 0$ for $B \in \mathbb{B}$. Since $h \in \mathfrak{H}$ was arbitrary, this implies $1_{f^{-1}B} \in \mathfrak{F}_0$, hence $f^{-1}B \in \mathcal{A}_0$.

(ii) Assume that $\hat\kappa_0 \in \mathcal{L}_\infty(X, \mathcal{A})$ is mean unbiased for $\kappa : \mathfrak{P} \to \mathbb{R}$ with minimal quadratic risk. For any $h \in \mathfrak{H}$ and $\lambda \in \mathbb{R}$, the estimator $\hat\kappa_0 + \lambda h$ is mean unbiased for κ, so that

$$P\left((\hat\kappa_0 - \kappa(P))^2\right) \le P\left(((\hat\kappa_0 + \lambda h) - \kappa(P))^2\right) \quad \text{for every } P \in \mathfrak{P}.$$

Hence $P(h\hat\kappa_0) = 0$ for every $P \in \mathfrak{P}$. Since $h \in \mathfrak{H}$ was arbitrary, this implies that $\hat\kappa_0 \in \mathfrak{F}_0$. Hence $\hat\kappa_0$ is $\mathcal{A}_0$-measurable.

(iii) If $\hat\kappa \in \mathcal{L}_2(X, \mathcal{A}, \mathfrak{P})$ is mean unbiased for κ, we have $P(\hat\kappa_0 - \hat\kappa) = 0$ for every $P \in \mathfrak{P}$, hence $\hat\kappa_0 - \hat\kappa \in \mathfrak{H}$. Therefore, $P((\hat\kappa_0 - \hat\kappa)f) = 0$ for $f \in \mathfrak{F}_0$, in particular:

$$P\left((\hat\kappa_0 - \hat\kappa)1_A\right) = 0 \qquad \text{for } A \in \mathcal{A}_0. \tag{3.5.2}$$

Since $\hat\kappa_0$ is $\mathcal{A}_0$-measurable, it is by (3.5.2) a conditional expectation of $\hat\kappa$, given $\mathcal{A}_0$. Hence

$$\int L(\hat\kappa_0(x), P) P(dx) \le \int L(\hat\kappa(x), P) P(dx)$$

for any convex loss function L by Jensen's inequality. $\square$

Chapter 4
Testing hypotheses

4.1 Basic concepts

Introduces basic concepts like critical region, critical function, power function, similar tests.

Let $(X, \mathcal{A})$ be a measurable space, $\mathfrak{P}$ a family of p–measures on $\mathcal{A}$, and $\mathfrak{P}_0 \subset \mathfrak{P}$ a distinguished subset, called "hypothesis". The problem is: Given a realization from a p–measure $P \in \mathfrak{P}$, one has to decide whether this p–measure is in $\mathfrak{P}_0$ or in $\mathfrak{P} - \mathfrak{P}_0$. Presuming that "abstention" is not allowed, this amounts to distinguishing a certain subset $C \in \mathcal{A}$ as "critical region" and to

$$\begin{matrix} \text{reject} \\ \text{accept} \end{matrix} \quad \text{the hypothesis } \mathfrak{P}_0 \text{ if } \quad x \begin{matrix} \in \\ \notin \end{matrix} C.$$

The term "region" is somewhat misleading, insinuating that C is a set with smooth boundary. In fact, there is no other condition on C except $C \in \mathcal{A}$.

This is what is done in practice. Statistical theory is based on the more general concept of a critical function.

Definition 4.1.1. A *cricitcal function* is an $\mathcal{A}$–measurable map $\varphi : X \to [0, 1]$. If φ is an indicator function, say $\varphi = 1_C$, then C is called *critical region*. The set of all critical functions will be denoted by Φ.

The concept of a critical function is connected with the following directions.

After x has been observed, the hypothesis is *rejected* with probability $\varphi(x)$ and *accepted* with probability $1 - \varphi(x)$. If φ is an indicator function, the decision is uniquely determined by the observation: The hypothesis is rejected if the observation falls into the critical region, and accepted otherwise.

With this interpretation in mind, we follow the common abuse of language and speak of the "test" φ (instead of the "critical function" φ).

Considering that randomized tests (i.e. critical functions other than those representing critical regions) are never used in practice, it is natural to ask why they have been introduced in statistical theory. This question will be discussed in Section 4.2.

Desirable are tests φ with

$$P(\varphi) = \begin{cases} \text{small} \\ \text{large} \end{cases} \quad \text{for} \quad P \begin{matrix} \in \\ \notin \end{matrix} \mathfrak{P}_0 \, .$$

Since the ideal

$$P(\varphi) = \begin{cases} 0 \\ 1 \end{cases} \quad \text{for} \quad P \begin{matrix} \in \\ \notin \end{matrix} \mathfrak{P}_0$$

is out of reach except for trivial cases, statisticians are usually content with

$$P(\varphi) \le \alpha \quad \text{for } P \in \mathfrak{P}_0,$$

for some small $\alpha \in (0,1)$, say $\alpha = 0.01$ or $\alpha = 0.05$.

Given the hypothesis $\mathfrak{P}_0$, we denote by

$$\Phi_\alpha := \{ \varphi \in \Phi : P(\varphi) \le \alpha \quad \text{for } P \in \mathfrak{P}_0 \} \tag{4.1.2}$$

the class of all level–α–tests. The desideratum that $P(\varphi)$ should be "as large as possible for all $P \notin \mathfrak{P}_0$" is intuitively plausible, but without a precise meaning, and this is the point where the general considerations come to an end.

Definition 4.1.3. (i) *Power function* of φ: $\quad P \to P(\varphi), \; P \in \mathfrak{P}$.
    (ii) *Envelope power function* of Φ_α: $\quad P \to \sup\{P(\varphi) : \varphi \in \Phi_\alpha\}, \quad P \in \mathfrak{P} - \mathfrak{P}_0$.

For any $P_1 \in \mathfrak{P} - \mathfrak{P}_0$ there exists (see Theorem 4.2.3) a test φ_1 which maximizes $P_1(\varphi)$ in the class Φ_α, i.e. $\varphi_1 \in \Phi_\alpha$, and $P_1(\varphi_1) = \sup\{P_1(\varphi) : \varphi \in \Phi_\alpha\}$.

A test φ is *uniformly most powerful* if its power function coincides with the envelope power function on $\mathfrak{P} - \mathfrak{P}_0$. Uniformly most powerful tests exist under special circumstances only. In general, the test φ_1 maximizing $P_1(\varphi)$ under the condition $\varphi \in \Phi_\alpha$ depends on P_1. The envelope power function can, however, be used as a standard for evaluating the power function of a particular test.

In connection with the general test problem, Wald (1943) introduced the concept of a "most stringent test", which minimizes $\varphi \to \sup\{\beta(P) - P(\varphi) : P \in \mathfrak{P} - \mathfrak{P}_0\}$ among all level–α–tests for $\mathfrak{P}_0$ (where $\beta(P)$ denotes the envelope power function). Since the difference between 0.50 and 0.51 is something else than the difference between 0.95 and 0.96, say, and since the power is of no interest where it is low, it is hard to understand why $\sup\{\beta(P) - P(\varphi) : P \in \mathfrak{P} - \mathfrak{P}_0\}$ should be a reasonable criterion for the choice of a critical function.

Example 4.1.4. For $\vartheta \in [1, \infty)$ let $P_\vartheta | \mathbb{B}$ be the normal distribution with mean ϑ and variance $1/2\,\vartheta$. It has Lebesgue density $x \rightarrow \sqrt{\vartheta/\pi} \exp[-\vartheta(x - \vartheta)^2]$. We consider tests of level $\alpha = 0.05$ for the hypothesis $\vartheta = 1$. The following Figure 4.1 refers to the sample size $n = 15$. It shows the envelope power function, and the power function of the test C_n which is most powerful against the alternative $\vartheta = 1.5$. This power function drops to almost 0 for alternatives greater 4, say, and is, therefore, of no practical use. A reasonable test is

$$C'_n = \{(x_1, \ldots, x_n) \in \mathbb{R}^n : \sum_1^n x_\nu > n + \sqrt{n/2}u_\alpha\},$$

which has a power function close to the envelope power function.

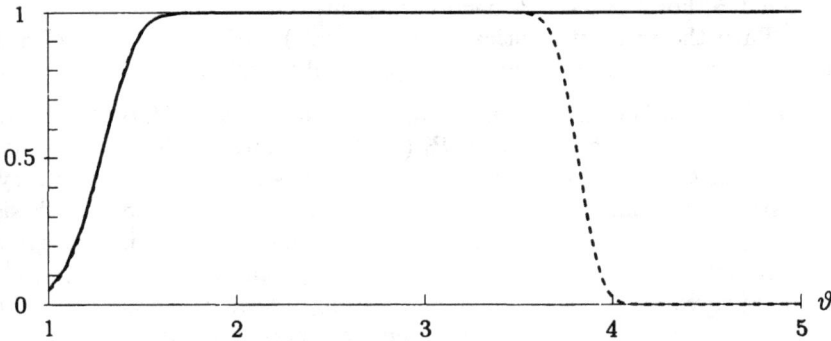

Figure 4.1 ——— envelope power function; $- - -$ power function of C_n.

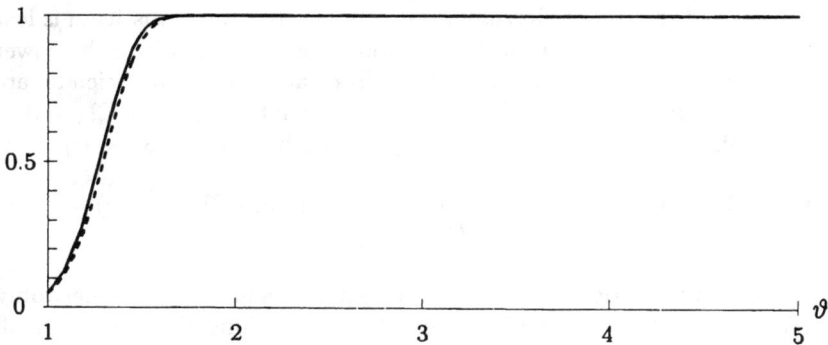

Figure 4.2 ——— envelope power function; $- - -$ power function of C'_n.

What we observe in this example is typical for the general case: In the absence of a uniformly most powerful test one has to find a practical compromise to which statistical theory cannot contribute much. The oversimplified distinction between the hypothesis $\mathfrak{P}_0$ and the alternative $\mathfrak{P} - \mathfrak{P}_0$ is of no great help. It embraces problems of different nature which require solutions in their own right. Here are a few examples.

(i) If $\mathfrak{P}$ is a family with one real parameter, say $\mathfrak{P} = \{P_\vartheta : \vartheta \in \mathbb{R}\}$, and the problem is to test the hypothesis $\vartheta \leq \vartheta_0$ against all alternatives $\vartheta > \vartheta_0$, asymptotic theory leads to reasonable results (see Chapter 10), and under special conditions there are even satisfying nonasymptotic results for every sample size (see Theorem 4.5.2). For the sake of curiosity, we mention the following example of a family $\mathfrak{P} = \{P_\vartheta : \vartheta > 0\}$ where there is a test for the hypothesis $\vartheta = \vartheta_0$ (i.e. $\mathfrak{P}_0 = \{P_{\vartheta_0}\}$) which is most powerful for every $\vartheta \neq \vartheta_0$. In general, a test which is good for alternatives $\vartheta > \vartheta_0$ will be poor for alternatives $\vartheta < \vartheta_0$ and vice versa.

Example 4.1.5. For $\vartheta > 0$ let P_ϑ denote the p–measure with Lebesgue density $\vartheta^{-1} 1_{(0,\vartheta)}$. Then the test with critical region $(0, \alpha\vartheta_0) \cup (\vartheta_0, \infty)$ is of level $\alpha \in (0,1)$ for the hypothesis $\vartheta = \vartheta_0$, and most powerful for all alternatives $\vartheta \neq \vartheta_0$.

(ii) The situation is of a different nature if the family $\mathfrak{P}$ is "large", and we consider tests for the hypothesis $P = P_0$ (i.e. $\mathfrak{P}_0 = \{P_0\}$) against all alternatives $P \neq P_0$. In this situation, the choice of a test becomes almost arbitrary. Even in a simple case like $\mathfrak{P} = \{N_{(\mu,\sigma^2)} : \mu \in \mathbb{R}, \sigma^2 > 0\}$ with the hypothesis $(\mu, \sigma^2) = (\mu_0, \sigma_0^2)$, the statistician will be helpless unless the experimenter gives him some advice. Alternatives (μ, σ_0^2) with $\mu \neq \mu_0$, and alternatives (μ_0, σ^2) with $\sigma^2 \neq \sigma_0^2$ have intrinsically different meanings. Needless to say that the problem becomes more difficult if $\mathfrak{P}$ is a "nonparametric" family.

(iii) The situation is even more difficult if both, $\mathfrak{P}$ and $\mathfrak{P}_0$, are large. As an instance consider the hypothesis of independence. Let $\mathfrak{P} \,|\, \mathbb{B}^2$ be the family of all p–measures with Lebesgue density, and $\mathfrak{P}_0$ the family of all product measures in $\mathfrak{P}$. In this case, the class of alternatives is so large that the statistician is lost unless the expert specifies a certain class of alternatives against which the power of the test is to be directed. Without the help of an expert, statisticians are inclined to select some functional $\kappa : \mathfrak{P} \to \mathbb{R}$ with $\kappa(P) = 0$ for $P \in \mathfrak{P}_0$ and to replace the original hypothesis $\mathfrak{P}_0$ by the larger hypothesis $\{P \in \mathfrak{P} : \kappa(P) = 0\}$.

Definition 4.1.6. A test φ is *similar* for the hypothesis $\mathfrak{P}_0$ if $P(\varphi) = \text{const}$ for $P \in \mathfrak{P}_0$.

To require a similar critical function for a hypothesis like $\vartheta \leq \vartheta_0$ is certainly not reasonable. Mostly $\varphi \equiv \alpha$ will be the only test fulfilling $P_\vartheta(\varphi) = \alpha$ for all $\vartheta \leq \vartheta_0$. Similarity may, however, be a meaningful requirement for the "boundary" of the hypothesis, for instance: If the hypothesis is $\{P \in \mathfrak{P} : \kappa(P) \leq r_0\}$, it appears not unreasonable to look for tests which are similar on $\{P \in \mathfrak{P} : \kappa(P) = r_0\}$.

The usual argument for supporting the requirement of similarity runs as follows. A level–α–test for the hypothesis $\mathfrak{P}_0$ is *unbiased* if $P(\varphi) \geq \alpha$ for $P \in \mathfrak{P} - \mathfrak{P}_0$. For the hypothesis $\mathfrak{P}_0 = \{P \in \mathfrak{P} : \kappa(P) \leq r_0\}$ this means

$$P(\varphi) \begin{smallmatrix} \leq \\ \geq \end{smallmatrix} \alpha \qquad \text{for} \qquad \kappa(P) \begin{smallmatrix} \leq \\ > \end{smallmatrix} r_0.$$

Presuming that κ is a continuous function of P, it suggests itself to look for tests φ fulfilling $P(\varphi) = \alpha$ for $\kappa(P) = r_0$, i.e. tests which are similar on $\{P \in \mathfrak{P} : \kappa(P) = r_0\}$.

One could ask, of course, why a test should be unbiased. The practically relevant question is as to which extent alternatives are rejected with reasonably high probability, say 0.8 at least. Whether a probability measure in the alternative, but close to the hypothesis, is rejected with probability $\alpha + 0.001$ or $\alpha - 0.001$ is irrelevant for all practical purposes. Hence the requirement of "unbiasedness" appears to be something like a deus ex machina, brought in for a single purpose: To restrict the class of tests to such an extent that, within this restricted class, there is a uniformly most powerful test. The reader interested in this branch of statistical theory is referred to Lehmann (1986, Chapter 4) or Witting (1985, Abschnitt 2.4).

It is the theory of median unbiased estimators or — more generally — of confidence bounds where the concept of a similar test is a useful auxiliary device. (See Sections 5.5 and 5.6).

Before we enter the mathematical problems connected with the construction of "good" tests, we ask the reader to keep in mind that all theoretical considerations can become irrelevant in view of certain misuses in the application of tests. These misuses can be characterized by the catchwords "multiple analysis" (the experimenter tries several different tests to find one which refutes the hypothesis), "optional stopping" (the experiment is repeated until there are "enough data" to refute the hypothesis) and "selective reporting" (only such results are published which refute a hypothesis).

To illustrate the impact of multiple analysis and optional stopping, we consider a series of independent observations, distributed as $N_{(\mu,1)}$. To refute the hypothesis $\mu = 0$, the experimenter applies after 20 observations two tests (multiple analysis!) with the critical regions $C_n = \{(x_1, \ldots, x_n) \in \mathbb{R}^n : |\sum_1^n x_\nu| > c_{n,\alpha}\}$ and $C'_n = \{(x_1, \ldots, x_n) \in \mathbb{R}^n : |\sum_1^n 1_{(0,\infty)}(x_\nu) - \frac{n}{2}| > c'_{n,\alpha}\}$ with rejection probability $\alpha = 0.05$. If neither of these tests leads to a refutation of the hypothesis, the experiment is continued until one of these tests refutes the hypothesis (optional stopping!). Figure 4.3 shows the probability that this procedure leads to rejecting the hypothesis for a sample of size less than or equal to n, as a function of n, if the hypothesis is, in fact, true. These probabilities were obtained as relative frequencies by a simulation experiment based on 10000 simulations.

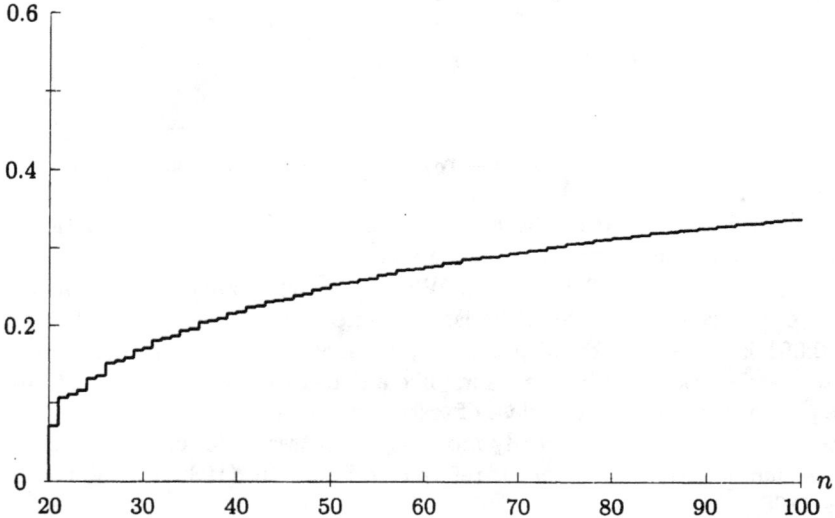

Figure 4.3   Shows the probability of rejecting the hypothesis
for $n = 20, 21, \ldots, 100$.

Historical Remark 4.1.7. Early examples of statistical tests occur in Arbuthnot (1710) and in D. Bernoulli (1734). The rejection probability of a test under alternatives as a criterion for evaluating its power was explicitly introduced by Neyman and Pearson (1928, 1933a). The idea to use randomized tests in order to fully exhaust the error probability α occurs in Anscombe (1948, p. 192) and in the thesis of Eudey (1949). It was further developed by Tocher (1950).

4.2 Critical functions, critical regions

Discusses the arguments given in literature for the use of critical functions as a generalization of critical regions.

In Section 4.1, the critical function φ was introduced as assigning to the observation x the probability $\varphi(x)$ for rejecting the hypothesis. We remark that this test procedure may also be interpreted as a test procedure based on the critical region $C := \{(x, u) \in X \times (0, 1) : \varphi(x) > u\}$.

Interpretation 1. Each p-measure P is replaced by the "smoothed" version $P \times U$ (where U is the uniform distribution over $(0, 1)$), and the critical function $\varphi : X \to [0, 1]$ is replaced by the critical region $C \subset X \times (0, 1)$.

Interpretation 2. For $u \in (0, 1)$ let

$$C_u := \{x \in X : \varphi(x) > u\}. \qquad (4.2.1)$$

Instead of rejecting the hypothesis with probability $\varphi(x)$, one selects first a critical region C_u by determining a realization u from the uniform distribution over $(0, 1)$, and rejects the hypothesis if $x \in C_u$.

These two interpretations are justified by the following relations:

$$\int_0^1 P(C_u)du = P \times U(C) = \int U\{u \in (0, 1) : \varphi(x) > u\}P(dx)$$

$$= \int \varphi(x)P(dx).$$

If the critical function attains only the values 0, γ and 1 as in (4.3.4), the family $\{C_u : u \in (0, 1)\}$ defined by (4.2.1) is most simple. We have

$$C_u = \begin{matrix} \{\varphi > 0\} \\ \{\varphi = 1\} \end{matrix} \quad \text{for} \quad \begin{matrix} 0 < u < \gamma \\ \gamma \leq u < 1, \end{matrix}$$

i.e. the critical region $\{\varphi > 0\}$ is chosen with probability γ, the critical region $\{\varphi = 1\}$ with probability $(1 - \gamma)$.

The original motivation for introducing critical functions was that an admitted error probability α cannot always be fully used by critical regions, which is somewhat disturbing for certain theoretical considerations. Think, as an example, of the problem of testing P_0 against the alternative P_1. If one wishes to compare the power functions of two different critical regions C_i, $i = 1, 2$, fulfilling $P_0(C_i) \leq \alpha$, one usually finds that $P_0(C_1) \neq P_0(C_2)$, so that there is no basis for comparing $P_1(C_1)$ and $P_1(C_2)$.

One could think that — for families of nonatomic measures — every critical function φ can be replaced by a critical region C with the same power function, i.e. $P(C) = P(\varphi)$ for $P \in \mathfrak{P}$. This holds true if $\mathfrak{P}$ is finite, a result due to Dvoretzky, Wald and Wolfowitz (1950, 1951) which follows easily from Ljapunov's Theorem (see Karlin and Studden, 1966, p. 266, Theorem 12.1). Regrettably, this result does not even extend to families $\mathfrak{P}$ which are countable. This would not yet exclude what is of interest for statistical theory, namely that the class of all critical regions is complete for families of nonatomic p-measures (in the sense that for any critical function φ there is a critical region C such that

$$P(C) \underset{\geq}{\leq} P(\varphi) \qquad \text{for} \qquad P \underset{\notin}{\in} \mathfrak{P}_0 .$$

To demonstrate that this is not the case it suffices to produce an example of a critical function φ_0 (not being an indicator function a.e.) which is most powerful in the sense that for any $\varphi \in \Phi$,

$$P(\varphi) \underset{\geq}{\leq} P(\varphi_0) \quad \text{for} \quad P \underset{\notin}{\in} \mathfrak{P}_0 \quad \text{implies} \quad \varphi = \varphi_0 \ \mathfrak{P}\text{-a.e.}$$

(The example uses a family of the power of the continuum, because this is more natural; it is obvious from the proof that the example also works with any countable dense subfamily.)

Example 4.2.2. Let $X = (-1,1)$ and let U be the uniform distribution over X. Let P_ϑ be the p-measure with U-density

$$p(x, \vartheta) = \begin{cases} \frac{1}{2} + 1_{(\vartheta-1,\vartheta)} & \text{for } \vartheta \in [0,1] \\ \frac{1}{2} + 1_{(-1,\vartheta)} + 1_{(\vartheta+1,1)} & \text{for } \vartheta \in [-1,0) \end{cases}$$

Let φ_0 be a critical function with $\varphi_0(x) = 0$ for $x \in (-1,0)$. We shall show that any other critical function which is at least as good as φ_0 for the test problem $P_0 : \{P_\vartheta : \vartheta \in [-1,1], \ \vartheta \neq 0\}$ (i.e. which fulfills $P_0(\varphi) \leq P_0(\varphi_0)$ and $P_\vartheta(\varphi) \geq P_\vartheta(\varphi_0)$ for $\vartheta \neq 0$) agrees with φ_0 a.e. In particular: There is no critical region at least as good as the critical function $\varphi_0 = 4\alpha 1_{[0,1)}$, with $\alpha \in (0, 1/4]$.

Proof. For any critical function φ we have

$$P_\vartheta(\varphi) = \frac{1}{2} U(\varphi) + U\big(\varphi 1_{(\vartheta-1,\vartheta)}\big) \qquad \text{for } \vartheta \in [0,1].$$

From $P_0(\varphi) \leq P_0(\varphi_0)$, we obtain $U(\varphi) \leq U(\varphi_0)$. From $P_1(\varphi) \geq P_1(\varphi_0)$, we obtain $U(\varphi) \geq U(\varphi_0)$. Hence we have

$$U(\varphi) = U(\varphi_0) \quad \text{and} \quad U\big(\varphi 1_{(-1,0)}\big) = 0,$$

where the second equality follows from the first since $P_0(\varphi) \leq P_0(\varphi_0)$.
This implies

$$P_\vartheta(\varphi) = \frac{1}{2} U(\varphi_0) + U\big(\varphi 1_{(0,\vartheta)}\big) \qquad \text{for } \vartheta \in (0,1]$$

and

$$P_\vartheta(\varphi) = \frac{1}{2} U(\varphi_0) + U\big(\varphi 1_{(\vartheta+1,1)}\big) \qquad \text{for } \vartheta \in [-1,0).$$

Hence $P_\vartheta(\varphi) \geq P_\vartheta(\varphi_0)$ for all $\vartheta \in [-1,1], \ \vartheta \neq 0$, implies

$$U\big(\varphi 1_{(0,\vartheta)}\big) \geq U\big(\varphi_0 1_{(0,\vartheta)}\big) \qquad \text{for } \vartheta \in (0,1],$$

and

$$U\big(\varphi 1_{(\vartheta+1,1)}\big) \geq U\big(\varphi_0 1_{(\vartheta+1,1)}\big) \qquad \text{for } \vartheta \in [-1,0).$$

Since $U\big(\varphi 1_{[0,1)}\big) = U\big(\varphi_0 1_{[0,1)}\big)$, this implies

$$U\big(\varphi 1_{[0,\vartheta)}\big) = U\big(\varphi_0 1_{[0,\vartheta)}\big) \qquad \text{for all } \vartheta \in (0,1].$$

Hence $\varphi = \varphi_0$ U–a.e., and $P_\vartheta(\varphi) = P_\vartheta(\varphi_0)$ for all $\vartheta \in [-1,1]$. $\qquad\square$

Another aspect which makes the use of critical functions mathematically more convenient (as compared to the use of critical regions) is that the function $\varphi \to P_*(\varphi)$ attains its supremum on $\Phi_\alpha = \{\varphi \in \Phi : P(\varphi) \leq \alpha \text{ for } P \in \mathfrak{P}_0\}$, whereas the function $C \to P_*(C)$ does not necessarily attain its supremum on $\{C \in \mathcal{A} : P(C) \leq \alpha \text{ for } P \in \mathfrak{P}_0\}$.

Theorem 4.2.3. *Let $\mathfrak{P}_0$ be dominated. For each $P \in \mathfrak{P}_0$ let I_P be a closed subset of $[0,1]$. If $\Phi_0 = \{\varphi \in \Phi : P(\varphi) \in I_P \text{ for } P \in \mathfrak{P}_0\}$ is nonempty, then for every p–measure $P_*|\mathcal{A}$ there exists $\varphi_* \in \Phi_0$ such that $P_*(\varphi_*) = \sup\{P_*(\varphi) : \varphi \in \Phi_0\}$.*

The relevant special cases are $I_P = [0,\alpha]$ and $I_P = \{\alpha\}$.

Proof. Let $\varphi_n \in \Phi_0$, $n \in \mathbb{N}$, be such that

$$\lim_{n\to\infty} P_*(\varphi_n) = \sup\{P_*(\varphi) : \varphi \in \Phi_0\}. \qquad (4.2.4)$$

By the Weak Compactness Lemma 4.2.6 there exists a subsequence $\mathbb{N}_0 \subset \mathbb{N}$ and a "limit" $\varphi_* \in \Phi$ such that

$$\lim_{n\in\mathbb{N}_0} P(\varphi_n) = P(\varphi_*) \qquad \text{for } P \in \mathfrak{P}_0 \text{ and } P = P_*. \qquad (4.2.5)$$

Relation (4.2.5) implies $P(\varphi_*) \in I_P$ for $P \in \mathfrak{P}_0$, hence $\varphi_* \in \Phi_0$. For $P = P_*$, (4.2.5), together with (4.2.4), implies $P_*(\varphi_*) = \sup\{P_*(\varphi) : \varphi \in \Phi_0\}$. $\qquad\square$

For the applied statistician such results are of no great concern. Consider, for instance, a sequence of alternatives P_n, $n \in \mathbb{N}$, such that $\sup\{P_n^n(\varphi) : \varphi \in \Phi_n, \ P^n(\varphi) \leq \alpha \text{ for } P \in \mathfrak{P}_0\}$ converges to some limit in $(0,1)$, where Φ_n denotes the family of all $\mathcal{A}^n$–measurable critical functions on X^n. Then, in "regular" cases, $\sup\{P_n^n(C) : C \in \mathcal{A}^n, \ P^n(C) \leq \alpha \text{ for } P \in \mathfrak{P}_0\}$ converges to the same limit. The difference will usually be of the order $O(n^{-1/2})$, and this order can even be made smaller if the p–measures are "smooth".

Weak Compactness Lemma 4.2.6. *If $\mathfrak{P}$ is dominated by a σ–finite measure, the following holds true: For every sequence $\varphi_n \in \Phi$, $n \in \mathbb{N}$, there exists a* *subsequence $\mathbb{N}_0$ and a "limit" $\varphi_* \in \Phi$ such that*

$$\lim_{n \in \mathbb{N}_0} P(\varphi_n) = P(\varphi_*) \qquad \text{for } P \in \mathfrak{P}.$$

This Lemma is, in fact, an easy consequence of Alaoglu's theorem. In the following we present the usual proof which uses measure theoretic tools only.

Proof. W.l.g. we assume that the dominating measure, say μ, is a p–measure. At first we prove the assertion under the assumption that $\mathcal{A}$ is countably generated. Let $\{A_m : m \in \mathbb{N}\}$ be a countable algebra generating $\mathcal{A}$.

Since $0 \le \mu(\varphi_n 1_{A_m}) \le \mu(A_m)$, there exists for every $m \in \mathbb{N}$ a subsequence $\mathbb{N}_m \subset \mathbb{N}$ with the following properties: $\mathbb{N}_m \subset \mathbb{N}_{m-1}$ for $m \ge 2$, and $\mu(\varphi_n 1_{A_m})$, $n \in \mathbb{N}_m$, converges. Let $\mathbb{N}_0$ denote the "diagonal" sequence, consisting of the m–th element of $\mathbb{N}_m$, for $m \in \mathbb{N}$. Since $\mathbb{N}_0$ is eventually a subset of $\mathbb{N}_m$, $\mu(\varphi_n 1_{A_m})$, $n \in \mathbb{N}_0$, converges for every $m \in \mathbb{N}$.

For every $A \in \mathcal{A}$ and every $\varepsilon > 0$ there exists A_m such that $\mu(A \triangle A_m) < \varepsilon$. From this and the convergence of $\mu(\varphi_n 1_{A_m})$, $n \in \mathbb{N}_0$, for every $m \in \mathbb{N}$ one concludes that $\mu(\varphi_n 1_A)$, $n \in \mathbb{N}_0$, converges for every $A \in \mathcal{A}$.

Let $\nu(A) := \lim_{n \in \mathbb{N}_0} \mu(\varphi_n 1_A)$, $A \in \mathcal{A}$. Since

$$0 \le \nu(A) \le \mu(A) \qquad \text{for } A \in \mathcal{A}, \tag{4.2.7}$$

the relation $B_k \downarrow \emptyset$ implies $\nu(B_k) \downarrow 0$. Since $\nu|\mathcal{A}$ is additive, it is, therefore, σ–additive. Since ν is absolutely continuous with respect to μ by (4.2.7), there exists a μ–density of ν, say φ:

$$\nu(A) = \mu(\varphi 1_A), \qquad \text{for } A \in \mathcal{A}, \tag{4.2.8}$$

which implies

$$\lim_{n \in \mathbb{N}_0} \mu(\varphi_n 1_A) = \mu(\varphi 1_A) \qquad \text{for } A \in \mathcal{A}. \tag{4.2.9}$$

As a consequence of (4.2.7) and (4.2.8) we have

$$0 \le \varphi \le 1 \quad \mu\text{–a.e., i.e. } \varphi \in \Phi.$$

Relation (4.2.9) implies

$$\lim_{n \in \mathbb{N}_0} \mu(\varphi_n f) = \mu(\varphi f) \qquad \text{for every } f \in \mathcal{L}_1(X, \mathcal{A}, \mu). \tag{4.2.10}$$

So far this result was obtained under the assumption that $\mathcal{A}$ is countably generated. It will now be extended to an arbitrary σ–algebra $\mathcal{A}$.

Let $\mathcal{A}_0$ denote the smallest sub-σ-algebra of $\mathcal{A}$ containing all sets $\varphi_n^{-1}(-\infty, r]$ with $n \in \mathbb{N}$, $r \in \mathbb{Q}$. Since $\mathcal{A}_0$ is countably generated, (4.2.10) holds for $f \in \mathcal{L}_1(X, \mathcal{A}_0, \mu)$, with an $\mathcal{A}_0$-measurable $\varphi \in \Phi$. To extend (4.2.10) to arbitrary $f \in \mathcal{L}_1(X, \mathcal{A}, \mu)$, let f_0 be a conditional expectation of f, given $\mathcal{A}_0$, with respect to μ. Since $\mu(gf) = \mu(gf_0)$ for arbitrary $\mathcal{A}_0$-measurable bounded functions g, we have $\mu(\varphi_n f) = \mu(\varphi_n f_0)$ and $\mu(\varphi f) = \mu(\varphi f_0)$. Since (4.2.10) holds for f_0, it holds for $f \in \mathcal{L}_1(X, \mathcal{A}, \mu)$.   □

4.3 The Neyman–Pearson Lemma

Introduces critical functions of Neyman–Pearson type and most powerful critical functions, and establishes the equivalence of these concepts by the fundamental lemma of Neyman and Pearson.

The problem of finding a critical function φ which maximizes $P_1(\varphi)$ under the condition $P_0(\varphi) \le \alpha$ has an explicit solution which is given by the so-called fundamental lemma of Neyman and Pearson (1933a, p. 300). In the following we present a slightly generalized version which is technically more convenient.

Throughout the following, $f_0 : X \to [0, \infty)$ is a measurable function fulfilling $\mu(f_0) = 1$, and $f_1 : X \to \mathbb{R}$ a μ-integrable function.

Definition 4.3.1. A critical function φ is of *Neyman–Pearson type* for $f_0 : f_1$ if there exists $c \in \mathbb{R}$ such that

$$\varphi(x) = \begin{cases} 1 \\ 0 \end{cases} \text{ if } f_1(x) \overset{>}{\underset{<}{}} c f_0(x).$$

This definition requires nothing about $\varphi(x)$ for $f_1(x) = c f_0(x)$ (except $\varphi(x) \in [0, 1]$, of course).

It is intuitively clear that the critical functions of Neyman–Pearson type are just those which maximize $\varphi \to \mu(f_1\varphi)$ under the side condition $\mu(f_0\varphi) \le \alpha$. This is made precise in the Neyman–Pearson Lemma 4.3.3.

Definition 4.3.2. (i) φ_0 is *most powerful* for $f_0 : f_1$ if for every $\varphi \in \Phi$, $\mu(f_0\varphi) \le \mu(f_0\varphi_0)$ and $\mu(f_1\varphi) \ge \mu(f_1\varphi_0)$, implies $\mu(f_0\varphi) = \mu(f_0\varphi_0)$ and $\mu(f_1\varphi) = \mu(f_1\varphi_0)$.
(ii) φ_0 is *most powerful* for $P_0 : P_1$ if (i) applies with $f_i \in dP_i/d\mu$.

Lemma 4.3.3 (Neyman–Pearson).
(i)   *For every $\alpha \in (0, 1)$ there exists a critical function φ_0 of Neyman–Pearson type such that $\mu(f_0\varphi_0) = \alpha$.*
(ii')   *A critical function of Neyman–Pearson type with $c \ge 0$ is most powerful.*

(ii'') *A most powerful critical function φ with $\mu(f_0\varphi) \in (0,1)$ is of Neyman–Pearson type μ–a.e.*

Proof. (i) For $c \in \mathbb{R}$ let $A(c) := \{f_1 > cf_0\}$ and $B(c) := \{f_1 \geq cf_0\}$.

It is straightforward to show for every $\alpha \in (0,1)$ the existence of c_α such that

$$\mu\big(f_0 1_{A(c_\alpha)}\big) \leq \alpha \leq \mu\big(f_0 1_{B(c_\alpha)}\big).$$

Hence there exists γ_α such that $\mu\big(f_0\big[(1-\gamma_\alpha)1_{A(c_\alpha)} + \gamma_\alpha 1_{B(c_\alpha)}\big]\big) = \alpha$.

$$\varphi_0 := (1-\gamma_\alpha)1_{A(c_\alpha)} + \gamma_\alpha 1_{B(c_\alpha)} \tag{4.3.4}$$

is a critical function of Neyman–Pearson type since

$$\varphi_0(x) = \begin{cases} 1 \\ 0 \end{cases} \quad \text{for} \quad f_1(x) \gtrless c_\alpha f_0(x).$$

(ii') Let φ_0 be of Neyman–Pearson type with $c \geq 0$. By Definition 4.3.1 the following relation holds for every critical function φ:

$$\big(\varphi_0(x) - \varphi(x)\big)\big(f_1(x) - cf_0(x)\big) \geq 0, \qquad x \in X. \tag{4.3.5}$$

Hence

$$\mu(f_1\varphi_0) - \mu(f_1\varphi) \geq c\big(\mu(f_0\varphi_0) - \mu(f_0\varphi)\big), \tag{4.3.6}$$

which implies the assertion.

(ii'') Assume that φ is most powerful with $\mu(f_0\varphi) \in (0,1)$. According to (i) there exists a critical function φ_0 of Neyman–Pearson type such that $\mu(f_0\varphi_0) = \mu(f_0\varphi)$. Hence (4.3.6) implies $\mu(f_1\varphi_0) \geq \mu(f_1\varphi)$. Since φ is most powerful, this implies $\mu(f_1\varphi_0) = \mu(f_1\varphi)$. Therefore, the μ–integral of the left hand side of (4.3.5) equals 0 μ–a.e. Hence $\varphi = \varphi_0$ μ–a.e. on $\{f_1 \neq cf_0\}$. Since φ_0 is of Neyman–Pearson type, the assertion follows.   □

Proposition 4.3.7. *If φ with $P_0(\varphi) \in (0,1)$ is most powerful for P_0:P_1, then*

$$P_0(\varphi) < P_1(\varphi) \qquad \text{unless } P_0 = P_1. \tag{4.3.8}$$

Proof. Let f_i be a μ–density of P_i, $i = 0,1$. Let $\psi \equiv \mu(f_0\varphi)$. We have $\mu(f_0\psi) = \mu(f_0\varphi)$. Since φ is most powerful, $P_0(\varphi) \geq P_1(\varphi)$, i.e. $\mu(f_1\psi) \geq \mu(f_1\varphi)$, implies $\mu(f_1\psi) = \mu(f_1\varphi)$, hence ψ is most powerful, too, and therefore by the Neyman–Pearson lemma of Neyman–Pearson type. Since $\psi(x) \in (0,1)$ for all $x \in X$ this implies $f_1 = cf_0$ μ–a.e. Since $\mu(f_i) = 1$ for $i = 0,1$, we obtain $c = 1$, i.e. $f_1 = f_0$ μ–a.e.   □

Lemma 4.3.9. *For $i = 0, 1$ let P_i be a p-measure with μ-density f_i. Let $A = \{f_1 > cf_0\}$ or $A = \{f_1 \geq cf_0\}$. If*

$$P_i(\varphi) = P_i(A), \qquad i = 0, 1, \tag{4.3.10}$$

for some $\varphi \in \Phi$, then

$$\varphi = 1_A \quad (P_0 + P_1)\text{-}a.e. \tag{4.3.11}$$

Proof. We consider the case $A = \{f_1 > cf_0\}$. Since $(1_A - \varphi)(f_1 - cf_0) \geq 0$, relation (4.3.10) implies

$$(1_A - \varphi)(f_1 - cf_0) = 0 \qquad (P_0 + P_1)\text{-}a.e.$$

Now $x \in A$ implies $f_1(x) - cf_0(x) \neq 0$, hence $\varphi(x) = 1 = 1_A(x)$ for $(P_0 + P_1)$-a.a. $x \in A$. Since $\varphi(x) \geq 0 = 1_A(x)$ for all $x \in A^c$, we have $\varphi(x) \geq 1_A(x)$ for $(P_0 + P_1)$-a.a. $x \in X$. Together with (4.3.10) this implies (4.3.11). $\square$

Lemma 4.3.12. *If $P_0(\varphi_0) \leq \alpha$ and $P_1(\varphi_0) = \sup\{P_1(\varphi) : \varphi \in \Phi, P_0(\varphi) \leq \alpha\} < 1$, then $P_0(\varphi_0) = \alpha$.*

Proof. If $\alpha_0 := P_0(\varphi_0) < \alpha$, define $\psi_0 = (1 - \varepsilon)\varphi_0 + \varepsilon$, with $\varepsilon := (\alpha - \alpha_0)/(1 - \alpha_0) \in (0, 1)$. We have $\psi_0 \in \Phi$, $P_0(\psi_0) = \alpha$, and

$$P_1(\psi_0) = (1 - \varepsilon)P_1(\varphi_0) + \varepsilon > P_1(\varphi_0). \qquad \square$$

Historical Remark 4.3.13. Starting with Neyman and Pearson (1936, p. 11) and Dantzig and Wald (1951) various generalizations of the Fundamental Lemma (to more than one side condition) have been obtained (see Lehmann, 1986, Section 3.6, pp. 96–101 or Witting, 1985, Abschnitt 2.4.1 , pp. 254–256).

4.4 Optimal tests for composite hypotheses

Theorem 4.2.3 establishes the existence of a test which is most powerful for testing a hypothesis $\mathfrak{P}_0$ against a particular alternative. In the present section it is shown how such a most powerful test can be obtained in certain cases by the use of "least favorable" prior distributions.

The Neyman–Pearson lemma shows that a test is most powerful for $P_0 : P_1$ if and only if it is of Neyman–Pearson type. Against that, Theorem 4.2.3 establishes the existence of a test which is most powerful for P_* in the class of all level-α-tests for $\mathfrak{P}_0$, but it says nothing about how to obtain such a most powerful test, and nothing about its power function.

Occasionally, these goals can be reached by the following device. It requires the concept of a mixture of p-measures in $\mathfrak{P}_0$. To present these considerations in a technically convenient way, we think of $\mathfrak{P}_0$ as a parametric family, say $\{P_\vartheta : \vartheta \in \Theta\}$, where the parameter space Θ is endowed with a σ-algebra $\mathcal{B}$. We assume that $\vartheta \to P_\vartheta(A)$ is measurable for every $A \in \mathcal{A}$. Then we may define for any p-measure $\Lambda|\mathcal{B}$ the mixture P_Λ by

$$P_\Lambda(A) := \int P_\vartheta(A)\Lambda(d\vartheta), \qquad A \in \mathcal{A}. \tag{4.4.1}$$

A straightforward proof shows that $P_\Lambda|\mathcal{B}$ is a p-measure. If $p(\cdot, \vartheta)$ is a μ-density of P_ϑ, then $x \to \int p(x,\vartheta)\Lambda(d\vartheta)$ is a μ-density of P_Λ provided $(x,\vartheta) \to p(x,\vartheta)$ is measurable.

Proposition 4.4.2. *Assume that a test φ_0 has the following properties.*
(a) $P_\vartheta(\varphi_0) \le \alpha$ *for every $P_\vartheta \in \mathfrak{P}_0$.* $\hfill (4.4.3)$
(b) φ_0 *maximizes $P_*(\varphi)$ under the condition $P_{\Lambda_0}(\varphi) \le \alpha$,*
      for some p-measure $\Lambda_0|\mathcal{B}$. $\hfill (4.4.4)$

   Then the following holds true.
(i)   *φ_0 maximizes $P_*(\varphi)$ under the condition $P_\vartheta(\varphi) \le \alpha$ for $P_\vartheta \in \mathfrak{P}_0$.*
(ii) *$\Lambda_0\{\vartheta \in \Theta : P_\vartheta(\varphi_0) = \alpha\} = 1$ if $P_*(\varphi_0) < 1$.*
(iii) *Λ_0 is least favorable in the following sense:*

$$\sup\{P_*(\varphi) : \varphi \in \Phi, \ P_\Lambda(\varphi) \le \alpha\} \ge P_*(\varphi_0) \ \text{for every p-measure $\Lambda|\mathcal{B}$.}$$

Proof. (i) $P_\vartheta(\varphi) \le \alpha$ for every $P_\vartheta \in \mathfrak{P}_0$ implies $P_{\Lambda_0}(\varphi) \le \alpha$, hence $P_*(\varphi) \le P_*(\varphi_0)$.

(ii) Since φ_0 maximizes $P_*(\varphi)$ under the side condition $P_{\Lambda_0}(\varphi_0) \le \alpha$, we have $P_{\Lambda_0}(\varphi_0) = \alpha$ by Lemma 4.3.12. Since $P_\vartheta(\varphi_0) \le \alpha$ for $\vartheta \in \Theta$, this implies $P_\vartheta(\varphi_0) = \alpha$ for Λ_0-a.a. $\vartheta \in \Theta$.

(iii) Let $\Lambda|\mathcal{B}$ be an arbitrary p-measure. $P_\vartheta(\varphi_0) \le \alpha$ for every $P_\vartheta \in \mathfrak{P}_0$ implies $P_\Lambda(\varphi_0) \le \alpha$. Hence

$$\sup\{P_*(\varphi) : \varphi \in \Phi, \ P_\Lambda(\varphi) \le \alpha\} \ge P_*(\varphi_0). \qquad \square$$

Part (iii) of Proposition 4.4.2 makes evident the intuitive idea underlying this device: To determine Λ such that the mixture P_Λ is as similar as possible to P_*, so that P_Λ and P_* are particularly difficult to distinguish by a test. It justifies the term "least favorable (prior) distribution" for Λ_0.

Whereas a test maximizing $P_*(\varphi)$ under the condition $P(\varphi) \le \alpha$ for $P \in \mathfrak{P}_0$ always exists (see Theorem 4.2.3), such a maximizing critical function can be obtained as a test for $P_{\Lambda_0} : P_*$ only under certain conditions (for instance if Θ is compact).

The basic idea underlying the use of least favorable distributions is due to Wald's decision theory (see Wald, 1945). Proposition 4.4.2 goes back to Lehmann and Stein (1948, Section 2). Among the papers studying the existence of least favorable prior distributions only a few are statistically relevant. (See Lehmann, 1952, Krafft and Witting, 1967).

Even if the least favorable prior distribution Λ_0 exists, it may not be representable in closed form, hence unfit for determining P_{Λ_0} and the most powerful test for $P_{\Lambda_0}:P_*$.

In spite of these difficulties, this device has proved applicable in several instances. (See, in particular, Lehmann and Stein, 1948.)

Example 4.4.5. Testing the hypothesis $\sigma = \sigma_0$ for the family $\{N^n_{(\mu,\sigma^2)} : \mu \in \mathbb{R}, \sigma^2 > 0\}$.

(i) We consider the alternative $P_1 = N^n_{(\mu_1,\sigma_1^2)}$ with $\mu_1 \in \mathbb{R}$ and $\sigma_1 > \sigma_0$.

Using the prior distribution $\Lambda_0 = N_{(\mu_1,\sigma_n^2)}$ over the parameter set $\mathbb{R} \times \{\sigma_0\}$, with $\sigma_n^2 = (\sigma_1^2 - \sigma_0^2)/n$, we obtain the following critical region C_n which is most powerful for testing $P_{\Lambda_0} = \int N^n_{(\mu,\sigma_0^2)}\Lambda_0(d\mu)$ against $N^n_{(\mu_1,\sigma_1^2)}$:

$$C_n := \{\sum_1^n (x_\nu - \bar{x}_n)^2 > \sigma_0^2 k_{n-1,1-\alpha}\} \tag{4.4.6}$$

(with $k_{m,\beta}$ denoting the β–quantile of χ_m^2).

It is easy to check that $N^n_{(\mu,\sigma^2)}(C_n) \leq \alpha$ for $\mu \in \mathbb{R}$ and $\sigma \leq \sigma_0$. Hence Λ_0 is least favorable, and C_n is most powerful for testing the (larger) hypothesis $\mu \in \mathbb{R}$, $\sigma \leq \sigma_0$ against the alternative (μ_1,σ_1). Since C_n does not depend on the special alternative, it is uniformly most powerful for all alternatives (μ,σ) with $\mu \in \mathbb{R}$, $\sigma > \sigma_0$.

(ii) Now we consider the alternative $P_1 = N^n_{(\mu_1,\sigma_1^2)}$ with $\mu_1 \in \mathbb{R}$, $\sigma_1 < \sigma_0$.

Using the (degenerate) prior distribution Λ_0, with $\Lambda_0(\{(\mu_1,\sigma_0)\}) = 1$, we obtain the following critical region $\hat{C}_n(\mu_1)$ which is most powerful for testing $P_{\Lambda_0} = N^n_{(\mu_1,\sigma_0^2)}$ against $N^n_{(\mu_1,\sigma_1^2)}$:

$$\hat{C}_n(\mu_1) := \{\sum_1^n (x_\nu - \mu_1)^2 < \sigma_0^2 k_{n,\alpha}\}. \tag{4.4.7}$$

It is easy to check that $N^n_{(\mu,\sigma^2)}(\hat{C}_n(\mu_1)) \leq \alpha$ for $\mu \in \mathbb{R}$ and $\sigma \geq \sigma_0$. Hence Λ_0 is least favorable, and $\hat{C}_n(\mu_1)$ is most powerful for testing the (larger) hypothesis $\mu \in \mathbb{R}$, $\sigma \geq \sigma_0$ against the alternative (μ_1,σ_1). Regrettably, the critical region $\hat{C}_n(\mu_1)$ depends through μ_1 on the alternative in a crucial way.

The conclusion of this example: Maximizing the power for a particular alternative may lead to a test which is useless for all practical purposes. No applied statistician would ever think of using a critical region $\hat{C}_n(\mu_1)$ instead of $C'_n := \{\sum_1^n (x_\nu - \bar{x}_n)^2 < \sigma_0^2 k_{n-1,\alpha}\}$. The critical region $\hat{C}_n(\mu_1)$ is superior to C'_n only in a neighborhood of μ_1 which is of the order $n^{-3/4}$, and C'_{n+1} is superior to $\hat{C}_n(\mu_1)$ for all alternatives $N^n_{(\mu,\sigma^2)}$ with $\mu \in \mathbb{R}$, $\sigma < \sigma_0$. That the power $N^n_{(\mu,\sigma^2)}(\hat{C}_n(\mu_1))$ is so low for $\mu \neq \mu_1$ finds its natural explanation in the fact that $N^n_{(\mu,\sigma_0^2)}(\hat{C}_n(\mu_1))$ drops down to 0 as μ moves away from μ_1.

Figure 4.4 shows the power functions of C'_n and $\hat{C}_n(0)$ for $\sigma_0^2 = 1$, $\alpha = 0.1$ and $n = 10$ for $N^n_{(\mu,\sigma^2)}$ with $\sigma_1^2 = k_{n-1,0.1}/k_{n-1,0.9}$.

(This value for σ_1 was chosen to obtain $N^n_{(\mu,\sigma^2)}(C'_n) = 0.9$, a probability for which the difference between $N^n_{(0,\sigma_1^2)}(C'_n)$ and $N^n_{(0,\sigma_1^2)}(\hat{C}_n(0))$ is still visible for $n = 10$.)

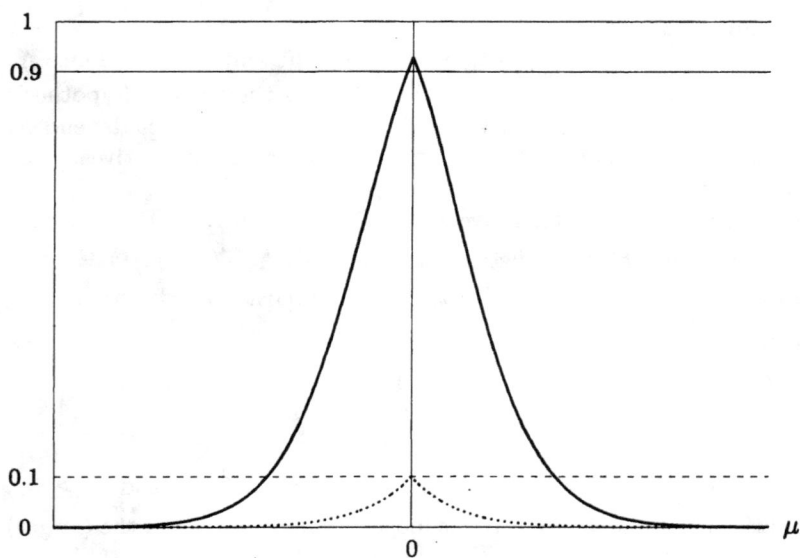

Figure 4.4   ———— $N^n_{(\mu,\sigma_1^2)}(C'_n)$;   ———— $N_{(\mu,\sigma^2)}(\hat{C}_n(0))$.
            ─ ─ ─ $N^n_{(\mu,\sigma_0^2)}(C'_n)$;   ········ $N^n_{(\mu,\sigma_0^2)}(\hat{C}_n(0))$.

4.5 Optimal tests for families with monotone likelihood ratios

For families with monotone likelihood ratios there exist tests for the hypothesis $\vartheta \leq \vartheta_0$ which are uniformly most powerful for alternatives $\vartheta > \vartheta_0$. If the family is exponential with one parameter, this holds true for all sample sizes. One–parameter exponential families are the only ones with this property.

Let $\Theta \subset \mathbb{R}$ be an interval, say $(\vartheta', \vartheta'')$ (including the cases $\vartheta' = -\infty$ and/or $\vartheta'' = \infty$). Let $\{P_\vartheta : \vartheta \in \Theta\}$ be a family of mutually absolutely continuous p-measures on $(X, \mathcal{A})$ admitting a sufficient statistic $S : (X, \mathcal{A}) \to (\mathbb{R}, \mathbb{B})$ with isotone likelihood ratios (see Definition 1.7.10). Since S is sufficient, it is clear that for any critical function $\varphi : (X, \mathcal{A}) \to ([0,1], \mathbb{B}_0)$ there exists a critical function, depending on x through $S(x)$, with the same power function. If the sufficient statistic S has isotone likelihood ratios, the critical functions which are isotone functions of S solve the problem of testing hypotheses $\vartheta \leq \vartheta_0$ in an ideal way. Theorem 4.5.2 is essentially due to Karlin and Rubin (1956, see in particular Theorems 2 and 4). Similar results for exponential families have earlier been obtained by Blackwell and Girshick (1954, p. 182, Lemma 7.4.1). See also Brown, Cohen and Strawderman (1976).

Let $\Psi_S \subset \Phi$ denote the class of all critical functions ψ such that, for some $c \in \mathbb{R}$,

$$S(x) \begin{array}{c} > \\ < \end{array} c \quad \text{implies} \quad \psi(x) = \begin{cases} 1 \\ 0 \end{cases} \quad \text{for } x \in X. \tag{4.5.1}$$

Theorem 4.5.2. *Let $\{P_\vartheta : \vartheta \in \Theta\}$ be a family of mutually absolutely continuous p–measures.*

(i) *For every $\vartheta \in \Theta$ and $\alpha \in (0,1)$ there exists $\psi \in \Psi_S$ with $P_\vartheta(\psi) = \alpha$.*

(ii) *If S is a sufficient statistic with isotone likelihood ratios for $\{P_\vartheta : \vartheta \in \Theta\}$, then every $\psi \in \Psi_S$ is most powerful for testing the hypothesis P_{ϑ_1} against the alternative P_{ϑ_2}, for all $\vartheta_i \in \Theta$ with $\vartheta_1 < \vartheta_2$. Moreover, the power function $\vartheta \to P_\vartheta(\psi)$ is strictly increasing if $\vartheta_1 \neq \vartheta_2$ implies $P_{\vartheta_1} \neq P_{\vartheta_2}$ (unless $P_\vartheta(\psi) \equiv 0$ or $P_\vartheta(\psi) \equiv 1$).*

Theorem 4.5.2 implies, in particular, that the class of tests Ψ_S is complete in the following sense: For every test $\varphi \in \Phi$ and every $\vartheta_0 \in \Theta$ there exists a test $\psi \in \Psi_S$ such that

$$P_\vartheta(\psi) \begin{array}{c} \leq \\ \geq \end{array} P_\vartheta(\varphi) \quad \text{for} \quad \vartheta \begin{array}{c} \leq \\ \geq \end{array} \vartheta_0. \tag{4.5.3}$$

Proof. (i) Apply Lemma 4.3.3(i) for $\mu = P_\vartheta$, $f_0 \equiv 1$, $f_1 = S$.

(ii) Since $\psi \in \Psi_S$, there exists $c \in \mathbb{R}$ fulfilling (4.5.1). Since S is sufficient with isotone likelihood ratios (see (1.7.11)), there exist densities $x \to g(x)h_\vartheta(S(x))$ with $t \to h_{\vartheta_2}(t)/h_{\vartheta_1}(t)$ isotone if $\vartheta_1 < \vartheta_2$. Hence

$$h_{\vartheta_2}(S(x))/h_{\vartheta_1}(S(x)) \gtrless h_{\vartheta_2}(c)/h_{\vartheta_1}(c) \quad \text{implies} \quad S(x) \gtrless c.$$

(If we exclude the trivial cases $P_{\vartheta_1}(\psi) = 1$ and $P_{\vartheta_2}(\psi) = 0$, this constant c is in the convex hull of $S(X)$. Hence $h_{\vartheta_i}(t)$ is defined for $t = c$.)

Together with (4.5.1) this implies that ψ is of Neyman–Pearson type for $P_{\vartheta_1}:P_{\vartheta_2}$. Hence it is most powerful for this test problem by Lemma 4.3.3(ii′).

By Proposition 4.3.7 this implies $P_{\vartheta_1}(\psi) < P_{\vartheta_2}(\psi)$ if $P_{\vartheta_1} \neq P_{\vartheta_2}$ and $P_{\vartheta_1}(\psi) \in (0,1)$. □

According to Theorem 4.5.2, the existence of a sufficient statistic with isotone likelihood ratios implies that for every $\vartheta_0 \in \Theta$ and every $\alpha \in (0,1)$ there exists a test φ with $P_{\vartheta_0}(\varphi) = \alpha$ which is most powerful for every test problem $P_{\vartheta_1}:P_{\vartheta_2}$ with $\vartheta_1 < \vartheta_2$. Theorem 4.5.4 shows that tests with such a strong optimum property cannot exist unless there exists a sufficient statistic with isotone likelihood ratios.

Theorem 4.5.4 is a simplified version (under the additional assumption of mutual absolute continuity) of Pfanzagl (1962, p. 110, Satz). See also Schmetterer (1974, p. 191, Theorem 5.3), Heyer (1982, p. 89, Theorem 13.7), Strasser (1985, pp. 52/3, Theorem 9.17). For a generalization to not necessarily dominated families of p–measures see Mussmann (1987).

Theorem 4.5.4. *Assume that $\{P_\vartheta : \vartheta \in \Theta\}$ is a family of mutually absolutely continuous p–measures with the following property:*

For every $\vartheta_0 \in \Theta$ and $\alpha \in (0,1)$ there exists $\varphi_0 \in \Phi$ with $P_{\vartheta_0}(\varphi_0) = \alpha$ which is most powerful for every test problem $P_{\vartheta_1}:P_{\vartheta_2}$ with $\vartheta_1 < \vartheta_2$.

Then there exists a sufficient statistic with isotone likelihood ratios.

Proof. Let P be a p–measure equivalent to P_ϑ, $\vartheta \in \Theta$, and let $p(\cdot, \vartheta)$ be a P–density of P_ϑ, $\vartheta \in \Theta$. Let $\mathcal{C}$ denote the class of all nondegenerate sets $\{p(\cdot, \tau) \leq cp(\cdot, \vartheta)\}$, $\vartheta, \tau \in \Theta$ with $\vartheta < \tau$, $c > 0$. (Nondegenerate means here: not of measure 0 or 1.) We shall show that $\mathcal{C}$ is totally ordered by inclusion (P). Let $A_0, A_1 \in \mathcal{C}$ be arbitrary, say $A_i = \{p(\cdot, \tau_i) \leq c_i p(\cdot, \vartheta_i)\}$, $i = 0, 1$. By assumption there exists $\psi_0 \in \Phi$ which is most powerful for $P_{\vartheta_1}:P_{\vartheta_2}$ for all $\vartheta_1 < \vartheta_2$ and fulfills

$$P_{\vartheta_0}(\psi_0) = P_{\vartheta_0}(A_0^c). \tag{4.5.5}$$

Since ψ_0 is most powerful for $P_{\vartheta_0} : P_{\tau_0}$, Lemma 4.3.3(ii″) implies

$$\psi_0 = 1_{A_0^c} \quad P\text{-a.e.} \tag{4.5.6}$$

hence

$$\{\psi_0 = 0\} = \{\psi_0 < 1\} = A_0 \quad (P).$$

Since ψ_0 is also most powerful for $P_{\vartheta_1} : P_{\tau_1}$, the Neyman–Pearson Lemma 4.3.3(ii′) implies the existence of $k \in \mathbb{R}$ such that

$$\{p(\cdot, \tau_1) < kp(\cdot, \vartheta_1)\} \subset \{\psi_0 = 0\} \quad (P) \tag{4.5.7′}$$

and

$$\{\psi_0 < 1\} \subset \{p(\cdot, \tau_1) \le kp(\cdot, \vartheta_1)\} \quad (P). \tag{4.5.7″}$$

Hence

$$A_1 \subset \{\psi_0 = 0\} \quad (P) \quad \text{if } c_1 < k \tag{4.5.8′}$$

respectively

$$\{\psi_0 < 1\} \subset A_1 \quad (P) \quad \text{if } c_1 \ge k. \tag{4.5.8″}$$

Together with (4.5.6) this implies $A_1 \subset A_0$ (P) or $A_0 \subset A_1$ (P).

By Lemma 1.7.27 there exists a measurable function $T : X \to [0,1]$ such that

$$A = \{T \le P(A)\} \quad (P) \quad \text{for every } A \in \mathcal{C}. \tag{4.5.9}$$

Let $\vartheta, \tau \in \Theta$ with $\vartheta < \tau$ be arbitrary. For $c > 0$ let

$$A_{\vartheta, \tau}(c) := \{p(\cdot, \tau) \le cp(\cdot, \vartheta)\}. \tag{4.5.10}$$

Since $c \to P(A_{\vartheta, \tau}(c))$ is isotone and continuous from the right, there exists (see Lemma 1.7.40) an isotone function $H_{\vartheta, \tau}$ such that $r \le P(A_{\vartheta, \tau}(c))$ iff $H_{\vartheta, \tau}(r) \le c$. Hence (4.5.9) implies $A_{\vartheta, \tau}(c) = \{T \le P(A_{\vartheta, \tau}(c))\} = \{H_{\vartheta, \tau} \circ T \le c\}$ (P) for every $c > 0$. Together with (4.5.10) this implies (notice that $P\{p(\cdot, \vartheta) = 0\} = 0$ and $P\{p(\cdot, \tau) = 0\} = 0$)

$$\{p(\cdot, \tau)/p(\cdot, \vartheta) \le c\} = \{H_{\vartheta, \tau} \circ T \le c\} \quad (P) \quad \text{for every } c \ge 0.$$

From this, $p(\cdot, \tau)/p(\cdot, \vartheta) = H_{\vartheta, \tau} \circ T$ (P). □

The statistical relevance of Theorem 4.5.4 is limited by the fact that it presumes more properties of φ than one would like to require of a "good" test: We would certainly be satisfied with a test for $\vartheta \le \vartheta_0$ which maximizes the power simultaneously for all $\vartheta > \vartheta_0$. That it minimizes, in addition, the power simultaneously for all $\vartheta < \vartheta_0$ is a welcome, but not indispensable attribute of a good test. The tests described in the assertion of Theorem 4.5.2 and in the assumption of Theorem 4.5.4 do much more: They are most powerful for every test problem $P_{\vartheta_1}\!:\!P_{\vartheta_2}$ with $\vartheta_1 < \vartheta_2$. Example 4.5.11 shows that the assertion of Theorem 4.5.4 is not true any more if we require only the existence of tests which are uniformly most powerful for alternatives $\vartheta > \vartheta_0$.

Example 4.5.11. Let $\mathbb{B}_0$ denote the Borel algebra of $(0,1)$, and let $P_\vartheta|\,\mathbb{B}_0$, $\vartheta \in (0,1)$, be defined by its Lebesgue density

$$p(x,\vartheta) = \begin{cases} (1+\frac{1}{\vartheta})/2 & 0 < x \le \vartheta \\ 1/2 & \vartheta < x < 1. \end{cases}$$

(i) For $\vartheta_0 \in (0,1)$ fixed, let $\Phi_\alpha := \{\varphi \in \Phi : P_\vartheta(\varphi) \le \alpha \text{ for } \vartheta \le \vartheta_0\}$. We shall show that there exists $\varphi_0 \in \Phi_\alpha$ such that $\vartheta > \vartheta_0$ implies $P_\vartheta(\varphi) \le P_\vartheta(\varphi_0)$ for every $\varphi \in \Phi_\alpha$.

Let

$$C_0 = \begin{cases} (\vartheta_0, \vartheta_0 + 2\alpha) \\ (2(1-\alpha)/(1+1/\vartheta_0), 1) \end{cases} \text{if} \quad \begin{array}{l} 0 < \alpha < (1-\vartheta_0)/2 \\ (1-\vartheta_0)/2 \le \alpha < 1. \end{array} \tag{4.5.12}$$

We shall show that $\varphi_0 = 1_{C_0}$ has the asserted properties. That $P_\vartheta(C_0) \le \alpha$ for $\vartheta \le \vartheta_0$ is easy to check. That $\vartheta > \vartheta_0$ implies $P_\vartheta(\varphi) \le P_\vartheta(C_0)$ for $\varphi \in \Phi_\alpha$ follows from the Neyman–Pearson Lemma 4.3.3 since φ_0 is of Neyman–Pearson type for $P_{\vartheta_0}\!:\!P_\vartheta$ for every $\vartheta > \vartheta_0$:

$$p(x,\vartheta) \mathop{\gtrless}^{>}_{<} c(\vartheta,\vartheta_0,\alpha)p(x,\vartheta_0) \quad \text{implies} \quad x \mathop{\in}^{\in}_{\notin} C_0$$

with

$$c(\vartheta,\vartheta_0,\alpha) = \begin{cases} 1 & \vartheta \le \vartheta_0 + 2\alpha < 1 \\ 1+1/\vartheta & \vartheta_0 + 2\alpha < \vartheta < 1 \\ (1+1/\vartheta)/(1+1/\vartheta_0) & (1-\vartheta_0)/2 \le \alpha < 1. \end{cases}$$

(ii) $\{P_\vartheta : \vartheta \in (0,1)\}$ admits no sufficient statistic with monotone likelihood ratios. If it does, then there exists a critical function φ_0 with $P_{\frac{1}{2}}(\varphi_0) = \frac{1}{2}$ which is most powerful for $P_{\vartheta_1}\!:\!P_{\vartheta_2}$ for arbitrary $\vartheta_1 < \vartheta_2$ (by Theorem 4.5.2). φ_0 is, in particular, most powerful for $P_{\frac{1}{2}}\!:\!P_\vartheta$ with $\vartheta > \frac{1}{2}$. According to (4.5.12), $\varphi_0 = 1_{(\frac{1}{3},1)}$ has this property, and it is unique λ–a.e. on $(0,1)$. It is straightforward to check that φ_0 is not of Neyman–Pearson type for $P_\vartheta\!:\!P_{\frac{1}{2}}$ with $\vartheta < 1/3$, hence not most powerful for $P_\vartheta\!:\!P_{\frac{1}{2}}$.

As a consequence of Proposition 1.7.15 and Theorem 4.5.2, the problem of testing the hypothesis $\vartheta \le \vartheta_0$ against $\vartheta > \vartheta_0$ has a satisfactory solution for one–parameter exponential families. Combining Theorems 4.5.4 and 1.7.16 we obtain the following converse

Corollary 4.5.13. *Assume that $\{P_\vartheta : \vartheta \in \Theta\}$ is a family of mutually absolutely continuous p–measures such that a test with the properties specified in Theorem 4.5.4 exists for infinitely many sample sizes, including the sample size 1.*
Then $\{P_\vartheta : \vartheta \in \Theta\}$ is a one–parameter exponential family.

Proof. By Theorem 4.5.4 there exists a sufficient statistic with isotone likelihood ratios for infinitely many sample sizes, including the sample size 1. According to Theorem 1.7.16 this implies exponentiality. □

As indicated above, the properties of the test specified in Theorem 4.5.4 are somewhat stronger than necessary from the statistical point of view. The existence of a test for $\vartheta \leq \vartheta_0$ which is uniformly most powerful for $\vartheta > \vartheta_0$ does not imply exponentiality (according to Example 4.5.11). The situation is different if this holds for infinitely many sample sizes.

We cite the following theorem which is a simplified version of Theorem 1 in Pfanzagl (1968).

Theorem 4.5.14. *Let P_0 be a p–measure and $\mathfrak{P}$ a family of p–measures on a measurable space $(X, \mathcal{A})$. Assume that $P_0 \neq P$ and that P_0, P are mutually absolutely continuous for every $P \in \mathfrak{P}$.*
Assume there exists $P_1 \in \mathfrak{P}$ with P_0–density p_1 such that $P_1((\log p_1)^+) < \infty$ and $P_0((\log p_1)^-) < \infty$.
Assume that for infinitely many sample sizes $\mathbb{N}_0 \subset \mathbb{N}$ (including the sample size $n = 1$) there exists a critical function $\varphi^{(n)} : (X^n, \mathcal{A}^n) \to ([0,1], \mathbb{B}_0)$ with the following properties
(i) $\lim_{n \in \mathbb{N}_0} P_0^n(\varphi^{(n)}) = \alpha \in (0,1)$;
(ii) *for each $n \in \mathbb{N}_0$, $\varphi^{(n)}$ is most powerful for testing the hypothesis P_0^n against the alternatives P^n, for any $P \in \mathfrak{P}$.*
Then $\mathfrak{P}$ is a one–parameter exponential family, i.e. there exists a function $S : (X, \mathcal{A}) \to (\mathbb{R}, \mathbb{B})$ and functions $C : \mathfrak{P} \to (0, \infty)$ and $a : \mathfrak{P} \to \mathbb{R}$ such that

$$x \to C(P) \exp\big[a(P)S(x)\big]$$

is a P_0–density of P.

4.6 Tests of Neyman structure

Conditional distributions, given a sufficient statistic, are applied for the construction of similar tests. If the sufficient statistic is complete, all similar tests are of this type.

Let $\mathfrak{P}_0$ be a dominated family of p–measures. The basic assumption of this section is the existence of a statistic $S : (X, A) \to (Y, B)$ which is sufficient for the hypothesis $\mathfrak{P}_0$, but not for the larger family including the alternative.

Definition 4.6.1. A critical function φ is of *Neyman structure* for the hypothesis $\mathfrak{P}_0$ if there exists a conditional expectation of φ, given S, under P, which is constant. (Since S is sufficient, this constant is the same for every $P \in \mathfrak{P}_0$.)

Proposition 4.6.2. (i) *If a critical function is of Neyman structure for $\mathfrak{P}_0$, it is similar for $\mathfrak{P}_0$.*

(ii) *If $\mathfrak{P}_0 \circ S$ is boundedly complete, then every critical function which is similar for $\mathfrak{P}_0$ is of Neyman structure for $\mathfrak{P}_0$.*

(iii) *If $\mathfrak{P}_0 \circ S$ is not boundedly complete, then there are similar tests which are not of Neyman structure.*

Proof. (i) If α is a constant conditional expectation of φ, given S, we have $P(\varphi) = \alpha$ for every $P \in \mathfrak{P}_0$.

(ii) Let φ be a similar critical function, say $P(\varphi) = \alpha$ for all $P \in \mathfrak{P}_0$. Since S is sufficient for $\mathfrak{P}_0$, there exists $\psi : Y \to [0, 1]$ such that $P(\psi \circ S) = P(\varphi)$ for all $P \in \mathfrak{P}_0$. Hence $P \circ S(\psi) = \alpha$ for all $P \in \mathfrak{P}_0$, and bounded completeness of $\mathfrak{P}_0 \circ S$ implies $\psi = \alpha$ $\mathfrak{P}_0 \circ S$–a.e.

(iii) If $\mathfrak{P}_0 \circ S$ is not boundedly complete, there exists a bounded $h : Y \to \mathbb{R}$ with $P \circ S\{h \neq 0\} > 0$ for some $P \in \mathfrak{P}_0$ such that $P \circ S(h) = 0$ for all $P \in \mathfrak{P}_0$. Let $c \neq 0$ be such that $-\alpha < ch < 1 - \alpha$. Then $\alpha + ch \circ S$ is a nonconstant critical function which is similar of level α for $\mathfrak{P}_0$. $\square$

Remark 4.6.3. If $\mathfrak{P}_0 \circ S$ is not boundedly complete, the restriction from "similar tests" to "similar tests of Neyman structure" may reduce the power. If the minimal sufficient statistic for $\mathfrak{P}_0$ is also sufficient for $\mathfrak{P}_0 \cup \{P_1\}$, then $\varphi \equiv \alpha$ is most powerful for testing $\mathfrak{P}_0$ against P_1 in the class of all critical functions which are of Neyman structure for $\mathfrak{P}_0$. In spite of this, there may exist nontrivial similar tests (see Example 4.7.15 and Exercise 4.7.18).

Tests of Neyman structure, so–called "conditional tests", have first been used by Neyman and Pearson (1933a, Sections IV, V). These authors also make plausible that under certain conditions all similar critical functions are of the conditional type. Completeness and bounded completeness were introduced by Lehmann and Scheffé (1950, p. 312) as a condition under which all critical functions are of Neyman structure, after it had been used implicitly by Scheffé (1943, p. 230, Theorem 3).

If there exists a conditional distribution of x, given S, it is the same for every $P \in \mathfrak{P}_0$, due to the sufficiency of S for $\mathfrak{P}_0$. Hence tests of Neyman structure reduce the test problem $\mathfrak{P}_0 : P_1$, involving a composite hypothesis, to the problem of testing a *simple* hypothesis against a simple alternative, a problem which is solved by the Neyman–Pearson lemma. More precisely, we have a family of such problems: one within each segment $S^{-1}\{y\}$, $y \in Y$.

Proposition 4.6.4. *Given a sufficient statistic S for $\mathfrak{P}_0$, we write the μ-density of $P \in \mathfrak{P}_0$ as $x \to g(x)h_P(S(x))$ (with g not depending on P).*

Assume that P_1 and $P \in \mathfrak{P}_0$ are mutually absolutely continuous. Let p_1 be a μ-density of P_1. Then the following holds true.

(i) For every $\alpha \in (0,1)$ there exists a level–α–test φ of Neyman structure for $\mathfrak{P}_0$ such that for some function $c_\alpha : (Y, \mathcal{B}) \to (\mathbb{R}, \mathbb{B})$

$$p_1(x) \underset{<}{\overset{>}{\gtrless}} c_\alpha(S(x))g(x) \quad \text{implies} \quad \varphi(x) = \begin{cases} 1 \\ 0 \end{cases}. \qquad (4.6.5)$$

(ii) A level–α–test φ of Neyman structure for $\mathfrak{P}_0$ fulfilling (4.6.5) is most powerful for P_1 in the class of all level–α–tests of Neyman structure for $\mathfrak{P}_0$.

Proof. (i) According to Proposition 1.3.1 there exists a Markov kernel $M|Y \times \mathbb{B}$ representing the conditional distribution of p_1/g, given S, with respect to $P \in \mathfrak{P}_0$. (Since P_1 and $P \in \mathfrak{P}_0$ are mutually absolutely continuous, $g(x) = 0$ implies $p_1(x) = 0$ for μ–a.a. $x \in X$.) By the Neyman–Pearson Lemma 4.3.3(i), applied with $\mu = M(y, \cdot)$, $f_1(t) = t$ and $f_0(t) \equiv 1$ there exist $c_\alpha(y)$ and $\gamma_\alpha(y)$ such that the critical function

$$\psi(t,y) = \begin{cases} 1 \\ \gamma_\alpha(y) \\ 0 \end{cases} \quad \text{for} \quad t \underset{<}{\overset{>}{=}} c_\alpha(y)$$

fulfills $\int \psi(t,y)M(y,dt) = \alpha$. It is easily seen from the proof of Lemma 4.3.3(i) that the functions c_α and γ_α are measurable. This implies measurability of ψ and, therefore, measurability of the critical function φ, defined by $\varphi(x) = \psi(p_1(x)/g(x), S(x))$. Since $\int \psi(t,y)M(y,dt)$ is a conditional expectation of φ, given $S = y$, for every $P \in \mathfrak{P}_0$ by Proposition 1.10.26, φ has the asserted properties.

(ii) If φ fulfills (4.6.5), the following relation holds for every critical function $\hat{\varphi}$:

$$(\varphi(x) - \hat{\varphi}(x))(p_1(x) - c_\alpha(S(x))g(x)) \geq 0 \quad \text{for } x \in X.$$

Integration with respect to μ yields

$$P_1(\varphi - \hat{\varphi}) \geq \int (\varphi(x) - \hat{\varphi}(x))c_\alpha(S(x))g(x)\mu(dx).$$

(Notice that $(\varphi - \hat{\varphi})c_\alpha(S(\cdot))g \leq p_1$, so that the integral on the right hand side exists.)

If both, φ and $\hat{\varphi}$, are level–α–tests of Neyman structure for $\mathfrak{P}_0$, we have by Proposition 1.10.13, $\int \big(\varphi(x) - \hat{\varphi}(x)\big) f\big(S(x)\big) P(dx) = 0$ for every function f (for which this integral exists). Applied with $f(y) = c_\alpha(y)/h_P(y)$ this implies $\int \big(\varphi(x) - \hat{\varphi}(x)\big) c_\alpha\big(S(x)\big) g(x) \mu(dx) = 0$, so that $P_1(\varphi) \geq P_1(\hat{\varphi})$. $\square$

Exercise 4.6.6. For $\vartheta > 0$ let Π_ϑ denote the Poisson distribution with parameter ϑ. Let (k_1, k_2) be a realization from $\Pi_{\vartheta_1} \times \Pi_{\vartheta_2}$. The problem is to test the hypothesis $\vartheta_1 = \vartheta_2$ (i.e. $\mathfrak{P}_0 = \{\Pi_\vartheta \times \Pi_\vartheta : \vartheta > 0\}$ against alternatives $\vartheta_1 > \vartheta_2$.

(i) We have $\Pi_\vartheta\{k_1\}\Pi_\vartheta\{k_2\} = B_{k_1+k_2,\,1/2}\{k_1\}\Pi_{2\vartheta}\{k_1 + k_2\}$, where $B_{k,p}$ denotes the binomial distribution. For $k = 0$ define $B_{0,p}\{0\} = 1$ for $p \in (0,1)$. Hence $S(k_1, k_2) = k_1 + k_2$ is sufficient for $\mathfrak{P}_0$, and the conditional distribution of k_1, given $S(k_1, k_2) = k$, under $\Pi_\vartheta \times \Pi_\vartheta$ is $B_{k,\,1/2}$ (independent of $\vartheta > 0$).

(ii) The following critical function φ is most powerful for $\vartheta_1 > \vartheta_2$ in the class of all similar level–α–tests for $\mathfrak{P}_0$

$$\varphi(k_1, k_2) = \begin{cases} 1 & > \\ \gamma_\alpha(k_1 + k_2) & k_1 = c_\alpha(k_1 + k_2), \\ 0 & < \end{cases}$$

where $c_\alpha(k)$ is the largest integer c such that $B_{k,\,1/2}\{c,\dots,k\} \geq \alpha$, and

$$\gamma_\alpha(k) := \big(\alpha - B_{k,\,1/2}\{c_\alpha(k) + 1,\dots,k\}\big)/B_{k,\,1/2}\{c_\alpha(k)\}.$$

Observe that $0 \leq \gamma_\alpha < 1$.

The test of Neyman structure for $\mathfrak{P}_0$ which is most powerful for $\Pi_{\vartheta_1} \times \Pi_{\vartheta_2}$ selects within each segment — consisting of the points $(0,k), (1,k-1), \dots, (k,0)$ — those points $(k_1, k-k_1)$, for which $\Pi_{\vartheta_1}\{k_1\}\Pi_{\vartheta_2}\{k-k_1\}/B_{k,\,1/2}\{k_1\}$ is largest. For $\vartheta_1 > \vartheta_2$ these are the points with k_1 large.

4.7 Most powerful similar tests for a real parameter in the presence of a nuisance parameter

The results of Section 4.6 on the construction of most powerful tests of Neyman structure are applied to obtain most powerful tests for families with a real structural parameter and an arbitrary nuisance parameter. Examples include one– and two–dimensional normal distributions and gamma distributions.

Theorem 4.7.1. *Let $\Theta \subset \mathbb{R}$ be an interval, and H an arbitrary set. Let $P_{\vartheta,\eta}$ be a p–measure with μ–density*

$$p(x, \vartheta, \eta) = g(x)H\big(T(x), \vartheta, \eta\big)G\big(S(x), \vartheta, \eta\big), \quad x \in X, \qquad (4.7.2)$$

with

$$H(\cdot,\vartheta,\eta) : (\mathbb{R},\mathbb{B}) \to ((0,\infty),\mathbb{B}_+),$$
$$G(\cdot,\vartheta,\eta) : (Y,\mathcal{B}) \to ((0,\infty),\mathbb{B}_+),$$

and

$$T : (X,\mathcal{A}) \to (\mathbb{R},\mathbb{B}),$$
$$S : (X,\mathcal{A}) \to (Y,\mathcal{B}).$$

Assume that for some $\vartheta_0 \in \Theta$

$$H(t,\vartheta_0,\eta) \equiv H_*(t,\vartheta_0) \quad \text{for } \eta \in H, \tag{4.7.3'}$$

and for every $\eta \in H$,

$$t \to H(t,\vartheta,\eta)/H_*(t,\vartheta_0) \quad \text{is} \quad \begin{matrix} \text{isotone} \\ \text{antitone} \end{matrix} \quad \text{for} \quad \vartheta \gtrless \vartheta_0. \tag{4.7.3''}$$

Then the following holds true.

(i) *For every* $\alpha \in (0,1)$ *there exists a level–α–test* φ_0 *of Neyman structure for the hypothesis* $\mathfrak{P}_0 = \{P_{\vartheta_0,\eta} : \eta \in H\}$ *such that, for some function* $c_\alpha : (Y,\mathcal{B}) \to (\mathbb{R},\mathbb{B})$,

$$T(x) \gtrless c_\alpha(S(x)) \quad \text{implies} \quad \varphi_0(x) = \begin{cases} 1 \\ 0 \end{cases}. \tag{4.7.4}$$

(ii) *If* $\mathfrak{P}_0 \circ S$ *is boundedly complete, any similar level–α–test for* $\mathfrak{P}_0$ *fulfilling (4.7.4) maximizes [minimizes] the power simultaneously for all alternatives* P_{ϑ_1,η_1} *with* $\vartheta_1 > \vartheta_0$ *[$\vartheta_1 < \vartheta_0$] and arbitrary* $\eta_1 \in H$, *in the class of all similar level–α–tests for the hypothesis* $\mathfrak{P}_0$.

Proof. (i) According to Proposition 1.3.1 there exists a Markov kernel $M|Y \times \mathbb{B}$, representing the conditional distribution of T, given S, with respect to $P_{\vartheta_0,\eta}$ for any $\eta \in H$. By the Neyman–Pearson Lemma 4.3.3(i), applied with $\mu = M(y,\cdot)$, $f_1(t) = t$ and $f_0(t) \equiv 1$ there exist $c_\alpha(y)$ and $\gamma_\alpha(y)$ such that the critical function

$$\psi(t,y) = \begin{cases} 1 \\ \gamma_\alpha(y) \\ 0 \end{cases} \quad \text{for} \quad t \begin{matrix} > \\ = c_\alpha(y) \\ < \end{matrix}$$

fulfills $\int \psi(t,y)M(y,dt) = \alpha$ for $y \in Y$.

It is easily seen from the proof of Lemma 4.3.3 that the functions c_α and γ_α are measurable. This implies measurability of ψ and therefore measurability of $\varphi_0 = \psi \circ (T,S)$. By Proposition 1.10.26, φ_0 is of Neyman structure for $\mathfrak{P}_0$.

(ii) Let

$$\hat{c}_\alpha(y) = \frac{H(c_\alpha(y), \vartheta_1, \eta_1)}{H_*(c_\alpha(y), \vartheta_0)} G(y, \vartheta_1, \eta_1).$$

Using (4.7.2) and (4.7.3'') we obtain for $\vartheta_1 > \vartheta_0$ that

$$p(x, \vartheta_1, \eta_1) \gtrless \hat{c}_\alpha(S(x)) g(x) H_*(T(x), \vartheta_0) \quad \text{implies} \quad T(x) \gtrless c_\alpha(S(x)).$$

Hence a level–α–test φ of Neyman structure fulfilling (4.7.4) fulfills (4.6.5) and is, therefore, most powerful for P_{ϑ_1, η_1} in the class of all level–α–tests of Neyman structure for $\mathfrak{P}_0$.

If $\mathfrak{P}_0 \circ S$ is boundedly complete, a test is similar for $\mathfrak{P}_0$ iff it is of Neyman structure for $\mathfrak{P}_0$ by Proposition 4.6.2. This implies the assertion for $\vartheta_1 > \vartheta_0$.

The proof for the case $\vartheta_1 < \vartheta_0$ runs analogously.   □

Remark 4.7.5. In the applications of Theorem 4.7.1 one usually thinks of η as a nuisance parameter attaining its values in $H \subset \mathbb{R}^k$. This is not necessarily so. If (4.7.2) holds with $H(\cdot, \vartheta, \eta)$ independent of η, say $H(\cdot, \vartheta)$, then the assertions of Theorem 4.7.1 extend easily to mixtures over η: Assume that H is endowed with a σ–algebra $\mathcal{C}$ and let $\mathcal{M}$ be a family of p–measures $M | \mathcal{C}$. Let $P_{\vartheta, M}$ denote the p–measure defined by $P_{\vartheta, M}(A) := \int P_{\vartheta, \eta}(A) M(d\eta)$. Then, under obvious measurability assumptions, $P_{\vartheta, M}$ has μ–density

$$p(x, \vartheta, M) = g(x) H(T(x), \vartheta) G(S(x), \vartheta, M),$$

with

$$G(y, \vartheta, M) := \int G(y, \vartheta, \eta) M(d\eta).$$

Sufficiency of S for $\{P_{\vartheta_0, \eta} : \eta \in H\}$ implies sufficiency for $\{P_{\vartheta_0, M} : M \in \mathcal{M}\}$. If the family $\mathcal{M}$ is boundedly complete and $\eta \to P_{\vartheta_0, \eta} \circ S(B)$ continuous for every $B \in \mathcal{B}$, then bounded completeness of $\{P_{\vartheta_0, \eta} \circ S : \eta \in H\}$ implies bounded completeness of $\{P_{\vartheta_0, M} \circ S : M \in \mathcal{M}\}$. Hence any level–$\alpha$–test fulfilling (4.7.4), which is similar for $\{P_{\vartheta_0, M} : M \in \mathcal{M}\}$, has the optimum property specified in Theorem 4.7.1(ii) with respect to all alternatives P_{ϑ_1, M_1}.

In order to determine the function c_α in (4.7.4) one needs the conditional distribution of T, given S, under $P_{\vartheta_0, \eta}$. In some instances this can be avoided by the following

Trick 4.7.6. Find a measurable function $q : \mathbb{R} \times Y \to \mathbb{R}$ which is increasing in the first component such that the distribution of $q \circ (T, S)$ under $P_{\vartheta_0, \eta}$ does not depend on $\eta \in H$.

Since $\{P_{\vartheta_0,\eta} \circ S : \eta \in H\}$ is boundedly complete, this implies by Basu's Theorem 1.8.2 that $q \circ (T, S)$ is stochastically independent of S under $P_{\vartheta_0,\eta}$, for every $\eta \in H$.

Since q is increasing in the first component, the relation $T(x) \underset{<}{\overset{>}{\gtrless}} c_\alpha(S(x))$ is equivalent to

$$q\big(T(x), S(x)\big) \underset{<}{\overset{>}{\gtrless}} q\big(c_\alpha(S(x)), S(x)\big) =: \hat{c}_\alpha(S(x)).$$

In principle, the function $\hat{c}_\alpha$ has to be determined such that

$$P_{\vartheta_0,\cdot}^S \big\{x \in X : q\big(T(x), S(x)\big) > \hat{c}_\alpha(S(x))\big\}$$
$$\leq \alpha \leq P_{\vartheta_0,\cdot}^S \big\{x \in X : q\big(T(x), S(x)\big) \geq \hat{c}_\alpha(S(x))\big\} \qquad \text{on } Y.$$

Since $q \circ (T, S)$ is stochastically independent of S, this amounts to determining a *constant* $\hat{c}_\alpha$ such that

$$P_{\vartheta_0,\cdot} \big\{x \in X : q\big(T(x), S(x)\big) > \hat{c}_\alpha\big\}$$
$$\leq \alpha \leq P_{\vartheta_0,\cdot} \big\{x \in X : q\big(T(x), S(x)\big) \geq \hat{c}_\alpha\big\}.$$

(In these relations we write $P_{\vartheta_0,\cdot}^S$ and $P_{\vartheta_0,\cdot}$ to stress that these probabilities do not depend on $\eta \in H$.)

Corollary 4.7.7. *Assume that $\{P_{\vartheta,\eta} : \vartheta \in \Theta,\ \eta \in H\}$ is an exponential family with μ–densities*

$$p(x, \vartheta, \eta) = C(\vartheta, \eta)g(x)\exp\Big[a_0(\vartheta, \eta)T_0(x) + \sum_{i=1}^{k} a_i(\vartheta, \eta)T_i(x)\Big].$$

Assume, moreover, that for some $\vartheta_0 \in \Theta$, the following holds true.

$$a_0(\vartheta_0, \eta) = a_*(\vartheta_0) \quad \text{for } \eta \in H; \tag{4.7.8'}$$

$$a_0(\vartheta, \eta) \underset{<}{\overset{>}{\gtrless}} a_*(\vartheta_0) \quad \text{for every } \eta \in H, \qquad \text{if} \quad \vartheta \underset{<}{\overset{>}{\gtrless}} \vartheta_0; \tag{4.7.8''}$$

$$\big\{(a_1(\vartheta_0, \eta), \dots, a_k(\vartheta_0, \eta)) : \eta \in H\big\} \text{ has a nonempty interior.} \tag{4.7.9}$$

Under these assumptions the following holds true, for every sample size n.

(i) For every $\alpha \in (0, 1)$ there exists a level-α-test φ_0 of Neyman structure for the hypothesis $\mathfrak{P}_0 = \{P_{\vartheta_0,\eta}^n : \eta \in H\}$ such that for some function $c_{n,\alpha} : (\mathbb{R}^n, \mathbb{B}^n) \to (\mathbb{R}, \mathbb{B})$

$$\sum_{\nu=1}^{n} T_0(x_\nu) \underset{<}{\overset{>}{\gtrless}} c_{n,\alpha}\Big(\sum_{\nu=1}^{n} T_1(x_\nu), \dots, \sum_{\nu=1}^{n} T_k(x_\nu)\Big) \text{ implies } \varphi_0(x) = \begin{cases} 1 \\ 0 \end{cases}. \tag{4.7.10}$$

(ii) Any similar level-α-test for $\mathfrak{P}_0$ fulfilling (4.7.10) maximizes [minimizes] the power simultaneously for all alternatives P_{ϑ_1,η_1} with $\vartheta_1 > \vartheta_0$ [$\vartheta_1 < \vartheta_0$]

and arbitrary $\eta_1 \in H$, in the class of all similar level–α–tests for the hypothesis $\mathfrak{P}_0$.

Proof. Assumption (4.7.2) of Theorem 4.7.1 is fulfilled with X^n in place of X, $T(x_1, \ldots, x_n) = \sum_1^n T_0(x_\nu)$ and $S(x_1, \ldots, x_n) = (\sum_1^n T_1(x_\nu), \ldots, \sum_1^n T_k(x_\nu))$. Assumptions (4.7.3) follow from (4.7.8). Moreover, $\mathfrak{P}_0 \circ S$ is complete by Theorem 1.6.10.   □

The following Examples 4.7.12 and 4.7.13 refer to the family $\{N^n_{(\mu, \sigma^2)} : \mu \in \mathbb{R}, \sigma^2 > 0\}$. It is therefore useful to write the Lebesgue density of $N_{(\mu, \sigma^2)}$ in such a way that the structure as an exponential family becomes manifest. This form is

$$x \to C(\mu, \sigma^2) \exp\left[\frac{\mu}{\sigma^2} x - \frac{1}{2\sigma^2} x^2\right]. \tag{4.7.11}$$

Example 4.7.12. Testing a hypothesis about the mean of a normal distribution. Wanted: a similar test for $\{N^n_{(\mu_0, \sigma^2)} : \sigma^2 > 0\} : N^n_{(\mu_1, \sigma_1^2)}$ with $\mu_1 > \mu_0$.

At first we restrict ourselves to the special case $\mu_0 = 0$, to which Corollary 4.7.7 applies immediately. The general case can be reduced to this case by the transformation $x \to x - \mu_0$.

It is easy to see from (4.7.11) that the assumptions of Corollary 4.7.7 are fulfilled for $\vartheta = \mu$, $\vartheta_0 = \mu_0 = 0$, $\eta = \sigma^2$ with

$$T_0(x) = x, \qquad\qquad T_1(x) = x^2,$$
$$a_0(\mu, \sigma^2) = \mu/\sigma^2, \qquad a_1(\mu, \sigma^2) = -1/2\sigma^2.$$

Hence the critical region $\{(x_1, \ldots, x_n) \in \mathbb{R}^n : \sum_1^n x_\nu > c_{n,\alpha}(\sum_1^n x_\nu^2)\}$ maximizes the power simultaneously for all alternatives $N^n_{(\mu, \sigma_1^2)}$ with $\mu_1 > 0$, $\sigma_1^2 > 0$, among all similar level–α–tests for the hypothesis $\{N^n_{(0, \sigma^2)} : \sigma^2 > 0\}$, if $c_{n,\alpha}$ is chosen such that the conditional probability of this critical region, given $\sum_1^n x_\nu^2$, equals α.

With $s_n^2 := (n-1)^{-1} \sum_1^n (x_\nu - \bar{x}_n)^2$, the function $\bar{x}_n/s_n$ can be expressed as a function of $\sum_1^n x_\nu$ and $\sum_1^n x_\nu^2$ which is increasing in $\sum_1^n x_\nu$. Its distribution under $N^n_{(0, \sigma^2)}$ does not depend on σ^2. Hence (see Trick 4.7.6) the most powerful critical region is equivalent to

$$\{(x_1, \ldots, x_n) \in \mathbb{R}^n : \bar{x}_n/s_n > c'_{n,\alpha}\}.$$

Expressed in terms of the original variables this critical region becomes

$$\{(x_1, \ldots, x_n) \in \mathbb{R}^n : (\bar{x}_n - \mu_0)/s_n > c'_{n,\alpha}\}.$$

This proves that the usual t-test is most powerful for $\{N^n_{(\mu_0,\sigma^2)} : \sigma^2 > 0\} : N^n_{(\mu_1,\sigma_1^2)}$ among all similar level–α–tests, simultaneously for all (μ_1,σ_1^2) with $\mu_1 > \mu_0$. (A corresponding result holds with $\mu_1 < \mu_0$).

Example 4.7.13. Testing the variance of a normal distribution. Wanted: a similar test for $\{N^n_{(\mu,\sigma_0^2)} : \mu \in \mathbb{R}\} : N^n_{(\mu_1,\sigma_1^2)}$ with $\sigma_1^2 > \sigma_0^2$.

It is easy to see from (4.7.11) that the assumptions of Corollary 4.7.7 are fulfilled for $\vartheta = \sigma^2$, $\eta = \mu$, with

$$T_0(x) = x^2, \qquad\qquad T_1(x) = x,$$
$$a_0(\sigma^2,\mu) = -1/2\sigma^2, \qquad a_1(\sigma^2,\mu) = \mu/\sigma^2.$$

Hence the critical region

$$\{(x_1,\ldots,x_n) \in \mathbb{R}^n : \sum_1^n x_\nu^2 > c_{n,\alpha}(\sum_1^n x_\nu)\}$$

is most powerful for all alternatives (μ_1,σ_1^2) with $\sigma_1^2 > \sigma_0^2$, among all similar level–α–tests for $\{N^n_{(\mu,\sigma_0^2)} : \mu \in \mathbb{R}\}$, if $c_{n,\alpha}$ is chosen in such a way that the conditional probability of this critical region, given $\sum_1^n x_\nu$, equals α. $\sum_1^n (x_\nu - \bar{x}_n)^2 = \sum_1^n x_\nu^2 - n^{-1}(\sum_1^n x_\nu)^2$ is a function of $\sum_1^n x_\nu^2$ and $\sum_1^n x_\nu$ which is increasing in $\sum_1^n x_\nu^2$. Its distribution under $N^n_{(\mu,\sigma_0^2)}$ does not depend on μ. Hence (see Trick 4.7.6) the most powerful critical region is equivalent to

$$\{(x_1,\ldots,x_n) \in \mathbb{R}^n : \sum_1^n (x_\nu - \bar{x}_n)^2 > c'_{n,\alpha}\}.$$

This proves that the usual χ^2–test is most powerful for $\{N^n_{(\mu,\sigma_0^2)} : \mu \in \mathbb{R}\} :$ $N^n_{(\mu_1,\sigma_1^2)}$ among all similar level–α–tests, simultaneously for all $\sigma_1^2 > \sigma_0^2$.

The corresponding result holds for similar level–α–tests for alternatives $\sigma_1^2 < \sigma_0^2$. There is, however, a remarkable difference: the critical region

$$\{(x_1,\ldots,x_n) \in \mathbb{R}^n : \sum_1^n (x_\nu - \bar{x}_n)^2 > c'_{n,\alpha}\}$$

is most powerful for alternatives $\sigma_1^2 > \sigma_0^2$ even within the class of all level–α–tests (without similarity), whereas $\{(x_1,\ldots,x_n) \in \mathbb{R}^n : \sum_1^n (x_\nu - \bar{x}_n)^2 < c'_{n,\alpha}\}$ lacks this stronger optimum property for alternatives $\sigma_1^2 < \sigma_0^2$ (see Example 4.4.5).

The following Examples 4.7.15, 4.7.17 and 4.7.18 refer to the family of *bivariate normal distributions*

$$\{N^n_{(\mu_1,\mu_2,\sigma_1^2,\sigma_2^2,\varrho)} : \mu_i \in \mathbb{R}, \ \sigma_i^2 > 0, \ \varrho \in (-1,1)\}.$$

It is therefore useful to write the Lebesgue density of $N_{(\mu_1,\mu_2,\sigma_1^2,\sigma_2^2,\varrho)}$ in such a way that the structure as an exponential family becomes manifest. This is

$$(x,y) \rightarrow C(\mu_1,\mu_2,\sigma_1^2,\sigma_2^2,\varrho) \exp\left[-\frac{1}{1-\varrho^2}\left(\left(\varrho\frac{\mu_2}{\sigma_2} - \frac{\mu_1}{\sigma_1}\right)\frac{x}{\sigma_1}\right.\right. \quad (4.7.14)$$
$$\left.\left. + \left(\varrho\frac{\mu_1}{\sigma_1} - \frac{\mu_2}{\sigma_2}\right)\frac{y}{\sigma_2} + \frac{1}{2}\frac{x^2}{\sigma_1^2} + \frac{1}{2}\frac{y^2}{\sigma_2^2} - \varrho\frac{xy}{\sigma_1\sigma_2}\right)\right].$$

Example 4.7.15. Testing for independence in a bivariate normal distribution. Wanted: a similar test for $\mathfrak{P}_0 = \{N^n_{(\mu_1,\mu_2,\sigma_1^2,\sigma_2^2,0)} : \mu_i \in \mathbb{R}, \sigma_i^2 > 0\} : N^n_{(\bar\mu_1,\bar\mu_2,\bar\sigma_1^2,\bar\sigma_2^2,\bar\varrho)}$ with $\bar\varrho > 0$.

It is easy to see from (4.7.14) that the assumptions of Corollary 4.7.7 are fulfilled for $\vartheta = \varrho$, $\vartheta_0 = 0$, and $\eta = (\mu_1,\mu_2,\sigma_1^2,\sigma_2^2)$ with

$$T_0(x,y) = xy, \quad T(x,y) = (x,y,x^2,y^2), \quad a_0(\mu_1,\mu_2,\sigma_1^2,\sigma_2^2,\varrho) = -\frac{\varrho}{\sigma_1\sigma_2(1-\varrho^2)},$$

$$a(\mu_1,\mu_2,\sigma_1^2,\sigma_2^2,\varrho) = -\frac{1}{1-\varrho^2}\left(\varrho\frac{\mu_2}{\sigma_2} - \frac{\mu_1}{\sigma_1}, \varrho\frac{\mu_1}{\sigma_1} - \frac{\mu_2}{\sigma_2}, \frac{1}{2\sigma_1^2}, \frac{1}{2\sigma_2^2}\right).$$

Hence the critical region

$$\left\{((x_1,y_1),\ldots,(x_n,y_n)) \in \mathbb{R}^{2n} : \sum_1^n x_\nu y_\nu > c_{n,\alpha}\left(\sum_1^n x_\nu, \sum_1^n y_\nu, \sum_1^n x_\nu^2, \sum_1^n y_\nu^2\right)\right\}$$

is most powerful for all alternatives $(\bar\mu_1,\bar\mu_2,\bar\sigma_1^2,\bar\sigma_2^2,\bar\varrho)$ with $\bar\varrho > 0$ among all similar level–α–tests for $\mathfrak{P}_0$, if the function $c_{n,\alpha}$ is chosen such that the conditional probability of this critical region, given $(\sum_1^n x_\nu, \sum_1^n y_\nu, \sum_1^n x_\nu^2, \sum_1^n y_\nu^2)$, equals α.

The sample correlation coefficient,

$$r_n((x_1,y_1),\ldots,(x_n,y_n)) = \frac{\sum_1^n (x_\nu - \bar x_n)(y_\nu - \bar y_n)}{\left(\sum_1^n (x_\nu - \bar x_n)^2 \sum_1^n (y_\nu - \bar y_n)^2\right)^{1/2}}$$

is a function of $\sum_1^n x_\nu y_\nu$ and $(\sum_1^n x_\nu, \sum_1^n y_\nu, \sum_1^n x_\nu^2, \sum_1^n y_\nu^2)$ which is increasing in $\sum_1^n x_\nu y_\nu$. The distribution of r_n under $N^n_{(\mu_1,\mu_2,\sigma_1^2,\sigma_2^2,0)}$ does not depend on $\mu_1,\mu_2,\sigma_1^2,\sigma_2^2$. Hence the critical region $\{((x_1,y_1),\ldots,(x_n,y_n)) \in \mathbb{R}^{2n} : r_n((x_1,y_1),\ldots,(x_n,y_n)) > c'_{n,\alpha}\}$ is most powerful for all alternatives with $\bar\varrho > 0$ in the class of all similar level–α–tests for the hypothesis $\mathfrak{P}_0$. (A corresponding result holds for alternatives with $\bar\varrho < 0$.)

We remark that Corollary 4.7.7 cannot be applied to the more general problem of testing the hypothesis $\{N^n_{(\mu_1,\mu_2,\sigma_1^2,\sigma_2^2,\varrho_0)} : \mu_i \in \mathbb{R}, \sigma_i^2 > 0\}$. If $\varrho_0 \neq 0$, the minimal sufficient statistic for this family, namely

$$T_*\big((x_1, y_1), \ldots, (x_n, y_n)\big) \tag{4.7.16}$$

$$= \Big(\sum_1^n x_\nu, \sum_1^n y_\nu, \sum_1^n x_\nu^2, \sum_1^n y_\nu^2, \sum_1^n x_\nu y_\nu\Big),$$

is also sufficient for the enlarged family $\{N^n_{(\mu_1, \mu_2, \sigma_1^2, \sigma_2^2, \varrho)} : \mu_i \in \mathbb{R}, \; \sigma_i^2 > 0, \; \varrho \in (0, 1)\}$. (This is due to the fact that the five coefficients in (4.7.14), if considered for fixed $\varrho_0 \neq 0$ as functions of $\mu_1, \mu_2, \sigma_1^2, \sigma_2^2$, are *not* affinely dependent.) Hence all tests of Neyman structure are trivial. On the other hand, the family $\{N^n_{(\mu_1, \mu_2, \sigma_1^2, \sigma_2^2, \varrho_0)} \circ T_* : \mu_i \in \mathbb{R}, \; \sigma_i^2 > 0\}$ is not boundedly complete if $\varrho_0 \neq 0$, so that similar tests do exist, so for instance the test based on the sample correlation coefficient. (Notice that r_n is a contraction of T_*.) Hence in this case the theory of tests of Neyman structure is not applicable. In particular: It does not answer the question whether the usual test based on the sample correlation coefficient is most powerful among all similar tests if the hypothetical value ϱ_0 is different from 0.

Since the family $\{N^n_{(\mu_1, \mu_2, \sigma_1^2, \sigma_2^2, \varrho)} \circ r_n : \varrho \in (-1, 1)\}$ has isotone likelihood ratios, the critical region $\{r_n > c_{n,\alpha}\}$ is most powerful for testing $\varrho = \varrho_0$ against any $\varrho > \varrho_0$. However: This optimum property holds only within the class of tests depending on the observations through r_n.

Exercise 4.7.17. Let

$$R_n\big((x_1, y_1), \ldots, (x_n, y_n)\big) = \frac{\sum\limits_1^n (x_\nu - \bar{x}_n)(y_\nu - \bar{y}_n)}{\frac{1}{2}\sum\limits_1^n (x_\nu - \bar{x}_n)^2 + \frac{1}{2}\sum\limits_1^n (y_\nu - \bar{y}_n)^2}.$$

Show that the critical region

$$\big\{((x_1, y_1), \ldots, (x_n, y_n)) \in \mathbb{R}^{2n} : R_n\big((x_1, y_1), \ldots, (x_n, y_n)\big) > c_{n,\alpha}\big\}$$

is most powerful for all alternatives $(\bar{\mu}_1, \bar{\mu}_2, \bar{\sigma}^2 \bar{\varrho})$ with $\bar{\varrho} > 0$ in the class of all level-α-tests for the hypothesis $\{N^n_{(\mu_1, \mu_2, \sigma^2, \sigma^2, 0)} : \mu_i \in \mathbb{R}, \; \sigma^2 > 0\}$.

Exercise 4.7.18. Show that the critical region

$$\big\{((x_1, y_1), \ldots, (x_n, y_n)) \in \mathbb{R}^{2n} : (\bar{y}_n - \bar{x}_n)(s_1^2 + s_2^2 - s_{12})^{-1/2} > c_{n,\alpha}\big\}$$

with $s_1 = \sum_1^n (x_\nu - \bar{x}_n)^2$, $s_2 = \sum_1^n (y_\nu - \bar{y}_n)^2$, $s_{12} = \frac{1}{n}\sum_1^n (x_\nu - \bar{x}_n)(y_\nu - \bar{y}_n)$, is most powerful for alternatives $N^n_{(\bar{\mu}_1, \bar{\mu}_2, \bar{\sigma}^2, \bar{\sigma}^2, \bar{\varrho})}$ with $\bar{\mu}_2 > \bar{\mu}_1$ among all similar level-α-tests for the hypothesis $\{N^n_{(\mu, \mu, \sigma^2, \sigma^2, \varrho)} : \mu \in \mathbb{R}, \; \sigma^2 > 0, \; \varrho \in (-1, 1)\}$. This critical region was suggested by Hsu (1940). [Hint: Use the transformation $u_\nu = x_\nu + y_\nu$, $v_\nu = x_\nu - y_\nu$.]

The more interesting problem is to obtain a similar test for the larger hypothesis $\{N^n_{(\mu,\mu,\sigma_1^2,\sigma_2^2,\varrho)} : \mu \in \mathbb{R},\ \sigma_i^2 > 0, \varrho \in (-1,1)\}$. Since the minimal sufficient statistic is (4.7.16), which is also sufficient for the alternative, all tests of Neyman structure are trivial. In spite of this, a similar critical region exists, namely

$$\left\{((x_1,y_1),\dots,(x_n,y_n)) \in \mathbb{R}^{2n} : (\bar{y}_n - \bar{x}_n)\left(\sum_1^n (y_\nu - x_\nu - (\bar{y}_n - \bar{x}_n))^2\right)^{-1/2} > c_{n,\alpha}\right\}.$$

Exercise 4.7.19. Show that the critical region

$$\left\{((x_1,y_1),\dots,(x_n,y_n)) \in \mathbb{R}^{2n} : (s_2^2 - s_1^2)((s_1^2 + s_2^2)^2 - 4s_{12}^2)^{-1/2} > c_{n,\alpha}\right\}$$

is most powerful for alternatives $N^n_{(\bar{\mu}_1,\bar{\mu}_2,\bar{\sigma}_1^2,\bar{\sigma}_2^2,\bar{\varrho})}$ with $\bar{\sigma}_2^2 > \bar{\sigma}_1^2$ among all similar level–α–tests for the hypothesis $\{N^n_{(\mu_1,\mu_2,\sigma^2,\sigma^2,\varrho)} : \mu_i \in \mathbb{R},\ \sigma^2 > 0,\ \varrho \in (-1,1)\}$.

This critical region was suggested by Pitman (1939) and Morgan (1939). [Hint: Use the transformation $u_\nu = x_\nu + y_\nu$, $v_\nu = x_\nu - y_\nu$.]

The following examples refer to the family $\{\Gamma^n_{a,b} : a,b > 0\}$ of *gamma distributions*. To stress the exponential nature of this family, we write the Lebesgue density of $\Gamma_{a,b}$ as

$$x \to \frac{1}{a^b \Gamma(b)} x^{-1} \exp\left[-\frac{x}{a} + b \log x\right], \qquad x > 0,$$

where $\Gamma(b)$ is the gamma function, defined by

$$\Gamma(b) = \int_0^\infty x^{b-1} \exp[-x]dx.$$

Example 4.7.20. Testing a hypothesis about the shape parameter. Wanted: A similar test for $\mathfrak{P}_0 = \{\Gamma^n_{a,b_0} : a > 0\} : \Gamma^n_{a_1,b_1}$ with $b_1 > b_0$.

The assumptions of Corollary 4.7.7 are fulfilled for $\vartheta = b$, $\vartheta_0 = b_0$, $\eta = a$, with

$$T_0(x) = \log x, \qquad T_1(x) = x,$$
$$a_0(b,a) = b, \qquad a_1(b,a) = -1/a.$$

Hence the critical region

$$\left\{(x_1,\dots,x_n) \in (0,\infty)^n : \sum_1^n \log x_\nu > c_{n,\alpha}\left(\sum_1^n x_\nu\right)\right\}$$

is most powerful for all alternatives (b_1, a_1) with $b_1 > b_0$, among all similar level–α–tests for $\mathfrak{P}_0$, if $c_{n,\alpha}$ is chosen in such a way that the conditional probability of this critical region, given $\sum_1^n x_\nu$, under Γ^n_{a,b_0}, equals α.

$\left(\prod_1^n x_\nu\right)^{1/n}/\bar{x}_n$ is a function of $\sum_1^n \log x_\nu$ and $\sum_1^n x_\nu$ which is increasing in $\sum_1^n \log x_\nu$. Its distribution under Γ^n_{a,b_0} does not depend on a. Hence (see Trick 4.7.6) the most powerful critical region is equivalent to

$$\left\{(x_1, \ldots, x_n) \in (0, \infty)^n : \left(\prod_1^n x_\nu\right)^{1/n}/\bar{x}_n > c'_{n,\alpha}\right\}.$$

Example 4.7.21. Testing a hypothesis about the scale parameter. Wanted: a similar test for $\mathfrak{P}_0 = \{\Gamma^n_{a_0,b} : b > 0\} : \Gamma_{a_1,b_1}$ with $a_1 > a_0$.

The assumptions of Corollary 4.7.7 are fulfilled for $\vartheta = a$, $\vartheta_0 = a_0$, $\eta = b$ with

$$T_0(x) = x, \qquad\qquad T_1(x) = \log x,$$
$$a_0(a, b) = -1/a, \qquad a_1(a, b) = b.$$

Hence the critical region

$$\left\{(x_1, \ldots, x_n) \in (0, \infty)^n : \sum_1^n x_\nu > c_{n,\alpha}\left(\prod_1^n x_\nu\right)\right\}$$

is most powerful for all alternatives (a_1, b_1) with $a_1 > a_0$ among all similar level–α–tests for $\mathfrak{P}_0$, if $c_{n,\alpha}$ is chosen in such a way that the conditional probability of this critical region, given $\prod_1^n x_\nu$, under $\Gamma^n_{a_0,b}$, equals α. Since the conditional distribution of $\sum_1^n x_\nu$, given $\prod_1^n x_\nu$, cannot be obtained in closed form, the function $c_{n,\alpha}$ has to be determined numerically (see Engelhardt and Bain, 1977, p. 78, Table 1).

The next example refers to the family of *inverse normal distributions*, $\{P^n_{\mu,\lambda} : \mu, \lambda > 0\}$, with Lebesgue density

$$x \to \sqrt{\frac{\lambda}{2\pi}} x^{-3/2} \exp\left[-\frac{\lambda(x - \mu)^2}{2\mu^2 x}\right], \qquad x > 0.$$

Example 4.7.22. Testing a hypothesis about μ. Wanted: a similar test for $\mathfrak{P}_0 = \{P^n_{\mu_0,\lambda} : \lambda > 0\} : P^n_{\mu_1,\lambda_1}$ with $\mu_1 > \mu_0$.

At first we restrict ourselves to the special case $\mu_0 = 1$, to which Corollary 4.7.7 applies immediately. The general case can be reduced to this case by the transformation $x \to x/\mu_0$.

The assumptions of Corollary 4.7.7 are fulfilled for $\vartheta = \mu$, $\vartheta_0 = \mu_0 = 1$, $\eta = \lambda$ with

$$T_0(x) = x \qquad\qquad T_1(x) = x + 1/x$$
$$a_0(\mu, \lambda) = \lambda(1 - \mu^{-2})/2 \qquad a_1(\mu, \lambda) = -\lambda/2.$$

Hence the critical region

$$\left\{ (x_1, \ldots, x_n) \in (0, \infty)^n : \sum_1^n x_\nu > c_{n,\alpha}\left(\sum_1^n (x_\nu + \frac{1}{x_\nu})\right) \right\}$$

maximizes the power simultaneously for all alternatives $P^n_{\mu_1, \lambda_1}$ with $\mu_1 > 1$, $\lambda_1 > 0$, in the class of all similar level-α-tests for the hypothesis $\mathfrak{P}_0$.

The function $c_{n,\alpha}$ is determined by the conditional distribution of $\sum_1^n x_\nu$, given $\sum_1^n (x_\nu + 1/x_\nu)$, under $P^n_{1,\lambda}$. Since $\sum_1^n (x_\nu + 1/x_\nu)$ is sufficient for $\{P^n_{1,\lambda} : \lambda > 0\}$, the function $c_{n,\alpha}$ is independent of λ.

Since

$$P^n_{1,\lambda}\left\{ (x_1, \ldots, x_n) \in (0, \infty)^n : \sum_1^n x_\nu > c_{n,\alpha}\left(\sum_1^n (x_\nu + \frac{1}{x_\nu})\right) \right\} = \alpha \qquad \text{for } \lambda > 0,$$

and $P_{\mu,\lambda} \circ (x \to x/\mu) = P_{1,\lambda/\mu}$, we obtain

$$P^n_{\mu_0,\lambda}\left\{ (x_1, \ldots, x_n) \in (0, \infty)^n : \frac{1}{\mu_0}\sum_1^n x_\nu > c_{n,\alpha}\left(\sum_1^n (\frac{x_\nu}{\mu_0} + \frac{\mu_0}{x_\nu})\right) \right\} = \alpha \text{ for } \lambda > 0.$$

Chhikara and Folks (1989, Chapter 6) suggest an equivalent critical region which can be determined by means of the t-distribution. Though this idea works, the conditional distribution, expressed by their relation (6.13), is not correct.

Chapter 5
Confidence procedures

5.1 Basic concepts

Introduces the concept of a confidence procedure and explains the relationship to hypothesis testing.

Let $\mathfrak{P}$ be a family of p-measures and $\kappa : \mathfrak{P} \to Y$ a functional. Our main interest will be in the case $Y = \mathbb{R}$.

A *confidence procedure* is a map assigning to each $x \in X$ a subset $K(x) \subset Y$ such that $\{x \in X : y \in K(x)\} \in \mathcal{A}$ for every $y \in Y$. The intuitive interpretation: $K(x)$ contains the "true" value of the functional with high probability. This interpretation is justified if the *confidence coefficient*

$$\inf_{P \in \mathfrak{P}} P\{x \in X : \kappa(P) \in K(x)\}$$

is large, say 0.9 at least. A more careful formulation of this interpretation reads as follows: The statement "$\kappa(P) \in K(x)$", made prior to the observation of x, will be true with high probability.

The interpretation of the confidence coefficient can be made more intuitive in terms of betting: If the confidence coefficient is β, one may bet $\beta{:}(1 - \beta)$ that the confidence set will contain the true $\kappa(P)$, whatever $P \in \mathfrak{P}$.

The interpretation of $K(x)$ as a set embracing $\kappa(P)$ requires implicitly that $K(x)$ is of a geometrically simple shape. If $\kappa : \mathfrak{P} \to \mathbb{R}$, the natural requirement is that $K(x)$ be an interval, perhaps a ray.

Confidence procedures can be used advantageously to supplement estimators by error bounds. Consider a functional $\kappa : \mathfrak{P} \to \mathbb{R}$ and an estimator $\hat{\kappa} : X \to \mathbb{R}$. Assume there exists an error bound $\Delta(P)$ such that

$$P\{x \in X : |\hat{\kappa}(x) - \kappa(P)| \leq \Delta(P)\} \geq \beta \quad \text{for } P \in \mathfrak{P}. \tag{5.1.1}$$

If $\Delta(P)$ can be replaced by an estimate $\hat{\Delta}(x)$, i.e. if

$$P\{x \in X : |\hat{\kappa}(x) - \kappa(P)| \leq \hat{\Delta}(x)\} \geq \beta \quad \text{for } P \in \mathfrak{P}, \tag{5.1.2}$$

then

$$P\{x \in X : \kappa(P) \in \big(\hat{\kappa}(x) - \hat{\Delta}(x), \ \hat{\kappa}(x) + \hat{\Delta}(x)\big)\} \geq \beta \quad \text{for } P \in \mathfrak{P}.$$

Hence $x \rightarrow \big(\hat{\kappa}(x) - \hat{\Delta}(x), \ \hat{\kappa}(x) + \hat{\Delta}(x)\big)$ is a confidence procedure with confidence coefficient $\geq \beta$, the confidence sets of which are intervals.

There seems to be a general agreement that a confidence coefficient should be large, say 0.9 at least. This was not always so. If $\hat{\Delta}(x)$ is an estimate of the "probable error" (which is surpassed with probability $1/2$) this leads to a confidence procedure with covering probability equal to $1/2$ for every $P \in \mathfrak{P}$, so that one may bet 1:1 that $\kappa(P) \in \big(\hat{\kappa}(x) - \hat{\Delta}(x), \ \hat{\kappa}(x) + \hat{\Delta}(x)\big)$. (This is exactly the same wording used by Gauß (1825, p. 425) in a letter to Olbers, in a special context.)

The usual theory views confidence procedures primarily in their relation to tests of hypotheses: $K(x)$ specifies those values of $\kappa(P)$ which are compatible with the observation x. Obviously, this can be turned into a test for the value of $\kappa(P)$: Incompatible with the observation x are all hypotheses $\{P \in \mathfrak{P} : \kappa(P) = r\}$ with $r \notin K(x)$.

More formally: To each confidence procedure there corresponds a family of critical regions, one for each hypothesis $\{P \in \mathfrak{P} : \kappa(P) = r\}$, defined by

$$C(r) := \big\{x \in X : r \notin K(x)\big\}.$$

For every $r \in Y$, we have

$$x \in C(r) \quad \text{iff} \quad r \notin K(x).$$

Hence $\kappa(P) = r$ implies $P\big(C(r)\big) = P\{x \in X : r \notin K(x)\} = P\{x \in X : \kappa(P) \notin K(x)\}$. If K has confidence coefficient β, this implies $P\big(C(r)\big) \leq 1 - \beta$ if the hypothesis $\kappa(P) = r$ holds true.

These relations can also be read the other way round: If we are given a family of critical regions $C(r)$, $r \in Y$, with $P\big(C(r)\big) \leq \alpha$ for $P \in \mathfrak{P}$ fulfilling $\kappa(P) = r$, then

$$K(x) := \big\{r \in Y : x \notin C(r)\big\}$$

defines a confidence procedure with covering probability

$$P\{x \in X : \kappa(P) \in K(x)\} \geq 1 - \alpha.$$

There is one weak point in this correspondence between confidence procedures and families of tests. Whereas the shape of the critical region is irrelevant for the interpretation of the test, the shape of a confidence set is essential for its meaning.

Our discussion of the relation between confidence procedures and families of tests was confined to tests based on critical regions. Now we consider a family of critical functions φ_r, $r \in Y$, such that $P(\varphi_r) \leq \alpha$ for $P \in \mathfrak{P}$ fulfilling $\kappa(P) = r$.

As outlined in Section 4.2, a test based on a critical function $\varphi : X \to [0,1]$ may also be considered as a test based on the critical region $C \subset X \times (0,1)$, defined by $C = \{(x,u) \in X \times (0,1) : \varphi(x) > u\}$, with u being a realization from the uniform distribution over $(0,1)$.

The application of this approach in the present context leads from the family of critical functions φ_r, $r \in Y$, to the confidence procedure K defined by

$$K(x,u) = \{r \in Y : \varphi_r(x) \leq u\}, \qquad (x,u) \in X \times (0,1).$$

This confidence procedure based on (x,u) may also be considered as a two–step procedure: A realization u from the uniform distribution over $(0,1)$ determines a confidence procedure $K(\cdot, u)$ which is then applied to the observation x.

One could think of more complex "randomized confidence procedures" where x determines a p–measure on a suitable σ–algebra on the space consisting of subsets of Y. However: Any such a randomized confidence procedure assigns to x the probability $\pi(x,r)$ that the confidence set covers $r \in Y$. Hence $\varphi_r(x) := 1 - \pi(x,r)$ may be considered as a test for the hypothesis $\kappa(P) = r$. The family of these critical functions is equivalent to the confidence procedure

$$K(x,u) := \{r \in Y : \pi(x,r) \geq u\}, \qquad (x,u) \in X \times (0,1).$$

In other words: Any confidence procedure based on the random selection of a subset of Y may be replaced by the following two–step procedure which has the same covering probabilities: Use a realization u from the uniform distribution over $(0,1)$ and apply then the confidence procedure $x \to K(x,u)$.

Historical Remark 5.1.3. The notion of a confidence interval as a set of parameter values containing the true parameter value with "high probability" is used in a vague form by several writers in the 19th century, the first one being perhaps Laplace (1812), who gives an asymptotic confidence interval for the parameter of the binomial distribution (2nd Book, Section 16). Other examples can be found in Gauß (1816), Fourier (1826), Cournot (1843, pp. 185/6) and Lexis (1875).

The first formally correct statement of confidence sets as random objects containing the fixed parameter with prescribed probability is given by Wilson (1927). Other examples of conceptually precise confidence statements prior to Neyman's general theory are Working and Hotelling (1929), Hotelling (1931) and Clopper and Pearson (1934). Fisher (1930) gives a method for obtaining exact confidence bounds for a real parameter, although the interpretation he later gives of this procedure is somewhat different. (See Neyman, 1941, for a detailed discussion). A first outline of a general theory of confidence procedures was given by Neyman (1934, Appendix, Note I). The discussion of Neyman's paper shows how many statisticians misunderstood this concept. A fuller treatment of confidence procedures in their relation to hypotheses testing was given in Neyman (1937, 1938).

5.2 The evaluation of confidence procedures

Suggests the concentration about the "true" value as a criterion for the quality of one–sided confidence bounds. The same criterion applies to median unbiased estimators (i.e., one–sided confidence bounds with covering probability $1/2$). Discusses criteria for the evaluation of confidence intervals. Points to the problem of conditional covering probabilities.

The shape of the confidence sets is a decisive feature of a confidence procedure. What $\kappa(P) \in K(x)$ really means is hard to say if $K(x)$ is a rather complex set (say one being the union of 13 intervals).

Example 5.2.1. Let P_ϑ, $\vartheta \in \mathbb{R}$, be the location parameter family of Cauchy distributions. Let C_ϑ be the critical region symmetric about ϑ, which is most powerful for testing the hypothesis ϑ against alternatives $\vartheta \pm 3$, say. For appropriate values of the level α we have $C_\vartheta = (\vartheta - b, \vartheta - a) \cup (\vartheta + a, \vartheta + b)$. From the family C_ϑ, $\vartheta \in \mathbb{R}$, we obtain the confidence sets $(-\infty, x - b] \cup [x - a, x + a] \cup [x + b, \infty)$.

To no purpose are confidence procedures where confidence sets occur which are empty or equal to $\kappa(\mathfrak{P})$.

Example 5.2.2. Let $\{P_\vartheta : \vartheta \geq 0\}$ denote the family of noncentral χ_k^2–distributions with k degrees of freedom and noncentrality parameter ϑ. Wanted: a lower confidence bound for ϑ. Since the family has isotone likelihood ratios, a lower confidence bound which is maximally concentrated in all intervals containing ϑ can be obtained by inverting the family of critical regions

$$C_\vartheta := (c_\alpha(\vartheta), \infty), \qquad \vartheta \in \Theta,$$

where $c_\alpha(\vartheta)$ is determined such that $P_\vartheta(C_\vartheta) = \alpha$.
    This leads to

$$K(x) := \{\vartheta \geq 0 : x \leq c_\alpha(\vartheta)\}.$$

Since $\{P_\vartheta : \vartheta \geq 0\}$ has isotone likelihood ratios, $\vartheta \to c_\alpha(\vartheta)$ is increasing, so that $c_\alpha(\vartheta) \geq c_\alpha(0)$ for all $\vartheta \geq 0$. Hence $K(x) = [0, \infty)$ if $x \leq c_\alpha(0)$. If ϑ is small, this will occur with a probability almost $1 - \alpha$. (See Lehmann, 1986, pp. 427/8, Problems 2 and 4 for details.)

The following example is taken from Kendall and Stuart (1961, p. 110, Example 20.5).

Example 5.2.3. For the family $\{N^n_{(\vartheta,1)} : \vartheta \in \mathbb{R}\}$, the critical region

$$C_\vartheta := \Big\{(x_1,\ldots,x_n) \in \mathbb{R}^n : \sum_{\nu=1}^{n}(x_\nu - \vartheta)^2 > c_{n,\alpha}\Big\}$$

is reasonable for testing the hypothesis ϑ against two-sided alternatives, with $c_{n,\alpha}$ chosen such that $N^n_{(\vartheta,1)}(C_\vartheta) = \alpha$. The pertaining confidence procedure is given by

$$K_n(x_1,\ldots,x_n) := \Big\{\vartheta \in \mathbb{R} : \sum_{\nu=1}^{n}(x_\nu - \vartheta)^2 \le c_{n,\alpha}\Big\}.$$

Since $\sum_{\nu=1}^{n}(x_\nu - \vartheta)^2 \ge \sum_{\nu=1}^{n}(x_\nu - \bar{x}_n)^2$, we have $K_n(x_1,\ldots,x_n) = \emptyset$ if $\sum_{\nu=1}^{n}(x_\nu - \bar{x}_n)^2 > c_{n,\alpha}$. This will happen with probability almost α, irrespectively of the value of ϑ.

In addition to shape, covering probability is an essential characteristic of the quality of a confidence procedure. Covering probability of K corresponds to the power of the pertaining family of tests: Assume that

$$P\{x \in X : \kappa(P) \in K_0(x)\} \ge P\{x \in X : \kappa(P) \in K_1(x)\} \quad \text{for } P \in \mathfrak{P},$$

and

$$P\{x \in X : r \in K_0(x)\} < P\{x \in X : r \in K_1(x)\}$$
$$\text{for } r \in Y,\ P \in \mathfrak{P},\ \text{with } r \neq \kappa(P).$$

Then the confidence procedure K_0 is better than K_1. Correspondingly: For every $r \in Y$, the critical region $C_0(r)$ is more powerful than the critical region $C_1(r)$, with

$$C_i(r) := \{x \in X : r \notin K_i(x)\}, \qquad i = 0, 1.$$

Regrettably, confidence procedures are hardly ever comparable in this strong sense. The situation becomes more favorable if we turn to a more special model. Assume now that the functional κ is real valued, and that the problem is to obtain one-sided confidence bounds.

Definition 5.2.4. $\hat{\kappa} : X \to \mathbb{R}$ is an *upper confidence bound* for the functional κ with *confidence coefficient* β if

$$\inf_{P \in \mathfrak{P}} P\{x \in X : \kappa(P) \le \hat{\kappa}(x)\} = \beta.$$

Lower confidence bounds are defined correspondingly.

In the terminology introduced in Section 5.1, $K(x) := (-\infty, \hat{\kappa}(x)]$ defines a confidence procedure with confidence coefficient β.

Remark 5.2.5. At a surface inspection it appears as an advantage if $P\{x \in X : \hat{\kappa}(x) = \kappa(P)\} > 0$. Since such a property cannot hold for all $P \in \mathfrak{P}$, it results in a certain inhomogeneity. Hence we prefer confidence bounds fulfilling $P\{x \in X : \hat{\kappa}(x) = \kappa(P)\} = 0$ for $P \in \mathfrak{P}$.

If Θ is an open subset of $\mathbb{R}^p$, and $\kappa : \Theta \to \mathbb{R}$ a map with continuous partial derivatives $\kappa^{(\cdot)}(\vartheta) \neq 0$, then $P_\vartheta\{x \in X : \hat{\kappa}(x) = \kappa(\vartheta)\} = 0$ for λ^p–a.a. $\vartheta \in \Theta$, provided P_ϑ, $\vartheta \in \Theta$, are mutually absolutely continuous. Notice that this carries over to arbitrary sample sizes. If

$$\overline{\lim_{n \to \infty}} P_\vartheta^{(n)}\{x \in X_n : \kappa(\vartheta) < \kappa^{(n)}(x)\} \leq \beta$$
$$\leq \underline{\lim_{n \to \infty}} P_\vartheta^{(n)}\{x \in X_n : \kappa(\vartheta) \leq \kappa^{(n)}(x)\} \qquad \text{for } \vartheta \in \Theta,$$

this relation holds for λ^p–a.a. $\vartheta \in \Theta$ with $\leq$ replaced by $=$, and $\overline{\lim}$, $\underline{\lim}$ replaced by lim. For further asymptotic results on atoms of $P_\vartheta^{(n)} \circ \kappa^{(n)}$ see Proposition 7.1.11 and the comments to condition (8.2.1').

With $\vartheta_0 \in \Theta$ fixed, let

$$B := \{t \in \mathbb{R} : P_{\vartheta_0}\{x \in X : \hat{\kappa}(x) = t\} > 0\}.$$

Since $P_\vartheta \ll P_{\vartheta_0}$ implies $P_\vartheta \circ \hat{\kappa} \ll P_{\vartheta_0} \circ \hat{\kappa}$, the relation $P_\vartheta\{x \in X : \hat{\kappa}(x) = \kappa(\vartheta)\} > 0$ implies $P_{\vartheta_0}\{x \in X : \hat{\kappa}(x) = \kappa(\vartheta)\} > 0$, hence $\kappa(\vartheta) \in B$. Let $N := \{\vartheta \in \Theta : \kappa(\vartheta) \in B\}$. Since $P_{\vartheta_0} \circ \hat{\kappa}$ has at most countably many atoms, B is countable, and since $\kappa^{(\cdot)}$ has continuous partial derivatives $\neq 0$, this implies $\lambda^p(N) = 0$ (see, e.g., Nöbeling, 1978, p. 16, Korollar 2).

For confidence bounds it is comparatively easy to introduce an appropriate optimality concept, and to specify conditions under which optimal confidence bounds exist.

The assertion $\kappa(P) \leq \hat{\kappa}_0(x)$ is more accurate than $\kappa(P) \leq \hat{\kappa}_1(x)$ if $\kappa(P) \leq \hat{\kappa}_0(x) < \hat{\kappa}_1(x)$. If $P\{x \in X : \kappa(P) \leq \hat{\kappa}_i(x)\} = \beta$ for $i = 0, 1$, we therefore prefer $\hat{\kappa}_0$ over $\hat{\kappa}_1$ if

$$P\{x \in X : \hat{\kappa}_0(x) > t\} \leq P\{x \in X : \hat{\kappa}_1(x) > t\} \quad \text{for } P \in \mathfrak{P} \text{ and } t > \kappa(P).$$

If $\hat{\kappa}_1(x) < \hat{\kappa}_0(x) < \kappa(P)$, the assertion $\kappa(P) \leq \hat{\kappa}_0(x)$ is less misleading than the assertion $\kappa(P) \leq \hat{\kappa}_1(x)$. Hence we prefer $\hat{\kappa}_0$ over $\hat{\kappa}_1$ if

$$P\{x \in X : \hat{\kappa}_0(x) < t\} \leq P\{x \in X : \hat{\kappa}_1(x) < t\} \quad \text{for } P \in \mathfrak{P} \text{ and } t < \kappa(P).$$

Taking both cases together this means that we prefer $\hat{\kappa}_0$ over $\hat{\kappa}_1$ if

$$P\{x \in X : t' \leq \hat{\kappa}_0(x) \leq t''\} \geq P\{x \in X : t' \leq \hat{\kappa}_1(x) \leq t''\} \qquad (5.2.6)$$
$$\text{for } P \in \mathfrak{P} \text{ and } t' < \kappa(P) < t''.$$

Hence an upper confidence bound is optimal if it is maximally concentrated in every interval containing $\kappa(P)$, for every $P \in \mathfrak{P}$.

Conditions for the existence and the optimality of upper confidence bounds with covering probability β will be given in Section 5.5. Applied with $\beta = 1/2$ this yields results on median unbiased estimators. The optimality concept (5.2.6) for confidence bounds corresponds to maximal concentration in the sense of Definition 2.3.1.

For *confidence intervals*, say $[\underline{\kappa}(x), \overline{\kappa}(x)]$, the desiderata are less obvious. Of course, the true parameter value should be covered with high probability, and the confidence intervals should be as small as possible. This could mean that their expected length is small. (Should the expectation also extend over those points $x \in X$ for which the confidence intervals fail to cover the true parameter value?) The duality to tests suggest another concept of "smallness": That they cover parameter values other than the true one with low probability.

In our opinion, both notions of "smallness" neglect some aspects of the problem. In many cases, two–sided confidence bounds are meant as error margins for an estimator. Having obtained an estimate $\hat{\kappa}(x)$, one would like to know something about its accuracy. How far off the true parameter value the estimate $\hat{\kappa}(x)$ might be? The appropriate answer to such a question are two–sided confidence bounds: The estimate $\hat{\kappa}(x)$ is supplemented by two other values, say $\underline{\kappa}(x)$, $\overline{\kappa}(x)$, marking off an interval which includes the true parameter value with high probability. In such a situation, neither covering probabilities nor expected length seem appropriate.

Consider the intervals represented by Figure 5.1.

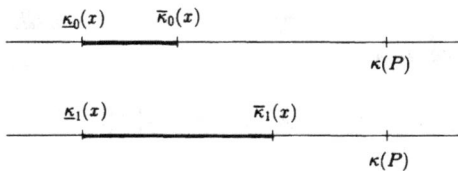

Figure 5.1   Shows two confidence intervals for $\kappa(P)$.

Probably, most statisticians would agree that the assertion

$$\kappa(P) \in [\underline{\kappa}_1(x), \overline{\kappa}_1(x)]$$

is less faulty than the assertion

$$\kappa(P) \in [\underline{\kappa}_0(x), \overline{\kappa}_0(x)].$$

Nevertheless, a confidence procedure which tends to produce two–sided bounds $[\underline{\kappa}_0(x), \overline{\kappa}_0(x)]$ of the kind indicated in the upper line has a fair chance to be preferred over a confidence procedure which tends to produce bounds $[\underline{\kappa}_1(x), \overline{\kappa}_1(x)]$, if evaluated by covering probabilities, because it covers values $t < \kappa(P)$ with smaller probability.

For $t < \kappa(P)$ we would prefer $P\{x \in X : \underline{\kappa}(x) \le t\}$ to be small, but there is no justification for requiring that, given $\underline{\kappa}(x) \le t$, the P–probability of $\overline{\kappa}(x) \le t$ should be large. On the contrary. Irrespectively of the value of $\underline{\kappa}(x)$, we would like to have $\overline{\kappa}(x)$ close to $\kappa(P)$. There is another aspect of this problem. After having obtained the confidence interval $[\underline{\kappa}(x), \overline{\kappa}(x)]$, we are confident that $\kappa(P)$ is in $[\underline{\kappa}(x), \overline{\kappa}(x)]$; we are a fortiori confident that $\kappa(P)$ is below $\overline{\kappa}(x)$. There is no reason for preferring procedure 0 over procedure 1 because of $P\{x \in X : t \in [\underline{\kappa}_0(x), \overline{\kappa}_0(x)]\} < P\{x \in X : t \in [\underline{\kappa}_1(x), \overline{\kappa}_1(x)]\}$ for $t < \kappa(P)$, if this is brought about by $P\{x \in X : \overline{\kappa}_0(x) < t\} > P\{x \in X : \overline{\kappa}_1(x) < t\}$.

Our conclusion: The event $\{x \in X : t \notin [\underline{\kappa}(x), \overline{\kappa}(x)]\}$ should be split up into $\{x \in X : \underline{\kappa}(x) > t\}$ and $\{x \in X : \overline{\kappa}(x) < t\}$, and both functions,

$$t \to P\{x \in X : \underline{\kappa}(x) > t\} \quad \text{and} \quad P\{x \in X : \overline{\kappa}(x) < t\},$$

should be considered. We will prefer the confidence procedure $[\underline{\kappa}_0, \overline{\kappa}_0]$ over $[\underline{\kappa}_1, \overline{\kappa}_1]$ if $\underline{\kappa}_0$ and $\overline{\kappa}_0$ are more closely concentrated around $\kappa(P)$ than $\underline{\kappa}_1$ and $\overline{\kappa}_1$, respectively.

The following result, due to Pratt (1961, p. 550, formula (2)) shows the close connection between covering probabilities and expected length. Notice, however, that expected length may not be a valid criterion in case of a scale parameter.

Proposition 5.2.7. *For nonrandomized confidence procedures with* $\{(x,t) \in X \times \mathbb{R} : t \in K(x)\} \in \mathcal{A} \times \mathbb{B}$, *we have*

$$\int \lambda(K(x)) P(dx) = \int P\{x \in X : t \in K(x)\} \lambda(dt).$$

Proof. By Fubini's theorem,

$$\int P\{x \in X : t \in K(x)\} \lambda(dt) = P \times \lambda\{(x,t) \in X \times \mathbb{R} : t \in K(x)\}$$

$$= \int \lambda\{t \in \mathbb{R} : t \in K(x)\} P(dx) = \int \lambda(K(x)) P(dx).$$

$\square$

The following exercise is due to Tate and Klett (1959). See also Pratt (1961, Section 8).

Exercise 5.2.8. For the family $\{N^n_{(\mu,\sigma^2)} : \mu \in \mathbb{R}, \sigma^2 > 0\}$, consider for σ^2 the confidence procedure

$$K_n(\underline{x}) := \left(b_n^{-1} \sum_1^n (x_\nu - \bar{x}_n)^2, \ a_n^{-1} \sum_1^n (x_\nu - \bar{x}_n)^2\right). \tag{5.2.9}$$

Since the distribution of $\sum_1^n (x_\nu - \bar{x}_n)^2 / \sigma^2$ under $N^n_{(\mu,\sigma^2)}$ is χ^2_{n-1}, we have

$$N^n_{(\mu,\sigma^2)}\{\underline{x} \in \mathbb{R}^n : \sigma^2 \in K_n(\underline{x})\} = 1 - \alpha \quad \text{iff} \quad \chi^2_{n-1}(a_n, b_n) = 1 - \alpha.$$

Show that the expected length for confidence intervals of type (5.2.9) is minimal if $h_{n+3}(a_n) = h_{n+3}(b_n)$, where h_m denotes the density of χ^2_m.

Hint: The expected length of K_n is $(a_n^{-1} - b_n^{-1})(n - 1)\sigma^2$. The problem is to minimize $a_n^{-1} - b_n^{-1}$ under the side condition $\int_{a_n}^{b_n} h_{n-1}(u)du = 1 - \alpha$. Since

$$\int_{a_n}^{b_n} h_{n-1}(u)du = (n^2 - 1) \int_{b_n^{-1}}^{a_n^{-1}} h_{n+3}(u^{-}\,)du,$$

this amounts to minimizing $a_n^{-1} - b_n^{-1}$ under the side condition

$$\int_{b_n^{-1}}^{a_n^{-1}} h_{n+3}(u^{-1})du = \text{const}.$$

For $m \geq 2$, χ^2_m is strongly unimodal, hence $u \to h_m(u^{-1})$ is superconvex. For any superconvex function h, $\beta - \alpha$ is minimized under the side condition $\int_\alpha^\beta h(v)dv = \text{const}$ if $h(\alpha) = h(\beta)$.

Exercise 5.2.10. For the family $\{N^n_{(\mu,1)} : \mu \in \mathbb{R}\}$ consider for μ the confidence procedure

$$K_n(\underline{x}) = (\bar{x}_n - n^{-1/2}c_\beta, \ \bar{x}_n + n^{-1/2}c_\beta), \tag{5.2.11}$$

with $c_\beta = u_{(1-\beta)/2}$. The length of the confidence intervals is $2n^{-1/2}c_\beta$. Another confidence procedure with the same covering probability β which works for the more general family $\{N^n_{(\mu,\sigma^2)} : \mu \in \mathbb{R}, \sigma^2 > 0\}$ is

$$\hat{K}_n(\underline{x}) = (\bar{x}_n - n^{-1/2}\hat{c}_{n,\beta}s_n, \ \bar{x}_n + n^{-1/2}\hat{c}_{n,\beta}s_n), \tag{5.2.12}$$

with $\hat{c}_{n,\beta} = t_{n-1,(1-\beta)/2}$ and $s_n^2 := (n-1)^{-1}\sum_1^n(x_\nu - \bar{x}_n)^2$. The length of these confidence intervals, $2n^{-1/2}\hat{c}_{n,\beta}s_n$, is now random. Show that the expected length of these confidence intervals is larger than $2n^{-1/2}c_\beta$.

Hint: Show that $a_n := \int s_n dN_{(0,1)}^n = \frac{\sqrt{2}}{\sqrt{n-1}}\Gamma(n/2)/\Gamma((n-1)/2)$. The assertion amounts to $a_n t_{n-1,\alpha} > u_\alpha$ for $\alpha \in (0, 1/2)$.

$a_n t_{n-1,\alpha}$ is the upper α–quantile of the distribution induced by t_{n-1} and the map $x \to a_n x$, say Q_n; u_α is the upper α–quantile of $N_{(0,1)}$. With both distributions symmetric about 0, it remains to be shown that $N_{(0,1)}$ is more concentrated about 0 than Q_n. This follows by Lemma 2.3.2 from the fact that $I := \{x \in \mathbb{R} : q_n(x) \leq \varphi(x)\}$ is an interval containing 0, and that $q_n(x) < \varphi(x)$ for $x \in I$, unless $x = \inf I$, $x = 0$ or $x = \sup I$ (q_n denotes the Lebesgue density of Q_n).

We complete the section about the evaluation of confidence procedures with a brief discussion of the so–called "ancillarity paradox". The problem is this: The fundamental property of a confidence procedure K with confidence coefficient $\geq \beta$ is that

$$P\{x \in X : \kappa(P) \in K(x)\} \geq \beta \qquad \text{for every } P \in \mathfrak{P}.$$

Hence one could bet $\beta{:}(1-\beta)$ that $\kappa(P) \in K(x)$ holds true. This interpretation becomes somewhat shaky if an opponent can always win in the long run provided he has the freedom to decide — based on the knowledge of x — whether he accepts the bet or not. This will be the case if there exists a subset $A \in \mathcal{A}$ such that the conditional expectation of $\kappa(P) \in K(x)$, given $x \in A$, under P, is less than β for every $P \in \mathfrak{P}$.

Example 5.2.13. $\hat{K}_n$, defined by (5.2.12), is a confidence procedure for μ with covering probability β for the family $\mathfrak{P} = \{N_{(\mu,1)}^n : \mu \in \mathbb{R}\}$. Since $\bar{x}_n$ and s_n are stochastically independent, the conditional probability of $\mu \in \hat{K}_n$, given $s_n < c_\beta/\hat{c}_{n,\beta}$, under $N_{(\mu,1)}^n$, is less than β. The confidence procedure $\hat{K}_n$ is not "optimal" for the family $\{N_{(\mu,1)}^n : \mu \in \mathbb{R}\}$; K_n, defined by (5.2.11), is a better one. Hence one might hope that a phenomenon like this is impossible for confidence procedures which are "optimal". This is, however, not the case. If $\mathfrak{P} = \{N_{(\mu,\sigma^2)}^n : \mu \in \mathbb{R}, \sigma^2 > 0\}$, then $\hat{K}_n$ is optimal in any reasonable sense. Yet the conditional probability of $\mu \in \hat{K}_n$, given $|\bar{x}_n|/s_n < t$, under $N_{(\mu,\sigma^2)}^n$ is less than β if $t > \hat{c}_{n,\beta}/\left(\sqrt{\hat{c}_{n,\beta}^2 + 1} - 1\right)$. (See L.D. Brown, 1967, p. 1068.)

We stop the discussion of this phenomenon since we have no reasonable answer to this problem.

5.3 The construction of one–sided confidence bounds and median unbiased estimators

Contains a procedure for the construction of one–sided confidence bounds for a real parameter under certain monotonicity conditions. Randomized confidence procedures are constructed for which the covering probability is the same for every p–measure in the family, and which can, therefore, be used as median unbiased estimators if the covering probability is chosen equal to $1/2$.

Throughout this section let $\Theta = (\vartheta_0, \vartheta_1)$ be an open interval (including the cases $\vartheta_0 = -\infty$ and/or $\vartheta_1 = \infty$).

The first result refers to a family of mutually absolutely continuous p–measures $\{P_\vartheta | \mathbb{B} : \vartheta \in \Theta\}$. Let $I \subset \mathbb{R}$ denote the smallest interval (closed, semi–closed or open; finite or infinite) such that $P_\vartheta(I) = 1$.

Let $t_0 := \inf I$, $t_1 := \sup I$. If P_ϑ is nonatomic, $I = (t_0, t_1)$. If t_i is finite, we may have $t_i \in I$ if $P_\vartheta\{t_i\} > 0$ (for some, and therefore for all, $\vartheta \in \Theta$).

We shall use the following conditions on the distribution function F_ϑ of P_ϑ.

$$\vartheta \to F_\vartheta(t) \text{ is continuous for every } t \in I; \qquad (5.3.1)$$

$$\vartheta \to F_\vartheta(t) \text{ is antitone for every } t \in I. \qquad (5.3.2)$$

Theorem 5.3.3. *Assume that $\{P_\vartheta : \vartheta \in \Theta\}$ is a family of mutually absolutely continuous p–measures $P_\vartheta | \mathbb{B}$ fulfilling conditions (5.3.1) and (5.3.2). For $\beta \in (0,1)$, let $\hat{\vartheta}_\beta : I \to [\vartheta_0, \vartheta_1]$ be defined by*

$$\hat{\vartheta}_\beta(t) := \sup\{\vartheta \in \Theta : F_\vartheta(t) \geq 1 - \beta\} \quad \text{for } t \in I, \qquad (5.3.4)$$

where $\sup \emptyset := \vartheta_0$.

Then the following holds true.

(i) $\hat{\vartheta}_\beta$ *is isotone.*

(ii) $\hat{\vartheta}_\beta$ *is an upper confidence bound with confidence coefficient $\geq \beta$, i.e.*

$$P_\vartheta\{t \in I : \vartheta \leq \hat{\vartheta}_\beta(t)\} \geq \beta \quad \text{for } \vartheta \in \Theta. \qquad (5.3.5)$$

(iii) *If P_ϑ is nonatomic and*

$$1 - \beta \in \{F_\vartheta(t) : \vartheta \in \Theta\}^\circ \quad \text{for } t \in (t_0, t_1), \qquad (5.3.6)$$

then equality holds in (5.3.5), the distribution of $\hat{\vartheta}_\beta$ under P_ϑ is nonatomic, and $\hat{\vartheta}_\beta$ attains only values in Θ.

Remark 5.3.7. Notice that definition (5.3.4) is not without a problem. The pertaining confidence set $(\vartheta_0, \hat{\vartheta}_\beta(t)]$ is empty or equal to Θ if $\hat{\vartheta}_\beta(t)$ equals ϑ_0

or ϑ_1, respectively. If we consider $\hat{\vartheta}_{1/2}$ as a median unbiased estimator, this estimator may attain the virtual values ϑ_0 and/or ϑ_1 with positive probability.

Proof of Theorem 5.3.3. (i) Follows since F_ϑ is isotone for $\vartheta \in \Theta$.

(ii) Follows from Lemma 5.3.14 since (5.3.4) implies for every $\vartheta \in \Theta$ that

$$\vartheta \le \hat{\vartheta}_\beta(t) \qquad \text{iff} \qquad F_\vartheta(t) \ge 1 - \beta. \tag{5.3.8}$$

Since F_ϑ is measurable, $\{t \in I : \vartheta \le \hat{\vartheta}_\beta(t)\}$ is measurable for every $\vartheta \in \Theta$. Since $\{t \in I : \vartheta_0 \le \hat{\vartheta}_\beta(t)\} = I$ and $\{t \in I : \vartheta_1 \le \hat{\vartheta}_\beta(t)\} = \bigcap_{n=1}^\infty \{t \in I : \tau_n \le \hat{\vartheta}_\beta(t)\}$ for a sequence $\tau_n \uparrow \vartheta_1$, $\{t \in I : \vartheta \le \hat{\vartheta}_\beta(t)\}$ is measurable for every $\vartheta \in [\vartheta_0, \vartheta_1]$.

(iii) If P_ϑ is nonatomic, then I is an open interval. Hence (5.3.6) implies that $\hat{\vartheta}_\beta$ attains only values in Θ. Moreover, we have $P_\vartheta \circ F_\vartheta = U$, which, together with (5.3.8), implies that equality holds in (5.3.5) and that the distribution of $\hat{\vartheta}_\beta$ under P_ϑ is nonatomic. □

Example 5.3.9. Let $\{B_{n,\vartheta} : \vartheta \in (0,1)\}$ be the family of binomial distributions. Since

$$B_{n,\vartheta}\{k, \ldots, n\} = k \binom{n}{k} \int_0^\vartheta t^{k-1}(1-t)^{n-k}dt, \qquad k = 1, \ldots, n,$$

the function $\vartheta \to B_{n,\vartheta}\{0, \ldots, k\}$ is decreasing for $k < n$, and equal to 1 for $k = n$. We have $I = [0, n]$. The definition of $\hat{\vartheta}_\beta$ according to (5.3.4) leads for $k_0 \in \{0, \ldots, n-1\}$ to a value $\hat{\vartheta}_\beta(k_0)$ such that $B_{n,\hat{\vartheta}_\beta(k_0)}\{0, \ldots, k_0\} = 1 - \beta$, and $\hat{\vartheta}_\beta(n) = 1$. Hence Theorem 5.3.3 implies that $\hat{\vartheta}_\beta$ is an upper confidence bound for ϑ with confidence coefficient β, since $\lim_{\vartheta \downarrow \hat{\vartheta}_\beta(k_0)} B_{n,\vartheta}\{k \in \{0, \ldots, n\} : \vartheta \le \hat{\vartheta}_\beta(k)\} = \beta$ for $k_0 \in \{0, \ldots, n-1\}$.

The following Figure 5.2 shows the covering probability $B_{n,\vartheta}\{k \in \{0, \ldots, n\} : \vartheta \le \hat{\vartheta}_\beta(k)\}$ as a function of ϑ for $n = 4$ and $\beta = 0.9$.

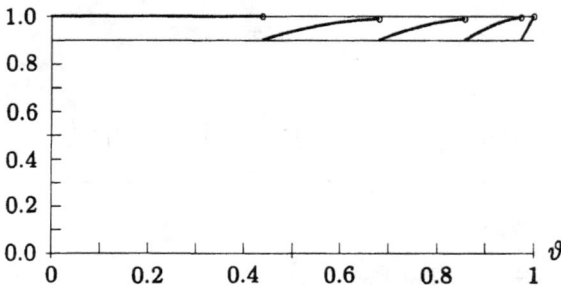

Figure 5.2

If P_ϑ has atoms, then strict inequality holds in (5.3.5) for most $\vartheta \in \Theta$. This is acceptable if one is interested in upper confidence bounds with a large covering probability. If the interest is in median unbiased estimators, equality in (5.3.5) with $\beta = 1/2$ is a necessity. For p–measures P_ϑ with atoms, equality in (5.3.5) can be achieved by randomized confidence bounds only.

For this purpose we introduce the following randomized version of a distribution function: Given a p–measure $P|\mathbb{B}$, let $F : \mathbb{R} \times (0,1) \to [0,1]$ be defined by

$$F(t, u) := P(-\infty, t) + uP\{t\}. \qquad (5.3.10)$$

By Lemma 5.3.15, $(P \times U) \circ F = U$ (with U being the uniform distribution over $(0,1)$).

Theorem 5.3.11. *Assume that $\{P_\vartheta|\mathbb{B} : \vartheta \in \Theta\}$ is a family of mutually absolutely continuous p–measures with distribution functions which are continuous and antitone functions of the parameter (see (5.3.1) and (5.3.2)).*

Using the randomized distribution function F_ϑ defined in (5.3.10), we define $\hat\vartheta_\beta : I \times (0,1) \to [\vartheta_0, \vartheta_1]$ for $\beta \in (0,1)$ as follows.

For $t \in I$ and $u \in (0,1)$ let

$$\hat\vartheta_\beta(t, u) := \sup\{\vartheta \in \Theta : F_\vartheta(t, u) \geq 1 - \beta\}, \qquad (5.3.12)$$

where $\sup \emptyset := \vartheta_0$.

Then the following holds true.

(i) $\hat\vartheta_\beta(t', u') \leq \hat\vartheta_\beta(t'', u'')$ *for $t' < t''$ and arbitrary $u', u'' \in (0,1)$.*

(ii) $\hat\vartheta_\beta$ *is a (randomized) upper confidence bound with covering probability equal to β, i.e.*

$$P_\vartheta \times U\{(t, u) \in I \times (0,1) : \vartheta \leq \hat\vartheta_\beta(t, u)\} = \beta \quad \text{for } \vartheta \in \Theta.$$

(iii) *The distribution of $\hat\vartheta_\beta$ under $P_\vartheta \times U$ is nonatomic on Θ.*

Observe that the confidence bound $\hat\vartheta_\beta$ attains the "virtual" value ϑ_i with positive probability if $t_i \in I$. Since $\vartheta \in (\vartheta_0, \vartheta_1)$, this means that the confidence interval is in these cases either empty or equal to the whole parameter set. $\hat\vartheta_{1/2}$, interpreted as a median unbiased estimator, fails to be proper.

Proof of Theorem 5.3.11. Relation (i) follows from

$$P_\vartheta(-\infty, t') + u'P_\vartheta\{t'\} \leq P_\vartheta(-\infty, t'] \leq P_\vartheta(-\infty, t'')$$
$$\leq P_\vartheta(-\infty, t'') + u''P_\vartheta\{t''\} \qquad \text{for } t' < t''.$$

Since $\vartheta \to P_\vartheta(-\infty, t]$ is antitone and continuous for every $t \in I$, these properties carry over to $\vartheta \to P_\vartheta(-\infty, t)$, and therefore to $\vartheta \to F_\vartheta(t, u)$. Together with (5.3.12), this implies for $\vartheta \in \Theta$ that

$$\vartheta \leq \hat{\vartheta}_\beta(t, u) \qquad \text{iff} \qquad F_\vartheta(t, u) \geq 1 - \beta.$$

Since $(P_\vartheta \times U) \circ F_\vartheta = U$ by Lemma 5.3.15, assertions (ii) and (iii) follow. Observe that measurability of F_ϑ implies measurability of $\hat{\vartheta}_\beta$. □

An upper confidence bound for the binomial family was given in Example 5.3.9. The following example gives a (randomized) median unbiased estimator for the binomial family.

Example 5.3.13. For $\vartheta \in (0, 1)$ let $B_{n,\vartheta}$ be the binomial distribution. It suffices to define the median unbiased estimator, say $\hat{\vartheta}(k, u)$, for $k \in \{0, 1, \ldots, n\}$. Definition (5.3.12) leads to

$$\hat{\vartheta}(k, u) := \sup\{\vartheta \in (0, 1) : B_{n,\vartheta}\{0, 1, \ldots, k - 1\} + u B_{n,\vartheta}\{k\} \geq \frac{1}{2}\}$$

(and $\hat{\vartheta}(0, u) = 0$ if $u \leq 1/2$, $\hat{\vartheta}(n, u) = 1$ if $u \geq 1/2$).
For the case $n = 2$ this becomes

$$\hat{\vartheta}(0, u) = \begin{cases} 0 & u \leq \frac{1}{2} \\ 1 - \frac{1}{\sqrt{2u}} & u > \frac{1}{2} \end{cases}$$

$$\hat{\vartheta}(1, u) = \begin{cases} \frac{1}{2} & u = \frac{1}{2} \\ \frac{1 - u - (u^2 - u + 1/2)^{1/2}}{1 - 2u} & u \neq \frac{1}{2} \end{cases}$$

$$\hat{\vartheta}(2, u) = \begin{cases} \frac{1}{\sqrt{2(1-u)}} & u < \frac{1}{2} \\ 1 & u \geq \frac{1}{2}. \end{cases}$$

If the true parameter value is ϑ, the median unbiased estimator attains the virtual value 0 with probability $\vartheta^2/2$, and the virtual value 1 with probability $(1 - \vartheta)^2/2$. The confidence interval $(0, \hat{\vartheta}(k, u))$ is empty or equal to $(0, 1)$ with probability $\vartheta^2/2$ and $(1 - \vartheta)^2/2$, respectively.

Lemma 5.3.14. *Let $Q | \mathbb{B}$ be a p–measure, and F its distribution function. Then the following holds true.*
(i) $Q\{t \in \mathbb{R} : F(t) \geq \alpha\} \geq 1 - \alpha$ *for every $\alpha \in (0, 1)$.*
(ii) $Q\{t \in \mathbb{R} : F(t) \geq \alpha\} = 1 - \alpha$ *if $Q\{t_\alpha\} = 0$, for $t_\alpha := \inf\{t \in \mathbb{R} : F(t) \geq \alpha\}$.*

Proof. Since F is isotone and continuous from the right, we have $F(t) \geq \alpha$ iff $t \geq t_\alpha$. Hence $Q\{t \in \mathbb{R} : F(t) \geq \alpha\} = Q[t_\alpha, \infty)$. Since $Q(-\infty, t_\alpha) \leq \alpha \leq Q(-\infty, t_\alpha]$, this implies the assertion. □

Lemma 5.3.15. *Let $Q|\mathbb{B}$ be a p–measure, and F its distribution function. Then the following holds true.*

$$Q \times U\{(t, u) \in \mathbb{R} \times (0, 1) : Q(-\infty, t) + uQ\{t\} \leq \alpha\} = \alpha \text{ for every } \alpha \in (0, 1),$$

with U the uniform distribution over $(0, 1)$.

Proof. It is straightforward to show that

$$\{(t, u) \in \mathbb{R} \times (0, 1) : Q(-\infty, t) + uQ\{t\} \leq \alpha\}$$
$$= \begin{cases} (-\infty, t_\alpha) \times (0, 1) + \{t_\alpha\} \times (0, (\alpha - Q(-\infty, t_\alpha))/Q\{t_\alpha\}) \\ (-\infty, t_\alpha] \times (0, 1) \end{cases} \text{ for } Q\{t_\alpha\} \overset{>}{=} 0.$$

$\square$

5.4 Optimal one–sided confidence bounds and median unbiased estimators

The confidence bounds and median unbiased estimators constructed in Section 5.3 are optimal if the family of p–measures has monotone likelihood ratios.

Let $\Theta = (\vartheta_0, \vartheta_1)$ be an interval in $\mathbb{R}$ (including the cases $\vartheta_0 = -\infty$ and/or $\vartheta_1 = \infty$), and $\{P_\vartheta : \vartheta \in \Theta\}$ a family of mutually absolutely continuous p–measures on $\mathcal{A}$. The basic assumption of this section is that this family admits a sufficient statistic $T : (X, \mathcal{A}) \to (\mathbb{R}, \mathbb{B})$.

According to Proposition 5.4.1, the construction described in Section 5.3 can be used to obtain confidence bounds which are isotone functions of T and have constant covering probability β. Theorem 5.4.3 establishes that such confidence bounds are maximally concentrated among all confidence bounds with constant covering probability β if $\{P_\vartheta \circ T : \vartheta \in \Theta\}$ has isotone likelihood ratios.

As to be seen from Section 5.3, the construction of confidence bounds with a given confidence level uses continuity and monotonicity of $\vartheta \to F_\vartheta$. In principle, such properties could be used to obtain families of critical functions with corresponding properties, and to convert such families into confidence procedures. The direct way to confidence bounds entered upon in Section 5.3 seems to be the simpler one. The situation is different with regard to optimum properties: Here it seems easier to make use of optimality results on tests for families with isotone likelihood ratios, as given in Theorem 4.5.2(ii).

For the case of nonrandomized confidence bounds, the results of Proposition 5.4.1 and Theorem 5.4.3 are due to Lehmann (1959, p. 80, Corollary 3; see also 1986, pp. 91/2). A related result occurs in Birnbaum (1961, p. 121, Lemma 2).

Let $I \subset \mathbb{R}$ be the smallest interval such that $P_\vartheta \circ T(I) = 1$. Let $t_0 := \inf I$, $t_1 = \sup I$.

Proposition 5.4.1. *Assume that $\{P_\vartheta \circ T : \vartheta \in \Theta\}$ is stochastically isotone and fulfills condition (5.3.1). Then the following holds true.*

· *For $\beta \in (0,1)$ there exists $\hat{\vartheta}_\beta : I \times (0,1) \to [\vartheta_0, \vartheta_1]$ such that*

(i) $\hat{\vartheta}_\beta(t', u') \leq \hat{\vartheta}_\beta(t'', u'')$ *for $t' < t''$ and arbitrary $u', u'' \in (0,1)$;*

(ii) $P_\vartheta \times U\{(x, u) \in X \times (0,1) : \vartheta \leq \hat{\vartheta}_\beta(T(x), u)\} = \beta$ *for $\vartheta \in \Theta$;*

(iii) *the distribution of $(x, u) \to \hat{\vartheta}_\beta(T(x), u)$ under $P_\vartheta \times U$ is nonatomic on Θ.*

Addendum. *If $P_\vartheta \circ T$, $\vartheta \in \Theta$, are nonatomic with distribution functions fulfilling condition (5.3.6), then there exists $\hat{\vartheta}_\beta : I \to (\vartheta_0, \vartheta_1)$ which has properties (i)–(iii) without randomization by U.*

Proof. Assertions (i)–(iii) follow from Theorem 5.3.11, applied for $P_\vartheta \circ T$ in place of P_ϑ. The addendum follows analogously from Theorem 5.3.3.   □

The following theorem establishes that confidence bounds with the properties specified in 5.4.1(i) and (ii) are maximally concentrated if $\{P_\vartheta : \vartheta \in \Theta\}$ has monotone likelihood ratios in T. Notice that this implies (by Proposition 1.7.7) that $\{P_\vartheta \circ T : \vartheta \in \Theta\}$ is stochastically ordered; hence Proposition 5.4.1 applies.

The formulation of Theorem 5.4.3 refers to the case of randomized confidence bounds. If $P_\vartheta \circ T$ is nonatomic, randomization can be dispensed with. Formulation and proof of the results for this case can be obtained by omitting the p–measure U.

The confidence bounds $\hat{\vartheta}_\beta$ described in Proposition 5.4.1 use a special type of randomization. The optimum property asserted in Theorem 5.4.3 holds in the class of arbitrary randomized confidence bounds.

Definition 5.4.2. A *randomized upper confidence bound* is a Markov kernel $K|X \times \overline{\mathbb{B}} \cap [\vartheta_0, \vartheta_1]$ with the following interpretation: If x is observed, the confidence bound is obtained as a realization from $K(x, \cdot)|\overline{\mathbb{B}} \cap [\vartheta_0, \vartheta_1]$.

Hence $K(x, [\vartheta, \vartheta_1])$ is the probability that the upper confidence bound will be greater or equal to ϑ, if x has been observed. The covering probability of the randomized upper confidence bound under P_ϑ is $\int K(x, [\vartheta, \vartheta_1]) P_\vartheta(dx)$.

The confidence bound $\hat{\vartheta}_\beta(T(x), u)$ can be described by the Markov kernel

$$K(x, B) = U\{u \in (0,1) : \hat{\vartheta}_\beta(T(x), u) \in B\}, \qquad B \in \overline{\mathbb{B}} \cap [\vartheta_0, \vartheta_1].$$

Recall Remark 2.1.1 about the representation of arbitrary Markov kernels by means of auxiliary random variables.

Theorem 5.4.3. *Let $\{P_\vartheta : \vartheta \in \Theta\}$ be a family of p–measures with isotone likelihood ratios in T. For $\beta \in (0, 1)$, assume that $\hat{\vartheta}_\beta : I \times (0, 1) \to [\vartheta_0, \vartheta_1]$ has the properties specified in 5.4.1(i) and (ii).*

If $K|X \times \overline{\mathbb{B}} \cap [\vartheta_0, \vartheta_1]$ is an arbitrary randomized confidence bound with the same covering probability β, then

$$P_\vartheta \times U\{(x, u) \in X \times (0, 1) : \vartheta' \le \hat{\vartheta}_\beta(T(x), u) \le \vartheta''\} \qquad (5.4.4)$$

$$\ge \int K(x, [\vartheta', \vartheta'']) P_\vartheta(dx)$$

for $\vartheta \in \Theta$ and $\vartheta', \vartheta'' \in [\vartheta_0, \vartheta_1]$ fulfilling $\vartheta' \le \vartheta \le \vartheta''$.

Proof. With $\vartheta' < \vartheta$ fixed, let

$$t' := \inf\{t \in I : \vartheta' \le \hat{\vartheta}_\beta(t, u) \text{ for some } u \in (0, 1)\}.$$

(Observe that $\hat{\vartheta}_\beta$ attains values greater than or equal to ϑ'.)

We shall show for $t \in I$ that

$$t \begin{smallmatrix} < \\ > \end{smallmatrix} t' \quad \text{implies} \quad \hat{\vartheta}_\beta(t, u) \begin{smallmatrix} < \\ \ge \end{smallmatrix} \vartheta' \qquad \text{for all } u \in (0, 1). \qquad (5.4.5)$$

By definition of t', the relation $t < t'$ implies $\vartheta' > \hat{\vartheta}_\beta(t, u)$ for all $u \in (0, 1)$. If $t > t'$, there exists by definition of t' some $t'' \in [t', t)$ with $\vartheta' \le \hat{\vartheta}_\beta(t'', u)$ for some $u \in (0, 1)$. Hence, as a consequence of 5.4.1(i), $\vartheta' \le \hat{\vartheta}_\beta(t, u)$ for all $u \in (0, 1)$. This proves (5.4.5).

Let

$$\varphi_0 := 1_{\{(x, u) \in X \times (0, 1) : \vartheta' \le \hat{\vartheta}_\beta(T(x), u)\}}.$$

Relation 5.4.1(ii) implies $P_{\vartheta'} \times U(\varphi_0) = \beta$. Because of (5.4.5),

$$T(x) \begin{smallmatrix} > \\ < \end{smallmatrix} t' \quad \text{implies} \quad \varphi_0(x, u) = \begin{cases} 1 \\ 0 \end{cases} \qquad \text{for } x \in X, \ u \in (0, 1).$$

Therefore, φ_0 is most powerful of level β for testing $P_{\vartheta'} \times U$ against $P_\vartheta \times U$ by Theorem 4.5.2(ii).

Let $\varphi(x) = K(x, [\vartheta', \vartheta_1])$. Since $\int K(x, [\vartheta', \vartheta_1]) P_{\vartheta'}(dx) = \beta$, we have

$$P_{\vartheta'} \times U(\varphi_0) = P_{\vartheta'} \times U(\varphi),$$

and therefore

$$P_\vartheta \times U(\varphi_0) \ge P_\vartheta \times U(\varphi).$$

Written explicitly, this is

$$P_\vartheta \times U\{(x,u) \in X \times (0,1) : \vartheta' \le \hat\vartheta_\beta(T(x),u)\}$$
$$\ge \int K(x,[\vartheta',\vartheta_1])\,P_\vartheta(dx),$$

which is one of the inequalities needed for (5.4.4). The corresponding inequality follows similarly.   □

Exercise 5.4.6. Show that the randomized median unbiased estimator obtained in Example 5.3.13 for the binomial distribution is maximally concentrated in the class of all randomized median unbiased estimators.

5.5 Optimal one–sided confidence bounds and median unbiased estimators in the presence of a nuisance parameter

The conditions for the existence of maximally concentrated confidence bounds and median unbiased estimators given in Section 5.4 refer to a family of p–measures with one real parameter. In the present section these results are applied to families with a real structural parameter, and an arbitrary nuisance parameter. Examples are given in Section 5.6.

Let $\Theta = (\vartheta_0, \vartheta_1)$, including the cases $\vartheta_0 = -\infty$ and/or $\vartheta_1 = \infty$, and H an arbitrary set. Throughout this section we assume that $P_{\vartheta,\eta}|\mathcal{A}$ is a p–measure with μ–density

$$x \to g(x)H\big(T(x),\vartheta\big)G\big(S(x),\vartheta,\eta\big), \qquad (5.5.1)$$

with

$$H(\cdot,\vartheta) : (\mathbb{R},\mathbb{B}) \to \big((0,\infty),\mathbb{B}_+\big),$$
$$G(\cdot,\vartheta,\eta) : (Y,\mathcal{B}) \to \big((0,\infty),\mathbb{B}_+\big),$$

and $T : (X,\mathcal{A}) \to (\mathbb{R},\mathbb{B})$, $S : (X,\mathcal{A}) \to (Y,\mathcal{B})$.

The results of this section are based on the following assumptions.

$$\vartheta \to H(t,\vartheta) \text{ is continuous for every } t \in \mathbb{R}. \qquad (5.5.2)$$
$$t \to H(t,\vartheta')/H(t,\vartheta) \text{ is isotone if } \vartheta < \vartheta'. \qquad (5.5.3)$$

Recall that $x \to (T(x), S(x))$ is sufficient for the family $\{P_{\vartheta,\eta} : \vartheta \in \Theta, \eta \in H\}$, and S is sufficient for each of the subfamilies $\{P_{\vartheta,\eta} : \eta \in H\}$, with ϑ fixed.

According to Proposition 1.3.1 there exists a conditional distribution of T, given S, with respect to $P_{\vartheta,\eta}$, say $M_\vartheta | Y \times \mathbb{B}$, which does not depend on η. Roughly speaking, the results of this section are obtained by applying the results of Section 5.3 to the families $\{M_\vartheta(y, \cdot) : \vartheta \in \Theta\}$, with $y \in Y$ fixed. Part of the regularity conditions needed for these families can be obtained from conditions on the densities (5.5.1).

Lemma 5.5.4. *Assume that $\{P_{\vartheta,\eta} : \vartheta \in \Theta, \eta \in H\}$ has μ–densities (5.5.1) fulfilling conditions (5.5.2) and (5.5.3).*

Then there exists a family of Markov kernels $\{M_\vartheta : \vartheta \in \Theta\}$ with the following properties.

(i)   *$M_\vartheta | Y \times \mathbb{B}$ is a conditional distribution of T, given S, under $P_{\vartheta,\eta}$ (for every $\eta \in H$).*

(ii)   *For every $y \in Y$, $\vartheta \to M_\vartheta(y, \cdot) | \mathbb{B}$ is continuous with respect to the sup-metric, and $\{M_\vartheta(y, \cdot) : \vartheta \in \Theta\}$ has isotone likelihood ratios.*

In more detail: There is some Markov kernel $M_* | Y \times \mathbb{B}$ such that, for every $\vartheta \in \Theta$ and every $y \in Y$, $M_\vartheta(y, \cdot) | \mathbb{B}$ has with respect to $M_*(y, \cdot) | \mathbb{B}$ a density

$$t \to C(y, \vartheta) h(t, \vartheta), \qquad t \in \mathbb{R}, \tag{5.5.5}$$

such that
(ii')   *$\vartheta \to C(y, \vartheta)$ and $\vartheta \to h(t, \vartheta)$ are continuous for $y \in Y$ and $t \in \mathbb{R}$, and*
(ii'') *$t \to h(t, \vartheta') / h(t, \vartheta)$ is isotone if $\vartheta < \vartheta'$.*

Notice that (ii') implies continuity of $\vartheta \to M_\vartheta(y, \cdot) | \mathbb{B}$ with respect to the sup–metric by Scheffé's lemma (see Lehmann, 1986, p. 573, Lemma 4).

Proof. With $(\vartheta_*, \eta_*) \in \Theta \times H$ fixed, let $\mu_* := P_{\vartheta_*, \eta_*}$, and let $M_* | Y \times \mathbb{B}$ be a Markov kernel representing the conditional distribution of T, given S, under μ_*. By (5.5.1), $P_{\vartheta,\eta}$ has μ_*–density

$$x \to h(T(x), \vartheta) \frac{G(S(x), \vartheta, \eta)}{G(S(x), \vartheta_*, \eta_*)}, \tag{5.5.6}$$

with

$$h(t, \vartheta) := H(t, \vartheta) / H(t, \vartheta_*).$$

This implies by Proposition 1.10.26 that

$$\int \left(\int h(t, \vartheta) \frac{G(y, \vartheta, \eta)}{G(y, \vartheta_*, \eta_*)} M_*(y, dt) \right) \mu_* \circ S(dy) = 1,$$

so that

$$\int h(t,\vartheta) M_*(y, dt) < \infty \qquad \text{for } \mu_* \circ S\text{-a.a. } y \in Y. \qquad (5.5.7)$$

The exceptional μ_*-null set in (5.5.7), say N_ϑ, still depends on ϑ. If Θ_0 is a countable dense subset of Θ, relation (5.5.7) holds for all $\vartheta \in \Theta_0$ if $y \notin N := \cup\{N_\vartheta : \vartheta \in \Theta_0\}$. Since $\vartheta \to h(t,\vartheta)$ is continuous and $t \to h(t,\vartheta')/h(t,\vartheta)$ is isotone for $\vartheta < \vartheta'$, the function $\vartheta \to \int h(t,\vartheta) M_0(y, dt)$ is continuous on Θ and $\int h(t,\vartheta) M_*(y, dt) < \infty$ for $\vartheta \in \Theta$ if $y \in N^c$. To see this, let $\hat\vartheta \in \Theta$ be an arbitrary element, and choose $\vartheta', \vartheta'' \in \Theta_0$ such that $\hat\vartheta \in (\vartheta', \vartheta'') \subset \Theta$. Since $t \to h(t,\vartheta')/h(t,\vartheta)$ is isotone for $\vartheta < \vartheta'$, we obtain for $\vartheta \in [\vartheta', \vartheta'']$ and an arbitrary $t_0 \in \mathbb{R}$,

$$0 \le h(t,\vartheta) \le \frac{h(t,\vartheta') h(t_0,\vartheta)/h(t_0,\vartheta')}{h(t,\vartheta'') h(t_0,\vartheta)/h(t_0,\vartheta'')} \qquad \text{for } t \gtrless t_0.$$

Since $h(t_0, \cdot)$ is continuous, it is bounded on $[\vartheta', \vartheta'']$. Therefore, $0 \le h(t,\vartheta) \le c_0(h(t,\vartheta') + h(t,\vartheta''))$ for $t \in \mathbb{R}$, $\vartheta \in [\vartheta', \vartheta'']$.

Since $\int h(t,\vartheta') M_*(y, dt)$ and $\int h(t,\vartheta'') M_*(y, dt)$ are finite for $y \in N^c$, we have $\int h(t,\hat\vartheta) M_*(y, dt) < \infty$ for $y \in N^c$. Continuity of $\vartheta \to \int h(t,\vartheta) M_*(y, dt)$ at $\hat\vartheta$ now follows by the bounded convergence theorem.

Let $M_\vartheta | Y \times \mathbb{B}$ be defined for $y \in N$ by $M_\vartheta(y, \cdot) := M_*(y, \cdot)$, and for $y \in N^c$ by the condition that $M_\vartheta(y, \cdot)$ has the density (5.5.5) with respect to $M_*(y, \cdot)$, with

$$C(y,\vartheta) := \left(\int h(t,\vartheta) M_*(y, dt)\right)^{-1}. \qquad (5.5.8)$$

Since $h(t,\vartheta) > 0$ for $t \in \mathbb{R}$, we have $C(y,\vartheta) < \infty$ for $y \in Y$. Moreover, $y \in N^c$ implies $C(y,\vartheta) > 0$.

Using that $y \to G(y,\vartheta,\eta)/C(y,\vartheta)G(y,\vartheta_*,\eta_*)$ is a density of $P_{\vartheta,\eta} \circ S$ with respect to $\mu_* \circ S$ according to Lemma 1.10.32(i), it is easy to check that M_ϑ is a conditional distribution of T, given S, under $P_{\vartheta,\eta}$ (for any $\eta \in H$). $\qquad \square$

According to Lemma 5.5.4, conditions (5.3.1) and (5.3.2) for the families $\{M_\vartheta(y, \cdot) : \vartheta \in \Theta\}$, $y \in Y$, follow from natural conditions on the functions occurring in the density (5.5.1). To make sure that the confidence bound defined below attains only values in Θ (and not the boundary values ϑ_0 and/or ϑ_1) we need, in addition, that this family of conditional distributions fulfills condition (5.3.6) for every $y \in Y$, and that $M_\vartheta(y, \cdot)$ is nonatomic. A version which is not restricted to the nonatomic case and requires, therefore, randomized confidence bounds, can be obtained in the same way; see Pfanzagl (1979) for details.

In most applications, the p-measures $M_\vartheta(y, \cdot)$ for $y \in Y$ are dominated by a σ-finite measure ν (usually the Lebesgue measure). In this case there exists a ν–density of $M_*(y,\cdot)$, say $q(y,\cdot)$, and we obtain

$$M_\vartheta(y,B) = C(y,\vartheta) \int h(t,\vartheta) q(y,t) 1_B(t) \nu(dt),$$

i.e. $M_\vartheta(y,\cdot)$ has ν–density $t \to C(y,\vartheta) q(y,t) h(t,\vartheta)$.

Theorem 5.5.9. *Assume that* $\{P_{\vartheta,\eta} : \vartheta \in \Theta, \ \eta \in H\}$ *is a family of p–measures with* μ*–densities (5.5.1) fulfilling conditions (5.5.2) and (5.5.3).*

Let $M_*|Y \times \mathbb{B}$ *be the Markov kernel representing the conditional distribution of* T*, given* S*, under* P_{ϑ_*,η_*}*, and let* $M_\vartheta|Y \times \mathbb{B}$ *be the Markov kernel representing the conditional distribution of* T*, given* S*, under* $P_{\vartheta,\eta}$ *(for any* $\eta \in H$*), defined in Lemma 5.5.4.*

For $\beta \in (0,1)$*, let* $\hat\vartheta_\beta : \mathbb{R} \times Y \to [\vartheta_0, \vartheta_1]$ *be defined by*

$$\hat\vartheta_\beta(t,y) := \sup\{\vartheta \in \Theta : M_\vartheta(y,(-\infty,t]) \geq 1-\beta\} \quad \text{for } t \in \mathbb{R}, \ y \in Y,$$

where $\sup \emptyset := \vartheta_0$*.*

Then the following holds true.

(i) $t \to \hat\vartheta_\beta(t,y)$ *is isotone for every* $y \in Y$*.*

(ii) $\hat\vartheta_\beta \circ (T,S)$ *is an upper confidence bound with confidence level* β*, i.e.*

$$P_{\vartheta,\eta}\{x \in X : \vartheta \leq \hat\vartheta_\beta(T(x),S(x))\} \geq \beta \quad \text{for } \vartheta \in \Theta, \ \eta \in H. \quad (5.5.10)$$

(iii)   *If* $M_*(y,\cdot)$ *is nonatomic for* $y \in Y$*, let* I_y *denote the smallest open interval such that* $M_*(y,I_y) = 1$*. If*

$$1-\beta \in \{M_\vartheta(y,(-\infty,t]) : \vartheta \in \Theta\}^\circ \quad \text{for } y \in Y, \ t \in I_y, \quad (5.5.11)$$

then equality holds in (5.5.10), the distribution of $x \to \hat\vartheta_\beta(T(x),S(x))$ *under* $P_{\vartheta,\eta}$ *is nonatomic, and* $\hat\vartheta_\beta(T(x),S(x))$ *attains for* $P_{\vartheta,\eta}$*–a.a.* $x \in X$ *only values in* Θ*.*

Proof. (i) Straightforward.

(ii) By Lemma 5.5.4 and Proposition 1.7.7, $\vartheta \to M_\vartheta(y,(-\infty,t])$ is continuous and antitone for $y \in Y$ and $t \in \mathbb{R}$. Hence, Theorem 5.3.3(ii), applied for $P_\vartheta = M_\vartheta(y,\cdot)$, yields that

$$M_\vartheta(y,\{t \in \mathbb{R} : \vartheta \leq \hat\vartheta_\beta(t,y)\}) \geq \beta \quad \text{for } \vartheta \in \Theta. \quad (5.5.12)$$

Integration over y with respect to $P_{\vartheta,\eta} \circ S$ yields

$$\beta \leq \int M_\vartheta(y,\{t \in \mathbb{R} : \vartheta \leq \hat\vartheta_\beta(t,y)\}) P_{\vartheta,\eta} \circ S(dy)$$

$$= P_{\vartheta,\eta}\{x \in X : \vartheta \leq \hat\vartheta_\beta(T(x),S(x))\}$$

by Proposition 1.10.26.

Since $\vartheta \to M_\vartheta(y, (-\infty, t])$ is continuous and antitone, the definition of $\hat{\vartheta}_\beta$ implies for every $\vartheta \in \Theta$ that

$$\vartheta \leq \hat{\vartheta}(t, y) \qquad \text{iff} \qquad M_\vartheta(y, (-\infty, t]) \geq 1 - \beta.$$

Therefore, $\hat{\vartheta}_\beta$ is measurable since $(t, y) \to M_\vartheta(y, (-\infty, t])$ is measurable by Lemma 6.7.3(ii).

(iii) If $M_*(y, \cdot)$ is nonatomic, then so is $M_\vartheta(y, \cdot)$ (for any $\vartheta \in \Theta$). Therefore, Theorem 5.3.3(iii), applied with $P_\vartheta = M_\vartheta(y, \cdot)$, yields that equality holds in (5.5.12), the distribution of $t \to \hat{\vartheta}_\beta(t, y)$, $t \in I_y$, under $M_\vartheta(y, \cdot)$ is nonatomic, and $\{\hat{\vartheta}_\beta(t, y) : t \in I_y\} \subset \Theta$. This implies by Proposition 1.10.26, applied with $f(t, y) = 1_{\{\vartheta \leq \hat{\vartheta}_\beta(\cdot, y)\}}(t)$, that equality holds in (5.5.10), the distribution of $x \to \hat{\vartheta}_\beta(T(x), S(x))$ under $P_{\vartheta, \eta}$ is nonatomic, and $\hat{\vartheta}_\beta(T(x), S(x))$ attains for $P_{\vartheta, \eta}$–a.a. $x \in X$ only values in Θ.   □

Theorem 5.5.13. *Assume that $\{P_{\vartheta, \eta} : \vartheta \in \Theta, \eta \in H\}$ is a family of p-measures with μ–densities (5.5.1) fulfilling condition (5.5.3). Assume that $\hat{\vartheta}_\beta :$ $\mathbb{R} \times Y \to [\vartheta_0, \vartheta_1]$ has, for $\beta \in (0, 1)$, the properties specified in 5.5.9(i) and (ii) with equality holding in (5.5.10).*

If $\{P_{\vartheta, \eta} \circ S : \eta \in H\}$ is boundedly complete for every $\vartheta \in \Theta$, then $\hat{\vartheta}_\beta \circ (T, S)$ is maximally concentrated about ϑ (in the sense of (5.4.4)) in the class of all randomized confidence bounds with the same covering probability β.

Proof. Let $K|X \times \overline{\mathbb{B}} \cap [\vartheta_0, \vartheta_1]$ be a randomized confidence bound for ϑ with covering probability β, i.e.

$$\int K(x, [\vartheta, \vartheta_1]) P_{\vartheta, \eta}(dx) = \beta \qquad \text{for } \vartheta \in \Theta, \ \eta \in H.$$

Then φ_ϑ, defined by $\varphi_\vartheta(x) = K(x, [\vartheta, \vartheta_1])$, is a similar test of level β for the hypothesis $\{P_{\vartheta, \eta} : \eta \in H\}$. Moreover, $1_{\{\hat{\vartheta}_\beta \circ (T, S) \geq \vartheta\}}$ is for this hypothesis a level–β–test of Neyman structure fulfilling (4.7.4). Therefore, Theorem 4.7.1 implies

$$P_{\vartheta, \eta}\{x \in X : \hat{\vartheta}_\beta(T(x), S(x)) \geq \vartheta'\} \overset{\leq}{\underset{\geq}{}} P_{\vartheta, \eta}(\varphi_{\vartheta'})$$

$$= \int K(x, [\vartheta', \vartheta_1]) P_{\vartheta, \eta}(dx) \qquad \text{for } \vartheta' \overset{>}{\underset{<}{}} \vartheta.$$

From this the assertion follows easily.   □

If $\{P_{\vartheta,\eta} : \vartheta \in \Theta, \ \eta \in H\}$ is an exponential family with density

$$x \rightarrow C(\vartheta,\eta)g(x)\exp\Big[a_0(\vartheta)T_0(x) + \sum_{i=1}^{k}a_i(\vartheta,\eta)T_i(x)\Big], \quad x \in X, \qquad (5.5.14)$$

it is of type (5.5.1) for every sample size n (see pp. 22/3).
The following is a useful Corollary to Theorem 5.5.13.

Corollary 5.5.15. *Assume that a family* $\{P_{\vartheta,\eta} : \vartheta \in \Theta, \ \eta \in H\}$ *of type (5.5.14) has the following properties.*
(a) *a_0 is isotone.*
(b) *$\{(a_1(\vartheta,\eta),\ldots,a_k(\vartheta,\eta)) : \eta \in H\}$ has a nonempty interior for every $\vartheta \in \Theta$.*
Assume that $\hat{\vartheta}_\beta : \mathbb{R} \times \mathbb{R}^k \rightarrow [\vartheta_0,\vartheta_1]$ has the following properties.
(i) *$t_0 \rightarrow \hat{\vartheta}_\beta(t_0,t_1,\ldots,t_k)$ is isotone for every $(t_1,\ldots,t_k) \in \mathbb{R}^k$.*
(ii) *$\vartheta_\beta^{(n)}(x_1,\ldots,x_n) := \hat{\vartheta}_\beta\big(\sum_1^n T_0(x_\nu),\sum_1^n T_1(x_\nu),\ldots,\sum_1^n T_k(x_\nu)\big)$ is an upper confidence bound for ϑ with covering probability β for the family $\{P_{\vartheta,\eta}^n : \vartheta \in \Theta, \ \eta \in H\}$.*
Then $\vartheta_\beta^{(n)}$ is maximally concentrated among all randomized upper confidence bounds with the same covering probability β.

Proof. $\{P_{\vartheta,\eta}^n : \vartheta \in \Theta, \ \eta \in H\}$ is for every $n \in \mathbb{N}$ of the type (5.5.1) with

$$\exp\Big[a_0(\vartheta)\sum_1^n T_0(x_\nu)\Big] \text{ in place of } H(T(x),\vartheta),$$

and

$$C(\vartheta,\eta)^n\exp\Big[\sum_{i=1}^{k}a_i(\vartheta,\eta)\sum_{\nu=1}^n T_i(x_\nu)\Big] \text{ in place of } G\big(S(x),\vartheta,\eta\big).$$

Since a_0 is isotone, this implies (5.5.3). Since $\{(a_1(\vartheta,\eta),\ldots,a_k(\vartheta,\eta)) : \eta \in H\}$ has a nonempty interior,

$$P_{\vartheta,\eta}^n \circ \Big((x_1,\ldots,x_n) \rightarrow \big(\sum_1^n T_1(x_\nu),\ldots,\sum_1^n T_k(x_\nu)\big)\Big)$$

is complete by Theorem 1.6.10. Hence the assertion follows from Theorem 5.5.13. $\qquad\square$

To establish the existence of confidence bounds fulfilling conditions 5.5.9(i) and (ii) with equality holding in (5.5.10), one has to solve the problem of verifying condition (5.5.11) on the conditional distributions.

5.6 Examples of maximally concentrated confidence bounds

The results of Section 5.5 on maximally concentrated confidence bounds are applied to some exponential families. Examples include the normal, the lognormal, the inverse normal and the gamma distributions.

The following Examples 5.6.2 and 5.6.3 refer to $\{N^n_{(\mu,\sigma^2)} : \mu \in \mathbb{R}, \sigma^2 > 0\}$. To apply the results of Section 5.5 we write the Lebesgue density of $N_{(\mu,\sigma^2)}$ as

$$x \to C(\mu,\sigma^2) \exp\left[\frac{\mu}{\sigma^2}x - \frac{1}{2\sigma^2}x^2\right], \qquad x \in \mathbb{R}. \tag{5.6.1}$$

Example 5.6.2. *Confidence bounds for μ.* With both parameters unknown, (5.6.1) is not of the type (5.5.14) (with μ in place of ϑ and σ^2 in place of η).

For the case $\beta = 1/2$, i.e. for median unbiased estimators for μ, we may proceed as follows. Since $\bar{x}_n$ is, obviously, median unbiased in the full family $\{N^n_{(\mu,\sigma^2)} : \mu \in \mathbb{R}, \sigma^2 > 0\}$, it suffices to show that $\bar{x}_n$ is maximally concentrated among all median unbiased estimators for each of the subfamilies $\{N^n_{(\mu,\sigma^2)} : \mu \in \mathbb{R}\}$, with $\sigma^2 > 0$ fixed. Since $\sum_1^n x_\nu$ is sufficient for each of these subfamilies and $\bar{x}_n$ is an increasing function of $\sum_1^n x_\nu$, this follows immediately from (the nonrandomized version of) Theorem 5.4.3.

The general case of an upper confidence bound with arbitrary covering probability β can be dealt with as follows.

$$\mu^{(n)}_\beta(\underline{x}) := \bar{x}_n + n^{-1/2}t_{n-1,1-\beta}s_n(\underline{x}),$$

with $s^2_n(\underline{x}) = (n-1)^{-1}\sum_1^n(x_\nu - \bar{x}_n)^2$ and $t_{n-1,\alpha}$ the upper α–quantile of the t–distribution with $(n-1)$ degrees of freedom, is an upper confidence bound for μ with covering probability β in the family $\{N^n_{(\mu,\sigma^2)} : \mu \in \mathbb{R}, \sigma^2 > 0\}$. That $\mu^{(n)}_\beta$ is maximally concentrated follows from the optimality of the t–test, in the class of all similar level–α–tests, established in Example 4.7.12.

Example 5.6.3. *Confidence bounds for σ^2.* The density (5.6.1) is of the type (5.5.14) for $\vartheta = \sigma^2$, $\eta = \mu$ with

$$\begin{aligned} T_0(x) &= x^2, & T_1(x) &= x, \\ a_0(\sigma^2) &= -1/2\sigma^2, & a_1(\sigma^2,\mu) &= \mu/\sigma^2. \end{aligned}$$

With $k_{m,\beta}$ denoting the upper β–quantile of χ^2_m, we have $N^n_{(\mu,\sigma^2)}\{(x_1,\dots,x_n) \in \mathbb{R}^n : \sigma^2 \leq \sum_1^n(x_\nu - \bar{x}_n)^2/k_{n-1,\beta}\} = \beta$ for all $\mu \in \mathbb{R}$, $\sigma^2 > 0$. Hence $(x_1,\dots,x_n) \to \sum_1^n(x_\nu - \bar{x}_n)^2/k_{n-1,\beta}$ is an upper confidence bound for σ^2 with covering probability β; the existence Theorem 5.5.9 is not needed here.

Since $\sum_1^n (x_\nu - \bar{x}_n)^2 = \sum_1^n x_\nu^2 - n^{-1}(\sum_1^n x_\nu)^2$, this confidence bound is an increasing function of $\sum_1^n x_\nu^2$ $(= \sum_1^n T_0(x_\nu))$ for every value of $\sum_1^n x_\nu$ $(= \sum_1^n T_1(x_\nu))$. It is, therefore, maximally concentrated among all confidence bounds with the same covering probability β, according to Corollary 5.5.15.

The following exercise refers to the family $\{P^n_{\mu,\sigma^2} : \mu \in \mathbb{R}, \ \sigma^2 > 0\}$ of *lognormal distributions*. Recall that P_{μ,σ^2} is defined by the property that $x \to \log x$ is under P_{μ,σ^2} normally distributed with mean μ and variance σ^2. Hence the Lebesgue density of P_{μ,σ^2} is given by

$$x \to C(\mu,\sigma^2)x^{-1}\exp\left[\frac{\mu}{\sigma^2}\log x - \frac{1}{2\sigma^2}(\log x)^2\right], \qquad x > 0. \tag{5.6.4}$$

Exercise 5.6.5. Let

$$\hat{s}_n^2(\underline{x}) := (n-1)^{-1}\sum_{\nu=1}^n \left(\log x_\nu - n^{-1}\sum_{i=1}^n \log x_i\right)^2.$$

Show that
(i) $(\prod_1^n x_\nu)^{1/n}\exp\left[n^{-1/2}t_{n-1,1-\beta}\hat{s}_n(\underline{x})\right]$ is maximally concentrated among all upper confidence bounds with covering probability β for the functional $\kappa(P_{\mu,\sigma^2}) = \exp[\mu]$, the median of P_{μ,σ^2}.
(ii) $(n-1)s_n^2(\underline{x})/k_{n-1,\beta}$ is maximally concentrated among all upper confidence bounds for σ^2 with covering probability β.

The following example refers to the family $\{P^n_{\mu,\lambda} : \mu, \lambda > 0\}$ of *inverse normal distributions*. To stress the exponential nature of this family, we write the Lebesgue density of $P_{\mu,\lambda}$ as

$$x \to C(\mu,\lambda)x^{-3/2}\exp\left[-\frac{\lambda}{2}x^{-1} - \frac{\lambda}{2\mu^2}x\right], \qquad x > 0. \tag{5.6.6}$$

Example 5.6.7. *Confidence bounds for λ.* The density (5.6.6) is of the type (5.5.14) for $\vartheta = \lambda$, $\eta = \mu$ with

$$T_0(x) = x^{-1}, \qquad T_1(x) = x,$$
$$a_0(\lambda,\mu) = -\lambda/2, \qquad a_1(\lambda,\mu) = -\lambda/2\mu^2.$$

By Tweedie (1957), $\lambda \sum_1^n (x_\nu^{-1} - \bar{x}_n^{-1})$ is under $P^n_{\mu,\lambda}$ distributed as χ^2_{n-1}. Hence, if $k_{m,\beta}$ denotes the upper β-quantile of χ^2_m,

$$(x_1,\ldots,x_n) \to k_{n-1,1-\beta} / \sum_1^n (x_\nu^{-1} - \bar{x}_n^{-1})$$

is an upper confidence bound for λ with covering probability β. Moreover, this confidence bound is a decreasing function of $\sum_1^n x_\nu^{-1}$ for every value of $\bar{x}_n^{-1}$.

It is, therefore, maximally concentrated among all confidence bounds with the same confidence probability β, in analogy to Corollary 5.5.15. (Notice that a_0 is decreasing in this case.)

Obtaining confidence bounds for μ seems to be an unsolved problem. See Chhikara and Folks (1989) for further results on tests and confidence bounds for the inverse normal distributions.

The following examples refer to the family $\{\Gamma_{a,b}^n : a, b > 0\}$ of *gamma distributions*. To apply the results of Section 5.5, we write the Lebesgue density of $\Gamma_{a,b}$ as

$$x \to C(a,b)x^{-1}\exp\left[-\frac{x}{a} + b\log x\right], \qquad x > 0. \tag{5.6.8}$$

Example 5.6.9. *Confidence bounds for the shape parameter* b. The density (5.6.8) is of the type (5.5.14) for $\vartheta = b$, $\eta = a$ with

$$\begin{aligned}
T_0(x) &= \log x, & T_1(x) &= x, \\
a_0(b) &= b, & a_1(b,a) &= -1/a.
\end{aligned}$$

According to Theorem 5.5.9 and Corollary 5.5.15, a maximally concentrated confidence bound for b can be obtained from the conditional distribution of $\sum_1^n \log x_\nu$, given $\sum_1^n x_\nu$, under $\Gamma_{a,b}^n$ (which depends on b only). The use of conditional distributions can be avoided by taking advantage of particular features of the present example:

Instead of using the conditional distribution of $\sum_1^n \log x_\nu$, given $\sum_1^n x_\nu$, one may as well use the conditional distribution of $\left(\prod_1^n x_\nu\right)^{1/n}/n^{-1}\sum_1^n x_\nu$, given $\sum_1^n x_\nu$. Since the distribution of $\left(\prod_1^n x_\nu\right)^{1/n}/n^{-1}\sum_1^n x_\nu$ under $\Gamma_{a,b}^n$ depends on b only, and since $\sum_1^n x_\nu$ is a complete sufficient statistic for each subfamily $\{\Gamma_{a,b}^n : a > 0\}$, the function $\left(\prod_1^n x_\nu\right)^{1/n}/n^{-1}\sum_1^n x_\nu$ is stochastically independent of $\sum_1^n x_\nu$ under $\Gamma_{a,b}^n$ by Basu's Theorem 1.8.2. Hence the optimal confidence bound can be simply based on the distribution of $\left(\prod_1^n x_\nu\right)^{1/n}/n^{-1}\sum_1^n x_\nu$ under $\Gamma_{1,b}^n$, say $Q_{n,b}$. Notice that $Q_{n,b}(0,1) = 1$ as a consequence of the inequality between the arithmetic and the geometric mean. Below we shall show that for $n \geq 2$

$$t \to C_n(b)t^{n(b-1)}, \qquad t \in (0,1), \tag{5.6.10}$$

with $C_n(b) := \Gamma(nb)/\Gamma(b)^n n^{n(b-1)}(n-1)!$, is a density of $Q_{n,b}$ with respect to $Q_{n,1}$. The family $\{Q_{n,b} : b > 0\}$ is of exponential type. Therefore, $b \to Q_{n,b}(0,t]$ is continuous and decreasing by Propositions 1.7.15 and 1.7.7.

Moreover,

$$\lim_{b \to 0} Q_{n,b}(0,t] = 1 \quad \text{and} \quad \lim_{b \to \infty} Q_{n,b}(0,t] = 0 \quad \text{for } t \in (0,1). \tag{5.6.11}$$

Hence the equation $Q_{n,b}(0, t] = 1 - \beta$ has a unique solution in b, say $b_\beta^{(n)}(t)$. By Theorem 5.5.9,

$$(x_1, \ldots, x_n) \to b_\beta^{(n)}\left(\left(\prod_1^n x_\nu\right)^{1/n} \bigg/ n^{-1} \sum_1^n x_\nu\right) \qquad (5.6.12)$$

is an upper confidence bound for b with covering probabiliy β. Since $b_\beta^{(n)}$ is increasing, this confidence bound is an increasing function of $\sum_1^n \log x_\nu$ (= $\sum_1^n T_0(x_\nu)$) for every value of $\sum_1^n x_\nu$ (= $\sum_1^n T_1(x_\nu)$). Hence it is maximally concentrated according to Corollary 5.5.15.

Relation (5.6.10) is equivalent to

$$\int 1_B(t) C_n(b) t^{n(b-1)} Q_{n,1}(dt) = \int 1_B(t) Q_{n,b}(dt) \qquad \text{for } B \in \mathbb{B}_+ \, .$$

This can be easily checked using that $(x_1, \ldots, x_n) \to \left(\prod_1^n x_\nu\right)^{b-1}/\Gamma(b)^n$ is a $\Gamma_{1,1}^n$–density of $\Gamma_{1,b}^n$, and that $\left(\prod_1^n x_\nu\right)^{1/n}/\bar{x}_n$ and $\bar{x}_n$ are stochastically independent under $\Gamma_{1,b}^n$.

It remains to prove (5.6.11). For $b < 1$, we have

$$Q_{n,b}(t, \infty) = C_n(b) \int_t^1 x^{n(b-1)} Q_{n,1}(dx) \leq C_n(b) t^{n(b-1)}.$$

This implies the first relation in (5.6.11) since $\Gamma(nb) < \Gamma(b)$ for $b < 1/n$ and $\lim_{b \to 0} \Gamma(b) = \infty$.

For $b > 1$, we have

$$Q_{n,b}(0, t] \leq C_n(b) t^{n(b-1)}.$$

Hence the second relation in (5.6.11) follows from (see Whittaker and Watson, 1958, pp. 248/9)

$$\log \Gamma(b) = \left(b - \frac{1}{2}\right) \log b - b + O(b^0) \qquad \text{for } b \to \infty.$$

This proof is based on an unpublished note of L. Schröder.

Example 5.6.13. *Confidence bounds for the scale parameter* a. The density (5.6.8) is of the type (5.5.14) for $\vartheta = a$, $\eta = b$ with

$$\begin{aligned} T_0(x) &= x, & T_1(x) &= \log x, \\ a_0(a) &= -1/a, & a_1(a, b) &= b. \end{aligned}$$

Since $\sum_1^n \log x_\nu$ is sufficient for $\{\Gamma_{a,b}^n : b > 0\}$, the conditional distribution of $\sum_1^n x_\nu$, given $\sum_1^n \log x_\nu$, under $\Gamma_{a,b}^n$ depends on a only. Equivalently: The distribution of $n^{-1}\sum_1^n x_\nu$, given $(\prod_1^n x_\nu)^{1/n}$, under $\Gamma_{a,b}^n$, depends on a only. Let $M_{n,a}|(0,\infty) \times \mathbb{B}_+$ denote the pertaining Markov kernel.

Let q_n be the Lebesgue density of the distribution of $n^{-1}\sum_1^n x_\nu / (\prod_1^n x_\nu)^{1/n}$ under $\Gamma_{a,1}^n$, $n \geq 2$ (which does not depend on a). Since $n^{-1}\sum_1^n x_\nu$ and $n^{-1}\sum_1^n x_\nu / (\prod_1^n x_\nu)^{1/n}$ are stochastically independent under $\Gamma_{a,1}^n$, we obtain that

$$t \to C_n(y,a)t^n \exp\left[-\frac{n}{a}t\right]q_n\left(\frac{t}{y}\right), \qquad t > y, \qquad (5.6.14)$$

is a Lebesgue density of $M_{n,a}(y,\cdot)|\mathbb{B}\cap(y,\infty)$. From this relation it is easily seen what we know from Lemma 5.5.4: that $a \to M_{n,a}\big(y,(y,t]\big)$ is continuous and decreasing.

Moreover,

$$\lim_{a\to 0} M_{n,a}\big(y,(y,t]\big) = 1 \quad \text{and} \quad \lim_{a\to\infty} M_{n,a}\big(y,(y,t]\big) = 0 \quad \text{for } t > y. \quad (5.6.15)$$

Hence the equation $M_{n,a}\big(y,(y,t]\big) = 1 - \beta$ has a unique solution in a, say $a_\beta^{(n)}(t,y)$. By Theorem 5.5.9

$$(x_1,\ldots,x_n) \to a_\beta^{(n)}\left(n^{-1}\sum_1^n x_\nu, (\prod_1^n x_\nu)^{1/n}\right) \qquad (5.6.16)$$

is an upper confidence bound for a with covering probability β. It is maximally concentrated according to Corollary 5.5.15.

It remains to prove (5.6.15). Since (see (5.6.14))

$$C_n(y,a) = \left(\int_y^\infty t^n \exp\left[-\frac{n}{a}t\right]q_n\left(\frac{t}{y}\right)dt\right)^{-1},$$

we obtain

$$M_{n,a}\big(y,(t,\infty)\big)$$

$$\leq \int_t^\infty x^n \exp\left[-\frac{n}{a}x\right]q_n\left(\frac{x}{y}\right)dx \Big/ \int_y^t x^n \exp\left[-\frac{n}{a}x\right]q_n\left(\frac{x}{y}\right)dx$$

$$\leq \int_t^\infty x^n \exp\left[-\frac{n}{a}(x-t)\right]q_n\left(\frac{x}{y}\right)dx \Big/ \int_y^t x^n q_n\left(\frac{x}{y}\right)dx,$$

and

$$M_{n,a}\big(y,(y,t]\big) \leq \int_y^t x^n q_n\Big(\frac{x}{y}\Big)\,dx \Big/ \int_y^\infty x^n \exp\Big[-\frac{n}{a}(x-y)\Big]q_n\Big(\frac{x}{y}\Big)\,dx.$$

The first relation in (5.6.15) follows by Fatou's lemma from

$$\lim_{a\to 0} \int_t^\infty x^n \exp\Big[-\frac{n}{a}(x-t)\Big]q_n\Big(\frac{x}{y}\Big)\,dx = 0,$$

the second from

$$\lim_{a\to\infty} \int_y^\infty x^n \exp\Big[-\frac{n}{a}(x-y)\Big]q_n\Big(\frac{x}{y}\Big)\,dx = \infty.$$

(For the second relation, use that

$$\int_y^\infty x^n q_n\Big(\frac{x}{y}\Big)\,dx = y^{n+1}\int \Big(n^{-1}\sum_1^n x_\nu\Big)^n\Big/\prod_1^n x_\nu \Gamma_{1,1}^n\big(d(x_1,\ldots,x_n)\big) = \infty. \;)$$

Notice a practically relevant difference between the confidence bound for b (given by (5.6.12)), and the confidence bound for a (given by (5.6.16)): Whereas a table with 3 entries (t, n, β) suffices for b, a table with 4 entries (t, y, n, β) is needed for a.

Chapter 6
Consistent estimators

6.1 Introduction

Introduces the concepts of "consistency" and "(locally) uniform consistency".

Though the main results of Chapter 6 refer to sequences of independent and identically distributed observations, we choose a more general framework for introducing certain basic concepts.

Let Θ be an abstract parameter set, (Y, d) a separable metric space, endowed with its Borel algebra, and $\kappa : \Theta \to Y$ a functional. For $\vartheta \in \Theta$ and $n \in \mathbb{N}$ let $P_\vartheta^{(n)}$ be a p–measure on a measurable space $(X_n, \mathcal{A}_n)$. The problem is to estimate $\kappa(\vartheta)$, based on a realization $x \in X_n$ of $P_\vartheta^{(n)}$.

This framework includes the case of estimating the functional κ on the basis of n independent, identically distributed observations from an unknown p–measure P in a given family $\mathfrak{P}$, if we identify $\Theta = \mathfrak{P}$, $(X_n, \mathcal{A}_n) = (X^n, \mathcal{A}^n)$ and $P_\vartheta^{(n)} = P^n$.

An estimator for the functional κ is a measurable map $\kappa^{(n)} : X_n \to Y$. If $X_n = X^n$, it is understood that $\kappa^{(n)}(\underline{x})$ means $\kappa^{(n)}(x_1, \dots, x_n)$.

Definition 6.1.1. (i) The estimator sequence $\kappa^{(n)}$, $n \in \mathbb{N}$, is *consistent* for κ in the family of sequences $P_\vartheta^{(n)}$, $n \in \mathbb{N}$, $\vartheta \in \Theta$, if $\kappa^{(n)}$ converges stochastically to $\kappa(\vartheta)$ for every $\vartheta \in \Theta$, i.e. if

$$\lim_{n \to \infty} P_\vartheta^{(n)} \{ x \in X_n : d(\kappa^{(n)}(x), \kappa(\vartheta)) > \varepsilon \} = 0 \qquad \text{for } \varepsilon > 0, \ \vartheta \in \Theta.$$

(ii) The estimator sequence is *consistent uniformly on* Θ if

$$\lim_{n \to \infty} \sup_{\vartheta \in \Theta} P_\vartheta^{(n)} \{ x \in X_n : d(\kappa^{(n)}(x), \kappa(\vartheta)) > \varepsilon \} = 0 \qquad \text{for } \varepsilon > 0.$$

From this we derive the concepts of "locally uniform consistency at ϑ" and "locally uniform consistency on Θ" as specified in (6.7.6). For such definitions, we need a Hausdorff topology on Θ.

Remark 6.1.2. Locally uniform consistency on Θ implies uniform consistency on every compact subset of Θ. If Θ is locally compact, "locally uniform consistency on Θ" and "uniform consistency on compact subsets of Θ" are equivalent.

Consistency of an estimator sequence is technically useful. It is an important first step towards the asymptotic distribution of an estimator sequence. It is the asymptotic distribution which can be used to obtain approximate results about the accuracy of estimators, about their efficiency compared with other estimators, etc. Consistency as such is an assertion about the asymptotic behavior of an estimator sequence which cannot be interpreted as an approximate assertion for finite sample sizes.

Many results in literature on the consistency of estimator sequences in the i.i.d. case refer to *strong consistency*, i.e. convergence of $\kappa^{(n)}(\underline{x})$, $n \in \mathbb{N}$, to $\kappa(\vartheta)$ for $P_\vartheta^{\mathbb{N}}$–a.a. $\underline{x} \in X^{\mathbb{N}}$. Using the characterization of a.e. convergence of $\kappa^{(n)}$, $n \in \mathbb{N}$, to $\kappa(\vartheta)$ by

$$\lim_{m \to \infty} P_\vartheta^{\mathbb{N}} \left(\bigcup_{n=m}^{\infty} \left\{ \underline{x} \in X^{\mathbb{N}} : d\big(\kappa^{(n)}(\underline{x}), \kappa(\vartheta)\big) > \varepsilon \right\} \right) = 0 \quad \text{for } \varepsilon > 0,$$

one may define *uniform strong consistency* of $\kappa^{(n)}$, $n \in \mathbb{N}$, by

$$\lim_{m \to \infty} \sup_{\vartheta \in \Theta} P_\vartheta^{\mathbb{N}} \left(\bigcup_{m=n}^{\infty} \left\{ \underline{x} \in X^{\mathbb{N}} : d\big(\kappa^{(n)}(\underline{x}), \kappa(\vartheta)\big) > \varepsilon \right\} \right) = 0 \quad \text{for } \varepsilon > 0.$$

The results which we present in the following for uniform consistency hold, in the i.i.d. case, under the same conditions for uniform *strong* consistency. We avoid the (modest) complications connected with this generalization, since "strong consistency" adds nothing to "consistency" which could be of use on the way to the asymptotic distributions of estimator sequences. This is in harmony with our conception of consistency as a technical tool. For all that, strong consistency is a mathematically interesting object to study. This interest results from thinking in terms of *limits* for estimator *sequences*. What counts for applications are approximations, not limits.

If Θ is compact, uniformly consistent estimator sequences exist under virtually no conditions (see Section 6.2). Starting from such estimator sequences, one may obtain consistent sequences of solutions to estimating equations (see Section 6.4). These results will be applied in Section 6.5 to obtain results on the consistency of ML (= maximum likelihood) estimators.

6.2 A general consistency theorem

Asserts the existence of estimator sequences which are consistent, uniformly on compact subsets.

Let now Θ be an open subset of $\mathbb{R}^p$, and $\{P_\vartheta : \vartheta \in \Theta\}$ a family of p–measures on $(X, \mathcal{A})$. Throughout the following we assume that ϑ is identifiable, i.e. $\vartheta' \neq \vartheta''$ implies $P_{\vartheta'} \neq P_{\vartheta''}$. To prepare the existence theorem for consistent estimators of ϑ, we need an auxiliary "nonparametric" result.

Let $\mathfrak{P}$ denote the family of all p–measures on $(X, \mathcal{A})$, and let $f_k : X \to [0, 1]$, $k \in \mathbb{N}$, be measurable functions such that

$$P(f_k) = Q(f_k) \quad \text{for } k \in \mathbb{N} \text{ implies } P = Q. \tag{6.2.1}$$

If $\mathcal{A}$ is countably generated, there exists a countable algebra generating $\mathcal{A}$, say $\{A_k : k \in \mathbb{N}\}$. Then (6.2.1) holds with $f_k = 1_{A_k}$. If X is a separable metric space, one may choose the functions f_k to be continuous (see Parthasarathy, 1967, p. 47, Theorem 6.6).

Let $\Delta : \mathfrak{P} \times \mathfrak{P} \to [0, 1]$ be defined by

$$\Delta(P, Q) := \sum_{k=1}^{\infty} 2^{-k} |P(f_k) - Q(f_k)|. \tag{6.2.2}$$

For $\underline{x} \in X^n$ let $Q_{\underline{x}}^{(n)}$ denote the *empirical p–measure* on $\mathcal{A}$, defined by

$$Q_{\underline{x}}^{(n)}(A) := n^{-1} \sum_{\nu=1}^{n} 1_A(x_\nu), \qquad A \in \mathcal{A}. \tag{6.2.3}$$

Lemma 6.2.4. (i) *The function Δ defined by (6.2.2) is a metric on $\mathfrak{P}$.*

(ii) *The estimator sequence $\underline{x} \to Q_{\underline{x}}^{(n)}$, $n \in \mathbb{N}$, is consistent for P, uniformly on $\mathfrak{P}$, i.e.*

$$\lim_{n \to \infty} \sup_{P \in \mathfrak{P}} P^n \{\underline{x} \in X^n : \Delta(Q_{\underline{x}}^{(n)}, P) > \varepsilon\} = 0 \quad \text{for } \varepsilon > 0.$$

Proof. (i) $\Delta(P, Q) = 0$ implies $P(f_k) = Q(f_k)$ for $k \in \mathbb{N}$, hence $P = Q$.

(ii) For $f : X \to [0, 1]$ we have $f^2 \leq f$, hence $P(f^2) - P(f)^2 \leq P(f)(1 - P(f)) \leq 1/4$ and, therefore, by Čebyshev's inequality

$$P^n \{\underline{x} \in X^n : |Q_{\underline{x}}^{(n)}(f) - P(f)| > \varepsilon\} \leq 1/4\varepsilon^2 n \quad \text{for } P \in \mathfrak{P}.$$

Let $\varepsilon \in (0,1)$ be arbitrary. With $K_\varepsilon = [-\log_2 \varepsilon/2]$ being the smallest integer $\geq -\log_2 \varepsilon/2$, we obtain

$$\sum_{k=K_\varepsilon+1}^{\infty} 2^{-k}|P(f_k) - Q(f_k)| \leq \varepsilon/2. \qquad (6.2.5)$$

Hence

$$P^n\{\underline{x} \in X^n : \sum_{k=1}^{\infty} 2^{-k}|Q_{\underline{x}}^{(n)}(f_k) - P(f_k)| > \varepsilon\}$$

$$\leq P^n\{\underline{x} \in X^n : \sum_{k=1}^{K_\varepsilon} 2^{-k}|Q_{\underline{x}}^{(n)}(f_k) - P(f_k)| > \varepsilon/2\}$$

$$\leq \sum_{k=1}^{K_\varepsilon} P^n\{\underline{x} \in X^n : |Q_{\underline{x}}^{(n)}(f_k) - P(f_k)| > \varepsilon/2\}$$

$$\leq \frac{K_\varepsilon}{\varepsilon^2}n^{-1} \qquad \text{for } P \in \mathfrak{P}.$$

$\square$

Based on an artificial metric, Lemma 6.2.4 itself is not of immediate interest. It can, however, be used to establish the existence of consistent estimator sequences for ϑ.

Remark 6.2.6. The following lemma requires some continuity conditions for $\vartheta \to P_\vartheta$. A natural requirement is continuity of $\vartheta \to P_\vartheta(A)$ for $A \in \mathcal{A}$, which is equivalent to continuity of $\vartheta \to P_\vartheta(f)$ for every bounded measurable function f. If X is a separable metric space, continuity of $\vartheta \to P_\vartheta(f)$ for all bounded continuous functions f suffices. What is really needed is continuity of $\vartheta \to P_\vartheta(f_k)$ for the functions f_k used in the definition of the distance function Δ (see (6.2.2)).

There is still one more difficulty to circumvent: Continuity of $\vartheta \to P_\vartheta(f_k)$ implies continuity of $\vartheta \to \Delta(P_\vartheta, Q)$ for every $Q \in \mathfrak{P}$. What we need is something different: namely, that $\Delta(P_{\vartheta_n}, P_{\vartheta_0}) \to 0$ implies $\vartheta_n \to \vartheta_0$. This follows if Θ is compact.

Lemma 6.2.7. *Assume that* (i) (Θ, d) *is a compact metric space, and* (ii) $\vartheta \to P_\vartheta(f_k)$ *is continuous for $k \in \mathbb{N}$.*

Then there exists an estimator sequence $\vartheta^{(n)} : X^n \to \Theta$ which is uniformly consistent on Θ, i.e.

$$\lim_{n \to \infty} \sup_{\vartheta \in \Theta} P_\vartheta^n\{\underline{x} \in X^n : d(\vartheta^{(n)}(\underline{x}), \vartheta) > \varepsilon\} = 0 \quad \text{for } \varepsilon > 0. \qquad (6.2.8)$$

Proof. (i) $\vartheta \to \Delta(P_\vartheta, Q)$ is continuous on Θ for every $Q \in \mathfrak{P}$. To see this, let $\vartheta \in \Theta$ be arbitrary. For every $\varepsilon > 0$ there exists a neighborhood U_ε of ϑ_0 such that $\vartheta \in U_\varepsilon$ implies $|P_\vartheta(f_k) - P_{\vartheta_0}(f_k)| < \varepsilon/2$ for $k = 1, \ldots, K_\varepsilon := [-\log_2 \varepsilon/2]$.

By (6.2.5), this implies $\Delta(P_\vartheta, P_{\vartheta_0}) < \varepsilon$ for $\vartheta \in U_\varepsilon$, hence $|\Delta(P_\vartheta, Q) - \Delta(P_{\vartheta_0}, Q)| < \varepsilon$ for $\vartheta \in U_\varepsilon$.

(ii) Since $\vartheta \to \Delta(P_\vartheta, Q)$ is continuous and Θ is compact, $\inf_{\vartheta \in \Theta} \Delta(P_\vartheta, Q)$ is attained. By the Selection Theorem 6.7.22, applied with $X = \mathfrak{P}$, $Y = \Theta$ and $f(Q, \vartheta) = \Delta(P_\vartheta, Q)$, there exists a measurable function $H : (\mathfrak{P}, \Delta) \to (\Theta, d)$ such that

$$\Delta(P_{H(Q)}, Q) = \inf_{\vartheta \in \Theta} \Delta(P_\vartheta, Q).$$

Therefore,

$$\Delta(P_{H(Q)}, P_\vartheta) \leq \Delta(P_{H(Q)}, Q) + \Delta(Q, P_\vartheta)$$
$$\leq 2\Delta(Q, P_\vartheta) \qquad \text{for } \vartheta \in \Theta, \; Q \in \mathfrak{P}.$$

Applied with $Q = Q_{\underline{x}}^{(n)}$ we obtain with $\vartheta^{(n)}(\underline{x}) := H(Q_{\underline{x}}^{(n)})$

$$P_\vartheta^n \{\underline{x} \in X^n : \Delta(P_{\vartheta^{(n)}(\underline{x})}, P_\vartheta) > \varepsilon\} \leq P_\vartheta^n \{\underline{x} \in X^n : \Delta(Q_{\underline{x}}^{(n)}, P_\vartheta) > \varepsilon/2\}.$$

By Lemma 6.2.4(ii) this implies

$$\lim_{n\to\infty} \sup_{\vartheta \in \Theta} P_\vartheta^n \{\underline{x} \in X^n : \Delta(P_{\vartheta^{(n)}(\underline{x})}, P_\vartheta) > \varepsilon\} = 0 \quad \text{for } \varepsilon > 0. \tag{6.2.9}$$

(iii) Since Θ is compact and $\vartheta \to P_\vartheta$ injective and continuous in the Δ–metric, the set $\{P_\vartheta : \vartheta \in \Theta\}$ is compact in the Δ–metric, and $P_\vartheta \to \vartheta$ is uniformly continuous (Kelley, 1955, p. 141, Theorem 8). Thus for every $\varepsilon > 0$ there exists $c_\varepsilon > 0$ such that $\Delta(P_{\vartheta'}, P_{\vartheta''}) < \delta_\varepsilon$ implies $d(\vartheta', \vartheta'') < \varepsilon$ for $\vartheta', \vartheta'' \in \Theta$. Therefore,

$$P_\vartheta^n \{\underline{x} \in X^n : d(\vartheta^{(n)}(\underline{x}), \vartheta) > \varepsilon\} \leq P_\vartheta^n \{\underline{x} \in X^n : \Delta(P_{\vartheta^{(n)}(\underline{x})}, P_\vartheta) > \delta_\varepsilon\}.$$

Hence (6.2.8) follows from (6.2.9) □

So far, the existence of uniformly consistent estimator sequences has been established for compact parameter spaces. Starting from such estimator sequences one can construct estimator sequences for σ–locally compact Θ. This is the content of Theorem 6.2.11. The proof of this theorem is based on Lemma 6.2.10 which is slightly more general than needed for this particular purpose. The reason: Lemma 6.2.10 will also be used in Section 6.3 to establish the existence of a consistent sequence of "approximate" solutions to estimating equations.

A Hausdorff space is called σ–*locally compact* if there exists a sequence of nonempty compact sets $K_i \subset \Theta$, $i \in \mathbb{N}$, such that $K_i \subset K_{i+1}^\circ$ and $\bigcup_1^\infty K_i^\circ = \Theta$.

Lemma 6.2.10. *Let Θ be a σ-locally compact metric space. For $\vartheta \in \Theta$ and $n \in \mathbb{N}$ let $P_\vartheta^{(n)}$ be a p-measure on some measurable space $(X_n, \mathcal{A}_n)$.*

Assume that for every $i \in \mathbb{N}$ there exists an estimator sequence $\vartheta_i^{(n)} : X_n \to \Theta$, $n \in \mathbb{N}$, which is consistent for $\vartheta \in K_i$, uniformly on K_i.

Then there exists a sequence $i_n \uparrow \infty$, $n \in \mathbb{N}$, such that $\vartheta_{i_n}^{(n)}$, $n \in \mathbb{N}$, is consistent for $\vartheta \in \Theta$, uniformly on compact subsets of Θ.

Proof. Since K_i is compact, $K_i \subset K_{i+1}^\circ$ implies $d(K_i, K_{i+1}^c) > 0$ (see Kelley, 1955, p. 155). Let ε_i, $i \in \mathbb{N}$, be a decreasing sequence with $\varepsilon_i \downarrow 0$, such that $0 < \varepsilon_i < \frac{1}{2} d(K_i, K_{i+1}^c)$.

Since $\vartheta_i^{(n)}$, $n \in \mathbb{N}$, is uniformly consistent on K_i, there exists n_i such that

$$P_\vartheta^{(n)} \{ x \in X_n : d(\vartheta_i^{(n)}(x), \vartheta) > \varepsilon_i \} < \varepsilon_i \quad \text{for } \vartheta \in K_i \text{ and } n \geq n_i.$$

W.l.g. we assume $n_i < n_{i+1}$ for $i \in \mathbb{N}$.

For $n \in \mathbb{N}$ sufficiently large, let i_n be such that $n_{i_n} \leq n < n_{i_n+1}$. Notice that $i_n \uparrow \infty$ and

$$P_\vartheta^{(n)} \{ x \in X_n : d(\vartheta_{i_n}^{(n)}(x), \vartheta) > \varepsilon_{i_n} \} < \varepsilon_{in} \quad \text{for } \vartheta \in K_{i_n}.$$

For every compact set K and every $\varepsilon > 0$, we have $K \subset K_{i_n}$ and $\varepsilon_{i_n} < \varepsilon$ for n sufficiently large, hence

$$\sup_{\vartheta \in K} P_\vartheta^{(n)} \{ x \in X_n : d(\vartheta_{i_n}^{(n)}(x), \vartheta) > \varepsilon \} < \varepsilon_{i_n}$$

for n sufficiently large. This implies the assertion. $\square$

Theorem 6.2.11. *Let Θ be a σ-locally compact metric space. For $\vartheta \in \Theta$ let P_ϑ be a p-measure on $(X, \mathcal{A})$.*

Assume there exist functions $f_k : X \to [0,1]$, $k \in \mathbb{N}$, fulfilling (6.2.1) such that $\vartheta \to P_\vartheta(f_k)$ is continuous for $k \in \mathbb{N}$. Then there exists an estimator sequence $\vartheta^{(n)} : X^n \to \Theta$, $n \in \mathbb{N}$, which is consistent for ϑ uniformly on compact subsets of Θ.

Proof. Since Θ is σ-locally compact, there exists an ascending sequence K_i, $i \in \mathbb{N}$, of compact sets fulfilling $\bigcup_1^\infty K_i^\circ = \Theta$. By Lemma 6.2.7, for every $i \in \mathbb{N}$ there exists an estimator sequence $\vartheta_i^{(n)}$, $n \in \mathbb{N}$, which is uniformly consistent on K_i. Hence the assertion follows from Lemma 6.2.10. $\square$

6.3 Consistency of M-estimators

Gives conditions under which M-estimators, defined as solutions in ϑ of $\sum_1^n f(x_\nu, \vartheta) = \sup$, are consistent if $P_\vartheta(f(\cdot, \tau) - f(\cdot, \vartheta)) < 0$ for $\tau \neq \vartheta$. In addition to regularity conditions on f, some sort of pseudo-compactness of Θ is needed.

Theorem 6.2.11 asserts the existence of consistent estimator sequences under rather general conditions. The construction upon which this assertion is based is, however, not applicable for practical purposes. In this section we present results on the consistency of some sort of minimum contrast estimators, and we study some of their properties which are needed later on to obtain their asymptotic distributions. These results will be applied to ML estimators in Section 6.5.

Let Θ be a separable metric space with distance function d. Let $f : X \times \Theta \to \mathbb{R}$ be a function with the following properties.

$$x \to f(x, \vartheta) \text{ is measurable for } \vartheta \in \Theta, \tag{6.3.1'}$$

$$\vartheta \to f(x, \vartheta) \text{ is continuous for } x \in X. \tag{6.3.1''}$$

$$P_\vartheta\big(f(\cdot, \tau) - f(\cdot, \vartheta)\big) < 0 \text{ for } \tau \in \Theta \text{ with } \tau \neq \vartheta. \tag{6.3.2}$$

Since $\tau \to P_\vartheta(f(\cdot, \tau))$ attains its maximum value at ϑ, (presuming that $P_\vartheta(f(\cdot, \vartheta)) \in \mathbb{R}$) it is not unreasonable to expect that $\tau \to n^{-1} \sum_1^n f(x_\nu, \tau)$ will attain its maximum value close to ϑ under P_ϑ^n.

Definition 6.3.3. (i) $\vartheta^{(n)} : X^n \to \Theta$ is an *M-estimator* if

$$n^{-1} \sum_1^n f\big(x_\nu, \vartheta^{(n)}(\underline{x})\big) = \sup_{\tau \in \Theta} n^{-1} \sum_1^n f(x_\nu, \tau) \qquad \text{for } \underline{x} \in X^n. \tag{6.3.4'}$$

(ii) $\vartheta^{(n)}$, $n \in \mathbb{N}$, is a *sequence of asymptotic M-estimators* if

$$\lim_{n \to \infty} \inf_{\vartheta \in \Theta} P_\vartheta^n \Big\{ \underline{x} \in X^n : n^{-1} \sum_1^n f(x_\nu, \vartheta^{(n)}(\underline{x})) \tag{6.3.4''}$$

$$\geq \sup_{\tau \in \Theta} n^{-1} \sum_1^n f(x_\nu, \tau) - \delta \Big\} = 1 \quad \text{for every } \delta > 0.$$

If $\tau \to P_\vartheta(f(\cdot, \tau))$ attains its *minimum* (rather than its maximum) at $\tau = \vartheta$, the function f in Definition 6.3.3 is replaced by $-f$, and the resulting estimators $\vartheta^{(n)}$ are called "minimum contrast" estimators. We deviate from this convention, since the most important application is to ML estimators, and this would require to change from minimization to maximization in the transition from Section 6.3 to Section 6.5.

Condition (6.3.4') gives a clear advice about how to obtain the estimate $\vartheta^{(n)}(\underline{x})$. It involves, however, some mathematical problems: Whether

$$\sup_{\tau \in \Theta} n^{-1} \sum_{1}^{n} f(x_\nu, \tau)$$

is attained for every $\underline{x} \in X^n$, and, if so, whether the choices of $\vartheta^{(n)}(\underline{x})$ for different values of $\underline{x} \in X^n$ can be coordinated in such a way that the function $\vartheta^{(n)} : X^n \to \Theta$ is measurable. Since consistency is an asymptotic property of an estimator sequence, it is clear that an asymptotic condition on the estimator sequence will suffice. (6.3.4'') is such an asymptotic condition. Estimator sequences fulfilling (6.3.4'') exist in general, and their measurability is easier to establish. That measurable versions of estimators fulfilling (6.3.4') or (6.3.4'') exist follows from the Selection Theorem 6.7.22 and Lemma 6.7.23.

Condition (6.3.2), i.e. $P_\vartheta\big(f(\cdot, \tau) - f(\cdot, \vartheta)\big) < 0$ for $\tau \neq \vartheta$, enables us to distinguish τ from ϑ by means of the family $f(\cdot, \tau)$, $\tau \in \Theta$. This condition is, however, not strong enough to guarantee consistency of M–estimators, since it does not prevent $P_\vartheta\big(f(\cdot, \tau) - f(\cdot, \vartheta)\big)$ from coming arbitrarily close to 0 for τ at a large distance from ϑ. An immediate strengthening of this condition is

$$P_\vartheta\big(\overline{f}(\cdot, V_\vartheta^c) - f(\cdot, \vartheta)\big) < 0 \tag{6.3.5}$$

for every neighborhood V_ϑ of ϑ. This condition is, however, too restrictive for most applications.

To establish consistency under less restrictive conditions we introduce the following

Condition 6.3.6. The function $f : X \times \Theta \to \mathbb{R}$ fulfills at ϑ_0 the following conditions: For some set $V \subset \Theta$,

$$\overline{f}(\cdot, V) - \pmb{f}(\cdot, \vartheta_0) \text{ is locally uniformly } P_\vartheta\text{–integrable at } \vartheta_0 \tag{6.3.7'}$$

(see Definition 6.7.8), and

$$P_{\vartheta_0}\big(\overline{f}(\cdot, V) - f(\cdot, \vartheta_0)\big) < 0. \tag{6.3.7''}$$

With this condition, we are able to formulate the following

Covering Condition 6.3.8. The function $f : X \times \Theta \to \mathbb{R}$ fulfills for $\Theta_0 \subset \Theta$ the [locally uniform] covering condition at ϑ_0 if there exists a cover of Θ_0 by finitely many sets V_i, $i = 1, \ldots, m$, fulfilling (6.3.7'') [and (6.3.7')].

In the following, it will be shown that [locally uniform] consistency of M–estimator sequences follows if the [locally uniform] covering condition holds for the complement of every neighborhood of any $\vartheta \in \Theta$. The covering condition for the complement of every neighborhood of ϑ_0 is, obviously, less restrictive than (6.3.5).

Lemma 6.3.9. *Let $f : X \times \Theta \to \mathbb{R}$ be a function fulfilling conditions (6.3.1) and (6.3.2). Assume that for every $\tau \neq \vartheta_0$ there exists a neighborhood V of τ such that $\overline{f}(\cdot, V) - f(\cdot, \vartheta_0)$ is uniformly P_ϑ–integrable at ϑ_0. Then the following holds true.*

(i) *The locally uniform covering condition holds for every compact subset of Θ which is disjoint from ϑ_0.*

(ii) *If the [locally uniform] covering condition holds at ϑ_0 for the complement of some compact neighborhood of ϑ_0, then it holds on the complement of every neighborhood of ϑ_0.*

Proof. (i) Assume that $\Theta_0 \subset \Theta$ is compact, and $\vartheta_0 \notin \Theta_0$. For $\tau \neq \vartheta_0$, there exists a neighborhood V_τ of τ such that $\overline{f}(\cdot, V_\tau) - f(\cdot, \vartheta_0)$ is locally uniformly P_ϑ–integrable at ϑ_0. By Lemma 6.7.5 there exists a neighborhood $V'_\tau \subset V_\tau$ of τ such that

$$P_{\vartheta_0}\big(\overline{f}(\cdot, V'_\tau) - f(\cdot, \vartheta_0)\big) < 0.$$

The relation $f(\cdot, \tau) \leq \overline{f}(\cdot, V'_\tau) \leq \overline{f}(\cdot, V_\tau)$ implies uniform P_ϑ–integrability at ϑ_0 of $\overline{f}(\cdot, V'_\tau) - f(\cdot, \vartheta_0)$.

Since $\{V'_\tau : \tau \in \Theta_0\}$ covers the compact set Θ_0, there exists a finite subcover, say $V'_{\tau_1}, \ldots, V'_{\tau_k}$, by sets fulfilling Condition 6.3.6.

(ii) Let $C \subset \Theta$ be a compact neighborhood of ϑ_0 such that the [locally uniform] covering condition holds true on C^c. By (i) there is a finite cover of the compact set $C \cap V^c$, for any neighborhood V of ϑ_0. Together with the finite cover of C^c, this forms a finite cover of V^c. $\qquad\qquad\square$

The intuitive idea of the consistency proof comes out more clearly if we present the result in the abstract framework of Theorem 6.3.10. For $\vartheta \in \Theta$ let $P_\vartheta^{(n)}$ be a p–measure on a measurable space $(X_n, \mathcal{A}_n)$, and let f_n be a function on $X_n \times \Theta$. Apart from the complication brought about by local uniformity, condition (6.3.11) requires that $\overline{f}_n(x, V_\vartheta^c)$ is for $n \to \infty$ stochastically smaller than $f_n(x, \vartheta)$, for every neighborhood V_ϑ of ϑ. By definition of $\vartheta^{(n)}$, $f_n(x, \vartheta^{(n)}(x))$ is for $n \to \infty$ stochastically not smaller than $f_n(x, \vartheta)$. This excludes that $\vartheta^{(n)}(x) \in V_\vartheta^c$ with positive probability, for $n \to \infty$.

That (6.3.11) holds for $f_n(\underline{x}, \vartheta) = n^{-1} \sum_1^n f(x_\nu, \vartheta)$ follows from the locally uniform Covering Condition 6.3.8, as will be shown in Proposition 6.3.15.

Theorem 6.3.10. *Assume that $f_n : X_n \times \Theta \to \mathbb{R}$ has properties (6.3.1), and that the following condition holds for every $\vartheta \in \Theta$.*

For every neighborhood V of ϑ there exists a neighborhood U of ϑ and $\delta_0 > 0$ such that

$$\lim_{n \to \infty} \sup_{\tau \in U} P_\tau^{(n)} \{ x \in X_n : \overline{f}_n(x, V^c) > f_n(x, \vartheta) - \delta_0 \} = 0. \tag{6.3.11}$$

Let $\vartheta^{(n)} : X_n \to \Theta$, $n \in \mathbb{N}$, be a sequence of asymptotic M-estimators, i.e.

$$\lim_{n \to \infty} \inf_{\vartheta \in \Theta} P_\vartheta^{(n)} \{ x \in X_n : f_n(x, \vartheta^{(n)}(x)) \geq \overline{f}_n(x, \Theta) - \delta \} = 1 \text{ for every } \delta > 0.$$

Then $\vartheta^{(n)}$, $n \in \mathbb{N}$, is consistent for ϑ, uniformly on compact subsets of Θ, i.e.

$$\lim_{n \to \infty} \sup_{\vartheta \in K} P_\vartheta^{(n)} \{ x \in X_n : d(\vartheta^{(n)}(x), \vartheta) > \varepsilon \} = 0$$

for every compact subset $K \subset \Theta$ and every $\varepsilon > 0$.

The conditions of Theorem 6.3.10 presume $\overline{f}_n(\cdot, \Theta) < \infty$. If $P_\vartheta^{(n)} \{ x \in X_n : f_n(x, \Theta) = \infty \} > 0$, apply the Theorem to $s \circ f_n$, where $s : \mathbb{R} \to \mathbb{R}$ is increasing and bounded from above. If (6.3.11) holds for f_n, it holds for $s \circ f_n$, provided $f_n(\cdot, \vartheta)$ is under $P_\vartheta^{(n)}$ stochastically bounded form above as $n \to \infty$, locally uniformly on Θ.

Proof. With $\varepsilon > 0$ fixed, let $V_\vartheta := \{ \tau \in \Theta : d(\tau, \vartheta) < \varepsilon/2 \}$. Using

$$f_n(x, \vartheta^{(n)}(x)) \leq \overline{f}_n(x, V_\vartheta^c) \quad \text{for } \vartheta^{(n)}(x) \in V_\vartheta^c$$

and $f_n(x, \vartheta) \leq \overline{f}_n(x, \Theta)$, we obtain

$$\{ x \in X_n : \vartheta^{(n)}(x) \in V_\vartheta^c \} \subset \{ x \in X_n : f_n(x, \vartheta^{(n)}(x)) < \overline{f}_n(x, \Theta) - \delta \}$$
$$\cup \{ x \in X_n : \overline{f}_n(x, V_\vartheta^c) \geq f_n(x, \vartheta) - \delta \}.$$

With $\delta \in (0, \delta_0)$, this implies

$$\lim_{n \to \infty} \sup_{\tau \in U_\vartheta} P_\tau^{(n)} \{ x \in X_n : \vartheta^{(n)}(x) \in V_\vartheta^c \} = 0, \tag{6.3.12}$$

for some open neighborhood U_ϑ of ϑ (depending on ε). We may assume w.l.g. that $\tau \in U_\vartheta$ implies $d(\tau, \vartheta) < \varepsilon/2$.

Since $\{ U_\vartheta : \vartheta \in K \}$ covers the compact set K, there exists a finite subcover, say $\{ U_{\vartheta_i} : i = 1, \ldots, m \}$. This implies

$$\sup_{\vartheta \in K} P_\vartheta^{(n)} \{ x \in X_n : d(\vartheta^{(n)}(x), \vartheta) \geq \varepsilon \} \tag{6.3.13}$$

$$\leq \max_{i=1,\ldots,m} \sup_{\vartheta \in U_{\vartheta_i}} P_\vartheta^{(n)} \{ x \in X_n : d(\vartheta^{(n)}, \vartheta) \geq \varepsilon \}.$$

For $\vartheta \in U_{\vartheta_i}$, the relation $d\big(\vartheta^{(n)}(x), \vartheta\big) \geq \varepsilon$ implies $d\big(\vartheta^{(n)}(x), \vartheta_i\big) \geq \varepsilon/2$, i.e. $\vartheta^{(n)}(x) \in V_{\vartheta_i}^c$. Hence

$$\sup_{\vartheta \in U_{\vartheta_i}} P_\vartheta^{(n)} \big\{ x \in X_n : d\big(\vartheta^{(n)}(x), \vartheta\big) \geq \varepsilon \big\} \tag{6.3.14}$$

$$\leq \sup_{\vartheta \in U_{\vartheta_i}} P_\vartheta^{(n)} \big\{ x \in X_n : \vartheta^{(n)}(x) \in V_{\vartheta_i}^c \big\}.$$

The assertion now follows from (6.3.13), (6.3.14), and (6.3.12) applied for $\vartheta = \vartheta_i$. $\qquad\square$

To apply Theorem 6.3.10 for the case $P_\vartheta^{(n)} = P_\vartheta^n$, we need conditions on the function $f : X \times \Theta \to \mathbb{R}$ which imply condition (6.3.11) for $f_n(\underline{x}, \vartheta) := n^{-1} \sum_1^n f(x_\nu, \vartheta)$.

Proposition 6.3.15. *Assume that $f : X \times \Theta \to \mathbb{R}$ fulfills conditions (6.3.1), and the locally uniform Covering Condition 6.3.8 for the complement of every neighborhood of ϑ.*
Then condition (6.3.11) holds true for $f_n(\underline{x}, \vartheta) = n^{-1} \sum_1^n f(x_\nu, \vartheta)$.

Proof. Let $\vartheta \in \Theta$ be fixed, and let V be an arbitrary neighborhood of ϑ. By assumption there exists a cover of V^c by sets V_i, $i = 1, \ldots, m$, such that for $i = 1, \ldots, m$

$$h_i(\cdot, \vartheta) := \overline{f}(\cdot, V_i) - f(\cdot, \vartheta)$$

is uniformly P_τ–integrable for τ in a neighborhood U of ϑ, and

$$P_\vartheta \big(h_i(\cdot, \vartheta) \big) < 0.$$

Let $\delta_i := -\frac{1}{2} P_\vartheta \big(h_i(\cdot, \vartheta) \big)$. From Proposition 6.7.15, applied with $\mathfrak{P} = \{ P_\tau : \tau \in U \}$ and $\varepsilon = \delta_i$, we obtain

$$\lim_{n \to \infty} \sup_{\tau \in U} P_\tau^n \big\{ \underline{x} \in X^n : n^{-1} \sum_1^n h_i(x_\nu, \vartheta) > -\delta_i \big\} = 0. \tag{6.3.16}$$

The relation

$$\overline{f}_n(\underline{x}, V^c) \leq \max_{i=1,\ldots,m} \overline{f}_n(\underline{x}, V_i)$$

implies

$$\big\{ \underline{x} \in X^n : \overline{f}_n(\underline{x}, V^c) > f_n(\underline{x}, \vartheta) - \delta \big\}$$

$$\subset \bigcup_{i=1}^m \big\{ \underline{x} \in X^n : n^{-1} \sum_1^n h_i(x_\nu, \vartheta) > -\delta_i \big\}$$

for $\delta := \min\{\delta_1, \ldots, \delta_m\} > 0$. Together with (6.3.16) this implies

$$\lim_{n \to \infty} \sup_{\tau \in U} P_\tau^n \{\underline{x} \in X^n : \overline{f}_n(\underline{x}, V^c) > f_n(\underline{x}, \vartheta) - \delta\} = 0.$$

$\square$

The following proposition is a slight generalization of Proposition 6.3.15, motivated by the fact that in certain cases the covering condition fails for $f|X \times \Theta$, but it holds true for $f_k : X^k \times \Theta \to \mathbb{R}$ defined by

$$f_k(x_1, \ldots, x_k, \vartheta) := \frac{1}{k} \sum_{\nu=1}^{k} f(x_\nu, \vartheta). \tag{6.3.17}$$

As an example, we mention location– and scale parameter families with $f = \log p$ (see Exercise 9.3.19 and Proposition 9.3.21).

For sample sizes which are multiples of k, say $n = km$, the Consistency Theorem 6.3.10 can be applied with $P_\vartheta^{(n)} = P_\vartheta^{km}$, and $X_n = X^{km}$, by writing $f_n(x_1, \ldots, x_n, \vartheta) = n^{-1} \sum_1^n f(x_\nu, \vartheta)$ for $n = km$ as

$$f_{km}(x_1, \ldots, x_n, \vartheta) = \frac{1}{m} \sum_{\nu=1}^{m} f_k(x_{(\nu-1)k+1}, \ldots, x_{\nu k}, \vartheta).$$

If the subsequence of ML estimators $\vartheta^{(n)}$ with $n = km$, $m = 1, 2, \ldots$, converges to ϑ, it is plausible that the whole sequence $\vartheta^{(n)}$, $n \in \mathbb{N}$, will do the same. This was proved by Perlman (1972) in a slightly different framework, using the law of large numbers for submartingales. For a more elementary approach see Pitman (1979, pp. 65/6, Theorem).

Proposition 6.3.18. *Assume that $f : X \times \Theta \to \mathbb{R}$ fulfills condition (6.3.1), and that $f_k : X^k \times \Theta \to \mathbb{R}$, defined by (6.3.17), fulfills the [locally uniform] covering condition for the complement of every neighborhood of ϑ. Then condition (6.3.11) holds true [locally uniformly] for the sequence $f_n(\cdot, \vartheta)$, $n \in \mathbb{N}$.*

Proof. Let V be an arbitrary neighborhood of ϑ. By assumption there exists a cover of V^c by sets $V_i \subset \Theta$, $i = 1, \ldots, q$, such that $\overline{f}_k(\cdot, V_i) - f_k(\cdot, \vartheta)$ is uniformly P_τ–integrable for τ in a neighborhood U of ϑ, and

$$P_\vartheta^k \big(\overline{f}_k(\cdot, V_i) - f_k(\cdot, \vartheta) \big) < 0.$$

Writing $n \in \mathbb{N}$ as $n = km + r$ with $0 \le r < k$, we have

$$f_n(x_1, \ldots, x_n, \vartheta') = \frac{k}{n} \sum_{\nu=1}^{m-1} f_k(x_{k(\nu-1)+1}, \ldots, x_{k\nu}, \vartheta') + \frac{1}{n} \sum_{\nu=k(m-1)+1}^{n} f(x_\nu, \vartheta').$$

This implies

$$\bar{f}_n(x_1, \ldots, x_n, V^c) - f_n(x_1, \ldots, x_n, \vartheta) \qquad (6.3.19)$$

$$\leq \frac{k(m-1)}{n} \sup_{\vartheta' \in V^c} \frac{1}{m-1} \sum_{\nu=1}^{m-1} \big(f_k(x_{k(\nu-1)+1}, \ldots, x_{k\nu}, \vartheta')$$

$$- f_k(x_{k(\nu-1)+1}, \ldots, x_{k\nu}, \vartheta) \big)$$

$$+ \frac{1}{n} \sup_{\vartheta' \in V^c} \sum_{\nu=k(m-1)+1}^{n} \big(f(x_\nu, \vartheta') - f(x_\nu, \vartheta) \big).$$

By Proposition 6.3.15, applied with X^k, P_ϑ^k and f_k in place of X, P_ϑ and f, there exists a neighborhood U of ϑ and $\delta > 0$ such that

$$\lim_{m \to \infty} \sup_{\tau \in U} P_\tau^{k(m-1)} \{ \underline{x} \in X^{k(m-1)} : \qquad (6.3.20)$$

$$\sup_{\vartheta' \in V^c} \frac{1}{m-1} \sum_{\nu=1}^{m-1} \big(f_k(x_{k(\nu-1)+1}, \ldots, x_{k\nu}, \vartheta')$$

$$- f_k(x_{k(\nu-1)+1}, \ldots, x_{k\nu}, \vartheta) \big) > -\delta \} = 0.$$

For any $\ell \geq k$,

$$\sum_{\nu=1}^{\ell} f(x_\nu, \vartheta') = \frac{1}{\binom{\ell}{k}} \sum_{(i_1, \ldots, i_k)} \sum_{j=1}^{k} f(x_{i_j}, \vartheta')$$

where $\sum_{(i_1, \ldots, i_k)}$ extends over all k-tuples $(i_1, \ldots, i_k)$ in $\{1, \ldots, \ell\}$ with $i_1 < \ldots < i_k$. This implies

$$\sup_{\vartheta' \in V^c} \sum_{\nu=1}^{\ell} f(x_\nu, \vartheta') \leq \frac{k}{\binom{\ell}{k}} \sum_{(i_1, \ldots, i_k)} \bar{f}_k(x_{i_1}, \ldots, x_{i_k}, V^c),$$

so that

$$\int \sup_{\vartheta' \in V^c} \sum_{\nu=1}^{\ell} \big(f(x_\nu, \vartheta') - f(x_\nu, \vartheta) \big) P_\vartheta^\ell \big(d(x_1, \ldots, x_\ell) \big)$$

$$\leq k P_\vartheta^k \big(\bar{f}_k(\cdot, V^c) - f_k(\cdot, \vartheta) \big).$$

Since $\bar{f}_k(\cdot, V^c) - f_k(\cdot, \vartheta)$ is uniformly P_τ-integrable for τ in a neighborhood U of ϑ, this implies for $\varepsilon > 0$,

$$\lim_{n \to \infty} \sup_{\tau \in U} P_\tau^{k+r} \{ \underline{x} \in X^{k+r} : \sup_{\vartheta' \in V^c} \frac{1}{n} \sum_{\nu=1}^{k+r} \big(f(x_\nu, \vartheta') - f(x_\nu, \vartheta) \big) > \varepsilon \} = 0. \quad (6.3.21)$$

From (6.3.19), (6.3.20) and (6.3.21) we obtain that

$$\lim_{n \to \infty} \sup_{\tau \in U} P_\tau^n \{ \underline{x} \in X^n : \overline{f}_n(x_1, \ldots, x_n, V^c) - f_n(x_1, \ldots, x_n, \vartheta) > -\delta' \} = 0$$

for $\delta' \in (0, \delta)$. $\qquad\qquad\qquad\qquad\qquad\qquad\qquad\qquad\qquad\qquad\qquad$ $\square$

6.4 Consistent solutions of estimating equations

If $\Theta \subset \mathbb{R}^p$ and if $\vartheta \to f(x, \vartheta)$ has a derivative $f^{(\cdot)}$, then consistent estimator sequences, which are solutions in ϑ of $\sum_1^n f^{(\cdot)}(x_\nu, \vartheta) = 0$, exist under mild regularity conditions.

Let $\Theta \subset \mathbb{R}^p$ be an open subset. Assume that $\vartheta \to f(x, \vartheta)$ has partial derivatives $f^{(i)}(x, \vartheta_1, \ldots, \vartheta_p) := \frac{\partial}{\partial \vartheta_i} f(x, \vartheta_1, \ldots, \vartheta_p)$. If convenient, we use $f^{(\cdot)}(x, \vartheta)$ to denote the row vector $f^{(i)}(x, \vartheta)$, $i = 1, \ldots, p$. As in Section 6.3 we assume that

$$P_\vartheta \big(f(\cdot, \tau) - f(\cdot, \vartheta) \big) < 0 \quad \text{for } \tau \in \Theta \text{ with } \tau \neq \vartheta. \tag{6.4.1}$$

Under the differentiability conditions stated above, estimators $\vartheta^{(n)}$ fulfilling

$$\sum_1^n f\big(x_\nu, \vartheta^{(n)}(\underline{x})\big) = \sup_{\vartheta \in \Theta} \sum_1^n f(x_\nu, \vartheta) \tag{6.4.2}$$

are also solutions to the estimating equation

$$\sum_1^n f^{(\cdot)}(x_\nu, \vartheta) = 0. \tag{6.4.3}$$

If sequences of such solutions are consistent, they are asymptotically normal under weak regularity conditions (see Section 7.4). The weak point in this argument is that it needs restrictive conditions (including the Covering Condition 6.3.8) to guarantee that estimator sequences fulfilling (6.4.2) are, in fact, consistent. The following theorem shows that consistent sequences of solutions to (6.4.3) exist under rather mild regularity conditions.

It is the particular type of this estimating equation (resulting from a contrast function) which accounts for this favorable result. A comparable theorem referring to arbitrary estimating equations $\sum_1^n g(x_\nu, \vartheta) = 0$ requires more restrictive conditions on g (like uniform integrability of the partial derivatives $g_i^{(j)}(x, \vartheta)$).

Theorem 6.4.4. *Assume that $f : X \times \Theta \to \mathbb{R}$ fulfills for every $\vartheta \in \Theta$ condition (6.3.1), and Condition 6.3.6 for some neighborhood V_τ of every $\tau \neq \vartheta$, and that $f^{(\cdot)}(x, \vartheta)$ exists for $x \in X$ and $\vartheta \in \Theta$.*

Then there exists an estimator sequence $\vartheta^{(n)}$, $n \in \mathbb{N}$, which is consistent for ϑ, uniformly on every compact subset K of Θ, and which fulfills

$$\lim_{n \to \infty} \inf_{\vartheta \in K} P_\vartheta^n \{ \underline{x} \in X^n : \sum_1^n f^{(\cdot)}(x_\nu, \vartheta^{(n)}(\underline{x})) = 0 \} = 1.$$

Theorem 6.4.4 implies in particular that

$$\lim_{n \to \infty} \inf_{\vartheta \in K} P_\vartheta^n \{ \underline{x} \in X^n : \sum_1^n f^{(\cdot)}(x_\nu, \tau) = 0 \quad \text{for some } \tau \in \Theta \} = 1.$$

Proof. Let K_i, $i \in \mathbb{N}$, be a sequence of nonempty compact subsets of Θ such that $K_i \subset K_{i+1}^\circ$ and $\Theta = \bigcup_1^\infty K_i^\circ$. By Theorem 6.3.10, applied with K_i in place of Θ, every estimator sequence $\vartheta_i^{(n)}$, $i \in \mathbb{N}$, fulfilling

$$\sum_1^n f(x_\nu, \vartheta_i^{(n)}(\underline{x})) = \sup_{\vartheta \in K_i} \sum_1^n f(x_\nu, \vartheta) \tag{6.4.5}$$

is uniformly consistent on K_i.

(Notice that this theorem applies, since condition (6.3.11) holds true as a consequence of Lemma 6.3.9(i) and Proposition 6.3.15.)

By Lemma 6.2.10 there exists a sequence $i_n \uparrow \infty$ such that $\vartheta_{i_n}^{(n)}$, $n \in \mathbb{N}$, is consistent, uniformly on compact subsets of Θ. If $K \subset \Theta$ is compact, we have $K \subset K_{i_n}^\circ$ for $n \geq n_0$, say, and therefore $\delta_0 := d(K, (K_{i_{n_0}}^\circ)^c) > 0$. This implies

$$\sup_{\vartheta \in K} P_\vartheta^n \{ \underline{x} \in X^n : \vartheta_{i_n}^{(n)}(\underline{x}) \in (K_{i_n}^\circ)^c \}$$

$$\leq \sup_{\vartheta \in K} P_\vartheta^n \{ \underline{x} \in X^n : d(\vartheta, \vartheta_{i_n}^{(n)}(\underline{x})) \geq \delta_0 \},$$

hence

$$\lim_{n \to \infty} \inf_{\vartheta \in K} P_\vartheta^n \{ \underline{x} \in X^n : \vartheta_{i_n}^{(n)}(\underline{x}) \in K_{i_n}^\circ \} = 1.$$

Since $\vartheta_{i_n}^{(n)}(\underline{x}) \in K_{i_n}^\circ$ implies $\sum_1^n f^{(\cdot)}(x_\nu, \vartheta_{i_n}^{(n)}(\underline{x})) = 0$ because of (6.4.5), this implies

$$\lim_{n \to \infty} \inf_{\vartheta \in K} P_\vartheta^n \{ \underline{x} \in X^n : \sum_1^n f^{(\cdot)}(x_\nu, \vartheta_{i_n}^{(n)}(\underline{x})) = 0 \} = 1.$$

Hence the assertion holds with $\vartheta^{(n)} = \vartheta_{i_n}^{(n)}$. $\square$

Theorem 6.4.4 asserts the existence of a consistent sequence of solutions to the equations (6.4.3). If one tries to construct such a sequence it becomes clear that one has first to fix a sequence of compact subsets of Θ, and a sequence $\varepsilon_i \downarrow 0$ as in the proof of Lemma 6.2.10 and to determine the smallest n for which

$$\sup_{\vartheta \in K_i} P_\vartheta^n \{\underline{x} \in X^n : d(\vartheta^{(n)}(\underline{x}), \vartheta) > \varepsilon_i\} < \varepsilon_i.$$

Obviously, such a procedure is not suitable for obtaining reasonably good estimators for a given sample size.

If the equations (6.4.3) have for every $n \in \mathbb{N}$ and every $\underline{x} \in X^n$ one solution only (as in Examples 9.1.1 and 9.1.10), then this yields a consistent estimator sequence (provided the regularity conditions on f specified in Theorem 6.4.4 are fulfilled). Recall that, in general, the number of solutions is a random variable, depending on $(x_1, \ldots, x_n)$. This number converges stochastically to 1 under certain conditions (see Mäkeläinen, Schmidt and Styan, 1981). There are, however, instances where the probability for more than one solution remains positive as n tends to infinity. As an example we mention a result of Reeds (1985). If $k_n(\underline{x})$ denotes the number of solutions of the likelihood equation for the location parameter of the Cauchy distribution, then $k_n + 1$ is asymptotically distributed according to the Poisson distribution with parameter $1/\pi$. With this in mind it is surprising that the situation becomes simpler for the more general case of a Cauchy family with location– and scale parameter: then the ML estimator is unique, according to Copas (1975).

If there are several solutions, the problem is to select one of these as an estimator. In mathematical terms: To select, among all possible sequences of solutions, a consistent one. The following lemma suggests that this can be achieved by projecting a sequence of auxiliary estimators into $\{\vartheta \in \Theta : \sum_1^n f^{(\cdot)}(x_\nu, \vartheta) = 0\}$.

Lemma 6.4.6. *For $n \in \mathbb{N}$ and $\underline{x} \in X^n$ let $\Theta_n(\underline{x}) \subset \Theta$ be a nonempty measurable set. If $\vartheta^{(n)}$, $n \in \mathbb{N}$, is a [locally uniformly] consistent estimator sequence for ϑ, then the projections of $\vartheta^{(n)}(\underline{x})$ into $\Theta_n(\underline{x})$ form a [locally uniformly] consistent estimator sequence in $\Theta_n(\underline{x})$, provided such a one exists.*

Proof. Let $\vartheta_0^{(n)}$, $n \in \mathbb{N}$, denote the (unknown) [locally uniformly] consistent estimator sequence fulfilling $\vartheta_0^{(n)}(\underline{x}) \in \Theta_n(\underline{x})$ for $\underline{x} \in X^n$, i.e. for every $\vartheta_0 \in \Theta$ there exists a neighborhood U_0 such that

$$\lim_{n \to \infty} \sup_{\vartheta \in U_0} P_\vartheta^n \{\underline{x} \in X^n : d(\vartheta_0^{(n)}(\underline{x}), \vartheta) > \varepsilon\} = 0 \quad \text{for } \varepsilon > 0. \tag{6.4.7}$$

Let $\vartheta^{(n)}$, $n \in \mathbb{N}$, be [locally uniformly] consistent, i.e. (6.4.7) holds also with $\vartheta^{(n)}(\underline{x})$ in place of $\vartheta_0^{(n)}(\underline{x})$.

For $n \in \mathbb{N}$ let $\hat{\vartheta}^{(n)}$ be a measurable function such that

$$\hat{\vartheta}^{(n)}(\underline{x}) \in \Theta_n(\underline{x}),$$

and

$$d\big(\hat{\vartheta}^{(n)}(\underline{x}), \vartheta^{(n)}(\underline{x})\big) \leq \inf\big\{d(\vartheta, \vartheta^{(n)}(\underline{x})) : \vartheta \in \Theta_n(\underline{x})\big\} + 1/n.$$

(For existence of $\hat{\vartheta}^{(n)}$ see Lemma 6.7.23.)

Since $\vartheta_0^{(n)}(\underline{x}) \in \Theta_n(\underline{x})$ we have

$$d\big(\hat{\vartheta}^{(n)}(\underline{x}), \vartheta^{(n)}(\underline{x})\big) \leq d\big(\vartheta_0^{(n)}(\underline{x}), \vartheta^{(n)}(\underline{x})\big) + 1/n$$
$$\leq d\big(\vartheta_0^{(n)}(\underline{x}), \vartheta\big) + d\big(\vartheta^{(n)}(\underline{x}), \vartheta\big) + 1/n.$$

Therefore,

$$d\big(\hat{\vartheta}^{(n)}(\underline{x}), \vartheta\big) \leq d\big(\hat{\vartheta}^{(n)}(\underline{x}), \vartheta^{(n)}(\underline{x})\big) + d\big(\vartheta^{(n)}(\underline{x}), \vartheta\big)$$
$$\leq d\big(\vartheta_0^{(n)}(\underline{x}), \vartheta\big) + 2d\big(\vartheta^{(n)}(\underline{x}), \vartheta\big) + 1/n.$$

Hence $\hat{\vartheta}^{(n)}$, $n \in \mathbb{N}$, is [locally uniformly] consistent by (6.4.7).   □

6.5 Consistency of maximum likelihood estimators

The results of Sections 6.3 and 6.4 are applied to maximum likelihood estimators.

Let $\Theta \subset \mathbb{R}^p$ be an open set, and $\{P_\vartheta : \vartheta \in \Theta\}$ a parametric family. Let $p(\cdot, \vartheta)$ denote a μ–density of P_ϑ. If ϑ is to be estimated on the basis of the observation $(x_1, \ldots, x_n)$, it suggests itself to consider the probability of $(x_1, \ldots, x_n)$, $\prod_{\nu=1}^n P_\vartheta\{x_\nu\}$, under different values of ϑ, and to choose as an estimator that value of ϑ for which $\prod_1^n P_\vartheta\{x_\nu\} = \max$. If the p–measures P_ϑ are nonatomic, it appears natural to take the densities as a substitute for the probabilites. Because of its intuitive appeal, the idea of ML estimators is so natural that it has been invented rather early (see the Historical Remark 6.5.9). The formal basis for the favorable asymptotic behavior of ML estimators is that $\tau \to P_\vartheta\big(\log p(\cdot, \tau)\big)$ attains its maximum at $\tau = \vartheta$, i.e. that $\log p(\cdot, \vartheta)$ fulfills condition (6.3.2). Together with some other technical conditions on $\log p(\cdot, \vartheta)$, and a covering condition corresponding to 6.3.8, this implies consistency of ML estimators.

Definition 6.5.1. (i) $\vartheta^{(n)} : X^n \to \Theta$ is a ML (= *maximum likelihood*) *estimator* if

$$\prod_1^n p(x_\nu, \vartheta^{(n)}(\underline{x})) = \sup_{\vartheta \in \Theta} \prod_1^n p(x_\nu, \vartheta) \quad \text{for every } \underline{x} \in X^n. \tag{6.5.2'}$$

(ii) $\vartheta^{(n)}$ is an *asymptotic ML estimator* if

$$\lim_{n \to \infty} \inf_{\vartheta \in \Theta} P_\vartheta^n \{\underline{x} \in X^n : n^{-1} \sum_1^n \ell(x_\nu, \vartheta^{(n)}(\underline{x})) \tag{6.5.2''}$$

$$\geq \sup_{\vartheta \in \Theta} n^{-1} \sum_1^n \ell(x_\nu, \vartheta) - \delta\} = 1 \text{ for every } \delta > 0.$$

Condition (6.5.2'') is even weaker than the condition that for some $a \in (0,1)$,

$$\prod_1^n p(x_\nu, \vartheta^{(n)}(\underline{x})) \geq a \sup_{\vartheta \in \Theta} \prod_1^n p(x_\nu, \vartheta) \quad \text{for every } \underline{x} \in X^n.$$

Definition 6.5.1 is, in fact, Definition 6.3.3, specialized for $f(x, \vartheta) = \log p(x, \vartheta)$.

Consistency Theorem 6.5.3. *Let* $\Theta \subset \mathbb{R}^p$ *be an open subset. For* $\vartheta \in \Theta$ *let* P_ϑ *be a p–measure on a measurable space* $(X, \mathcal{A})$ *such that* $\vartheta' \neq \vartheta''$ *implies* $P_{\vartheta'} \neq P_{\vartheta''}$.

(i) *Assume that there exist densities* $p(\cdot, \vartheta) \in dP_\vartheta/d\mu$ *such that* $\vartheta \to p(x, \vartheta)$ *is continuous on* Θ *and that the function* $\log p$ *fulfills the locally uniform Covering Condition 6.3.8 for the complement of every neighborhood of* ϑ.

Then every sequence of asymptotic ML estimators is consistent for ϑ, *uniformly on compact subsets of* Θ.

(ii) *If* $\log p$ *fulfills Condition 6.3.6 for some neighborhood* V_τ *of every* $\tau \neq \vartheta$, *and* $\vartheta \to p(x, \vartheta)$ *is differentiable for every* $x \in X$, *then there exists an estimator sequence* $\vartheta^{(n)}$, $n \in \mathbb{N}$, *which is consistent for* ϑ, *uniformly on compact subsets of* Θ, *and which fulfills for every compact* $K \subset \Theta$

$$\lim_{n \to \infty} \inf_{\vartheta \in K} P_\vartheta^n \{\underline{x} \in X^n : \sum_1^n p^{(\cdot)}(x_\nu, \vartheta^{(n)}(\underline{x})) \Big/ p(x_\nu, \vartheta^{(n)}(\underline{x})) = 0\} = 1.$$

Proof. Apply Proposition 6.3.15, Theorem 6.3.10 and Theorem 6.4.4 for $f(x, \vartheta) = \log p(x, \vartheta)$. □

Part (ii) of Theorem 6.5.3 is in the spirit of Cramér's original theorem (1946, p. 500) which asserts that consistent sequences *exist*, without identifying one. It is usually proved under much more restrictive regularity conditions. (For an elaborate proof using a fixed point theorem see Schmetterer, 1974, pp. 295ff., for a short proof using the inverse function theorem see Foutz, 1977.)

It is, of course, the covering condition which severely limits the applicability of Theorem 6.5.3(i). If the ML estimators can be obtained in explicit form, it is usually easier to use this for a consistency proof, than to apply Theorem 6.5.3. Lemma 6.3.9 establishes the covering condition for compact subsets, provided that for every $\tau \neq \vartheta$ there exists a neighborhood V_τ such that $P_\vartheta\big(\log \overline{p}(\cdot, V_\tau) - \log p(\cdot, \vartheta)\big) < \infty$. The condition $P_\vartheta\big(\log p(\cdot, \tau) - \log p(\cdot, \vartheta)\big) < 0$, is always true by Lemma 6.5.4. How the covering condition can be verified in a case where ϑ is not compact is illustrated in Example 6.6.5 and Exercise 6.6.7.

Without some substitute for the compactness of Θ, ML estimators are not necessarily consistent. This is demonstrated by Example 6.6.2. More examples of this type can be found in Bahadur (1958) and Pfanzagl (1969b). There are, of course, many modes of skilfull compactifications (other than the covering condition) to establish consistency for noncompact parameter sets Θ. The reader is referred to Pfanzagl (1969b) and Bahadur (1971).

It might look nicer at first sight to base conditions for the consistency of ML estimators on $P_\vartheta\big(\log p(\cdot, \tau)\big)$ or $P_\vartheta\big(\log \overline{p}(\cdot, V_\tau)\big)$ rather than $P_\vartheta\big(\log(p(\cdot, \tau) - \log p(\cdot, \vartheta)\big)$ or $P_\vartheta\big(\log \overline{p}(\cdot, V_\tau) - \log p(\cdot, \vartheta)\big)$. However, the more complicated condition is the more natural one: It is $\log p(\cdot, \tau) - \log(\cdot, \vartheta)$ which is independent of the dominating measure, which remains invariant under transformations on X, and which may be integrable if $\log p(\cdot, \tau)$ and $\log p(\cdot, \vartheta)$, taken separately, are not.

Lemma 6.5.4. *Assume that $\tau \neq \vartheta$ implies $P_\tau \neq P_\vartheta$. If $P_\vartheta\big(\log p(\cdot, \vartheta)\big) > -\infty$ for $\vartheta \in \Theta$ and $P_\vartheta\big(\log p(\cdot, \tau)\big) < \infty$ for $\vartheta, \tau \in \Theta$ with $\vartheta \neq \tau$, then*

$$P_\vartheta\big(\log p(\cdot, \tau) - \log p(\cdot, \vartheta)\big) < 0 \quad \text{if } \tau \neq \vartheta. \tag{6.5.5}$$

Proof. If $P_\vartheta\big(\log p(\cdot, \vartheta)\big) = \infty$ or $P_\vartheta\big(\log p(\cdot, \tau)\big) = -\infty$, relation (6.5.5) is trivially true. If both values are finite, strict concavity of $u \to \log u$ implies for $\vartheta \neq \tau$

$$P_\vartheta\big(\log p(\cdot, \tau) - \log p(\cdot, \vartheta)\big) < \log P_\vartheta\big(p(\cdot, \tau)/p(\cdot, \vartheta)\big) = \log \mu\big(p(\cdot, \tau)\big) = 0.$$

□

Definition 6.5.1 of the ML estimator refers to a particular version of the density. Straightforward examples show that the choice of the density influences existence, measurability, and even consistency of the resulting ML estimator. We exclude this problem from our considerations, since the Consistency Theorem 6.5.3 requires continuity of $\vartheta \to p(x, \vartheta)$, and such versions are unique μ–a.e. (Hint: Let $p_i(\cdot, \vartheta)$, $i = 1, 2$, be two continuous versions of the density. Let Θ_0 be a countable dense subset of Θ. We have $\mu\{p_1(\cdot, \vartheta) \neq p_2(\cdot, \vartheta)\} = 0$ for $\vartheta \in \Theta_0$. By continuity of $\vartheta \to p_i(x, \vartheta)$, $p_1(x, \vartheta) = p_2(x, \vartheta)$ for $\vartheta \in \Theta_0$ implies $p_1(x, \vartheta) = p_2(x, \vartheta)$ for $\vartheta \in \Theta$.) The reader interested in the problem of choosing the density is referred to Pfanzagl (1969b, pp. 257f.) and Landers (1972).

Remark 6.5.6. The continuity assumption for $\vartheta \to p(x, \vartheta)$ establishes an inherent relationship between the topology of Θ, a subset of $\mathbb{R}^p$, and topologies on the family $\{P_\vartheta : \vartheta \in \Theta\}$ defined in terms of the p–measures. First of all, $\vartheta_n \to \vartheta_0$ implies that $d(P_{\vartheta_n}, P_{\vartheta_0}) \to 0$ for $d(P, Q) := \sup_{A \in \mathcal{A}} |P(A) - Q(A)|$ by Scheffé's lemma (see Lehmann, 1986, p. 573, Lemma 4). The converse is true if the covering condition holds for the complement of every neighborhood V of ϑ_0. Under this condition, there exists a cover of V^c by a finite number of sets $V_1, \ldots, V_m$ such that

$$P_{\vartheta_0}\big(\log \overline{p}(\cdot, V_i) - \log p(\cdot, \vartheta_0)\big) < 0 \quad \text{for } i = 1, \ldots, m. \tag{6.5.7}$$

Assume now that for some sequence ϑ_n, $n \in \mathbb{N}$, we have $d(P_{\vartheta_n}, P_{\vartheta_0}) \to 0$, but $\vartheta_n \in V^c$ for infinitely many $n \in \mathbb{N}$. Then there exists $i_0 \in \{1, \ldots, m\}$ and an infinite subsequence $\mathbb{N}_0 \subset \mathbb{N}$ such that $\vartheta_n \in V_{i_0}$ for $n \in \mathbb{N}_0$. Since $d(P_{\vartheta_n}, P_{\vartheta_0}) \xrightarrow[n \in \mathbb{N}_0]{} 0$ implies $p(\cdot, \vartheta_n) \xrightarrow[n \in \mathbb{N}_0]{} p(\cdot, \vartheta_0)$ in μ–mean, there exists by a theorem of F. Riesz (see Ash, 1972, p. 93, Theorem 2.5.3) a subsequence $\mathbb{N}_1 \subset \mathbb{N}_0$ such that $p(\cdot, \vartheta_n) \xrightarrow[n \in \mathbb{N}_1]{} p(\cdot, \vartheta_0)$ μ–a.e. Since $\vartheta_n \in V_{i_0}$ for $n \in \mathbb{N}_1$, this implies $p(\cdot, \vartheta_0) \leq \overline{p}(\cdot, V_{i_0})$ μ–a.e., hence $P_{\vartheta_0}\big(\log \overline{p}(\cdot, V_{i_0}) - p(\cdot, \vartheta_0)\big) \geq 0$, in contradiction to (6.5.7).

Hence $d(P_{\vartheta_n}, P_{\vartheta_0}) \to 0$ if and only if $\vartheta_n \to \vartheta_0$. As a particular consequence we obtain for families $\{P_\vartheta : \vartheta \in \Theta\}$ with continuous densities fulfilling the covering condition for the complement of every neighborhood of ϑ_0, that $d(P_{\vartheta_n}, P_{\vartheta_0}) \to 0$ implies pointwise convergence of the densities.

The intuitive idea leading to the concept of a ML estimator is so convincing that ML estimators are considered as optimal by many statisticians. Leaving aside all mythology, one is left with the problem whether ML estimators have favorable properties if judged by their concentration about the true parameter value. The essential results are:

(i) There are no optimum properties of ML estimators for finite sample sizes.

(ii) Under suitable regularity conditions, sequences of asymptotic ML estimators are asymptotically maximally concentrated (see Section 8.5).

(iii) The asymptotic optimum properties depend in an essential way on regularity conditions. There are families of p–measures where the intuitive idea underlying Definition 6.5.1 is as convincing as in other cases, yet the ML estimator sequences fail to be consistent. (See Example 6.6.2.)

To obtain an assertion about the asymptotic distribution of ML estimator sequences, we assume that $\ell^{(i)}(x,\vartheta) := \frac{\partial}{\partial \vartheta_i} \log p(x,\vartheta_1,\ldots,\vartheta_p)$ exists. We write $\ell^{(\cdot)}(x,\vartheta)$ for the column vector $\left(\ell^{(1)}(x,\vartheta),\ldots,\ell^{(p)}(x,\vartheta)\right)^{\top}$. The starting point is that consistent sequences of ML estimators $\vartheta^{(n)}$ fulfill the *likelihood equations*

$$\sum_{\nu=1}^{n} \ell^{(\cdot)}\left(x_\nu, \vartheta^{(n)}(\underline{x})\right) = 0. \tag{6.5.8}$$

Since the Consistency Theorem 6.5.3 requires the covering condition for the complement of every neighborhood of ϑ_0, it is worthwhile to recall Theorem 6.4.4 which asserts the existence of consistent solutions under mild regularity conditions, and Lemma 6.4.6 which asserts that such a sequence can be obtained by projecting an auxiliary consistent estimator sequence into the set of solutions of the likelihood equations (6.5.8). In view of Lemma 6.5.4, the only remaining condition on the densities is for every $\tau \neq \vartheta$ the existence of a neighborhood V_τ such that $P_\vartheta\left(\log \overline{p}(\cdot, V_\tau) - \log p(\cdot, \vartheta)\right) < \infty$.

*Historical Remark 6.5.9.*The first suggestion of the ML method known to us is by Lambert (1760), §303. The ML method was applied by D. Bernoulli (1777) (see in particular Section 11 and Recapitulation) to determine the location "... qui maxima gaudet probabilitate", assuming the density function to be semicircular. He already ran into the trouble frequently occurring with ML estimators, namely that they cannot be obtained in closed form.

Later the ML method was used by Laplace (1781, pp. 383–385) and Gauß (1809, Section 176), who characterizes the normal distribution by the property that it has the arithmetic mean as ML estimator (Section 177), but he motivated the ML principle by a Bayesian argument (Section 176). Later, Gauß abandoned the ML principle in favour of the minimization of the risk for the quadratic loss function. In a letter to Bessel he writes (1839): "Ich muß es nämlich in alle Wege für weniger wichtig halten, denjenigen Wert einer unbekannten Größe auszumitteln, dessen Wahrscheinlichkeit die größte ist, die ja doch immer nur unendlich klein bleibt, als vielmehr denjenigen, an welchen sich haltend man das am wenigsten nachteilige Spiel hat; oder wenn $f(a)$ die Wahrscheinlichkeit des Wertes a für die Unbekannte x bezeichnet, so ist weniger daran gelegen, daß $f(a)$ ein Maximum werde, als daran, daß $\int f(x)F(x-a)dx\ldots$ ein Minimum werde, indem für F eine Funktion gewählt wird, die immer positiv und für größere Argumente auf eine schickliche Art immer größer wird."

Throughout the 19th century the ML method has been used by several authors though it is not always easy to discern whether the ML principle is used as such or as a consequence of Bayes' theorem with uniform prior. Encke (1832) mentions that the arithmetic mean is the ML estimator of the location of a normal distribution (p. 276), and determines the ML estimator for $1/\sqrt{2}\sigma$ (p. 280). De Morgan (1864) gives explicitly a Bayesian argument for the ML estimator (p. 421). Glaisher (1872) mentions that the arithmetic mean is the ML estimator of the location parameter of the normal distribution and shows that the median is the ML estimator of the location parameter of the Laplace distribution (pp. 121–123). K. Pearson (1896, pp. 262–265) obtains the sample correlation coefficient as ML estimator for the correlation coefficient of the population.

An untimely independent invention of the ML method is due to Zermelo (1929) who — seemingly unaware of the statistical literature — uses the ML method to estimate the power of chess–players, starting from the assumption that $u/(u+v)$ is the probability that a player with power u will beat a player with power v (p. 347: "Unser Verfahren kommt darauf hinaus, daß die relativen Spielstärken ... so bestimmt werden, daß die Wahrscheinlichkeit für das Eintreten des beobachteten Turnier–Ergebnisses eine möglichst große wird".)

K. Pearson and Filon (1898, pp. 231ff.) prove the asymptotic normality of the posterior distribution of a Bayes estimator of a location parameter for a uniform prior, which coincides with the distribution of the ML estimator. (They probably considered their result to be true also for the method of moments.) Edgeworth (1908) provides a more satisfactory proof for this result, and claims that this estimator has minimal variance. It is clear from the context (see pp. 505ff. and p. 662) that Edgeworth, at this point, considers the behavior of the estimator in repeated samples. In (1909, pp. 82ff.) he explicitly announces a proof "free from the speculative character which attaches to inverse probability", and shows for the case of a location parameter that the ML estimator is of minimal variance in the class of all solutions of estimating equations.

Fisher (1922) derives the asymptotic normality of the ML estimator (pp. 328/9) and claims its asymptotic efficiency without a valid proof. (He obviously intends to derive the asymptotic efficiency of the ML estimator from its sufficiency, which does not hold true (see pp. 330/1). A first proof of efficiency at a moderate level of rigor is contained in Fisher (1925). A second proof at the same moderate level of rigor is given in Fisher (1935), together with the statement that the ML estimator minimizes the variance in the class of all solutions of estimating equations (pp. 44ff.), thus extending Edgeworth's result from location to arbitrary parameters (without giving credit to Edgeworth).

The first mathematically satisfactory proof of asymptotic normality of the ML estimator is due to Doob (1934). Attempts by Dugué (1936) at giving a rigorous proof of the optimum property failed. Finally, an example produced by Hodges Jr. in 1951 (see LeCam, 1953, p. 280) revealed the existence of asymptotically normal estimator sequences the asymptotic variance of which exceeds the asymptotic variance of the ML estimator at a finite number of points.

We have discussed the history of this question in more detail, because Fisher usually obtains unlimited credit for these achievements. A fuller discussion of this question can be found in Pratt (1976).

6.6 Examples of ML estimators

Contains examples which contribute to clarifying the conditions for consistency of ML estimators. Examples concerning routine applications will be given in Section 9.1, together with the asymptotic distributions.

Without continuity of $\vartheta \to P_\vartheta$, ML estimator sequences will not be consistent in general. Here is a straightforward example.

Example 6.6.1. For $\vartheta \in \mathbb{R}$ let $P_\vartheta = N_{(\mu(\vartheta),1)}$, with

$$\mu(\vartheta) = \begin{cases} \vartheta \\ -\vartheta \end{cases} \quad \text{for} \quad \vartheta \underset{\in}{\notin} \{-1,1\}.$$

The ML estimator $\vartheta^{(n)}(x)$ is uniquely determined as the solution in ϑ of $\mu(\vartheta) = \bar{x}_n$, so that

$$\vartheta^{(n)}(x) := \begin{cases} \bar{x}_n \\ -\bar{x}_n \end{cases} \quad \text{if} \quad \bar{x}_n \underset{\in}{\notin} \{-1,1\}.$$

Hence we have $\vartheta^{(n)}(x) = \bar{x}_n$ for P_ϑ^n-a.a. $x \in \mathbb{R}^n$. Therefore, $\vartheta^{(n)}(x) \to \mu(\vartheta)$ (P_ϑ^n). Hence $\vartheta^{(n)}$, $n \in \mathbb{N}$, is not consistent if $\vartheta \in \{-1,1\}$.

The following example shows that the ML estimator sequence may fail to be consistent if the Covering Condition 6.3.8 is not fulfilled.

Example 6.6.2. Let $X = (0,1)$, $\Theta = (0,1]$. Let $p(\cdot,\vartheta)$ be a Lebesgue density with the following properties:
(i) $\vartheta \to p(x,\vartheta)$ is continuous for every $x \in (0,1)$,
(ii) $\frac{1}{2} \le p(x,\vartheta) \le p(\vartheta,\vartheta)$ for $x \in (0,1)$, $\vartheta \in (0,1]$,
(iii) $p(\vartheta,\vartheta) = 2^{\vartheta^{-3}-1}$ for $\vartheta \in (0,1]$,
(iv) $p(x,1) = 1$ for $x \in (0,1)$.
We shall show that any sequence of ML estimators fulfills $\lim_{n\to\infty} \vartheta^{(n)}(x) = 0$ for P_1^N-a.a. $x \in (0,1)^N$.

Proof. Let $\varepsilon \in (0,1)$ be arbitrary. Then $\vartheta \in [\varepsilon,1]$ implies

$$p(x,\vartheta) \le p(\vartheta,\vartheta) \le p(\varepsilon,\varepsilon),$$

hence

$$\sup_{\vartheta \in [\varepsilon,1]} \prod_{\nu=1}^{n} p(x_\nu,\vartheta) \le p(\varepsilon,\varepsilon)^n. \tag{6.6.3}$$

By the 1st Lemma of Borel–Cantelli (see Ash, 1972, p. 66),

$$P_\vartheta^{\mathbf{N}}\{\underline{x} \in (0,1)^{\mathbf{N}} : x_{1:n} > n^{-1/2}\} = (P_\vartheta(n^{-1/2},1))^n$$

implies that

$$P_\vartheta^{\mathbf{N}}\{\underline{x} \in (0,1)^{\mathbf{N}} : x_{1:n} > n^{-1/2} \text{ infinitely often}\} = 0$$

if $\sum_{n=1}^{\infty}(P_\vartheta(n^{-1/2},1))^n < \infty$.

By (iv),

$$\sum_{n=1}^{\infty}(P_1(n^{-1/2},1))^n = \sum_{n=1}^{\infty}(1 - n^{-1/2})^n < \infty$$

since $1 - x \le e^{-x}$.

Therefore, for $P_1^{\mathbf{N}}$-a.a. $\underline{x} \in (0,1)^{\mathbf{N}}$ there exists $n(\underline{x})$ such that $x_{1:n} \le n^{-1/2}$ for $n \ge n(\underline{x})$.

$x_{1:n} \le n^{-1/2}$ implies

$$\prod_{\nu=1}^{n} p(x_\nu, x_{1:n}) \ge \left(\frac{1}{2}\right)^{n-1} p(x_{1:n}, x_{1:n})$$

$$\ge \left(\frac{1}{2}\right)^{n-1} p(n^{-1/2}, n^{-1/2}) = \left(\frac{1}{2}\right)^{n-1} 2^{n^{3/2}-1} = 2^{n(n^{1/2}-1)}.$$

Hence for $P_1^{\mathbf{N}}$-a.a. $\underline{x} \in (0,1)^{\mathbf{N}}$,

$$\prod_{\nu=1}^{n} p(x_\nu, x_{1:n}) \ge 2^{n(n^{1/2}-1)} \text{ and } x_{1:n} \le n^{-1/2} \text{ for all } n \ge n(\underline{x}). \qquad (6.6.4)$$

Let n_ε be such that $n \ge n_\varepsilon$ implies

$$2^{n^{1/2}-1} > p(\varepsilon, \varepsilon) \qquad \text{and} \qquad n^{-1/2} < \varepsilon.$$

Hence, $n \ge \max\{n(\underline{x}), n_\varepsilon\}$ implies $x_{1:n} < \varepsilon$. Using (6.6.3) and (6.6.4), we obtain

$$\sup_{\vartheta \in (0,\varepsilon)} \prod_{\nu=1}^{n} p(x_\nu, \vartheta) > \sup_{\vartheta \in [\varepsilon,1]} \prod_{\nu=1}^{n} p(x_\nu, \vartheta),$$

hence $\vartheta^{(n)}(\underline{x}) \in (0,\varepsilon)$ for any ML estimator. Therefore, $\lim_{n\to\infty} \vartheta^{(n)}(\underline{x}) = 0$ for $P_1^{\mathbf{N}}$-a.a. $\underline{x} \in (0,1)^{\mathbf{N}}$. □

The following example illustrates how the covering condition underlying Consistency Theorem 6.5.3 can be verified in a case where Θ is not compact.

Example 6.6.5. Let $p : \mathbb{R} \to (0, \infty)$ be a continuous p–density with

$$\lim_{x \to \pm\infty} p(x) = 0.$$

Assume that

$$\int (\log p(x)) p(x) dx > -\infty. \qquad (6.6.6)$$

For $\vartheta \in \mathbb{R}$ let P_ϑ denote the p–measure with Lebesgue density $x \to p(x-\vartheta)$, $x \in \mathbb{R}$. Then the conditions of the Consistency Theorem 6.5.3 are fulfilled. According to Exercise 9.3.18, there are equivariant ML estimators. For these estimators, the consistency is automatically locally uniform. Conditions for the consistency of ML estimators for location– and scale parameter families will be given in Proposition 9.3.21.

Proof. Since p is continuous and $\lim_{x \to \pm\infty} p(x) = 0$, p is bounded on $\mathbb{R}$, say $p(x) \leq b$ for $x \in \mathbb{R}$.
   (i) We have

$$\int \sup_{\vartheta \in \mathbb{R}} \left[\log p(x - \vartheta) - \log p(x - \vartheta_0) \right] P_{\vartheta_0}(dx)$$

$$\leq \log b - \int \log p(x) P(dx) < \infty.$$

Hence, the covering condition is fulfilled for arbitrary compact subsets of $\mathbb{R}$ which are disjoint from ϑ_0 by Lemma 6.3.9.
   (ii) It remains to establish the covering condition for the complement of some compact subset of ϑ_0.
   Since $\lim_{x \to \pm\infty} p(x) = 0$, for every $\varepsilon > 0$ there exists $c_\varepsilon > 0$ such that $p(x) < \varepsilon$ for $x < -c_\varepsilon$. Hence

$$\vartheta > x + c_\varepsilon \quad \text{implies} \quad x - \vartheta < -c_\varepsilon \quad \text{implies} \quad p(x - \vartheta) < \varepsilon,$$

so that

$$\sup\{p(x - \vartheta) : \vartheta > x + c_\varepsilon\} \leq \varepsilon.$$

Hence we obtain for every $x \in \mathbb{R}$

$$\lim_{c \to \infty} \sup_{\vartheta > c} p(x - \vartheta) = 0,$$

and therefore

$$\lim_{c \to \infty} \sup_{\vartheta > c} \log p(x - \vartheta) = -\infty.$$

Since

$$\sup_{\vartheta > c} \log p(x - \vartheta) \leq \log b < \infty \qquad \text{for all } c \in \mathbb{R},$$

we obtain by Fatou's lemma

$$\overline{\lim_{c \to \infty}} \int \left(\sup_{\vartheta > c} \log p(x - \vartheta) \right) P_{\vartheta_0}(dx)$$

$$\leq \int \overline{\lim_{c \to \infty}} \left(\sup_{\vartheta > c} \log p(x - \vartheta) \right) P_{\vartheta_0}(dx) = -\infty.$$

Hence

$$\int \sup_{\vartheta > c'} \left[\log p(x - \vartheta) - \log p(x - \vartheta_0) \right] P_{\vartheta_0}(dx)$$

$$= \int \sup_{\vartheta > c'} \log p(x - \vartheta) P_{\vartheta_0}(dx) - \int \log p(x) P(dx) < 0$$

for c' sufficiently large.

Similarly, for c'' sufficiently large,

$$\int \sup_{\vartheta < -c''} \left[\log p(x - \vartheta) - \log p(x - \vartheta_0) \right] P_{\vartheta_0}(dx) < 0.$$

$C_0 := [\vartheta_0 - c, \vartheta_0 + c]$ is a compact neighborhood of ϑ_0. The complement of C_0 is covered by the sets $V_1 := (-\infty, \vartheta_0 - c)$ and $V_2 := (\vartheta_0 + c, \infty)$. If $c > \max\{c' - \vartheta_0, c'' + \vartheta_0\}$, these sets fulfill condition (6.3.7''), i.e.

$$\int \sup_{\vartheta \in V_i} \left[\log p(x - \vartheta) - \log p(x - \vartheta_0) \right] P_{\vartheta_0}(dx) < 0. \qquad \square$$

Exercise 6.6.7. Show that every sequence of asymptotic ML estimators is consistent for scale parameter families, based on a continuous Lebesgue density $p : (0, \infty) \to (0, \infty)$ with $\lim_{x \to 0} x p(x) = 0$ and $\lim_{x \to \infty} x p(x) = 0$, if

$$\int \log [x p(x)] p(x) dx > -\infty.$$

Hint: Apply the result of Example 6.6.5 for the location parameter family $P \circ (x \to \vartheta + \log x)$.

In the following example (due to Deemer and Votaw, 1955) the likelihood equation has no solution on a subset of $\mathbb{R}^n$, the probability of which converges to 0. This phenomenon occurs in connection with various truncated distributions (see Mittal and Dahiya, 1989, and the references cited there).

Example 6.6.8. For $\vartheta \in (0, \infty)$ let P_ϑ denote the exponential distribution, truncated at 1. The Lebesgue density of P_ϑ is

$$p(x, \vartheta) = \vartheta(1 - e^{-\vartheta})^{-1} \exp[-\vartheta x], \quad 0 \le x \le 1.$$

The likelihood equation for the sample size n becomes $\bar{x}_n = f(\vartheta)$, with

$$f(\vartheta) = \vartheta^{-1} - e^{-\vartheta}/(1 - e^{-\vartheta}).$$

Since $f(\vartheta) \in (0, 1/2)$ for $\vartheta \in (0, \infty)$, the likelihood equation has no solution if $\bar{x}_n \ge 1/2$. This is, however, not as terrifying as it sounds: $P_\vartheta^n\{\underline{x} \in (0, \infty)^n : \bar{x}_n \ge 1/2\}$ converges to 0 quickly.

6.7 Appendix: Uniform integrability, stochastic convergence and measurable selection

Collects auxiliary results on uniform integrability and uniform stochastic convergence, including a uniform version of the weak law of large numbers. Presents a "measurable selection theorem".

Let $(Y, \mathcal{V})$ be a Hausdorff space with countable base (in most applications an open subset of $\mathbb{R}^p$), $(X, \mathcal{A})$ a measurable space, and $f : X \times Y \to \mathbb{R}$ a function with the following properties:

$$x \to f(x, y) \quad \text{is measurable for every } y \in Y, \qquad (6.7.1')$$
$$y \to f(x, y) \quad \text{is continuous on } Y \text{ for every } x \in X. \qquad (6.7.1'')$$

We write $\overline{f}(x, B) := \sup\{f(x, y) : y \in B\}$, and $\underline{f}(x, B) := \inf\{f(x, y) : y \in B\}$.

Lemma 6.7.2. *If (6.7.1) holds true, the functions* $x \to \overline{f}(x, B)$ *and* $x \to \underline{f}(x, B)$ *are measurable for any set* $B \subset Y$.

Proof. Since $(Y, \mathcal{V})$ has a countable base, there exists a countable set, say $B_0 \subset B$, which is dense in $(B, B \cap \mathcal{V})$. Since B_0 is countable, $x \to \overline{f}(x, B_0)$ is measurable. It remains to be shown that $\overline{f}(x, B_0) = \overline{f}(x, B)$ for $x \in X$.

$B_0 \subset B$ implies $\overline{f}(x, B_0) \le \overline{f}(x, B)$. For $r < \overline{f}(x, B)$, the set $\{y \in B : r < f(x, y)\}$ is non–empty, and open in $(B, B \cap \mathcal{V})$ by continuity of $y \to f(x, y)$. Hence it contains an element of B_0, say y_0, so that $r < f(x, y_0) \le \overline{f}(x, B_0)$. $\square$

Lemma 6.7.3. (i) *If (6.7.1) holds true, then* $f : X \times Y \to \mathbb{R}$ *is* $\mathcal{A} \times \mathcal{B}$–*measurable, where* $\mathcal{B}$ *is the Borel algebra of* $(Y, \mathcal{V})$.

(ii) *If $f : X \times \mathbb{R} \to \mathbb{R}$ fulfills (6.7.1') and if $f(x, \cdot)$ is isotone and right continuous for $x \in X$, then f is $\mathcal{A} \times \mathbb{B}$–measurable.*

Proof. (i) We shall show that $S := \{(x,y) \in X \times Y : f(x,y) < r\} \in \mathcal{A} \times \mathcal{B}$ for every $r \in \mathbb{R}$.

Let $\mathcal{V}_0$ be a countable base of $\mathcal{V}$. We shall show that

$$S = \bigcup_{V \in \mathcal{V}_0} (\{x \in X : \overline{f}(x, V) < r\} \times V). \tag{6.7.4}$$

Since $x \to \overline{f}(x, V)$ is $\mathcal{A}$–measurable by Lemma 6.7.2, and $V \in \mathcal{B}$ (since $\mathcal{B}$ is the Borel algebra), the right hand side of (6.7.4) is in $\mathcal{A} \times \mathcal{B}$.

To prove (6.7.4), we remark that $(x_0, y_0) \in \{x \in X : \overline{f}(x, V) < r\} \times V$ for some $V \in \mathcal{V}_0$ implies $f(x_0, y_0) < r$, hence $(x_0, y_0) \in S$. Conversely, if $(x_0, y_0) \in S$, we have $f(x_0, y_0) < r$, hence $f(x_0, y_0) < r_0$ for some $r_0 < r$. Since $f(x_0, \cdot)$ is continuous, $\{y \in Y : f(x_0, y) < r_0\}$ is an open set, which contains y_0. Hence there exists $V \in \mathcal{V}_0$ such that $y_0 \in V \subset \{y \in Y : f(x_0, y) < r_0\}$. Since $f(x_0, y) < r_0$ for $y \in V$, we have $\overline{f}(x_0, V) \le r_0 < r$, hence $x_0 \in \{x \in X : \overline{f}(x, V) < r\}$. Therefore, $(x_0, y_0) \in \{x \in X : \overline{f}(x, V) < r\} \times V$ for some $V \in \mathcal{V}_0$.

(ii) It suffices to prove that

$$\{(x, y) \in X \times \mathbb{R} : f(x, y) < r\} = \bigcup_{z \in \mathbb{Q}} \{(x, y) \in X \times (-\infty, z] : f(x, z) < r\}.$$

Let $f(x, y) < r$. Since $f(x, \cdot)$ is right continuous, there exists $z \in \mathbb{Q}$, $z \ge y$, such that $f(x, z) < r$. Conversely, $y \le z \in \mathbb{Q}$ and $f(x, z) < r$ imply $f(x, y) < r$. $\square$

Lemma 6.7.5. *Assume (6.7.1') and (6.7.1'') at y_0. Let $f(\cdot, y_0)$ be P_0–integrable. If there exists a neighborhood V of y_0 such that $P_0(\overline{f}(\cdot, V)) < \infty$, then for every $\varepsilon > 0$ there exists a neighborhood V_ε of y_0 such that*

$$P_0(\overline{f}(\cdot, V_\varepsilon)) < P_0(f(\cdot, y_0)) + \varepsilon.$$

Proof. Let V_n, $n \in \mathbb{N}$, be a nonincreasing local base at y_0. Then $\overline{f}(x, V_n) \downarrow f(x, y_0)$ for every $x \in X$. (Since $\{y \in Y : f(x, y) < f(x, y_0) + \varepsilon\}$ is an open set containing y_0, it contains V_n for $n \ge n_{x, \varepsilon}$, say. Therefore, $\overline{f}(x, V_n) \le f(x, y_0) + \varepsilon$ for $n \ge n_{x, \varepsilon}$.)

Let $V \ni y_0$ be such that $P(\overline{f}(\cdot, V)) < \infty$. For n sufficiently large, we have $V_n \subset V$ and therefore $\overline{f}(\cdot, V_n) \le \overline{f}(\cdot, V)$. Hence Fatou's lemma implies

$$P(\overline{f}(\cdot, V_n)) \downarrow P(f(\cdot, y_0)). \qquad \square$$

In the following it will be convenient to formulate the definitions and results for p–measures, or sequences of p–measures, which are indexed by a parameter, say $\vartheta \in \Theta$. Throughout the following, Θ will be assumed to be a Hausdorff space, in most applications an open subset of $\mathbb{R}^p$.

Various properties are needed in a uniform or locally uniform version. From the definition of "uniformity on Θ" we derive the localized concepts according to the following rule.

(6.7.6′) "locally uniformly at ϑ_0" means "uniformly on some neighborhood of ϑ_0".

(6.7.6″) "locally uniformly on Θ" means "locally uniformly at ϑ_0 for every $\vartheta_0 \in \Theta$".

Remark 6.7.7. If "uniformly on Θ_1" and "uniformly on Θ_2" implies "uniformly on $\Theta_1 \cup \Theta_2$", then "locally uniformly on Θ" implies "uniformly on every compact subset of Θ." If Θ is locally compact, "uniformly on every compact subset of Θ" implies "locally uniformly on Θ".

Definition 6.7.8. A family of functions $f(\cdot, \vartheta) : (X, \mathcal{A}) \to (\mathbb{R}, \mathbb{B})$, $\vartheta \in \Theta$, is *uniformly integrable relative to* $\{P_\vartheta : \vartheta \in \Theta\}$ (*on* Θ, for short) if

$$\lim_{M \to \infty} \sup_{\vartheta \in \Theta} P_\vartheta\left(|f(\cdot, \vartheta)|1_{\{|f(\cdot, \vartheta)| > M\}}\right) = 0.$$

If $f(\cdot, \vartheta)$, $\vartheta \in \Theta$, is uniformly integrable relative to $\{P_\vartheta : \vartheta \in \Theta\}$, then

$$\sup_{\vartheta \in \Theta} P_\vartheta\left(|f(\cdot, \vartheta)|\right) < \infty.$$

(There exists $M > 0$ such that $P_\vartheta\left(|f(\cdot, \vartheta)|1_{\{|f(\cdot, \vartheta)| > M\}}\right) < 1$ for $\vartheta \in \Theta$. Hence $P_\vartheta(|f(\cdot, \vartheta)|) \le M + 1$ for $\vartheta \in \Theta$.)

If $\sup_{\vartheta \in \Theta} P_\vartheta\left(|f(\cdot, \vartheta)|^{1+\delta}\right) < \infty$ for some $\delta > 0$, then $f(\cdot, \vartheta)$, $\vartheta \in \Theta$, is uniformly integrable relative to $\{P_\vartheta : \vartheta \in \Theta\}$ (since $P_\vartheta\left(|f(\cdot, \vartheta)|1_{\{|f(\cdot, \vartheta)| > M\}}\right) \le M^{-\delta} P_\vartheta\left(|f(\cdot, \vartheta)|^{1+\delta}\right)$).

The following results have to do with continuity of $\vartheta \to P_\vartheta(f)$. They extend continuity for bounded functions $f : X \to \mathbb{R}$ to continuity for locally uniformly integrable functions. Depending on the context, the functions f are just measurable, or continuous.

Lemma 6.7.9. *Assume that* $\vartheta \to P_\vartheta(f)$ *is continuous for all bounded measurable [continuous] functions* f. *If a measurable [continuous] function* f *is uniformly integrable relative to* $\{P_\vartheta : \vartheta \in \Theta_0\}$, *then* $P_{\vartheta_0}(|f|) < \infty$ *if* ϑ_0 *is an accumulation point of* Θ_0. *Hence* f *is also uniformly integrable relative to* $\{P_\vartheta : \vartheta \in \Theta_0 \cup \{\vartheta_0\}\}$.

Proof. If $P_{\vartheta_0}(|f|) = \infty$, there exists $M > 0$ such that $P_{\vartheta_0}(|f| \wedge M) > \sup_{\vartheta \in \Theta_0} P_\vartheta(|f|) + 1$. Since $|f| \wedge M$ is bounded [and continuous], there exists a neighborhood U of ϑ_0, $U \cap \Theta_0 \neq \emptyset$, such that

$$P_{\vartheta_0}(|f| \wedge M) < P_\vartheta(|f| \wedge M) + 1 \qquad \text{for } \vartheta \in U,$$

which is contradictory. Hence, $P_{\vartheta_0}(|f|) < \infty$. $\square$

Lemma 6.7.10. *Assume that $\vartheta \to P_\vartheta(f)$ is continuous at ϑ_0 for every bounded measurable [continuous] function f.*

(i) Then $\vartheta \to P_\vartheta(f)$ is continuous at ϑ_0 for every measurable [continuous] function f which is locally uniformly integrable at ϑ_0.

(ii) If $\vartheta \to P_\vartheta(f)$ is continuous at ϑ_0 for some measurable [continuous] function $f \geq 0$, then f is locally uniformly integrable at ϑ_0.

As a particular consequence we obtain that $P_n(f) \to P_0(f)$ for bounded measurable [continuous] functions f implies

(i') $P_n(f) \to P_0(f)$ if f is [continuous and] uniformly integrable on $\{P_n : n \in \mathbb{N}\}$, and

(i'') if $P_n(f) \to P_0(f)$ for some nonnegative measurable [continuous] function f, then f is uniformly integrable on $\{P_n : n \in \mathbb{N}\}$.

Proof. (i) For $M > 0$ let f^M be defined by

$$f^M(x) = \begin{cases} f(x) \\ M \text{ sgn } f(x) \end{cases} \qquad \text{if} \quad |f(x)| \begin{matrix} \leq \\ > \end{matrix} M.$$

Notice that f^M is continuous if f is so.

We have

$$|P_\vartheta(f) - P_{\vartheta_0}(f)| \leq |P_\vartheta(f^M) - P_{\vartheta_0}(f^M)|$$
$$+ P_\vartheta(|f|1_{\{|f|>M\}}) + P_{\vartheta_0}(|f|1_{\{|f|>M\}}).$$

By uniform local integrability of f there exists a neighborhood U of ϑ_0 with the following property:

For every $\varepsilon > 0$ there is $M_\varepsilon > 0$ such that

$$\sup_{\vartheta \in U} P_\vartheta(|f|1_{\{|f|>M_\varepsilon\}}) < \varepsilon/3.$$

Since f^M is bounded [and continuous], there exists a neighborhood U_ε of ϑ_0 such that $|P_\vartheta(f^M) - P_{\vartheta_0}(f^M)| < \varepsilon/3$ for $\vartheta \in U_\varepsilon$. Hence $|P_\vartheta(f) - P_{\vartheta_0}(f)| < \varepsilon$ for $\vartheta \in U \cap U_\varepsilon$.

(ii) We have

$$P_\vartheta(f) = P_\vartheta(f1_{\{f \leq M\}}) + P_\vartheta(f1_{\{f > M\}}).$$

Let M_ε be such that $P_{\vartheta_0}(f1_{\{f > M_\varepsilon\}}) < \varepsilon/3$ [and $P_{\vartheta_0} \circ f\{M_\varepsilon\} = 0$].

$\vartheta \to P_\vartheta(f)$ is continuous at ϑ_0 by assumption, and $\vartheta \to P_\vartheta(f1_{\{f \le M_\epsilon\}})$ is continuous at ϑ_0 since $f1_{\{f \le M_\epsilon\}}$ is bounded [and continuous P_{ϑ_0}-a.e.]. Hence there exists a neighborhood U_ϵ at ϑ_0 such that for $\vartheta \in U_\epsilon$,

$$|P_\vartheta(f) - P_{\vartheta_0}(f)| < \epsilon/3$$

and

$$\left|P_\vartheta\left(f1_{\{f \le M_\epsilon\}}\right) - P_{\vartheta_0}\left(f1_{\{f \le M_\epsilon\}}\right)\right| < \epsilon/3.$$

This implies

$$P_\vartheta\left(f1_{\{f > M_\epsilon\}}\right) < \epsilon \quad \text{for } \vartheta \in U_\epsilon. \qquad \square$$

Corollary 6.7.11. *Assume that $P_n \Rightarrow P_0$. Let $q \ge 0$ be a continuous function such that $P_n(q) = 1$ for $n \in \mathbb{N}$ and $n = 0$. Then $P_n' \Rightarrow P_0'$, where P_n' denotes the p-measure with P_n-density q.*

Addendum. (LeCam's 3rd Lemma; see also Lemma 8.4.26.)
If $p_n' \in dP_n'/dP_n$ and $P_n \circ \log p_n' \Rightarrow N_{(-\frac{1}{2}\sigma^2, \sigma^2)}$, then

$$P_n' \circ \log p_n' \quad \Rightarrow \quad N_{(\frac{1}{2}\sigma^2, \sigma^2)}.$$

Proof. By Lemma 6.7.10(ii'), q is uniformly integrable relative to $\{P_n : n \in \mathbb{N}\}$. Hence hq is uniformly integrable for any bounded and continuous function h. This implies $P_n(hq) \to P_0(hq)$ by Lemma 6.7.10(i').

The addendum follows with P_n replaced by $P_n \circ \log p_n'$, $P_0 = N_{(-\frac{1}{2}\sigma^2, \sigma^2)}$ and $q(u) = \exp[u]$, $u \in \mathbb{R}$, since $\exp \in dP_n' \circ \log p_n'/dP_n \circ \log p_n'$ and $\exp \in dN_{(\frac{1}{2}\sigma^2, \sigma^2)}/dN_{(-\frac{1}{2}\sigma^2, \sigma^2)}$. $\qquad \square$

In the following definitions, f_n is a measurable function from $(X_n \times \Theta, \mathcal{A}_n \times \mathcal{B})$ to $(\mathbb{R}^p, \mathbb{B}^p)$.

Definition 6.7.12. f_n, $n \in \mathbb{N}$, *converges stochastically* to 0, *uniformly relative to* $\{P_\vartheta^{(n)} : \vartheta \in \Theta\}$ (on Θ, for short), if

$$\lim_{n \to \infty} \sup_{\vartheta \in \Theta} P_\vartheta^{(n)}\{x \in X_n : \|f_n(x, \vartheta)\| > \epsilon\} = 0 \qquad \text{for } \epsilon > 0.$$

Definition 6.7.13. f_n, $n \in \mathbb{N}$, *is stochastically bounded,* uniformly relative to $\{P_\vartheta^{(n)} : \vartheta \in \Theta\}$ (on Θ, for short), if for every $\epsilon > 0$ there exists $M_\epsilon > 0$ such that

$$\sup_{\vartheta \in \Theta} P_\vartheta^{(n)}\{x \in X_n : \|f_n(x, \vartheta)\| > M_\epsilon\} < \epsilon \qquad \text{for } n \in \mathbb{N}.$$

Lemma 6.7.14. *Assume that f_n, $n \in \mathbb{N}$, is stochastically bounded, uniformly on Θ, and g_n, $n \in \mathbb{N}$, converges stochastically to 0, uniformly on Θ. Then $f_n g_n$, $n \in \mathbb{N}$, converges stochastically to 0, uniformly on Θ.*

Proof. By assumption, for $\delta > 0$ there exists $M_\delta > 0$ such that

$$\sup_{\vartheta \in \Theta} P_\vartheta^{(n)} \{ x \in X_n : \| f_n(x, \vartheta) \| > M_\delta \} < \delta \qquad \text{for } n \in \mathbb{N}.$$

For $\varepsilon > 0$,

$$\sup_{\vartheta \in \Theta} P_\vartheta^{(n)} \{ x \in X_n : \| f_n(x, \vartheta) g_n(x, \vartheta) \| > \varepsilon \}$$
$$\leq \sup_{\vartheta \in \Theta} P_\vartheta^{(n)} \{ x \in X_n : \| f_n(x, \vartheta) \| > M_\delta \}$$
$$+ \sup_{\vartheta \in \Theta} P_\vartheta^{(n)} \{ x \in X_n : \| g_n(x, \vartheta) \| > \varepsilon / M_\delta \},$$

hence

$$\varlimsup_{n \to \infty} \sup_{\vartheta \in \Theta} P_\vartheta^{(n)} \{ x \in X_n : \| f_n(x, \vartheta) g_n(x, \vartheta) \| > \varepsilon \} \leq \delta.$$

Since $\delta > 0$ was arbitrary, this implies

$$\lim_{n \to \infty} \sup_{\vartheta \in \Theta} P_\vartheta^{(n)} \{ x \in X_n : \| f_n(x, \vartheta) g_n(x, \vartheta) \| > \varepsilon \} = 0.$$

$\square$

The following uniform version of the weak law of large numbers occurs in Chung (1951).

Proposition 6.7.15. *If $f(\cdot, \vartheta) : (X, \mathcal{A}) \to (\mathbb{R}, \mathbb{B})$, $\vartheta \in \Theta$, is uniformly integrable relative to $\{ P_\vartheta : \vartheta \in \Theta \}$, then $n^{-1} \sum_1^n f(x_\nu, \vartheta)$ converges stochastically to $P_\vartheta (f(\cdot, \vartheta))$, uniformly on $\{ P_\vartheta^n : \vartheta \in \Theta \}$, i.e.*

$$\lim_{n \to \infty} \sup_{\vartheta \in \Theta} P_\vartheta^n \{ \underline{x} \in X^n : | n^{-1} \sum_1^n f(x_\nu, \vartheta) - P_\vartheta (f(\cdot, \vartheta)) | > \varepsilon \} = 0 \quad \text{for } \varepsilon > 0.$$

Proof. We prove the assertion for a function f not depending on ϑ. The general case follows by transition from P_ϑ to $P_\vartheta \circ f(\cdot, \vartheta)$.

Let $f_n := f 1_{\{|f| \leq M_n\}}$ and $\hat{f}_n := f 1_{\{|f| > M_n\}}$.

From $f - P_\vartheta(f) = f_n - P_\vartheta(f_n) + \hat{f}_n - P_\vartheta(\hat{f}_n)$ we obtain

$$\left| n^{-1} \sum_{\nu=1}^n f(x_\nu) - P_\vartheta(f) \right| \le \left| n^{-1} \sum_{\nu=1}^n f_n(x_\nu) - P_\vartheta(f_n) \right|$$
$$+ \left| n^{-1} \sum_{\nu=1}^n \hat{f}_n(x_\nu) \right| + P_\vartheta(|\hat{f}_n|).$$

By Markov's inequality,

$$P_\vartheta^n \left\{ \underline{x} \in X^n : \left| n^{-1} \sum_{\nu=1}^n \hat{f}_n(x_\nu) \right| > \varepsilon \right\} \le \varepsilon^{-1} P_\vartheta(|\hat{f}_n|)$$

and

$$P_\vartheta^n \left\{ \underline{x} \in X^n : \left| n^{-1} \sum_{\nu=1}^n (f_n(x_\nu) - P_\vartheta(f_n)) \right| > \varepsilon \right\}$$
$$\le \varepsilon^{-2} n^{-1} P_\vartheta((f_n - P_\vartheta(f_n))^2) \le \varepsilon^{-2} n^{-1} P_\vartheta(f_n^2) \le \varepsilon^{-2} n^{-1} M_n^2.$$

Since $\lim_{n\to\infty} \sup_{\vartheta \in \Theta} P_\vartheta(|\hat{f}_n|) = 0$ for $M_n \to \infty$, the assertion follows.   □

Lemma 6.7.16. *Let Θ be a Hausdorff space and $\{P_\vartheta : \vartheta \in \Theta\}$ a family of p-measures on $(X, \mathcal{A})$. Let (Y, d) be a metric space, and $f : (X \times Y, \mathcal{A} \times \mathcal{B}) \to (\mathbb{R}, \mathbb{B})$ a function fulfilling conditions (6.7.1). Assume that for every $y \in Y$ there exists a neighborhood V of y such that $\overline{f}(\cdot, V)$ and $\underline{f}(\cdot, V)$ are locally uniformly integrable on Θ.*

Let $\Theta_0 \subset \Theta$ and $Y_0 \subset Y$ be compact subsets. Then for every $\varepsilon > 0$ there exists $\delta_\varepsilon > 0$ such that (with $S(y, \delta)$ denoting the open ball with center y and radius δ)

$$\lim_{n\to\infty} \sup_{\vartheta \in \Theta_0} P_\vartheta^n \left\{ \underline{x} \in X^n : \sup_{y \in Y_0} \left(n^{-1} \sum_1^n \overline{f}(x_\nu, S(y, \delta_\varepsilon)) \right. \right. \tag{6.7.17}$$
$$\left. \left. - P_\vartheta(f(\cdot, y)) \right) > \varepsilon \right\} = 0.$$

Together with the corresponding relation for $\underline{f}$ this implies

$$\lim_{n\to\infty} \sup_{\vartheta \in \Theta_0} P_\vartheta^n \left\{ \underline{x} \in X^n : \sup_{y \in Y_0} \left| n^{-1} \sum_1^n \overline{f}(x_\nu, S(y, \delta_\varepsilon)) \right. \right. \tag{6.7.18}$$
$$\left. \left. - P_\vartheta(f(\cdot, y)) \right| > \varepsilon \right\} = 0.$$

(since $\left| n^{-1} \sum_1^n \overline{f}(x_\nu, S(y, \delta_\varepsilon)) - P_\vartheta(f(\cdot, y)) \right| > \varepsilon$ implies

$$n^{-1} \sum_1^n \overline{f}(x_\nu, S(y, \delta_\varepsilon)) - P_\vartheta(f(\cdot, y)) > \varepsilon$$

or

$$n^{-1} \sum_{1}^{n} \underline{f}(x_\nu, S(y, \delta_\varepsilon)) - P_\vartheta(f(\cdot, y)) < -\varepsilon.$$

Proof. Let $\varepsilon > 0$ and $\vartheta_0 \in \Theta$ be fixed. By Lemma 6.7.5, for every $y \in Y$ there exists an open neighborhood $V_{y,\varepsilon}$ of y such that $\overline{f}(\cdot, V_{y,\varepsilon})$ is locally uniformly integrable at ϑ_0, and

$$P_{\vartheta_0}(\overline{f}(\cdot, V_{y,\varepsilon}) - f(\cdot, y)) < \varepsilon/2.$$

By Lemma 6.7.10(i) there exists a neighborhood $U_{y,\varepsilon}$ of ϑ_0 such that

$$P_\vartheta(\overline{f}(\cdot, V_{y,\varepsilon}) - f(\cdot, y)) < \varepsilon/2 \qquad \text{for } \vartheta \in U_{y,\varepsilon}. \tag{6.7.19}$$

Since $\{V_{y,\varepsilon} : y \in Y_0\}$ covers the compact set Y_0, there exists a finite subcover, say $V_{i,\varepsilon} \,(= V_{y_i,\varepsilon}), i = 1, \ldots, m$. By Lebesgue's covering lemma (see Kelley, 1955, p. 154, Theorem 26) there exists $\delta_\varepsilon > 0$ such that for every $y \in Y_0$, the ball $S(y, \delta_\varepsilon)$ is contained in some $V_{i,\varepsilon}, i = 1, \ldots, m$. Therefore,

$$\sup_{y \in Y_0} \left(n^{-1} \sum_{1}^{n} \overline{f}(x_\nu, S(y, \delta_\varepsilon)) - P_\vartheta(f(\cdot, y)) \right) > \varepsilon$$

implies for $\vartheta \in U_\varepsilon := \bigcap_{i=1}^{m} U_{y_i,\varepsilon}$ (use (6.7.19)) that

$$n^{-1} \sum_{1}^{n} \overline{f}(x_\nu, V_{i,\varepsilon}) - P_\vartheta(\overline{f}(\cdot, V_{i,\varepsilon})) > \varepsilon/2 \qquad \text{for some } i = 1, \ldots, m.$$

Hence

$$P_\vartheta^n \big\{ \underline{x} \in X^n : \sup_{y \in Y_0} \left(n^{-1} \sum_{1}^{n} \overline{f}(x_\nu, S(y, \delta_\varepsilon)) - P_\vartheta(f(\cdot, y)) \right) > \varepsilon \big\}$$

$$\leq \sum_{i=1}^{m} P_\vartheta^n \big\{ \underline{x} \in X^n : n^{-1} \sum_{1}^{n} \overline{f}(x_\nu, V_{i,\varepsilon}) - P_\vartheta(\overline{f}(\cdot, V_{i,\varepsilon})) > \varepsilon/2 \big\} \qquad \text{for } \vartheta \in U_\varepsilon.$$

By Proposition 6.7.15, this implies

$$\lim_{n \to \infty} \sup_{\vartheta \in U_\varepsilon} P_\vartheta^n \big\{ \underline{x} \in X^n : \sup_{y \in Y_0} \left(n^{-1} \sum_{1}^{n} \overline{f}(x_\nu, S(y, \delta_\varepsilon)) \right. \tag{6.7.20}$$

$$\left. - P_\vartheta(f(\cdot, y)) \right) > \varepsilon \big\} = 0.$$

So far, $\vartheta_0 \in \Theta$ was fixed. To indicate the dependence of U_ε and δ_ε on ϑ_0, we now write $U_{\vartheta_0,\varepsilon}$ and $\delta_{\vartheta_0,\varepsilon}$. Since $\{U_{\vartheta_0,\varepsilon} : \vartheta_0 \in \Theta_0\}$ covers the compact set

Θ_0, there exists a finite subcover, say $U_{\vartheta_j,\varepsilon}$, $j = 1, \ldots, k$. Hence (6.7.17) follows from (6.7.20) with $\delta_\varepsilon := \min\{\delta_{\vartheta_j,\varepsilon} : j = 1, \ldots, k\}$.   □

Corollary 6.7.21. *Assume the conditions of Lemma 6.7.16 with $Y = \Theta$. Then for every compact subset $\Theta_0 \subset \Theta$ and $\varepsilon > 0$ there exists $\delta_\varepsilon > 0$ such that*

$$\lim_{n \to \infty} \sup_{\tau \in \Theta_0} P_\tau^n \Big\{ \underline{x} \in X^n : \sup_{\vartheta \in \Theta_0} \sup_{\|a\| \leq \delta_\varepsilon} \Big| n^{-1} \sum_1^n \int_0^1 f(x_\nu, \vartheta + ua) du$$
$$- P_\tau\big(f(\cdot, \vartheta)\big) \Big| > \varepsilon \Big\} = 0.$$

Proof. Follows from Lemma 6.7.16. Hint: Use

$$\sup_{\|a\| \leq \delta} n^{-1} \sum_1^n \int_0^1 f(x_\nu, \vartheta + au) du - P_\tau\big(f(\cdot, \vartheta)\big)$$
$$\leq n^{-1} \sum_1^n \overline{f}\big(x_\nu, S(\vartheta, \delta)\big) - P_\tau\big(f(\cdot, \vartheta)\big)$$

and the corresponding lower inequality.   □

The following theorem occurs in the literature in numerous versions (for a survey see Wagner, 1977 and 1980, and Ioffe, 1978).

Measurable Selection Theorem 6.7.22. *Let $(Y, \mathcal{B})$ be a Polish space, and let $f : X \times Y \to \mathbb{R}$ be a function fulfilling conditions (6.7.1). Assume that for every $x \in X$ there exists $y \in Y$ such that $f(x, y) = \overline{f}(x, Y)$.*
Then there exists a measurable function $\psi : X \to Y$ such that

$$f\big(x, \psi(x)\big) = \overline{f}(x, Y) \qquad \text{for } x \in X.$$

Proof. Given a dense subset $\{y_k : k \in \mathbb{N}\} \subset Y$ and a sequence $0 < \varepsilon_n \downarrow 0$, let

$$S_n(k) := \{y \in Y : d(y, y_k) \leq \varepsilon_n\}, \qquad k \in \mathbb{N}, \ n \in \{0\} \cup \mathbb{N}.$$

Let $B_0(x) := \{y \in Y : f(x, y) = \overline{f}(x, Y)\}$. $B_0(x)$ is closed for every $x \in X$. If B is closed, we have

$$\{x \in X : B_0(x) \cap B = \emptyset\} = \{x \in X : \overline{f}(x, B) < \overline{f}(x, Y)\} \in \mathcal{A}$$

(since $\overline{f}(\cdot, B)$ and $\overline{f}(\cdot, Y)$ are $\mathcal{A}$–measurable by Lemma 6.7.2).
Starting from $B_0(x)$, we define inductively a sequence $B_n(x)$, $n \in \mathbb{N}$. Assume that $B_n(x)$ is nonempty and closed for every $x \in X$, and that $\{x \in X : B_n(x) \cap B = \emptyset\} \in \mathcal{A}$ for every closed set $B \in \mathcal{B}$. Since $\bigcup_{k=1}^{\infty} S_n(k) = Y$, the

set $\{k \in \mathbb{N} : B_n(x) \cap S_n(k) \neq \emptyset\}$ is nonempty. Let

$$k_n(x) := \inf\{k \in \mathbb{N} : B_n(x) \cap S_n(k) \neq \emptyset\}.$$

Since

$$\{x \in X : k_n(x) = k\} = \{x \in X : B_n(x) \cap S_n(k) \neq \emptyset\}$$
$$\cap \bigcap_{i=1}^{k-1}\{x \in X : B_n(x) \cap S_n(i) = \emptyset\},$$

the function k_n is $\mathcal{A}$–measurable.

Now we define

$$B_{n+1}(x) := B_n(x) \cap S_n(k_n(x)).$$

By definition of $k_n(x)$, we have $B_{n+1}(x) \neq \emptyset$. $B_{n+1}(x)$ is closed, and $\{x \in X : B_{n+1}(x) \cap B = \emptyset\} \in \mathcal{A}$ for every closed set B, since

$$\{x \in X : B_{n+1}(x) \cap B = \emptyset\} = \{x \in X : B_n(x) \cap S_n(k_n(x)) \cap B = \emptyset\}$$
$$= \bigcup_{k \in \mathbb{N}}\Big(\{x \in X : B_n(x) \cap S_n(k) \cap B = \emptyset\} \cap \{x \in X : k_n(x) = k\}\Big).$$

Since $B_{n+1}(x) \subset B_n(x)$ and the diameter of $B_n(x)$ is $\leq \varepsilon_n$, the set $\bigcap_{n \in \mathbb{N}} B_n(x)$ contains exactly one point, say $\psi(x)$. Since $\psi(x) \in B_0(x)$, we have $f(x, \psi(x)) = \overline{f}(x, Y)$. Since $\psi^{-1}(B) = \bigcap_{n=1}^{\infty}\{x \in X : B_n(x) \cap B \neq \emptyset\} \in \mathcal{A}$ for every closed set B, the function ψ is measurable. $\square$

If $\overline{f}(x, Y)$ is not attained for every $x \in X$, the following lemma is useful in connection with estimator sequences.

Lemma 6.7.23. *Let $(Y, \mathcal{B})$ be a separable Hausdorff space, endowed with its Borel algebra. Let $f : X \times Y \to \mathbb{R}$ be a function fulfilling conditions (6.7.1).*

Then for any $\varepsilon > 0$ there exists a measurable function $\psi_\varepsilon : X \to Y$ such that

$$f(x, \psi_\varepsilon(x)) > \overline{f}(x, Y) - \varepsilon.$$

Proof. Given a dense subset $\{y_k : k \in \mathbb{N}\} \subset Y$, let

$$k(x) := \inf\{k \in \mathbb{N} : f(x, y_k) > \overline{f}(x, Y) - \varepsilon\}.$$

Since

$$\{x \in X : k(x) = k\} = \{x \in X : f(x, y_k) > \overline{f}(x, Y) - \varepsilon\}$$
$$\cap \bigcap_{i=1}^{k-1} \{x \in X : f(x, y_i) \leq \overline{f}(x, Y) - \varepsilon\},$$

the function k is $\mathcal{A}$–measurable. The assertion holds with $\psi_\varepsilon(x) = y_{k(x)}$. $\square$

Chapter 7
Asymptotic distributions of estimator sequences

7.1 Limit distributions

Emphasizes the need for estimator sequences which converge locally uniformly to their limit distributions, discusses various modes of locally uniform convergence, and establishes special properties of limit distributions which are attained locally uniformly.

To judge the accuracy of an estimator one has to know its distribution. If this distribution is difficult to describe, one may be content with an approximation. It is the purpose of asymptotic theory to provide such approximations for large sample sizes.

To write the following in a convenient way which does not restrict the presentation ab ovo to the case of independent and identically distributed observations, we introduce a parameter set Θ, endowed with a Hausdorff topology. For $\vartheta \in \Theta$ and $n \in \mathbb{N}$ let $P_{\vartheta}^{(n)}$ be a p–measure on a measurable space $(X_n, \mathcal{A}_n)$. The functional to be estimated is $\kappa : \Theta \to \mathbb{R}^q$. This framework includes the i.i.d. case for $\Theta = \mathfrak{P}$ a family of p–measures P on $(X, \mathcal{A})$, with $(X_n, \mathcal{A}_n)$ replaced by $(X^n, \mathcal{A}^n)$, $P_{\vartheta}^{(n)}$ replaced by P^n, and $\kappa : \mathfrak{P} \to \mathbb{R}^q$.

Consistency of the estimator sequence $\kappa^{(n)}$, $n \in \mathbb{N}$, means convergence of $\kappa^{(n)}$, $n \in \mathbb{N}$, under $P_{\vartheta}^{(n)}$ to $\kappa(\vartheta)$, and this is equivalent to weak convergence of $P_{\vartheta}^{(n)} \circ \kappa^{(n)}$, $n \in \mathbb{N}$, to the degenerate distribution, concentrated in the point $\{\kappa(\vartheta)\}$. Hence consistency as such says nothing that could be used for judging the accuracy of an estimator, even for large sample sizes.

A practically useful approximate assertion on the distribution of $\kappa^{(n)}$ about $\kappa(\vartheta)$ can be obtained if it is possible to "rescale" the difference between $\kappa^{(n)}$ and $\kappa(\vartheta)$ by multiplying $\kappa^{(n)} - \kappa(\vartheta)$ by an appropriate factor, say c_n with $c_n \uparrow \infty$, such that the distributions of $c_n\big(\kappa^{(n)} - \kappa(\vartheta)\big)$ under $P_{\vartheta}^{(n)}$ form a sequence which converges weakly to some limit distribution, say $Q_{\vartheta} | \mathbb{B}^q$. Abusing language we shall also say that the estimator sequence $\kappa^{(n)}$, $n \in \mathbb{N}$, converges to a limit distribution (if the rescaling factor c_n for $c_n(\kappa^{(n)} - \kappa(\vartheta))$ is understood).

For technical reasons, weak convergence is defined by the convergence of integrals of bounded and continuous functions. There are, of course, more natural convergence concepts for probability measures, based on the convergence of $Q_\vartheta^{(n)}(B)$ to $Q_\vartheta(B)$, at least for sets B of simple geometric structure. Such concepts are easier to interpret than the convergence of integrals of bounded and continuous functions. The obstacle: A general concept of "convergence" which applies to arbitrary p–measures (including p–measures with atoms) cannot easily be based on the convergence of $Q_\vartheta^{(n)}(B)$, $n \in \mathbb{N}$. However: What occurs in statistical theory are limit distributions of estimator sequences. Under mild regularity conditions, such limit distributions have a Lebesgue density (see Proposition 7.1.11). For such limit distributions, weak convergence can be expressed in the following more intuitive form, based on convergence on $\mathcal{C}$, the class of all convex Borel sets in $\mathbb{R}^q$.

Because of its fundamental role for asymptotic estimation theory, we repeat Theorem 7.7.10 for the special case of limit distributions of estimator sequences.

Theorem 7.1.1. *Assume that*

(i) Θ *is a locally compact Hausdorff space fulfilling the first axiom of countability.*

(ii) $\vartheta \to Q_\vartheta$ *is continuous on Θ (with respect to weak convergence), and $Q_\vartheta(C^b) = 0$ for $\vartheta \in \Theta$ and $C \in \mathcal{C}$.*

Then the following assertions are equivalent:

$$P_\vartheta^{(n)} \circ c_n(\kappa^{(n)} - \kappa(\vartheta)) \Rightarrow Q_\vartheta \quad [\text{locally uniformly}] \text{ on } \Theta, \qquad (7.1.2)$$

$$\sup_{C \in \mathcal{C}} |P_\vartheta^{(n)}\{x \in X_n : c_n(\kappa^{(n)} - \kappa(\vartheta)) \in C\} - Q_\vartheta(C)| \to 0 \qquad (7.1.3)$$

$$[\text{locally uniformly}] \text{ on } \Theta.$$

Addendum. *The condition $Q_\vartheta(C^b) = 0$ for $C \in \mathcal{C}$ is, in particular, true if $Q_\vartheta \ll \lambda^q$. If $q = 1$, the condition $Q_\vartheta\{t\} = 0$ for $t \in \mathbb{R}$ suffices.*

To discuss some basic questions, we assume for the moment that $q = 1$, and that the limit distribution $Q_\vartheta | \mathbb{B}$ is nonatomic for every $\vartheta \in \Theta$. Let

$$\Delta_n(\vartheta) := \sup_{I \in \mathcal{I}} |P_\vartheta^{(n)}\{x \in X_n : c_n(\kappa^{(n)}(x) - \kappa(\vartheta)) \in I\} - Q_\vartheta(I)|, \qquad (7.1.4)$$

where $\mathcal{I}$ denotes the class of all intervals in $\mathbb{R}$. According to Theorem 7.1.1, weak convergence of $P_\vartheta^{(n)} \circ c_n(\kappa^{(n)} - \kappa(\vartheta))$ to Q_ϑ on Θ is equivalent to

$$\lim_{n \to \infty} \Delta_n(\vartheta) = 0 \quad \text{for every } \vartheta \in \Theta. \qquad (7.1.5)$$

The limit distribution Q_ϑ could be used to find an interval $[-t_\beta(\vartheta), t_\beta(\vartheta)]$ such that $Q_\vartheta[-t_\beta(\vartheta), t_\beta(\vartheta)] = \beta$. Then the probability under $P_\vartheta^{(n)}$ of

$$\kappa(\vartheta) - c_n^{-1} t_\beta(\vartheta) \le \kappa^{(n)}(x) \le \kappa(\vartheta) + c_n^{-1} t_\beta(\vartheta) \tag{7.1.6}$$

will be close to β if n is large. Hence one may use the limit distributions of two different estimator sequences for an approximate comparison of their accuracy. Even if it may be difficult, it is possible in principle to determine $\Delta_n(\vartheta)$ for every $\vartheta \in \Theta$ and to judge whether the difference between

$$P_\vartheta^{(n)}\{x \in X_n : \kappa(\vartheta) - c_n^{-1} t_\beta(\vartheta) \le \kappa^{(n)}(x) \le \kappa(\vartheta) + c_n^{-1} t_\beta(\vartheta)\}$$

and β, bounded by $\Delta_n(\vartheta)$, is small enough so that a comparison between the concentration of two estimators, based on the limit distributions, is accurate enough for a certain sample size. With the approximation error, $\Delta_n(\vartheta)$, depending on ϑ, this usage requires, strictly speaking, that the convergence of $\Delta_n(\vartheta)$ to zero is uniform on Θ. As a substitute we shall be content with uniform convergence on compact subsets of Θ.

The need for uniformity in ϑ is even more imperative if the limit distribution is used for obtaining a confidence interval. This problem will be discussed in Section 7.3.

Wolfowitz (1965, pp. 249/50) seems to have been the first one to justify the requirement of "convergence uniformly in ϑ" from the operational point of view. Other authors seem to be inclined to use such a uniformity condition as some sort of deus ex machina for ruling out superefficiency. (See Sections 8.2 and 8.3.)

The following Proposition 7.1.8 shows, in particular, that locally uniform convergence to a limit distribution depending continuously on the parameter implies continuous convergence (see Schmetterer, 1966). Notice that continuous convergence (i.e. $Q_{\vartheta_n}^{(n)} \Rightarrow Q_{\vartheta_0}$ for every sequence $\vartheta_n \to \vartheta_0$) is a property much stronger than the "regularity"

$$Q_{\vartheta_n}^{(n)} \Rightarrow Q_\vartheta \text{ for sequences } \vartheta_n = \vartheta_0 + c_n^{-1} a, \tag{7.1.7}$$

required as a regularity condition for the Convolution Theorem 8.4.1.

Proposition 7.1.8. *Assume that Θ is a locally compact Hausdorff space fulfilling the first axiom of countability. Then the following assertions are equivalent for arbitrary sequences $Q_\vartheta^{(n)}$ of p-measures.*

(i) $Q_\vartheta^{(n)} \Rightarrow Q_\vartheta$ locally uniformly on Θ, and $\vartheta \to Q_\vartheta$ is continuous on Θ (with respect to weak convergence).

(ii) $Q_{\vartheta_n}^{(n)} \Rightarrow Q_{\vartheta_0}$ for every $\vartheta_0 \in \Theta$ and every sequence $\vartheta_n \to \vartheta_0$.

Proof. Follows immediately from Corollary 7.7.14 applied for the functions $f_n(\vartheta) := Q_\vartheta^{(n)}(h)$, $h \in \mathfrak{C}$. □

Whereas continuous convergence of $Q_\vartheta^{(n)}$ to Q_ϑ at ϑ_0 implies continuity of $\vartheta \to Q_\vartheta$ at ϑ_0, this is not necessarily true for locally uniform convergence of $Q_\vartheta^{(n)}$ to Q_ϑ at ϑ_0. We need continuity of $\vartheta \to Q_\vartheta$ at ϑ_0 as an extra condition to conclude from locally uniform convergence at ϑ_0 to continuous convergence at ϑ_0 (by means of Lemma 7.7.13(i)). The situation is different if $Q_\vartheta^{(n)}$, $n \in \mathbb{N}$, is not an arbitrary sequence of p-measures, but of the special type $P_\vartheta^{(n)} \circ c_n(\kappa^{(n)} - \kappa(\vartheta))$. Under the regularity conditions specified in Propositions 7.1.8 and 7.1.9, $P_\vartheta^{(n)} \circ c_n(\kappa^{(n)} - \kappa(\vartheta)) \Rightarrow Q_\vartheta$ locally uniformly on Θ is equivalent to $P_{\vartheta_n}^{(n)} \circ c_n(\kappa^{(n)} - \kappa(\vartheta_n)) \Rightarrow Q_{\vartheta_0}$ for all sequences $\vartheta_n \to \vartheta_0$, $\vartheta_0 \in \Theta$.

Earlier versions of Proposition 7.1.9 occur in C.R. Rao (1963, p. 196, Lemma 2(i)) and Wolfowitz (1965, p. 254, Lemma 2).

Proposition 7.1.9. *Let Θ be a Hausdorff space. Assume that*
(i) $\vartheta \to P_\vartheta^{(n)}$ *is continuous with respect to the sup–metric for every $n \in \mathbb{N}$.*
(ii) $\kappa : \Theta \to \mathbb{R}^q$ *is continuous at ϑ_0.*
(iii) $\kappa^{(n)}$, $n \in \mathbb{N}$, *is an estimator sequence such that $P_\vartheta^{(n)} \circ c_n(\kappa^{(n)} - \kappa(\vartheta)) \Rightarrow Q_\vartheta$, locally uniformly at ϑ_0.*
Then $\vartheta \to Q_\vartheta$ is continuous at ϑ_0 with respect to weak convergence, and $P_{\vartheta_n}^{(n)} \circ c_n(\kappa^{(n)} - \kappa(\vartheta_n)) \Rightarrow Q_{\vartheta_0}$ for every sequence $\vartheta_n \to \vartheta_0$.

Corollary 7.1.10. *If $Q_\vartheta = N_{(0,\Sigma(\vartheta))}$, then $\vartheta \to \Sigma(\vartheta)$ is continuous at ϑ_0 (since $N_{(0,\Sigma_n)} \Rightarrow N_{(0,\Sigma_0)}$ implies $\Sigma_n \to \Sigma_0$).*

Proof. Let $h : \mathbb{R}^q \to [0,1]$ be uniformly continuous. It suffices to show that $\vartheta \to Q_\vartheta(h)$ is continuous at ϑ_0. With $Q_\vartheta^{(n)} := P_\vartheta^{(n)} \circ c_n(\kappa^{(n)} - \kappa(\vartheta))$ we have

$$|Q_\vartheta(h) - Q_{\vartheta_0}(h)| \le |Q_\vartheta^{(n)}(h) - Q_\vartheta(h)|$$
$$+ |Q_\vartheta^{(n)}(h) - Q_{\vartheta_0}^{(n)}(h)| + |Q_{\vartheta_0}^{(n)}(h) - Q_{\vartheta_0}(h)|.$$

Since $Q_\vartheta^{(n)}(h) \underset{n \to \infty}{\longrightarrow} Q_\vartheta(h)$ uniformly on a neighborhood U_0 of ϑ_0, it remains to be shown that $\vartheta \to Q_\vartheta^{(n)}(h)$ is continuous at ϑ_0 for every $n \in \mathbb{N}$. Let $n \in \mathbb{N}$ be fixed. We have

$$|Q_\vartheta^{(n)}(h) - Q_{\vartheta_0}^{(n)}(h)|$$
$$= \left| \int h(c_n(\kappa^{(n)}(x) - \kappa(\vartheta))) P_\vartheta^{(n)}(dx) - \int h(c_n(\kappa^{(n)}(x) - \kappa(\vartheta_0))) P_{\vartheta_0}^{(n)}(dx) \right|$$

$$\leq |\int h\big(c_n(\kappa^{(n)}(x)-\kappa(\vartheta))\big)P_\vartheta^{(n)}(dx) - \int h\big(c_n(\kappa^{(n)}(x)-\kappa(\vartheta_0))\big)P_\vartheta^{(n)}(dx)|$$

$$+|\int h\big(c_n(\kappa^{(n)}(x)-\kappa(\vartheta_0))\big)P_\vartheta^{(n)}(dx) - \int h\big(c_n(\kappa^{(n)}(x)-\kappa(\vartheta_0))\big)P_{\vartheta_0}^{(n)}(dx)|.$$

Since κ is continuous and h uniformly continuous, there exists $U_\varepsilon' \ni \vartheta_0$ such that

$$\sup_{t\in\mathbb{R}^q} |h(t) - h\big(t + c_n(\kappa(\vartheta) - \kappa(\vartheta_0))\big)| < \varepsilon/2 \quad \text{for } \vartheta \in U_\varepsilon'.$$

This implies that

$$|\int h\big(c_n(\kappa^{(n)}(x) - \kappa(\vartheta))\big)P_\vartheta^{(n)}(dx) - \int h\big(c_n(\kappa^{(n)}(x) - \kappa(\vartheta_0))\big)P_\vartheta^{(n)}(dx)| < \varepsilon/2$$

for $\vartheta \in U_\varepsilon'$.

Since $h : \mathbb{R}^q \to [0,1]$, we have

$$|\int h\big(c_n(\kappa^{(n)}(x) - \kappa(\vartheta_0))\big)P_\vartheta^{(n)}(dx) - \int h\big(c_n(\kappa^{(n)}(x) - \kappa(\vartheta_0))\big)P_{\vartheta_0}^{(n)}(dx)|$$

$$\leq \sup_{A\in\mathcal{A}_n} |P_\vartheta^{(n)}(A) - P_{\vartheta_0}^{(n)}(A)| \leq \varepsilon/2$$

if ϑ is sufficiently close to ϑ_0, say $\vartheta \in U_\varepsilon''$. Hence $|Q_\vartheta^{(n)}(h) - Q_{\vartheta_0}^{(n)}(h)| < \varepsilon$ for $\vartheta \in U_\varepsilon' \cap U_\varepsilon''$.

This implies continuity of $\vartheta \to Q_\vartheta$ at ϑ_0. The convergence of $P_{\vartheta_n}^{(n)} \circ c_n\big(\kappa^{(n)} - \kappa(\vartheta_n)\big)$, $n \in \mathbb{N}$, to Q_{ϑ_0} now follows from Lemma 7.7.13(i). $\qquad\square$

The next proposition shows, for the case $\Theta \subset \mathbb{R}^p$ and open, that a limit distribution which is approached locally uniformly in ϑ has always a Lebesgue density. The proposition presumes that $P_{\vartheta_0+c_n^{-1}a}^{(n)}$ is contiguous with respect to $P_{\vartheta_0}^{(n)}$. This is, in particular, the case under the LAN Condition 8.1.1 (see p. 259). Under this — stronger — condition, the existence of a Lebesgue density for the limit distribution follows immediately from the convolution theorem.

An earlier version of Proposition 7.1.11 was obtained by Kaufman (1966, p. 174, Lemma 5.4), following a weaker result of Wolfowitz (1965, p. 253, Lemma 1).

Proposition 7.1.11. *Let $\Theta \subset \mathbb{R}^p$ be open. Assume that*

(i) $P_{\vartheta_n}^{(n)}$ is contiguous with respect to $P_{\vartheta_0}^{(n)}$, $n \in \mathbb{N}$, for every sequence $\vartheta_n = \vartheta_0 + c_n^{-1}a$, $a \in \mathbb{R}^p$.

(ii) $\kappa : \Theta \to \mathbb{R}^q$, $q \leq p$, admits partial derivatives which are continuous in a neighborhood of ϑ_0. The $q \times p$-matrix of the derivatives at ϑ_0 is of rank q.

(iii) The estimator sequence $\kappa^{(n)}$, $n \in \mathbb{N}$, for κ is "regular" for sequences $\vartheta_n = \vartheta_0 + c_n^{-1}a$, $n \in \mathbb{N}$ (in the sense of condition (7.1.7)).

Then the limit distribution of $P_{\vartheta_0}^{(n)} \circ c_n (\kappa^{(n)} - \kappa(\vartheta_0))$ has a Lebesgue density.

Recall Theorem 7.1.1 according to which this implies convergence to the limit distribution uniformly over all convex Borel sets.

Proof. Let Q_{ϑ_0} denote the limit distribution. We have to show that $Q_{\vartheta_0} \ll \lambda^q$. Assume there exists $B \in \mathbb{B}^q$ such that $\lambda^q(B) = 0$ and $Q_{\vartheta_0}(B) > 0$.

Since the measures $\lambda^q | \mathbb{B}^q$ and $Q_{\vartheta_0} | \mathbb{B}^q$ are regular, there exists a closed set $C \subset B$ and a sequence of open sets $U_m \supset B$, $m \in \mathbb{N}$, such that $Q_{\vartheta_0}(C) > 0$, and $\lim_{m \to \infty} \lambda^q(U_m) = \lambda^q(B) = 0$. Let h_m, $m \in \mathbb{N}$, be a sequence of continuous functions fulfilling $1_C \le h_m \le 1_{U_m}$. This implies $Q_{\vartheta_0}(h_m) \ge Q_{\vartheta_0}(C) > 0$, and $\lim_{m \to \infty} \lambda^q(h_m) = 0$.

By assumption,

$$P_{\vartheta_0 + c_n^{-1}a}^{(n)} \circ c_n \left(\kappa^{(n)} - \kappa(\vartheta_0 + c_n^{-1}a) \right) \Rightarrow Q_{\vartheta_0}. \qquad (7.1.12)$$

Since

$$c_n \left(\kappa(\vartheta_0 + c_n^{-1}a) - \kappa(\vartheta_0) \right) = K(\vartheta_0)a + \Delta_n(a)$$

with $\Delta_n(a) \to 0$ (and $K_{ij}(\vartheta) = \kappa_i^{(j)}(\vartheta)$), we have

$$c_n \left(\kappa^{(n)} - (\kappa(\vartheta_0) + c_n^{-1}K(\vartheta_0)a) \right) = c_n \left(\kappa^{(n)} - \kappa(\vartheta_0 + c_n^{-1}a) \right) + \Delta_n(a).$$

Hence we obtain from (7.1.12) by Slutzky's Lemma 7.7.8 that

$$P_{\vartheta_0 + c_n^{-1}a}^{(n)} \circ c_n \left(\kappa^{(n)} - (\kappa(\vartheta_0) + c_n^{-1}K(\vartheta_0)a) \right) \Rightarrow Q_{\vartheta_0}$$

for every $a \in \mathbb{R}^p$.

This implies

$$\lim_{n \to \infty} \int h_m \left(c_n (\kappa^{(n)}(x) - (\kappa(\vartheta_0) + c_n^{-1}K(\vartheta_0)a)) \right) P_{\vartheta_0 + c_n^{-1}a}^{(n)}(dx)$$
$$= Q_{\vartheta_0}(h_m) \ge Q_{\vartheta_0}(C) > 0.$$

Since $P_{\vartheta_0 + c_n^{-1}a}^{(n)}$ is contiguous with respect to $P_{\vartheta_0}^{(n)}$, $n \in \mathbb{N}$, this implies by Lemma 7.1.13 the existence of $H(a) > 0$ such that

$$\varliminf_{n \to \infty} \int h_m \left(c_n (\kappa^{(n)}(x) - (\kappa(\vartheta_0) + c_n^{-1}K(\vartheta_0)a)) \right) P_{\vartheta_0}^{(n)}(dx) \ge H(a)$$

for $m \in \mathbb{N}$, $a \in \mathbb{R}^p$. Since

$$P_{\vartheta_0}^{(n)} \circ c_n \left(\kappa^{(n)} - (\kappa(\vartheta_0) + c_n^{-1}K(\vartheta_0)a) \right) \Rightarrow Q_{\vartheta_0} \circ (u \to u - K(\vartheta_0)a),$$

we obtain

$$\int h_m (u - K(\vartheta_0)a) Q_{\vartheta_0}(du) \ge H(a) \quad \text{for } m \in \mathbb{N}, a \in \mathbb{R}^p.$$

Since $K(\vartheta_0)$ has rank q, this implies that

$$\hat{H}(v) := \inf_{m \in \mathbb{N}} \int h_m(u - v)Q_{\vartheta_0}(du) > 0 \quad \text{for } v \in \mathbb{R}^q,$$

hence

$$0 < \int \hat{H}(v)\lambda^q(dv) \le \int h_m(u - v)Q_{\vartheta_0}(du)\lambda^q(dv) = \lambda^q(h_m) \quad \text{for } m \in \mathbb{N}.$$

Since $\lim_{m \to \infty} \lambda^q(h_m) = 0$, this is impossible. Hence $\lambda^q(B) = 0$ implies $Q_{\vartheta_0}(B) = 0$. □

Lemma 7.1.13. *Assume that $Q'_n|\mathcal{B}$, $n \in \mathbb{N}$, is contiguous with respect to $Q_n|\mathcal{B}$, $n \in \mathbb{N}$. Let h_m, $m \in \mathbb{N}$, be a sequence of bounded nonnegative functions such that $\inf_{m \in \mathbb{N}} \lim_{n \to \infty} Q'_n(h_m) > 0$. Then $\inf_{m \in \mathbb{N}} \underline{\lim}_{n \to \infty} Q_n(h_m) > 0$.*

Proof. If $\inf_{m \in \mathbb{N}} \lim_{n \to \infty} Q'_n(h_m) > 0$, then there exists $\delta > 0$ and $N_m \in \mathbb{N}$ for $m \in \mathbb{N}$, such that $Q'_n(h_m) > \delta$ for $n \ge N_m$. If $\inf_{m \in \mathbb{N}} \underline{\lim}_{n \to \infty} Q_n(h_m) = 0$, then for every k there exists m_k such that $\mathbb{N}_k := \{n \in \mathbb{N} : Q_n(h_{m_k}) < 1/k\}$ is infinite. For $n_k \in \mathbb{N}_k$, $n_k \ge N_{m_k}$, we have $Q'_{n_k}(h_{m_k}) > \delta$, and $Q_{n_k}(h_{m_k}) < 1/k$. For $k \to \infty$ this leads to $\lim_{k \to \infty} Q_{n_k}(h_{m_k}) = 0$. Since Q'_{n_k}, $k \in \mathbb{N}$, is contiguous with respect to Q_{n_k}, $k \in \mathbb{N}$, this yields $\lim_{k \to \infty} Q'_{n_k}(h_{m_k}) = 0$ by Exercise 8.1.16, in contradiction to $Q'_{n_k}(h_{m_k}) > \delta$ for $k \in \mathbb{N}$ sufficiently large. □

7.2 How to deal with limit distributions

Gives the limit distribution of a smooth function of an estimator sequence. Warns against using the limit distribution for approximating bias and risk for large samples.

From the asymptotic distribution of an estimator sequence one can easily obtain the asymptotic distribution of a smooth function of this estimator sequence.

Let Θ be a locally compact Hausdorff space, T an open subset of $\mathbb{R}^q$, and $\kappa : \Theta \to T$ a functional. For $\vartheta \in \Theta$ and $n \in \mathbb{N}$ let $P_\vartheta^{(n)}$ be a p-measure on $(X_n, \mathcal{A}_n)$.

Proposition 7.2.1. *Assume that $\kappa^{(n)} : X_n \to T$, $n \in \mathbb{N}$, is an estimator sequence such that, for some sequence $c_n \to \infty$,*

$$P_\vartheta^{(n)} \circ c_n(\kappa^{(n)} - \kappa(\vartheta)) \Rightarrow Q_\vartheta, \quad [\textit{locally uniformly}] \text{ at } \vartheta_0.$$

Assume that $h : T \to \mathbb{R}^p$, $p \leq q$, has partial derivatives which are continuous on Θ, say $H_{ij}(t) := h_i^{(j)}(t_1, \ldots, t_q)$, $i = 1, \ldots, p$, $j = 1, \ldots, q$, where $H(\kappa(\vartheta_0))$ has rank p.

Then

$$P_\vartheta^{(n)} \circ c_n \big(h \circ \kappa^{(n)} - h(\kappa(\vartheta)) \big) \Rightarrow Q_\vartheta \circ \big(t \to H(\kappa(\vartheta))t \big),$$

[locally uniformly] at ϑ_0.

Remark 7.2.2. Recall that

$$N_{(0,\Sigma(\vartheta))} \circ \big(t \to H(\kappa(\vartheta))t \big) = N_{(0, H(\kappa(\vartheta))\Sigma(\vartheta)H(\kappa(\vartheta))^\top)}.$$

Proof. It is straightforward to see that the proof goes through [locally uniformly] at ϑ_0.

$$h_i(s) = h_i(t) + \sum_{j=1}^q (s_j - t_j) \int_0^1 H_{ij}\big((1-u)t + us\big)\, du$$

implies

$$c_n \big(h_i(\kappa^{(n)}(x)) - h_i(\kappa(\vartheta)) \big)$$

$$= \sum_{j=1}^q c_n \big(\kappa_j^{(n)}(x) - \kappa_j(\vartheta) \big) \int_0^1 H_{ij}\big((1-u)\kappa(\vartheta) + u\kappa^{(n)}(x)\big)\, du \quad (7.2.3)$$

$$= \sum_{j=1}^q H_{ij}\big(\kappa(\vartheta)\big) c_n \big(\kappa_j^{(n)}(x) - \kappa_j(\vartheta) \big) + r_i^{(n)}(x, \vartheta)$$

with

$$r_i^{(n)}(x, \vartheta) :=$$

$$\sum_{j=1}^q \Big[\int_0^1 H_{ij}\big((1-u)\kappa(\vartheta) + u\kappa^{(n)}(x)\big)\, du - H_{ij}\big(\kappa(\vartheta)\big) \Big] c_n \big(\kappa_j^{(n)}(x) - \kappa_j(\vartheta) \big).$$

Since $t \to \int_0^1 H_{ij}\big((1-u)\kappa(\vartheta) + ut\big)\, du$ is continuous and $c_n \big(\kappa^{(n)} - \kappa(\vartheta) \big)$, $n \in \mathbb{N}$, is stochastically bounded by Corollary 7.7.7, we obtain that

$$\int_0^1 H_{ij}\big((1-u)\kappa(\vartheta) + u\kappa^{(n)}\big)\, du \to H_{ij}\big(\kappa(\vartheta)\big) \quad (P_\vartheta^{(n)}).$$

Lemma 6.7.14 implies $r_i^{(n)}(\cdot, \vartheta) \to 0$ $(P_\vartheta^{(n)})$. Moreover, $P_\vartheta^{(n)} \circ c_n(\kappa^{(n)} - \kappa(\vartheta)) \Rightarrow Q_\vartheta$ implies

$$P_\vartheta^{(n)} \circ H(\kappa(\vartheta))c_n(\kappa^{(n)} - \kappa(\vartheta)) \Rightarrow Q_\vartheta \circ (t \to H(\kappa(\vartheta))t).$$

Hence the assertion follows from (7.2.3) by Slutzky's Lemma 7.7.8.   □

We conclude this section on possible uses of limit distributions of estimator sequences by emphasizing what a limit distribution cannot be used for, namely: for approximating integrals of unbounded functions.

Weak convergence of Q_n, $n \in \mathbb{N}$, to Q_0 is defined by convergence of $Q_n(h)$, $n \in \mathbb{N}$, to $Q_0(h)$ for every bounded and continuous function $h : \mathbb{R}^q \to \mathbb{R}$. This convergence property does not extend to functions h which are continuous, but unbounded. Hence a property of the limit distribution Q_0, expressed by an integral of an unbounded function cannot be interpreted as being "approximately true" for Q_n.

To be more explicit, let

$$P_\vartheta^{(n)} \circ c_n(\kappa^{(n)} - \kappa(\vartheta)) \quad \Rightarrow \quad Q_\vartheta.$$

If $Q_\vartheta(-\infty, 0] = Q_\vartheta[0, \infty) = 1/2$, this implies $Q_\vartheta\{0\} = 0$, hence also

$$\lim_{n \to \infty} P_\vartheta^{(n)}\{x \in X_n : \kappa^{(n)}(x) \le \kappa(\vartheta)\} \qquad (7.2.4)$$
$$= \lim_{n \to \infty} P_\vartheta^{(n)}\{x \in X_n : \kappa^{(n)}(x) \ge \kappa(\vartheta)\} = 1/2,$$

a property which may fairly be called *asymptotic median unbiasedness* of the estimator sequence $\kappa^{(n)}$, $n \in \mathbb{N}$.

The natural definition of *asymptotic mean unbiasedness* is, perhaps,

$$\lim_{n \to \infty} n^{1/2} \int (\kappa^{(n)} - \kappa(\vartheta))dP_\vartheta^{(n)} = 0. \qquad (7.2.5)$$

This relation may be an important property of $\kappa^{(n)}$, $n \in \mathbb{N}$, in certain applications. From the technical point of view it is in no inherent relation with $\int u Q_\vartheta(du) = 0$, nor is $\int u^2 Q_{\vartheta,1}(du) / \int u^2 Q_{\vartheta,2}(du)$ (for the limit distributions of two estimator sequences $\kappa_i^{(n)}$, $n \in \mathbb{N}$) an approximation to

$$\int (\kappa_1^{(n)} - \kappa(\vartheta))^2 dP_\vartheta^{(n)} / \int (\kappa_2^{(n)} - \kappa(\vartheta))^2 dP_\vartheta^{(n)}.$$

Proposition 7.2.6. *If $Q_\vartheta^{(n)} \Rightarrow Q_\vartheta$, then*

$$\varliminf_{n \to \infty} Q_\vartheta^{(n)}(L(\cdot, \vartheta)) \ge Q_\vartheta(L(\cdot, \vartheta)) \qquad (7.2.7')$$

for any function $L : \mathbb{R}^q \times \Theta \to \mathbb{R}$ for which $u \to L(u, \vartheta)$ is lower semicontinuous and bounded from below. If $u \to L(u, \vartheta)$ is bounded and continuous, then

$$\lim_{n \to \infty} Q_\vartheta^{(n)}(L(\cdot, \vartheta)) = Q_\vartheta(L(\cdot, \vartheta)). \qquad (7.2.7'')$$

Proof. Since $m \wedge L(\cdot, \vartheta)$ is bounded and lower semicontinuous, we have by Lemma 7.7.4(ii),

$$\varliminf_{n \to \infty} Q_\vartheta^{(n)}(L(\cdot, \vartheta)) \geq \varliminf_{n \to \infty} Q_\vartheta^{(n)}(m \wedge L(\cdot, \vartheta)) \geq Q_\vartheta(m \wedge L(\cdot, \vartheta)).$$

This implies the assertion by the monotone convergence theorem. $\qquad \square$

A relation like (7.2.7''), written in terms of the estimator sequence, reads as follows.

$$\lim_{n \to \infty} P_\vartheta^{(n)}(L(c_n(\kappa^{(n)} - \kappa(\vartheta)), \vartheta)) = \int L(u, \vartheta) Q_\vartheta(du).$$

If this relation holds for two estimator sequences $\kappa_i^{(n)}$, $n \in \mathbb{N}$, $i = 1, 2$, then $\int L(u, \vartheta) Q_{\vartheta,1}(du) / \int L(u, \vartheta) Q_{\vartheta,2}(du)$ can be used as an approximation to $\int L(c_n(\kappa_1^{(n)} - \kappa(\vartheta)), \vartheta) dP_\vartheta^{(n)} / \int L(c_n(\kappa_2^{(n)} - \kappa(\vartheta)), \vartheta) dP_\vartheta^{(n)}$ for large sample sizes. If the loss function $L(\cdot, \vartheta)$ is homogeneous, say $L(u, \vartheta) = |u|$, then $\int |u| Q_{\vartheta,1}(du) / \int |u| Q_{\vartheta,2}(du)$ is an approximation to

$$\int |\kappa_1^{(n)} - \kappa(\vartheta)| dP_\vartheta^{(n)} / \int |\kappa_2^{(n)} - \kappa(\vartheta)| dP_\vartheta^{(n)}.$$

With nonhomogeneous loss functions, it is hard to see why the comparison of two estimator sequences should be based on the losses for rescaled values $c_n(\kappa_i^{(n)} - \kappa(\vartheta))$, a paradigm which is hardly ever under discussion.

In the following we give examples showing that strict inequality in (7.2.7') may occur with unbounded loss functions. To set out the domain for such examples we mention that strict inequality for the quadratic loss function is impossible in the simplest case:

Let $P(f) = 0$. If the asymptotic distribution of $n^{-1/2} \sum_1^n f(x_\nu)$ under P^n converges weakly to $N_{(0,\sigma^2)}$, this implies that f^2 is P–integrable, and $P(f^2) = \sigma^2$ (by the normal convergence criterion, see Loève, 1977, p. 328). Therefore,

$$\int \left(n^{-1/2} \sum_1^n f(x_\nu) \right)^2 P^n(d(x_1, \ldots, x_n)) = \sigma^2$$

for every $n \in \mathbb{N}$. With P replaced by P_ϑ, and f replaced by $f - P_\vartheta(f)$, this shows that $P_\vartheta^n \circ n^{1/2}(\kappa^{(n)} - \kappa(\vartheta)) \Rightarrow N_{(0, \sigma^2(\vartheta))}$ implies $n \int (\kappa^{(n)} - \kappa(\vartheta))^2 dP_\vartheta^n = \sigma^2(\vartheta)$

for $\kappa(\vartheta) = P_\vartheta(f)$ and $\kappa^{(n)}(\underline{x}) = n^{-1} \sum_1^n f(x_\nu)$. For results on the convergence of moments of ML estimators see Pfaff (1982).

It is easier to invent estimator sequences where strict inequality holds in (7.2.7') for $L(u, \vartheta) = (u - \vartheta)^2$, than to prove that they live up to our expectations.

Example 7.2.8. Let P_ϑ be the p–measure with Lebesgue density

$$x \to \frac{1}{2}|x - \vartheta| \exp\left[-(x - \vartheta)^2/2\right],$$

and $\vartheta^{(n)}(x_1, \ldots, x_n) := \frac{1}{2}(x_{1:n} + x_{n:n})$. Similarly as in Galambos (1978, p. 109, Example 2.9.1) it can be shown that the distribution of $(8\log n)^{1/2}(\vartheta^{(n)} - \vartheta)$ under P_ϑ^n converges to the logistic distribution with variance $\pi^2/3$, whereas $8\log n \int (\vartheta^{(n)} - \vartheta)^2 dP_\vartheta^n$, $n \in \mathbb{N}$, behaves asymptotically like $8(\log n)^2$. (Hint: Use McCord, 1964, p. 1739, Theorem 3.)

In the following more complex example the sequence of (standardized) variances converges to a finite limit different from the variance of the limit distribution.

Example 7.2.9. Let P_ϑ be the p–measure with Lebesgue density

$$x \to \left(1 + |x - \vartheta|\right)^{-3}.$$

The following estimator excludes observations x_ν which are too far off a preliminary estimator, say the sample median $\tilde{x}_n$:

$$\vartheta^{(n)}(x_1, \ldots, x_n) := \sum_{\nu=1}^n x_\nu 1_{[-n,n]}(x_\nu - \tilde{x}_n) \bigg/ \sum_{\nu=1}^n 1_{[-n,n]}(x_\nu - \tilde{x}_n).$$

The distribution of $\left(\frac{n}{\log n}\right)^{1/2}(\vartheta^{(n)} - \vartheta)$ under P_ϑ^n converges weakly to $N_{(0,1)}$, whereas

$$\lim_{n\to\infty} \frac{n}{\log n} \int (\vartheta^{(n)} - \vartheta)^2 dP_\vartheta^n = 2.$$

(Hint: Show that $\left(\frac{n}{\log n}\right)^{1/2}(\vartheta^{(n)} - \vartheta)$ behaves asymptotically like

$$(n\log n)^{-1/2} \sum_1^n x_\nu 1_{[-n,n]}(x_\nu)$$

and use the normal convergence criterion, see Loève, 1977, p. 328.)

7.3 Asymptotic confidence bounds

Shows how to obtain asymptotic confidence bounds for a real functional, and stresses the need for locally uniform convergence. Examples of asymptotic confidence bounds are given in Section 9.3. The optimality of confidence bounds and confidence intervals will be discussed in Sections 8.2 and 8.3.

Let Θ be a Hausdorff space. For $\vartheta \in \Theta$ and $n \in \mathbb{N}$ let $P_\vartheta^{(n)}$ be a p–measure on a measurable space $(X_n, \mathcal{A}_n)$. Let $\kappa : \Theta \to \mathbb{R}$ be a real functional and $\kappa_\beta^{(n)} : X_n \to \mathbb{R}$ a sequence of upper confidence bounds for κ.

$\kappa_\beta^{(n)} : X_n \to \mathbb{R}$ is an upper confidence bound for κ with covering probability β if

$$P_\vartheta^{(n)}\{x \in X_n : \kappa(\vartheta) \le \kappa_\beta^{(n)}(x)\} = \beta \qquad \text{for } \vartheta \in \Theta. \tag{7.3.1}$$

From asymptotic theory we cannot expect more than sequences of confidence bounds which obey an asymptotic version of (7.3.1).

Definition 7.3.2. (i) A sequence of upper confidence bounds has *asymptotic covering probability* β at ϑ_0 if

$$\lim_{n\to\infty} P_{\vartheta_0}^{(n)}\{x \in X_n : \kappa(\vartheta_0) \le \kappa_\beta^{(n)}(x)\} = \beta.$$

(ii) The asymptotic covering probability is kept *uniformly on* Θ if

$$\lim_{n\to\infty} \sup_{\vartheta\in\Theta} |P_\vartheta^{(n)}\{x \in X_n : \kappa(\vartheta) \le \kappa_\beta^{(n)}(x)\} - \beta| = 0.$$

Local uniformity at ϑ_0 and local uniformity on Θ are defined according to (6.7.6).

Example 7.3.3 demonstrates that an asymptotic property like 7.3.2(i), holding at ϑ_0 (yet not locally uniformly at ϑ_0) fails to reflect the performance of a confidence bound for large sample sizes in a proper way.

Example 7.3.3. For $P_\vartheta^{(n)} = N_{(\vartheta,1)}^n$, $n \in \mathbb{N}$, $\vartheta \in \mathbb{R}$, let

$$\vartheta^{(n)}(\underline{x}) := \begin{cases} \overline{x}_n \\ 2^{-1/2}(s_n^2(\underline{x}) - 1) \end{cases} \quad \text{if } |\overline{x}_n| \begin{matrix} > \\ \le \end{matrix} n^{-1/4},$$

with $s_n^2(\underline{x}) = (n-1)^{-1} \sum_1^n (x_\nu - \overline{x}_n)^2$.

It is easy to see that

$$N_{(\vartheta,1)}^n \circ n^{1/2}(\vartheta^{(n)} - \vartheta) \Rightarrow N_{(0,1)} \quad \text{for every } \vartheta \in \mathbb{R}. \tag{7.3.4}$$

Hence it is natural to use

$$\left(\vartheta^{(n)} - n^{-1/2}u_{(1-\beta)/2}, \ \vartheta^{(n)} + n^{-1/2}u_{(1-\beta)/2}\right) \tag{7.3.5}$$

as a confidence interval for ϑ. In fact, the asymptotic covering probability, $N^n_{(\vartheta,1)}\left\{\underline{x} \in \mathbb{R}^n : \vartheta \in \left(\vartheta^{(n)}(\underline{x}) - n^{-1/2}u_{(1-\beta)/2}, \ \vartheta^{(n)}(\underline{x}) + n^{-1/2}u_{(1-\beta)/2}\right)\right\}$ converges to β, for every $\vartheta \in \mathbb{R}$. The convergence of the covering probability is, however, not locally uniform at $\vartheta = 0$. This is responsible for the fact that the covering probability, considered — for n fixed — as a function of ϑ, falls below the intended value β in a neighborhood of 0 of order $n^{-1/2}$.

The following figure shows the covering probability for different sample sizes.

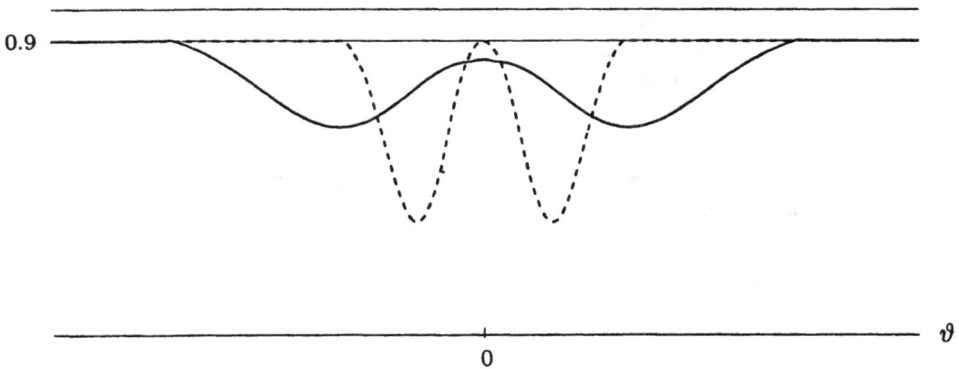

Figure 7.1. Covering probability of the confidence interval (7.3.5) for different sample sizes.
$\underline{\hspace{1.5cm}}$ $n = 10$, $- - -$ $n = 100$.

As in Section 5.2 (see pp. 160/1), we judge the quality of an upper confidence bound by the probability with which it covers values $\kappa(\vartheta) + t$, i.e.

$$P_\vartheta^{(n)}\left\{x \in X_n : \kappa(\vartheta) + t \le \kappa_\beta^{(n)}(x)\right\}.$$

This probability should be large for $t < 0$, and small for $t > 0$ (and equal to β for $t = 0$).

Definition 7.3.6. The *asymptotic power function* of the sequence of confidence bounds $\kappa_\beta^{(n)}$, $n \in \mathbb{N}$, is

$$t \to \lim_{n\to\infty} P_\vartheta^{(n)}\left\{x \in X_n : \kappa(\vartheta) + c_n^{-1}t \le \kappa_\beta^{(n)}(x)\right\}$$

(provided this limit exists for $t \in \mathbb{R}$).

To turn relation (7.1.6), i.e.

$$\kappa(\vartheta) - c_n^{-1} t_\beta(\vartheta) \leq \kappa^{(n)}(x) \leq \kappa(\vartheta) + c_n^{-1} t_\beta(\vartheta),$$

into a confidence interval for $\kappa(\vartheta)$, one has to replace $t_\beta(\vartheta)$ by an estimate, say $t_\beta^{(n)}(x)$.

Proposition 7.3.7. *Let Θ be a Hausdorff space, $\kappa : \Theta \to \mathbb{R}$ a functional and*

$$P_\vartheta^{(n)} \circ c_n\big(\kappa^{(n)} - \kappa(\vartheta)\big) \quad \Rightarrow \quad Q_\vartheta \quad [\text{locally uniformly}] \text{ on } \Theta.$$

Assume that the limit distribution Q_ϑ is nonatomic, and $\vartheta \to Q_\vartheta$ is continuous (with respect to weak convergence).

Let $t_\gamma(\vartheta)$ be the γ-quantile of Q_ϑ, defined by $Q_\vartheta\big(-\infty, t_\gamma(\vartheta)\big] = \gamma$, and let $t_\gamma^{(n)} : X_n \to \mathbb{R}$ be an estimator sequence for the functional t_γ which is consistent [locally uniformly] on Θ.

Then the following holds true.

$\kappa_\beta^{(n)}(x) := \kappa^{(n)}(x) - c_n^{-1} t_{1-\beta}^{(n)}(x)$ *is a sequence of upper confidence bounds for κ with [locally uniform] asymptotic covering probability β. The asymptotic power function of $\kappa_\beta^{(n)}$ is $t \to Q_\vartheta\big[t + t_{1-\beta}(\vartheta), \infty\big)$, $t \in \mathbb{R}$.*

The most important special case is $c_n = n^{1/2}$ and $Q_\vartheta = N_{(0, \sigma^2(\vartheta))}$, in which case $t_{1-\beta}(\vartheta) = -u_{1-\beta} \sigma(\vartheta)$, and $Q_\vartheta\big[t + t_{1-\beta}(\vartheta), \infty\big) = \Phi\big(u_{1-\beta} - t/\sigma(\vartheta)\big)$.

Proof.

$$P_\vartheta^{(n)}\big\{x \in X_n : \kappa(\vartheta) + c_n^{-1} t \leq \kappa^{(n)}(x) - c_n^{-1} t_{1-\beta}^{(n)}(x)\big\}$$
$$= P_\vartheta^{(n)}\big\{x \in X_n : c_n\big(\kappa^{(n)}(x) - \kappa(\vartheta)\big) + \big(t_{1-\beta}(\vartheta) - t_{1-\beta}^{(n)}(x)\big)$$
$$\geq t + t_{1-\beta}(\vartheta)\big\}, \qquad n \in \mathbb{N},$$

converges to $Q_\vartheta\big[t + t_{1-\beta}(\vartheta), \infty\big\}$ [locally uniformly] on Θ by Slutzky's Lemma 7.7.8, since

$$t_{1-\beta}^{(n)} - t_{1-\beta}(\vartheta) \to 0 \quad (P_\vartheta^{(n)}) \quad [\text{locally uniformly}] \text{ on } \Theta.$$

$\square$

The accuracy of the confidence bound depends primarily on the accuracy of the estimator sequence for $\kappa(\vartheta)$, which finds its expression in the standardizing sequence c_n, $n \in \mathbb{N}$, and in the concentration of the limit distribution Q_ϑ. (See Sections 8.2 and 8.3 for asymptotic optimality of confidence bounds.)

The asymptotic power function of the confidence bound,

$$t \to Q_\vartheta\big[t + t_{1-\beta}(\vartheta), \infty\big),$$

does not depend on the quality of the estimator sequence $t_{1-\beta}^{(n)}$ for $t_{1-\beta}(\vartheta)$: consistency is enough. This does not mean, of course, that the quality of the confidence bound $\kappa^{(n)} - c_n^{-1} t_{1-\beta}^{(n)}$ is independent of the quality of $t_{1-\beta}^{(n)}$. It just means that this dependence is comparatively weak so that it does not show up in an approximation by limit distributions. (It does show up in higher order approximations based on Edgeworth expansions.)

The following exercise gives the power function for an irregular case.

Exercise 7.3.8. For $\vartheta \in (0, \infty)$ let P_ϑ denote the uniform distribution with Lebesgue density $x \to \vartheta^{-1} 1_{(0,\vartheta)}(x)$. Show that

$$\vartheta_\beta^{(n)}(x_1, \ldots, x_n) := x_{n:n}\left(1 + n^{-1}|\log(1-\beta)|\right)$$

is an upper confidence bound with asymptotic power function

$$t \to 1 - (1-\beta)\exp[t/\vartheta], \quad -\infty < t \leq \vartheta|\log(1-\beta)|.$$

(Hint: $\lim_{n \to \infty} P_\vartheta^n\{n(x_{n:n} - \vartheta) \leq t\} = \exp[t/\vartheta]$ for $-\infty < t \leq 0$.)

Proposition 7.3.7 presumes, among other things, that c_n, $n \in \mathbb{N}$, is the same for every $\vartheta \in \Theta$, and that $t_\gamma(\vartheta)$ can be consistently estimated. The following example illustrates which difficulties may occur if these conditions are violated.

Exercise 7.3.9. For $\vartheta \in (0, 1)$ let P_ϑ be the binomial distribution, defined by $P_\vartheta\{x\} = \vartheta^x(1-\vartheta)^{1-x}$, $x \in \{0, 1\}$.

For $\kappa(\vartheta) = \vartheta(1-\vartheta)$ a reasonable estimator is $\kappa^{(n)}(\underline{x}) = \vartheta^{(n)}(\underline{x})(1-\vartheta^{(n)}(\underline{x}))$, with $\vartheta^{(n)}(\underline{x}) = n^{-1}\sum_1^n 1_{\{1\}}(x_\nu)$. It is easy to see that

$$P_\vartheta^n \circ n^{1/2}\left(\kappa^{(n)} - \kappa(\vartheta)\right) \Rightarrow N_{(0,\sigma^2(\vartheta))},$$

with $\sigma^2(\vartheta) = \vartheta(1-\vartheta)(1-2\vartheta)^2$. This leads to the upper confidence bound $\underline{x} \to \kappa^{(n)}(\underline{x}) + n^{-1/2}u_{1-\beta}\sigma(\vartheta^{(n)}(\underline{x}))$.

The expectation that

$$\lim_{n \to \infty} P_\vartheta^n\left\{\underline{x} \in \{0,1\}^n : \kappa(\vartheta) \leq \kappa^{(n)}(\underline{x}) + n^{-1/2}u_{1-\beta}\sigma(\vartheta^{(n)}(\underline{x}))\right\} = \beta$$

does, however, not materialize for $\vartheta = 1/2$. The reason: This parameter value is singular in the sense that $\sigma(1/2) = 0$. (To obtain a nondegenerate limit distribution in the case $\vartheta = 1/2$, one has to consider $n(\kappa^{(n)} - \kappa(1/2))$.)

We have

$$P_{1/2}^n\left\{\underline{x} \in \{0,1\}^n : \kappa(1/2) \leq \kappa^{(n)}(\underline{x}) + n^{-1/2}u_{1-\beta}\sigma(\vartheta^{(n)}(\underline{x}))\right\}$$

$$= B_{n,1/2}\left\{k \in \{0,\ldots,n\} : \frac{1}{4} \leq \frac{k}{n}\left(1 - \frac{k}{n}\right)\right.$$

$$\left. + n^{-1/2}u_{1-\beta}\left(\frac{k}{n}\left(1 - \frac{k}{n}\right)\right)^{1/2}\left|1 - 2\frac{k}{n}\right|\right\}$$

$$= B_{n,1/2}\{k \in \{0,\ldots,n\} : n(\frac{1}{4} - \frac{k}{n}(1 - \frac{k}{n}))$$
$$\leq 2u_{1-\beta}(\frac{k}{n}(1 - \frac{k}{n}))^{1/2}n^{1/2}|\frac{1}{2} - \frac{k}{n}|\}.$$

Since

$$n(\frac{1}{4} - \frac{k}{n}(1 - \frac{k}{n})) = n(\frac{1}{2} - \frac{k}{n})^2,$$

and since the distribution of $n^{1/2}(\frac{1}{2} - \frac{k}{n})$ under $B_{n,1/2}$ converges to $N_{(0,1/4)}$, we obtain

$$\lim_{n \to \infty} B_{n,1/2}\{k \in \{0,\ldots,n\} : n(\frac{1}{4} - \frac{k}{n}(1 - \frac{k}{n}))$$
$$\leq 2u_{1-\beta}(\frac{k}{n}(1 - \frac{k}{n}))^{1/2}n^{1/2}|\frac{1}{2} - \frac{k}{n}|\}$$
$$= N_{(0,1/4)}\{t \in \mathbb{R} : t^2 \leq u_{1-\beta}|t|\} = 2\Phi(2u_{1-\beta}) - 1.$$

From the intuitive point of view one might expect that the asymptotic covering probability for $\vartheta = 1/2$ will be larger than β. This is, in fact, true for large β (say $\beta > 0.7$). Though other values of β are hardly of interest from the practical point of view, it seems to be of methodological interest that the asymptotic covering probability is smaller than β for $\beta < 0.69$.

7.4 Solutions to estimating equations

The results of this section are useful for determining the limit distributions of (asymptotic) solutions to estimating equations. These results are first presented in a general framework, and then specialized to the i.i.d. case.

If $g : X \times \Theta \to \mathbb{R}^p$ fulfills $P_\vartheta(g(\cdot, \vartheta)) = 0$ for $\vartheta \in \Theta$, it is natural to determine an estimator for ϑ as a solution of $\sum_1^n g(x_\nu, \vartheta) = 0$, or as an approximation to such a solution.

Our considerations start in the following more general framework: Let Θ be an open subset of $\mathbb{R}^p$. For $\vartheta \in \Theta$ and $n \in \mathbb{N}$ let $P_\vartheta^{(n)}$ be a p–measure on a measurable space (X_n, A_n), and $\vartheta^{(n)} : X_n \to \mathbb{R}^p$ an estimator.

The following propositions are based on a sequence of functions $g_n : X_n \times \Theta \to \mathbb{R}^p$ such that

$$P_\vartheta^{(n)} \circ g_n(\cdot, \vartheta) \quad \Rightarrow \quad N_{(0,G(\vartheta))} \quad \text{[locally uniformly] at } \vartheta_0, \qquad (7.4.1)$$

with nonsingular matrix $G(\vartheta)$.

Theorem 7.4.3 gives the asymptotic distribution of consistent estimator sequences $\vartheta^{(n)}$, $n \in \mathbb{N}$, which are asymptotic solutions of the estimating equations $g_n(x, \vartheta) = 0$.

In the theorems which yield the distribution of certain estimator sequences we shall use the following representation:

$$g_n(x, \vartheta + c_n^{-1}a) = g_n(x, \vartheta) - D(\vartheta)a + r_n(x, \vartheta, a)a, \qquad (7.4.2)$$

with $a \in \mathbb{R}^p$, a nonsingular $p \times p$ matrix $D(\vartheta)$, and with varying conditions on the remainder matrix r_n.

Theorem 7.4.3. *Assume that for every $M > 0$*

$$\sup_{\|a\| \leq M} \|r_n(\cdot, \vartheta, a)a\| \to 0 \quad (P_\vartheta^{(n)}) \; [\text{locally uniformly}] \; \text{at} \; \vartheta_0 \qquad (7.4.4')$$

and, for every $\varepsilon > 0$ and some $\delta_\varepsilon > 0$,

$$P_\vartheta^{(n)}\{x \in X_n : \sup_{M < \|a\| \leq c_n \delta_\varepsilon} \|r_n(\cdot, \vartheta, a)\| > \varepsilon\} \to 0 \qquad (7.4.4'')$$

$$[\text{locally uniformly}] \; \text{at} \; \vartheta_0.$$

Assume that $\vartheta \to D(\vartheta)$ is continuous at ϑ_0, and $D(\vartheta_0)$ is nonsingular.

Let $\vartheta^{(n)}$, $n \in \mathbb{N}$, be an estimator sequence which is consistent for ϑ [locally uniformly] at ϑ_0 and which fulfills

$$g_n(x, \vartheta^{(n)}(x)) \to 0 \quad (P_\vartheta^{(n)}), \quad [\text{locally uniformly}] \; \text{at} \; \vartheta_0. \qquad (7.4.5)$$

Then, with $H(\vartheta) = D(\vartheta)^{-1}$,

$$c_n(\vartheta^{(n)}(x) - \vartheta) - H(\vartheta)g_n(x, \vartheta) \to 0 \quad (P_\vartheta^{(n)}) \qquad (7.4.6)$$

[locally uniformly] at ϑ_0, which implies

$$P_\vartheta^{(n)} \circ c_n(\vartheta^{(n)} - \vartheta) \Rightarrow N_{(0, H(\vartheta)G(\vartheta)H(\vartheta)^\top)} \qquad (7.4.7)$$

[locally uniformly] at ϑ_0.

Addendum. *If $\vartheta^{(n)}$, $n \in \mathbb{N}$, is c_n-consistent, then the assertion holds without condition (7.4.4'').*

$$P_\vartheta^{(n)}\{x \in X_n : \sup_{\|a\| \leq c_n \delta_\varepsilon} \|r_n(\cdot, \vartheta, a)\| > \varepsilon\} \to 0 \; [\text{locally uniformly}] \; \text{at} \; \vartheta_0 \quad (7.4.4''')$$

implies both, (7.4.4') and (7.4.4'').

In irregular cases, say $g_n(x, \vartheta) = n^{-1/2} \sum_1^n g(x_\nu, \vartheta)$ with a function g which is not differentiable for every $\vartheta \in \Theta$, like $g(x, \vartheta) = 1 - 2\,1_{(-\infty, \vartheta)}(x)$, relation (7.4.4''') may fail, whereas (7.4.4') and (7.4.4'') hold true.

Proof. We shall prove (7.4.6). From this and (7.4.1), relation (7.4.7) follows immediately by Slutzky's Lemma 7.7.8.

It is easy to check that all asymptotic relations in the proof hold [locally uniformly] at ϑ_0. Relation (7.4.2), applied with $a = c_n\big(\vartheta^{(n)}(x)-\vartheta\big)$, yields with $R_n(x,\vartheta) := r_n\big(x,\vartheta,c_n(\vartheta^{(n)}(x)-\vartheta)\big)$, the relation

$$D(\vartheta)c_n\big(\vartheta^{(n)}(x)-\vartheta\big) - g_n(x,\vartheta) \qquad (7.4.8)$$
$$= R_n(x,\vartheta)c_n\big(\vartheta^{(n)}(x)-\vartheta\big) - g_n\big(x,\vartheta^{(n)}(x)\big).$$

With $M > 0$ and $\varepsilon > 0$ fixed, let

$$A_n(\vartheta) := \big\{x \in X_n : c_n\|\vartheta^{(n)}(x)-\vartheta\| \le M\big\}$$
$$B_n(\vartheta) := \big\{x \in X_n : M < c_n\|\vartheta^{(n)}(x)-\vartheta\| \le c_n\delta_\varepsilon\big\}$$
$$C_n(\vartheta) := \big\{x \in X_n : \|\vartheta^{(n)}(x)-\vartheta\| > \delta_\varepsilon\big\}.$$

From (7.4.4'),

$$P_\vartheta^{(n)}\big\{x \in A_n(\vartheta) : \|R_n(x,\vartheta)c_n\big(\vartheta^{(n)}(x)-\vartheta\big)\| > \varepsilon\big\} \to 0 \quad \text{for } \varepsilon > 0.$$

Together with (7.4.5) and (7.4.8) this implies that

$$\big(c_n(\vartheta^{(n)}-\vartheta) - H(\vartheta)g_n(\cdot,\vartheta)\big)1_{A_n(\vartheta)} \to 0 \quad (P_\vartheta^{(n)}). \qquad (7.4.9)$$

In case $\vartheta^{(n)}$, $n \in \mathbb{N}$, is c_n–consistent, $M \ (= M_\varepsilon)$ can be chosen such that $P_\vartheta^{(n)}\big(A_n(\vartheta)^c\big) < \varepsilon$ for $n \in \mathbb{N}$, and this implies (7.4.6).

Now we consider the case $x \in B_n(\vartheta)$. With $D_n(x,\vartheta) := D(\vartheta) - R_n(x,\vartheta)$, let

$$\hat{B}_n(\vartheta) := \big\{x \in B_n(\vartheta) : \det D_n(x,\vartheta) \ne 0\big\}.$$

We shall show that $P_\vartheta^{(n)}\big(B_n(\vartheta) - \hat{B}_n(\vartheta)\big) \to 0$. Since $\det D(\vartheta_0) \ne 0$, there is $\varepsilon > 0$ such that any $p \times p$–matrix A is nonsingular if $\|A - D(\vartheta_0)\| < 2\varepsilon$. Since $\vartheta \to D(\vartheta)$ is continuous at ϑ_0, there exists a neighborhood U_ε of ϑ_0 such that

$$\|D(\vartheta) - D(\vartheta_0)\| < \varepsilon \qquad \text{for } \vartheta \in U_\varepsilon.$$

$x \in B_n(\vartheta) - \hat{B}_n(\vartheta)$ implies $\|D_n(x,\vartheta) - D(\vartheta_0)\| > 2\varepsilon$. For $\vartheta \in U_\varepsilon$, this implies $\|D_n(x,\vartheta) - D(\vartheta)\| > \varepsilon$, hence $\|R_n(x,\vartheta)\| > \varepsilon$.

Since, by (7.4.4''),

$$P_\vartheta^{(n)}\big\{x \in B_n(\vartheta) : \|R_n(x,\vartheta)\| > \varepsilon\big\} \to 0,$$

we obtain

$$P_\vartheta^{(n)}\big(B_n(\vartheta) - \hat{B}_n(\vartheta)\big) \to 0. \qquad (7.4.10)$$

For $x \in \hat{B}_n(\vartheta)$, there exists $H_n(x, \vartheta) := D_n(x, \vartheta)^{-1}$, and we obtain from (7.4.8)

$$c_n\big(\vartheta^{(n)}(x) - \vartheta\big) - H_n(x, \vartheta)g_n(x, \vartheta) \qquad (7.4.11)$$
$$= -H_n(x, \vartheta)g_n\big(x, \vartheta^{(n)}(x)\big).$$

Since matrix inversion is continuous, we have

$$H_n(\cdot, \vartheta)1_{\hat{B}_n(\vartheta)} - H(\vartheta)1_{\hat{B}_n(\vartheta)} \to 0 \qquad (P_\vartheta^{(n)}).$$

This implies in particular that $H_n(\cdot, \vartheta)1_{\hat{B}_n(\vartheta)}$ is stochastically bounded so that, by (7.4.5),

$$H_n(\cdot, \vartheta)g_n\big(\cdot, \vartheta^{(n)}(\cdot)\big)1_{\hat{B}_n(\vartheta)} \to 0 \qquad (P_\vartheta^{(n)}).$$

Since $g_n(\cdot, \vartheta)$ is stochastically bounded by (7.4.1), we have by Slutzky's Lemma 7.7.8.

$$H_n(\cdot, \vartheta)g_n(\cdot, \vartheta)1_{\hat{B}_n(\vartheta)} - H(\vartheta)g_n(\cdot, \vartheta)1_{\hat{B}_n(\vartheta)} \to 0 \qquad (P_\vartheta^{(n)}).$$

Hence we obtain from (7.4.10) and (7.4.11) that

$$\big(c_n(\vartheta^{(n)} - \vartheta) - H(\vartheta)g_n(\cdot, \vartheta)\big)1_{B_n(\vartheta)} \to 0 \qquad (P_\vartheta^{(n)}). \qquad (7.4.12)$$

Finally, we consider the case $x \in C_n(\vartheta)$. Since $\vartheta^{(n)}$, $n \in \mathbb{N}$, is consistent, $P_\vartheta^{(n)}\big(C_n(\vartheta)\big) \to 0$. Together with (7.4.9) and (7.4.12) this implies (7.4.6). $\quad\square$

The following proposition requires a less restrictive condition on g_n. It shows how an arbitrary c_n-consistent estimator sequence can be improved to obtain an estimator sequence which has the same limit distribution as consistent solutions to the estimating equations $g_n(x, \vartheta) = 0$.

Proposition 7.4.13. *Assume that the representation (7.4.2) holds with a remainder term fulfilling (7.4.4').*
Let $\vartheta^{(n)}$, $n \in \mathbb{N}$, be an estimator sequence for ϑ which is c_n-consistent [locally uniformly] at ϑ_0.
For $x \in X_n$, let $H_n(x)$ be a $p \times p$-matrix such that

$$H_n \to H(\vartheta) := D(\vartheta)^{-1} \quad (P_\vartheta^{(n)}) \quad [locally\ uniformly]\ at\ \vartheta_0. \qquad (7.4.14)$$

Then the estimator sequence

$$\hat{\vartheta}^{(n)}(x) := \vartheta^{(n)}(x) + c_n^{-1}H_n(x)g_n\big(x, \vartheta^{(n)}(x)\big) \qquad (7.4.15)$$

fulfills (7.4.6) and (7.4.7).

Proof. It is easy to check that all asymptotic relations in the proof hold [locally uniformly] at ϑ_0. We have

$$c_n\big(\hat{\vartheta}^{(n)}(x) - \vartheta\big) = c_n\big(\vartheta^{(n)}(x) - \vartheta\big) \tag{7.4.16}$$
$$+ H_n(x)g_n(x,\vartheta) - H_n(x)D(\vartheta)c_n\big(\vartheta^{(n)}(x) - \vartheta\big) + H_n(x)R_n(x,\vartheta),$$

with $R_n(x,\vartheta) := r_n\big(x,\vartheta,c_n(\vartheta^{(n)}(x) - \vartheta)\big)$.

$g_n(\cdot,\vartheta)$, $n \in \mathbb{N}$, is stochastically bounded because of (7.4.1) (see Corollary 7.7.7). Moreover, $c_n(\vartheta^{(n)} - \vartheta)$ is stochastically bounded by assumption. Therefore,

$$H_n g_n(\cdot,\vartheta) - H(\vartheta)g_n(\cdot,\vartheta) \to 0 \quad (P_\vartheta^{(n)}) \tag{7.4.17}$$

and

$$H_n D(\vartheta)c_n(\vartheta^{(n)} - \vartheta) - c_n(\vartheta^{(n)} - \vartheta) \to 0 \quad (P_\vartheta^{(n)}). \tag{7.4.18}$$

We shall show that

$$H_n R_n(\cdot,\vartheta) \to 0 \quad (P_\vartheta^{(n)}). \tag{7.4.19}$$

Since $c_n(\vartheta^{(n)} - \vartheta)$ is stochastically bounded, for every $\delta > 0$ there exists M_δ such that for some neighborhood U of ϑ_0,

$$\sup_{\vartheta \in U} P_\vartheta^{(n)}\big\{x \in X_n : c_n\|\vartheta^{(n)}(x) - \vartheta\| > M_\delta\big\} < \delta \qquad \text{for } n \in \mathbb{N}.$$

We have

$$\sup_{\vartheta \in U} P_\vartheta^{(n)}\big\{x \in X_n : \|R_n(x,\vartheta)\| > \varepsilon\big\}$$
$$\leq \sup_{\vartheta \in U} P_\vartheta^{(n)}\big\{x \in X_n : c_n\|\vartheta^{(n)}(x) - \vartheta\| > M_\delta\big\}$$
$$+ \sup_{\vartheta \in U} P_\vartheta^{(n)}\big\{x \in X_n : \sup_{\|a\| \leq M_\delta} \|r_n(x,\vartheta,a)a\| > \varepsilon\big\}.$$

By (7.4.4') this implies

$$\varlimsup_{n\to\infty} \sup_{\vartheta \in U} P_\vartheta^{(n)}\big\{x \in X_n : \|R_n(x,\vartheta)\| > \varepsilon\big\} \leq \delta.$$

Since $\delta > 0$ was arbitrary, this relation, together with (7.4.14), implies (7.4.19). The assertion now follows from (7.4.16)–(7.4.19). □

Remark 7.4.20. It is essential that in definition (7.4.15) the same auxiliary estimator $\vartheta^{(n)}$ is used in both places. Follow the proof of Proposition 7.4.13 to see that defining $\hat{\vartheta}^{(n)}(x) := \vartheta_1^{(n)}(x) + c_n^{-1}H^{(n)}(x)g_n\big(x,\vartheta_2^{(n)}(x)\big)$ with two different c_n–consistent estimator sequences $\vartheta_1^{(n)}$, $\vartheta_2^{(n)}$ fails.

Notice that the asymptotic distribution of the auxiliary estimator sequence $\vartheta^{(n)}$, $n \in \mathbb{N}$, is of no influence on the asymptotic distribution of the improved estimator sequence $\hat{\vartheta}^{(n)}$, $n \in \mathbb{N}$. Any influence the accuracy of $\vartheta^{(n)}$ might have on the accuracy of $\hat{\vartheta}^{(n)}$ escapes an asymptotic analysis of first order.

If we apply (7.4.2) with $a = c_n\big(\hat{\vartheta}^{(n)}(x) - \vartheta\big)$, we obtain from (7.4.16) that

$$g_n\big(\cdot, \hat{\vartheta}^{(n)}(\cdot)\big) \to 0 \quad (P_\vartheta^{(n)}). \tag{7.4.21}$$

This shows the origin of the improvement procedure: $\hat{\vartheta}^{(n)}(x)$, defined by (7.4.15), is an approximate solution to the estimating equation

$$g_n(x, \vartheta) = 0.$$

This idea was first suggested by Fisher (1925, Section 5) in connection with the likelihood equation for the location parameter family of Cauchy distributions. It was used by LeCam (1956, p. 139) as a general approach to the construction of asymptotically efficient estimator sequences.

In principle, the improvement procedure can be applied iteratively. However: Since already the first application leads to an estimator sequence $\hat{\vartheta}^{(n)}$, $n \in \mathbb{N}$, fulfilling (7.4.5), the second application of (7.4.15) starting now with $\hat{\vartheta}^{(n)}$ in place of $\vartheta^{(n)}$, leads to an estimator sequence $\hat{\hat{\vartheta}}^{(n)}$ fulfilling $c_n\big(\hat{\hat{\vartheta}}^{(n)} - \hat{\vartheta}^{(n)}\big) \to 0$ $(P_\vartheta^{(n)})$. Hence a second application of the improvement procedure proves ineffectual: It produces an estimator sequence $\hat{\hat{\vartheta}}^{(n)}$ with the same asymptotic distribution as $\hat{\vartheta}^{(n)}$. This does not mean that a second application of the improvement procedure has no effect for finite sample sizes. It just means that an asymptotic analysis of first order (working with approximations by limit distributions only) gives no information about whether $\hat{\hat{\vartheta}}^{(n)}$ is superior (or inferior) to $\hat{\vartheta}^{(n)}$ for finite sample sizes.

Of different nature is the question whether for fixed n the sequence of estimators $\vartheta^{(n)}, \hat{\vartheta}^{(n)}, \hat{\hat{\vartheta}}^{(n)}, \ldots$, obtained by applying the improvement procedure iteratively, converges to the exact solution of $g_n(x, \vartheta) = 0$. This problem is studied by several authors in connection with the system of likelihood equations. (As an example we mention Kale, 1961.) Such studies are motivated by the belief that the ML estimator is superior to estimators obtained by application of the improvement procedure to the likelihood equations. This belief has, however, no rational base. Both estimator sequences have the same asymptotic distribution (provided the ML estimator is consistent at all), and further properties (in particular such ones relevant for moderate sample sizes) are not known.

Remark 7.4.22. The difficulty with conditions (7.4.4) on the remainder term is the uniformity in a. A startling, if purely theoretical, remedy, due to LeCam (1960, Appendix 1) is the use of *discretized estimators*. If $\vartheta^{(n)}$, $n \in \mathbb{N}$, is c_n–consistent, condition (7.4.4'),

$$\sup_{\|a\| \leq M} \|R_n(\cdot, \vartheta, a)\| \to 0 \quad (P_\vartheta^{(n)})$$

(with $R_n(x, \vartheta, a) = r_n(x, \vartheta, a)a$) suffices for proving (7.4.6). The uniformity in a enables the replacement of a by $c_n(\vartheta^{(n)}(x) - \vartheta)$, i.e. the conclusion from $\sup_{\|a\| \leq M} \|R_n(\cdot, \vartheta, a)\| \to 0$ $(P_\vartheta^{(n)})$ to $R_n(\cdot, \vartheta, c_n(\vartheta^{(n)} - \vartheta)) \to 0$ $(P_\vartheta^{(n)})$ if $c_n(\vartheta^{(n)} - \vartheta)$, $n \in \mathbb{N}$, is stochastically bounded. If $\vartheta^{(n)}$ is replaced by the "discretized" estimator $\tilde{\vartheta}^{(n)}$, $n \in \mathbb{N}$, defined, for instance, by

$$\tilde{\vartheta}_i^{(n)}(x) = c_n^{-1}[c_n \vartheta_i^{(n)}(x)], \qquad i = 1, \ldots, p,$$

the condition on the remainder term can be weakened to "$R_n(\cdot, \vartheta, a_n) \to 0$ for any sequence $a_n \in \mathbb{R}$ fulfilling $\sup_{n \in \mathbb{N}} \|a_n\| < \infty$." The underlying reason is that the number of values attained by $c_n(\tilde{\vartheta}^{(n)}(x) - \vartheta)$ in $\{u \in \mathbb{R}^p : \|u\| \leq M\}$ remains bounded as n tends to infinity. The applied statistician may have a problem to decide whether a given estimate results from a discretized estimator or not.

In the i.i.d. case, the same purpose can be achieved by "splitting" (see, e.g., Klaassen, 1987).

In the i.i.d. case condition (7.4.1) holds under suitable regularity conditions (see Central Limit Theorem 7.7.11). The next proposition gives regularity conditions under which the conditions (7.4.2) and (7.4.4) are fulfilled in the i.i.d. case with $c_n = \sqrt{n}$. Most of these conditions are needed already for specifying the limit distribution.

Proposition 7.4.23. *Let $\{P_\vartheta : \vartheta \in \Theta\}$ be a family of p–measures on $(X, \mathcal{A})$. Assume that $\vartheta \to P_\vartheta$ is weakly continuous.*

Let $g : X \times \Theta \to \mathbb{R}^p$ be a function with the following properties.

(i) $x \to g_i(x, \vartheta)$ is measurable for every $\vartheta \in \Theta$, and $P_\vartheta(g_i(\cdot, \vartheta)) = 0$ for $\vartheta \in \Theta$, $i = 1, \ldots, p$.

(ii) There exists a neighborhood V of ϑ_0 such that $\vartheta \to g_i^{(j)}(x, \vartheta)$ is continuous on V and $\sup_{\tau \in V} |g_i^{(j)}(\cdot, \tau)|$ is P_ϑ–integrable, locally uniformly at ϑ_0.

(iii) The $p \times p$–matrix $D(\vartheta)$ with elements

$$D_{ij}(\vartheta) := -P_\vartheta(g_i^{(j)}(\cdot, \vartheta)), \quad i, j = 1, \ldots, p, \tag{7.4.24}$$

is nonsingular at ϑ_0.

Then the matrix $D(\vartheta)$ is continuous at ϑ_0, and conditions (7.4.2) and (7.4.4) are fulfilled for $P_\vartheta^{(n)} = P_\vartheta^n$ with $c_n = \sqrt{n}$, $\hat{g}_n(\underline{x}, \vartheta) := n^{-1/2} \sum_1^n g(x_\nu, \vartheta)$ and the matrix $D(\vartheta)$ defined by (7.4.24).

Proof. a) Since $\sup_{\tau \in V} |g_i^{(j)}(\cdot, \tau)|$ is P_ϑ-integrable, locally uniformly at ϑ_0, by condition (ii), continuity of $D(\vartheta)$ at ϑ_0 follows from Lemmas 6.7.5 and 6.7.10.

b) We have

$$g_i(x, \vartheta + n^{-1/2}a) = g_i(x, \vartheta)$$

$$+ n^{-1/2} \sum_{j=1}^p a_j \int_0^1 g_i^{(j)}(x, \vartheta + n^{-1/2}au)du,$$

hence

$$n^{-1/2} \sum_1^n g_i(x_\nu, \vartheta + n^{-1/2}a) \qquad (7.4.25)$$

$$= n^{-1/2} \sum_1^n g_i(x_\nu, \vartheta) - \sum_{j=1}^p D_{ij}^{(n)}(\underline{x}, \vartheta, a)a_j,$$

with

$$D_{ij}^{(n)}(\underline{x}, \vartheta, a) := -n^{-1} \sum_1^n \int_0^1 g_i^{(j)}(x_\nu, \vartheta + n^{-1/2}au)du.$$

By Corollary 6.7.21 (obtained from a uniform version of the Law of Large Numbers), there exists a neighborhood U of ϑ_0 such that for $\varepsilon > 0$ there exists $\delta_\varepsilon > 0$ with

$$\lim_{n \to \infty} \sup_{\vartheta \in U} P_\vartheta^n \{\underline{x} \in X^n : \sup_{\|a\| \le n^{1/2}\delta_\varepsilon} \|D^{(n)}(\underline{x}, \vartheta, a) - D(\vartheta)\| > \varepsilon\} = 0.$$

Hence (7.4.2) holds with $r_n(\underline{x}, \vartheta, a) = D(\vartheta) - D^{(n)}(\underline{x}, \vartheta, a)$, and r_n fulfills (7.4.4''') which implies (7.4.4') and (7.4.4''). □

7.5 The limit distribution of ML estimator sequences

The results of Sections 6.3, 6.4 and 7.4 are applied to obtain the limit distribution of ML estimator sequences and other asymptotically equivalent estimator sequences. The optimality of this limit distribution will be established in Section 8.4.

Let Θ be an open subset of $\mathbb{R}^p$, and $\{P_\vartheta : \vartheta \in \Theta\}$ a family of p–measures dominated by some σ–finite measure μ. Let $p(\cdot, \vartheta)$ be a μ–density of P_ϑ. We denote $\ell(\cdot, \vartheta) = \log p(\cdot, \vartheta)$.

$\ell^{(i)}(x, \vartheta)$ and $\ell^{(ij)}(x, \vartheta)$ denote the first and second partial derivatives of $\ell(x, \vartheta)$. If convenient we write $\ell^{(\cdot)}(x, \vartheta)$ for the column vector with elements $\ell^{(i)}(x, \vartheta)$, $i = 1, \ldots, p$. Let $L(\vartheta)$ denote the $p \times p$–matrix with elements

$$L_{ij}(\vartheta) := P_\vartheta\big(\ell^{(i)}(\cdot, \vartheta)\ell^{(j)}(\cdot, \vartheta)\big), \qquad i, j = 1, \ldots, p.$$

Throughout the following we assume

$$P_\vartheta\big(\ell^{(i)}(\cdot, \vartheta)\big) = 0 \quad \text{for } i, j = 1, \ldots, p \text{ and } \vartheta \in \Theta, \tag{7.5.1}$$

$$P_\vartheta\big(\ell^{(ij)}(\cdot, \vartheta)\big) + P_\vartheta\big(\ell^{(i)}(\cdot, \vartheta)\ell^{(j)}(\cdot, \vartheta)\big) = 0, \quad \text{for } i, j = 1, \ldots, p \tag{7.5.2}$$

$$\text{and } \vartheta \in \Theta.$$

The functions $\ell^{(i)}(\cdot, \vartheta)$, $i = 1, \ldots, p$, are linearly P_ϑ–independent. (7.5.3)

Relations (7.5.1) and (7.5.2) follow from $\mu\big(p(\cdot, \vartheta)\big) = 1$ for $\vartheta \in \Theta$ if the order between integration with respect to μ and partial differentiation with respect to ϑ can be interchanged.

Condition (7.5.3) implies that the matrix $L(\vartheta)$ is nonsingular: Since

$$u^\top L(\vartheta)u = P_\vartheta\big((u^\top \ell^{(\cdot)}(\cdot, \vartheta))^2\big),$$

the relation $u_0^\top L(\vartheta_0)u_0 = 0$ for some ϑ_0 and some $u_0 \in \mathbb{R}^p$ implies $u_0^\top \ell^{(\cdot)}(x, \vartheta_0) = 0$ for P_{ϑ_0}–a.a. $x \in X$.

Theorem 7.5.5 specifies regularity conditions under which a consistent sequence of solutions to the system of likelihood equations

$$\sum_1^n \ell^{(i)}(x_\nu, \vartheta) = 0 \quad \text{for } i = 1, \ldots, p \tag{7.5.4}$$

is asymptotically normal with covariance matrix $L(\vartheta)^{-1}$.

Recall Theorem 6.4.4 and Lemma 6.3.9 according to which consistent solutions to these estimating equations exist under rather general conditions.

Theorem 7.5.5. *Assume in addition to (7.5.1)–(7.5.3) the following regularity conditions at ϑ_0.*

(i) *There exists a neighborhood V of ϑ_0 such that*

$$\vartheta \to \ell^{(ij)}(x, \vartheta) \text{ is continuous on } V \text{ for } i, j = 1, \ldots, p \text{ for every } x \in X$$

and

$$\sup_{\tau \in V} |\ell^{(ij)}(\cdot, \tau)| \text{ is } P_\vartheta\text{–integrable,} \quad [\textit{locally uniformly}] \text{ at } \vartheta_0.$$

(ii) For $i = 1, \ldots, p$ the family of functions $x \to \ell^{(i)}(x, \vartheta)^2$, $\vartheta \in \Theta$, is P_ϑ–integrable [locally uniformly] at ϑ_0.

Let $\vartheta^{(n)} : X^n \to \mathbb{R}^p$, $n \in \mathbb{N}$, be an estimator sequence which is consistent for ϑ [locally uniformly] at ϑ_0, and which fulfills

$$n^{-1/2} \sum_1^n \ell^{(\cdot)}\big(x_\nu, \vartheta^{(n)}(\underline{x})\big) \to 0 \quad (P_\vartheta^n) \quad [locally \ uniformly] \ at \ \vartheta_0. \quad (7.5.6)$$

Then, with $\Lambda(\vartheta) := L(\vartheta)^{-1}$,

$$n^{1/2}\big(\vartheta^{(n)}(\underline{x}) - \vartheta\big) - \Lambda(\vartheta) n^{-1/2} \sum_1^n \ell^{(\cdot)}(x_\nu, \vartheta) \to 0 \quad (P_\vartheta^{(n)}) \quad (7.5.7)$$

$$[locally \ uniformly] \ at \ \vartheta_0$$

which implies

$$P_\vartheta^n \circ n^{1/2}(\vartheta^{(n)} - \vartheta) \Rightarrow N_{(0, \Lambda(\vartheta))} \quad [locally \ uniformly] \ at \ \vartheta_0. \quad (7.5.8)$$

Proof. According to Proposition 7.4.23, applied with $g_i(\cdot, \vartheta) = \ell^{(i)}(\cdot, \vartheta)$, the matrix $L(\vartheta)$ is continuous at ϑ_0, and the functions $\underline{x} \to n^{-1/2} \sum_1^n \ell^{(\cdot)}(x_\nu, \vartheta)$ fulfill conditions (7.4.2) and (7.4.4). Furthermore, the Central Limit Theorem 7.7.11 implies condition (7.4.1) with $G(\vartheta) = L(\vartheta)$. Hence Theorem 7.4.3 applies with $H(\vartheta) = \Lambda(\vartheta)$, and the asymptotic variance becomes $H(\vartheta)G(\vartheta)H(\vartheta)^\top = \Lambda(\vartheta)$. □

Let $\Theta \subset \mathbb{R}^p$ be open. If $\vartheta \to \sum_1^n \ell(x_\nu, \vartheta)$ attains its supremum in Θ, then $\sum_1^n \ell^{(\cdot)}(x_\nu, \vartheta) = 0$ has a solution in ϑ. If $\vartheta^{(n)}$, $n \in \mathbb{N}$, is a consistent auxiliary estimator sequence, choose among the solutions to $\sum_1^n \ell^{(\cdot)}(x_\nu, \vartheta) = 0$ the one which is closest to $\vartheta^{(n)}(\underline{x})$, say $\hat{\vartheta}^{(n)}(\underline{x})$. The estimator sequence $\hat{\vartheta}^{(n)}$, $n \in \mathbb{N}$, thus obtained fulfills the assumptions of Theorem 7.5.5 on the estimator sequence $\vartheta^{(n)}$. (Recall Lemma 6.4.6, according to which $\hat{\vartheta}^{(n)}$, $n \in \mathbb{N}$, is consistent.)

If the auxiliary estimator sequence is not just consistent, but even $\sqrt{n}$–consistent, the following improvement procedure is a convenient alternative in view of the fact that solving the system of likelihood equations may be troublesome. Recall Proposition 7.4.13 according to which the improvement procedure works under conditions even more general than those given below.

Theorem 7.5.9. Assume that the family $\{P_\vartheta : \vartheta \in \Theta\}$ fulfills the regularity conditions specified in Theorem 7.5.5.

Let $\vartheta^{(n)} : X^n \to \mathbb{R}^p$, $n \in \mathbb{N}$, be an estimator sequence which is $\sqrt{n}$–consistent for ϑ [locally uniformly] at ϑ_0.

Let $\Lambda^{(n)}$, $n \in \mathbb{N}$, be an estimator sequence which is consistent for $\Lambda(\vartheta)$ [locally uniformly] at ϑ_0.

Then the estimator sequence $\hat{\vartheta}^{(n)}$, $n \in \mathbb{N}$, defined by

$$\hat{\vartheta}^{(n)}(\underline{x}) := \vartheta^{(n)}(\underline{x}) + \Lambda^{(n)}(\underline{x})n^{-1}\sum_{1}^{n}\ell^{(\cdot)}\big(x_\nu, \vartheta^{(n)}(\underline{x})\big) \tag{7.5.10}$$

fulfills (7.5.7) and (7.5.8).

Proof. Follows immediately from Proposition 7.4.13 together with Proposition 7.4.23 and Central Limit Theorem 7.7.11.   □

Recall the discussion on p. 245 which shows that asymptotic theory gives no support to the idea that an iterative application of the improvement procedure leads to better estimators.

Even in the context of regular parametric families, there are various possibilities of estimating the matrix $L(\vartheta)$. Such estimators are needed for the computation of asymptotic confidence bounds, and they are needed in the improvement procedure (7.5.10) in order to obtain by inversion an estimator for $\Lambda(\vartheta)$.

If L is a continuous function of ϑ, and $\vartheta^{(n)}$ is consistent, then $L^{(n)}(\underline{x}) := L(\vartheta^{(n)}(\underline{x}))$, $n \in \mathbb{N}$, is a consistent estimator sequence. (If L admits continuous partial derivatives , and $\vartheta^{(n)}$ is $\sqrt{n}$–consistent, this estimator sequence is even $\sqrt{n}$–consistent.) Since $L(\vartheta)$ is defined by integration, it often occurs that it cannot be expressed in closed form (see Examples 9.1.7 and 9.1.10). In this case, a reasonable alternative is to estimate the integral by the sample mean, i.e. to use (7.5.12) or (7.5.13) as estimates for $L_{ij}(\vartheta)$.

Exercise 7.5.11. Assume that $\vartheta^{(n)}$, $n \in \mathbb{N}$, is consistent [locally uniformly] at ϑ_0.

a) Show that $L^{(n)} : X^n \to \mathbb{R}^p \times \mathbb{R}^p$ defined by

$$L_{ij}^{(n)}(\underline{x}) = n^{-1}\sum_{\nu=1}^{n}\ell^{(i)}\big(x_\nu, \vartheta^{(n)}(\underline{x})\big)\ell^{(j)}\big(x_\nu, \vartheta^{(n)}(\underline{x})\big) \tag{7.5.12}$$

is consistent for $L(\vartheta)$ [locally uniformly] at ϑ_0 if there exists a neighborhood V of ϑ_0 such that the following conditions hold for $i = 1, \ldots, p$.
(i) $\vartheta \to \ell^{(i)}(x, \vartheta)$ is continuous on V for $x \in X$, and
(ii) $\sup_{\tau \in V} \ell^{(i)}(\cdot, \tau)^2$ is P_ϑ–integrable [locally uniformly] at ϑ_0.

   Observe that condition (i) is always fulfilled under the assumptions of Theorem 7.5.5.

b) Show that $\hat{L}^{(n)} : X^n \to \mathbb{R}^p \times \mathbb{R}^p$ defined by

$$\hat{L}_{ij}^{(n)}(\underline{x}) = -n^{-1} \sum_{\nu=1}^{n} \ell^{(ij)}\left(x_\nu, \vartheta^{(n)}(\underline{x})\right) \tag{7.5.13}$$

is consistent for $L(\vartheta)$ [locally uniformly] at ϑ_0 if condition 7.5.5(i) holds true.

7.6 Stochastic approximations to estimator sequences

Introduces the concept of asymptotically linear estimator sequences and shows that ML estimators are asymptotically optimal in the class of all asymptotically linear estimator sequences.

For $\Theta \subset \mathbb{R}^p$ let $\kappa : \Theta \to \mathbb{R}^q$ be a functional with partial derivatives $K_{ij} = \kappa_i^{(j)}$, $i = 1, \ldots, q$, $j = 1, \ldots, p$. The usual way to prove asymptotic normality of $c_n(\kappa^{(n)} - \kappa(\vartheta))$ under $P_\vartheta^{(n)}$ is to approximate $c_n(\kappa^{(n)} - \kappa(\vartheta))$ by some asymptotically normal function, say $k_n(\cdot, \vartheta) : X_n \to \mathbb{R}^q$, i.e.

$$c_n\left(\kappa^{(n)}(x) - \kappa(\vartheta)\right) = k_n(x, \vartheta) + r_n(x, \vartheta), \tag{7.6.1}$$

with

$$r_n(\cdot, \vartheta + c_n^{-1}a) \to 0 \quad (P_\vartheta^{(n)}) \text{ for } a \in \mathbb{R}^p \tag{7.6.2}$$

and

$$P_\vartheta^{(n)} \circ k_n(\cdot, \vartheta) \Rightarrow N_{(0, \Sigma(\vartheta))}. \tag{7.6.3}$$

Condition (7.6.1) implies

$$k_n(x, \vartheta + c_n^{-1}a) = k_n(x, \vartheta) - K(\vartheta)a + r_n(x, \vartheta, a) \tag{7.6.4}$$

with $r_n(\cdot, \vartheta, a) \to 0$ $(P_\vartheta^{(n)})$ for $a \in \mathbb{R}^p$.

For solutions of estimating equations $g_n(x, \vartheta) = 0$, a stochastic approximation (7.6.1) holds with $\kappa(\vartheta) \equiv \vartheta$ and $k_n(x, \vartheta) = H(\vartheta)g_n(x, \vartheta)$ (see (7.4.6)). If H is continuous and $g_n(\cdot, \vartheta)$ stochastically bounded, relation (7.6.4) implies (use $D(\vartheta) = H(\vartheta)^{-1}$)

$$g_n(x, \vartheta + c_n^{-1}a) - g_n(x, \vartheta) \to -D(\vartheta)a \quad (P_\vartheta^{(n)}).$$

This is condition (7.4.2), which was needed to prove (7.4.6). It now turns out to be almost necessary for having the solution asymptotically linear.

We now turn to the i.i.d. case and assume that (7.6.1) holds with $c_n = n^{1/2}$ and $k_n(\underline{x}, \vartheta) = n^{-1/2} \sum_1^n k(x_\nu, \vartheta)$, i.e. that

$$n^{1/2} \left(\kappa^{(n)}(\underline{x}) - \kappa(\vartheta) \right) = n^{-1/2} \sum_1^n k(x_\nu, \vartheta) + r_n(\underline{x}, \vartheta), \qquad (7.6.5)$$

with

$$r_n(\cdot, \vartheta + n^{-1/2}a) \to 0 \quad (P_\vartheta^n) \text{ for } a \in \mathbb{R}^p. \qquad (7.6.6)$$

Relation (7.6.3) holds true if

$$P_\vartheta \left(k(\cdot, \vartheta) \right) = 0 \qquad (7.6.7')$$

and

$$\Sigma(\vartheta) = P_\vartheta \left(k(\cdot, \vartheta) k(\cdot, \vartheta)^\top \right) \qquad (7.6.7'')$$

is positive definite.

An estimator sequence $\vartheta^{(n)}$, $n \in \mathbb{N}$, with a representation (7.6.5) is *asymptotically linear* with *influence function* $k(\cdot, \vartheta)$. If $k_i(\cdot, \vartheta)^2$, $i = 1, \dots, q$, is uniformly P_ϑ–integrable relative to $\{P_\vartheta : \vartheta \in \Theta\}$ and $r_n(\cdot, \vartheta) \to 0$ (P_ϑ^n) uniformly on Θ, then the estimator sequence $\kappa^{(n)}$ is *asymptotically linear uniformly on* Θ.

Proposition 7.6.8. *If the estimator sequence* $\kappa^{(n)} : X^n \to \mathbb{R}^q$, $n \in \mathbb{N}$, *is asymptotically linear [uniformly on* Θ], *then*

$$P_\vartheta^n \circ n^{1/2} \left(\kappa^{(n)} - \kappa(\vartheta) \right) \Rightarrow N_{(0, \Sigma(\vartheta))} \quad [\text{uniformly on } \Theta].$$

Proof. By the uniform version of the Central Limit Theorem 7.7.11

$$P_\vartheta^n \circ n^{-1/2} \sum_1^n k(x_\nu, \vartheta) \Rightarrow N_{(0, \Sigma(\vartheta))} \quad [\text{uniformly on } \Theta].$$

Hence the assertion follows from Slutzky's Lemma 7.7.8.   □

Remark 7.6.9. If $\kappa^{(n)}$ is asymptotically linear with influence function k and $h : \mathbb{R}^q \to \mathbb{R}^r$ has continuous partial derivatives $H_{ij} = h_i^{(j)}$, $i = 1, \dots, r$, $j = 1, \dots, q$, then $h \circ \kappa^{(n)}$ is asymptotically linear with influence function $H\left(\kappa(\vartheta) \right) k(\cdot, \vartheta)$. Hence Proposition 7.2.1 is immediate for asymptotically linear estimator sequences.

The regularity of representation (7.6.5), expressed by (7.6.6), is indispensable. If (7.6.5) holds at ϑ_0 with $r_n(\cdot, \vartheta_0) \to 0$ $(P_{\vartheta_0}^n)$, this has no influence on the asymptotic performance of $n^{1/2} \left(\kappa^{(n)} - \kappa(\vartheta) \right)$ for $\vartheta \neq \vartheta_0$. (If $\tilde{\kappa}^{(n)}$ is any other estimator sequence, define an estimator sequence

$$\hat{\kappa}^{(n)}(\underline{x}) = \begin{cases} \kappa^{(n)}(\underline{x}) \\ \tilde{\kappa}^{(n)}(\underline{x}) \end{cases} \quad \text{if} \quad |\kappa^{(n)}(\underline{x}) - \kappa(\vartheta_0)| \begin{matrix} \leq \\ > \end{matrix} n^{-1/4}.$$

Then $\hat{\kappa}^{(n)}$ has at ϑ_0 a representation (7.6.5) with the same influence function $k(\cdot, \vartheta_0)$, whereas $n^{1/2}(\hat{\kappa}^{(n)} - \kappa(\vartheta))$ behaves under P_ϑ^n asymptotically like $n^{1/2}(\tilde{\kappa}^{(n)} - \kappa(\vartheta))$ for $\vartheta \neq \vartheta_0$.)

If $k_i^{(j)}(x, \vartheta_1, \ldots, \vartheta_p) := \frac{\partial}{\partial \vartheta_j} k_i(x, \vartheta_1, \ldots, \vartheta_p)$ is continuous in ϑ, one obtains from (7.6.4) by a Taylor expansion that

$$n^{-1} \sum_1^n \int_0^1 k_i^{(\cdot)}(x_\nu, \vartheta + n^{-1/2}au)a\,du = -\kappa_i^{(\cdot)}(\vartheta)a + o_p(n^0) \quad \text{for } a \in \mathbb{R}^p.$$

Since the left hand side converges to

$$P_\vartheta(k_i^{(\cdot)}(\cdot, \vartheta)a)$$

(provided $\sup_{\tau \in U} |k_i^{(j)}(\cdot, \tau)|$ is P_ϑ-integrable for some neighborhood U of ϑ), this implies

$$P_\vartheta(k_i^{(j)}(\cdot, \vartheta)) = -\kappa_i^{(j)}(\vartheta). \tag{7.6.10}$$

If integration with respect to μ and differentiation with respect to ϑ may be interchanged, we have by (7.6.7′)

$$P_\vartheta(k_i^{(j)}(\cdot, \vartheta)) + P_\vartheta(k_i(\cdot, \vartheta)\ell^{(j)}(\cdot, \vartheta)) = 0.$$

Together with (7.6.10) this implies

$$P_\vartheta(k_i(\cdot, \vartheta)\ell^{(j)}(\cdot, \vartheta)) = \kappa_i^{(j)}(\vartheta) \quad \text{for } i = 1, \ldots, q, \ j = 1, \ldots, p. \tag{7.6.11}$$

This is a condition which influence functions are bound to fulfill, if the representation (7.6.1) holds locally uniformly. Relation (7.6.11) has thorough consequences for the asymptotic distribution of asymptotically linear estimator sequences. Since

$$P_\vartheta(k_i^*(\cdot, \vartheta)\ell^{(j)}(\cdot, \vartheta)) = \kappa_i^{(j)}(\vartheta) \tag{7.6.12}$$

for $k_i^*(x, \vartheta) = \sum_{k,m=1}^p \kappa_i^{(k)}(\vartheta)\Lambda_{km}(\vartheta)\ell^{(m)}(x, \vartheta)$, relation (7.6.11) implies

$$P_\vartheta((k_i(\cdot, \vartheta) - k_i^*(\cdot, \vartheta))\ell^{(j)}(\cdot, \vartheta)) = 0 \quad \text{for } i = 1, \ldots, q, \ j = 1, \ldots, p.$$

For every $i = 1, \ldots, q$, the function $k_i(\cdot, \vartheta) - k_i^*(\cdot, \vartheta)$ is orthogonal to the "tangent space at ϑ", i.e. the subspace of $\mathcal{L}_2(X, \mathcal{A}, P_\vartheta)$ spanned by $\ell^{(1)}(\cdot, \vartheta), \ldots$ $\ldots, \ell^{(p)}(\cdot, \vartheta)$, say $[\ell^{(1)}(\cdot, \vartheta), \ldots, \ell^{(p)}(\cdot, \vartheta)]$. Since $k_i^*(\cdot, \vartheta)$ is an element of this subspace, it is the projection of $k_i(\cdot, \vartheta)$ into $[\ell^{(1)}(\cdot, \vartheta), \ldots, \ell^{(p)}(\cdot, \vartheta)]$. This implies that (see (7.6.7))

$$K(\vartheta)\Lambda(\vartheta)K(\vartheta)^{\mathsf{T}} = P_\vartheta\big(k^*(\cdot,\vartheta)k^*(\cdot,\vartheta)^{\mathsf{T}}\big) \leq_L P_\vartheta\big(k(\cdot,\vartheta)k(\cdot,\vartheta)^{\mathsf{T}}\big) = \Sigma(\vartheta).$$

This can also be seen directly from

$$0 \leq P_\vartheta\Big(\big(\sum_{i=j}^{p} u_i(k_i(\cdot,\vartheta) - k_i^*(\cdot,\vartheta))\big)^2\Big)$$

$$= \sum_{i,j=1}^{p} u_i u_j P_\vartheta\big((k_i(\cdot,\vartheta) - k_i^*(\cdot,\vartheta))k_j(\cdot,\vartheta)\big)$$

$$= \sum_{i,j=1}^{p} u_i u_j \big(\Sigma_{ij}(\vartheta) - \sum_{k,m=1}^{p} \kappa_i^{(k)}(\vartheta)\Lambda_{km}(\vartheta)\kappa_j^{(m)}(\vartheta)\big).$$

Hence the estimator sequence $\kappa^{(n)}$ is asymptotically optimal in the class of all asymptotically linear estimator sequences iff $k_i(\cdot,\vartheta) = k_i^*(\cdot,\vartheta)$, i.e. if the influence function is in the tangent space $[\ell^{(1)}, \ldots, \ell^{(p)}]$. Since ML estimators are asymptotically linear with influence function $\hat{\Lambda}_i(\cdot,\vartheta) = \sum_{j=1}^{p}\Lambda_{ij}(\vartheta)\ell^{(j)}(\cdot,\vartheta)$ (hence asymptotically normal with covariance matrix $\Lambda(\vartheta)$), they are asymptotically optimal in the class of all asymptotically linear estimator sequences. This is a comparatively weak optimum property of ML estimators, since it is optimality within a special class of estimator sequences. In Section 8.4 it will be shown that ML estimator sequences are asymptotically optimal in the class of all "regular" estimator sequences.

Example 7.6.13. For the family $\{N^n_{(\mu,\sigma^2)} : \mu \in \mathbb{R},\ \sigma > 0\}$ the tangent space at (μ,σ) is $[\ell^{(1)}(\cdot,\mu,\sigma), \ell^{(2)}(\cdot,\mu,\sigma)]$ with

$$\ell^{(1)}(x,\mu,\sigma) = \frac{1}{\sigma}\cdot\frac{x-\mu}{\sigma} \quad\text{and}\quad \ell^{(2)}(x,\mu,\sigma) = \frac{1}{\sigma}\big((\frac{x-\mu}{\sigma})^2 - 1\big).$$

$\sqrt{\frac{\pi}{2}}n^{-1}\sum_1^n |x_\nu - \bar{x}_n|$, $n \in \mathbb{N}$, is an estimator sequence for σ which is asymptotically linear with influence function

$$k(x,\mu,\sigma) = \sigma\Big(\sqrt{\frac{\pi}{2}}\frac{|x-\mu|}{\sigma} - 1\Big).$$

The sequence of ML estimators for σ, $(n^{-1}\sum_1^n(x_\nu - \bar{x}_n)^2)^{1/2}$, is asymptotically linear with influence function (see (9.4.6))

$$x \to \frac{\sigma}{2}\big((\frac{x-\mu}{\sigma})^2 - 1\big).$$

It is straightforward to see that $\sqrt{\frac{\pi}{2}}\frac{|x-\mu|}{\sigma} - 1 - \frac{1}{2}((\frac{x-\mu}{\sigma})^2 - 1)$ is orthogonal to the tangent space.

7.7 Appendix: Weak convergence

Collects auxiliary results on uniform weak convergence, with particular emphasis on weak convergence of p–measures on $\mathbb{B}^q$, and some auxiliary results on continuous convergence.

Let $(\Theta, \mathcal{U})$ be a Hausdorff space, in most applications an open subset of $\mathbb{R}^p$. For $\vartheta \in \Theta$ and $n \in \mathbb{N}$ let $Q_\vartheta^{(n)}$ be a p–measure on $(\mathbb{R}^q, \mathbb{B}^q)$. (The change of notation from $P_\vartheta^{(n)}$ to $Q_\vartheta^{(n)}$ is useful, since in the applications to estimation theory, the auxiliary results on weak convergence will be applied to $Q_\vartheta^{(n)} = P_\vartheta^{(n)} \circ c_n(\kappa^{(n)} - \kappa(\vartheta))$.)

Definition 7.7.1. The sequence $Q_\vartheta^{(n)}$, $n \in \mathbb{N}$, *converges weakly to* Q_ϑ (say $Q_\vartheta^{(n)} \Rightarrow Q_\vartheta$) *uniformly on* Θ if

$$\lim_{n \to \infty} \sup_{\vartheta \in \Theta} |Q_\vartheta^{(n)}(h) - Q_\vartheta(h)| = 0 \quad \text{for every } h \in \mathfrak{C},$$

where $\mathfrak{C}$ denotes the class of all bounded and continuous functions $h : \mathbb{R}^q \to \mathbb{R}$.

This concept of *uniform* weak convergence is technically useful only in connection with limit distributions Q_ϑ which depend continuously on ϑ.

Notice that the definition of weak convergence "locally uniformly at ϑ_0" requires the neighborhood of ϑ_0 to be the same for every $h \in \mathfrak{C}$. However: If $K \subset \Theta$ is compact, and if for every $\vartheta_0 \in K$ and every $h \in \mathfrak{C}$ there is a neighborhood $U_{\vartheta_0, h}$ such that

$$\lim_{n \to \infty} \sup_{\vartheta \in U_{\vartheta_0, h}} |Q_\vartheta^{(n)}(h) - Q_\vartheta(h)| = 0,$$

then this implies

$$\lim_{n \to \infty} \sup_{\vartheta \in K} |Q_\vartheta^{(n)}(h) - Q_\vartheta(h)| = 0 \quad \text{for every } h \in \mathfrak{C}.$$

(Since $\{U_{\vartheta_0, h} : \vartheta_0 \in K\}$ covers K, there exists a finite subcover.)

Lemma 7.7.2. *Assume that* $\vartheta \to Q_\vartheta$ *is weakly continuous. If* $h_m \downarrow 0$ *is a sequence of bounded continuous functions such that* $Q_\vartheta(h_1) < \infty$ *for* $\vartheta \in K$, *a compact subset of* Θ, *then*

$$\sup_{\vartheta \in K} Q_\vartheta(h_m) \downarrow 0. \tag{7.7.3}$$

Proof. For $m \in \mathbb{N}$, the function $\vartheta \to Q_\vartheta(h_m)$ is continuous. Since $Q_\vartheta(h_m) \downarrow 0$ for every $\vartheta \in \Theta$ by the monotone convergence theorem, the uniform convergence expressed by (7.7.3) follows from Dini's theorem (see, e.g., Kelley, 1955, p. 239, Problem E). □

Lemma 7.7.4. (i) *Let $K \subset \Theta$ be compact. If*

$$\lim_{n \to \infty} \sup_{\vartheta \in K} |Q_\vartheta^{(n)}(h) - Q_\vartheta(h)| = 0 \qquad (7.7.5)$$

for every bounded and uniformly continuous function h, then (7.7.5) holds for every bounded and continuous function h, i.e. $Q_\vartheta^{(n)} \Rightarrow Q_\vartheta$, uniformly on K.

(ii) $Q_\vartheta^{(n)} \Rightarrow Q_\vartheta$ *iff for all bounded and lower semicontinuous functions h,*

$$\varliminf_{n \to \infty} Q_\vartheta^{(n)}(h) \geq Q_\vartheta(h).$$

Proof. (i) For $h \in \mathfrak{C}$ there exists a sequence of bounded uniformly continuous functions $h_m \downarrow h$ (see Ash, 1972, p. 390, proof of Theorem A6.6). We have

$$\sup_{\vartheta \in K} \left(Q_\vartheta^{(n)}(h) - Q_\vartheta(h) \right) \leq \sup_{\vartheta \in K} |Q_\vartheta^{(n)}(h_m) - Q_\vartheta(h_m)|$$
$$+ \sup_{\vartheta \in K} Q_\vartheta(h_m - h).$$

Since (7.7.5) holds for $h = h_m$, this implies

$$\varlimsup_{n \to \infty} \sup_{\vartheta \in K} \left(Q_\vartheta^{(n)}(h) - Q_\vartheta(h) \right) \leq \sup_{\vartheta \in K} Q_\vartheta(h_m - h)$$

for every $m \in \mathbb{N}$. Since $h_m - h \downarrow 0$, Lemma 7.7.2 implies

$$\varlimsup_{n \to \infty} \sup_{\vartheta \in K} \left(Q_\vartheta^{(n)}(h) - Q_\vartheta(h) \right) \leq 0.$$

Together with the corresponding lower inequality this proves the assertion.
(ii) Ash (1972), p. 196, Theorem 4.5.1. □

Lemma 7.7.6. *Assume that*
(i) $Q_\vartheta^{(n)} \Rightarrow Q_\vartheta$ *uniformly on compact subsets of Θ,*
(ii) $\vartheta \to Q_\vartheta$ *is continuous on Θ with respect to weak convergence.*
Then, for every compact subset K of Θ,

$$\lim_{M \to \infty} \sup_{n \in \mathbb{N}} \sup_{\vartheta \in K} Q_\vartheta^{(n)} \{ u \in \mathbb{R}^q : \|u\| > M \} = 0.$$

Proof. Let $h_m \downarrow 0$ be a sequence of continuous functions such that $h_m(u) = 1$ for $\|u\| \geq m$. By Lemma 7.7.2, for $\varepsilon > 0$ there exists m_ε such that

$$\sup_{\vartheta \in K} Q_\vartheta(h_{m_\varepsilon}) < \varepsilon/2.$$

Since h_{m_ε} is bounded and continuous,

$$\sup_{\vartheta \in K} |Q_\vartheta^{(n)}(h_{m_\varepsilon}) - Q_\vartheta(h_{m_\varepsilon})| < \varepsilon/2 \qquad \text{for } n \geq n_\varepsilon,$$

hence

$$\sup_{\vartheta \in K} Q_\vartheta^{(n)}(h_{m_\varepsilon}) < \varepsilon \qquad \text{for } n \geq n_\varepsilon.$$

This implies

$$\sup_{\vartheta \in K} Q_\vartheta^{(n)} \{ u \in \mathbb{R}^q : \|u\| \geq m_\varepsilon \} < \varepsilon \qquad \text{for } n \geq n_\varepsilon.$$

Hence there exists $M_\varepsilon \geq m_\varepsilon$ such that

$$\sup_{\vartheta \in K} Q_\vartheta^{(n)} \{ u \in \mathbb{R}^q : \|u\| \geq M_\varepsilon \} < \varepsilon \qquad \text{for } n \in \mathbb{N}. \qquad \square$$

Corollary 7.7.7. *Assume that $P_\vartheta^{(n)} \circ f_n(\cdot, \vartheta)$, $n \in \mathbb{N}$, converges to a limit distribution Q_ϑ [uniformly on compact subsets of Θ] such that $\vartheta \to Q_\vartheta$ is continuous on Θ with respect to weak convergence. Then $f_n(\cdot, \vartheta)$, $n \in \mathbb{N}$, is stochastically bounded [uniformly on compact subsets of Θ].*

Proof. Apply Lemma 7.7.6 for $Q_\vartheta^{(n)} = P_\vartheta^{(n)} \circ f_n(\cdot, \vartheta)$. $\qquad \square$

Lemma 7.7.8 (Slutzky). *Assume that $P_\vartheta^{(n)} \circ f_n(\cdot, \vartheta) \Rightarrow Q_\vartheta$ [uniformly on compact subsets of Θ], and $g_n(\cdot, \vartheta) \to 0$ [uniformly on compact subsets of Θ]. Then*

$$P_\vartheta^{(n)} \big(f_n(\cdot, \vartheta) + g_n(\cdot, \vartheta) \big) \Rightarrow Q_\vartheta \qquad \text{[uniformly on compact subsets of } \Theta \text{]}.$$

Proof. Let $K \subset \Theta$ be compact. According to Lemma 7.7.4, it suffices to prove

$$I_n(h) := \sup_{\vartheta \in K} \Big| \int h\big(f_n(x, \vartheta) + g_n(x, \vartheta)\big) P_\vartheta^{(n)}(dx) - Q_\vartheta(h) \Big| \to 0$$

for any uniformly continuous function $h : \mathbb{R}^q \to [0, 1]$. We have

$$I_n(h) \leq \sup_{\vartheta \in K} \int |h\big(f_n(x, \vartheta) + g_n(x, \vartheta)\big) - h\big(f_n(x, \vartheta)\big)| P_\vartheta^{(n)}(dx)$$

$$+ \sup_{\vartheta \in K} \Big| \int h\big(f_n(x, \vartheta)\big) P_\vartheta^{(n)}(dx) - Q_\vartheta(h) \Big|,$$

hence

$$\varlimsup_{n\to\infty} I_n(h) \le \varlimsup_{n\to\infty} \sup_{\vartheta\in K} \int |h\big(f_n(x,\vartheta) + g_n(x,\vartheta)\big) \tag{7.7.9}$$

$$- h\big(f_n(x,\vartheta)\big)|P_\vartheta^{(n)}(dx).$$

Since h is uniformly continuous, for every $\varepsilon > 0$ there exists $\delta_\varepsilon > 0$ such that $|h(u + a) - h(u)| \le \varepsilon$ for $u \in \mathbb{R}^q$ and $\|a\| \le \delta_\varepsilon$. This implies

$$|h\big(f_n(x,\vartheta) + g_n(x,\vartheta)\big) - h\big(f_n(x,\vartheta)\big)| \le \varepsilon + 1_{\{|g_n(\cdot,\vartheta)|>\delta_\varepsilon\}}(x),$$

hence

$$|\int h\big(f_n(x,\vartheta) + g_n(x,\vartheta)\big) - h\big(f_n(x,\vartheta)\big)|P_\vartheta^{(n)}(dx)$$

$$\le \varepsilon + P_\vartheta^{(n)}\{x \in X_n : |g_n(x,\vartheta)| > \delta_\varepsilon\}.$$

Since

$$\lim_{n\to\infty} \sup_{\vartheta\in K} P_\vartheta^{(n)}\{x \in X_n : |g_n(x,\vartheta)| > \delta_\varepsilon\} = 0,$$

the assertion follows from (7.7.9)   □

Theorem 7.7.10. *Assume that*
   (i) Θ *is a locally compact Hausdorff space fulfilling the 1st axiom of countability,*
   (ii) $\vartheta \to Q_\vartheta$ *is continuous on* Θ *(with respect to weak convergence), and* $Q_\vartheta(C^b) = 0$ *for* $\vartheta \in \Theta$ *and* $C \in \mathcal{C}$.
   Then the following assertions are equivalent.
a) $Q_\vartheta^{(n)} \Rightarrow Q_\vartheta$ *locally uniformly on* Θ.
b) $\sup_{C\in\mathcal{C}} |Q_\vartheta^{(n)}(C) - Q_\vartheta(C)| \to 0$ *locally uniformly on* Θ.

Without local uniformity in ϑ, Theorem 7.7.10 is due to R.R. Rao (1962). See Fabian (1970, p. 142, Theorem 4.1) for an alternative proof.

Proof. According to R.R. Rao (1962, p. 665, Theorem 4.2), the relation $Q_n \Rightarrow Q_0$ is equivalent to $\sup_{C\in\mathcal{C}} |Q_n(C)-Q_0(C)| \to 0$ for any sequence of p–measures Q_n, $n \in \mathbb{N}$, if $Q_0(C^b) = 0$ for $C \in \mathcal{C}$.
   Written explicitly, this is the equivalence between
a') $Q_n(h) \to Q_0(h)$ for every $h \in \mathfrak{C}$, and
b') $\sup_{C\in\mathcal{C}} |Q_n(C) - Q_0(C)| \to 0$.
   To simplify our notations, we now write $Q_\vartheta^{(0)}$ for Q_ϑ. For $n \in \{0\} \cup \mathbb{N}$ let

$$f_{n,h}(\vartheta) := Q_\vartheta^{(n)}(h), \quad n \in \{0\} \cup \mathbb{N}, \quad \text{map } \Theta \text{ into } [0,1],$$

and

$$\hat{f}_n(\vartheta) := \left(Q^{(n)}_\vartheta(C)\right)_{C \in \mathcal{C}}, \quad \text{map } \Theta \text{ into } [0,1]^{\mathcal{C}},$$

endowed with the sup–distance.

From the equivalence between a') and b'), applied with Q_n replaced by $Q^{(n)}_{\vartheta_n}$, we obtain that "continuous convergence of $f_{n,h}$ to $f_{0,h}$ for every $h \in \mathfrak{C}$" is equivalent to "continuous convergence of $\hat{f}_n$ to $\hat{f}_0$". By Corollary 7.7.14, applied for the functions $f_{n,h}$ and $\hat{f}_n$, continuous convergence on Θ is equivalent to locally uniform convergence on Θ.   □

Central Limit Theorem 7.7.11. *Let Θ be an arbitrary parameter set. For $\vartheta \in \Theta$ let P_ϑ be a p–measure on $(X, \mathcal{A})$ and $f(\cdot, \vartheta)$ a measurable function from $(X, \mathcal{A})$ to $(\mathbb{R}^q, \mathbb{B}^q)$ such that*

$$P_\vartheta\big(f(\cdot, \vartheta)\big) = 0 \quad \text{for } \vartheta \in \Theta$$

and $\|f(\cdot, \vartheta)\|^2$, $\vartheta \in \Theta$, is uniformly integrable relative to $\{P_\vartheta : \vartheta \in \Theta\}$.
    Let

$$\Sigma(\vartheta) := P_\vartheta\big(f(\cdot, \vartheta) f(\cdot, \vartheta)^\top\big).$$

Then

$$P^n_\vartheta \circ n^{-1/2} \sum_1^n f(x_\nu, \vartheta) \Rightarrow N_{(0, \Sigma(\vartheta))},$$

uniformly on Θ, i.e.

$$\lim_{n \to \infty} \sup_{\vartheta \in \Theta} \sup_{C \in \mathcal{C}} \left| P^n_\vartheta\{\underline{x} \in X^n : n^{-1/2} \sum_1^n f(x_\nu, \vartheta) \in C\} - N_{(0, \Sigma(\vartheta))}(C) \right| = 0.$$

The matrix $\Sigma(\vartheta)$ is nonsingular iff the functions $f_i(\cdot, \vartheta)$, $i = 1, \ldots, q$, are linearly P_ϑ–independent.

Proof. Bhattacharya and Rao (1976), p. 184, Corollary 18.3, applied for the i.i.d. case with $s = 2$, $k_n = n$, $\varepsilon_n = n^{-1/4}$. (Notice a misprint in (18.23): read $\leq$ for $>$ in 184$_3$.)   □

We conclude this section with some lemmas on the relationship between uniform and continuous convergence. For a congenial result under different conditions see Bahadur's Lemma 8.4.19.

In the following, $(X, \mathcal{A})$ is a Hausdorff space, endowed with its Borel algebra, (Y, d) a metric space, with Borel algebra $\mathcal{B}$, and $f_n : (X, \mathcal{A}) \to (Y, \mathcal{B})$, $n \in \{0\} \cup \mathbb{N}$ a sequence of measurable functions.

Definition 7.7.12. (i) f_n, $n \in \mathbb{N}$, converges to f_0 *continuously* at x_0, if $f_n(x_n) \to f_0(x_0)$ for any sequence $x_n \to x_0$.

(ii) f_n, $n \in \mathbb{N}$, converges to f_0 continuously on X if it converges continuously at every $x_0 \in X$.

Lemma 7.7.13. (i) $f_n \to f_0$ *locally uniformly at* x_0 *implies* $f_n \to f_0$ *continuously at* x_0, *if* f_0 *is continuous at* x_0.

(ii) $f_n \to f_0$ *continuously on* X *implies* $f_n \to f_0$ *locally uniformly on* X, *provided* X *is compact and* f_0 *is continuous on* X.

(iii) *If* $f_n \to f_0$ *in a neighborhood of* x_0 *and if* $f_n \to f_0$ *continuously at* x_0, *then* f_0 *is continuous at* x_0, *provided* X *has at* x_0 *a countable local base*.

Corollary 7.7.14. *If* X *is a locally compact Hausdorff space fulfilling the 1st axiom of countability, then the following assertions are equivalent.*

(i) $f_n \to f_0$ *locally uniformly on* X, *and* f_0 *is continuous on* X.

(ii) $f_n \to f_0$ *continuously on* X.

Proof of Lemma 7.7.13. (i) By assumption there exists a neighborhood U_0 of x_0 with the following property: For every $\varepsilon > 0$ there exists n_ε such that $d\big(f_n(x), f_0(x)\big) < \varepsilon/2$ for $x \in U_0$ and $n \geq n_\varepsilon$. Since $x_n \to x_0$ implies $x_n \in U_0$ for $n \geq n_0$ and $d\big(f_0(x_n), f_0(x_0)\big) < \varepsilon/2$ for $n \geq n'_\varepsilon$, we obtain

$$d\big(f_n(x_n), f_0(x_0)\big) < \varepsilon \quad \text{for} \quad n \geq \max\{n_0, n'_\varepsilon, n''_\varepsilon\}.$$

(ii) If the convergence of f_n, $n \in \mathbb{N}$, to f_0 fails to be uniform on X, there exists an infinite sequence $x_n \in X$, $n \in \mathbb{N}_0$, such that

$$\lim_{n \in \mathbb{N}_0} d\big(f_n(x_n), f_0(x_n)\big) > 0.$$

Since X is compact, there exists a subsequence $x_n \in X$, $n \in \mathbb{N}_1 \subset \mathbb{N}_0$ converging to some $x_0 \in X$. Since f_0 is continuous, this implies $\lim_{n \in \mathbb{N}_1} f_0(x_n) = f_0(x_0)$, hence also $\lim_{n \in \mathbb{N}_1} d\big(f_n(x_n), f_0(x_0)\big) > 0$, i.e.: f_n, $n \in \mathbb{N}$, does not converge continuously to f_0 on X.

(iii) First we shall show that for every $\varepsilon > 0$ there exists a neighborhood U_ε of x_0 and n_ε such that

$$d\big(f_n(x), f_0(x_0)\big) < \varepsilon \quad \text{for } x \in U_\varepsilon \text{ and } n \geq n_\varepsilon. \tag{7.7.15}$$

Let U_m, $m \in \mathbb{N}$, be a local base at x_0. Assume that, in contradiction to (7.7.15), there exists $\varepsilon_0 > 0$ such that every U_m contains an element x_m with $d\big(f_n(x_m), f_0(x_0)\big) \geq \varepsilon_0$ for infinitely many $n \in \mathbb{N}$. Then there exists an increasing sequence n_m, $m \in \mathbb{N}$, such that

$$d\big(f_{n_m}(x_m), f_0(x_0)\big) \geq \varepsilon \qquad \text{for } m \in \mathbb{N}.$$

Since $x_m \to x_0$, this contradicts the assumption of continuous convergence.

Since $f_n(x) \to f_0(x)$ for $x \in U$ say, continuity of f_0 at x_0 follows from (7.7.15). $\qquad\qquad \square$

Chapter 8
Asymptotic bounds for the concentration of estimators and confidence bounds

8.1 Introduction

Introduces the concept of LAN (local asymptotic normality) as an asymptotic regularity condition on the family of p–measures. The LAN condition is used in Sections 8.2–8.4 to obtain asymptotic bounds for the concentration of estimators and confidence bounds.

Sections 3.3 and 5.6 contain several instances of estimators which are maximally concentrated about the true parameter value, for every sample size. Optimality results of this kind hold within a certain class of estimators, say mean– or median unbiased ones, and they hold for particular families of p–measures, mainly exponential ones. It is the purpose of the present chapter to obtain comparable *asymptotic* results for *arbitrary* families of p–measures (subject to regularity conditions). Such asymptotic results refer to limit distributions of estimator sequences. In regular cases these are limits of sequences $P_\vartheta^n \circ n^{1/2}\big(\kappa^{(n)} - \kappa(\vartheta)\big)$, $n \in \mathbb{N}$. They describe the distribution of $\kappa^{(n)}$ about $\kappa(\vartheta)$ in a neighborhood of order $n^{-1/2}$. Correspondingly, we need *local* properties of the family $\{P_\vartheta^n : \vartheta \in \Theta\}$ in a neighborhood of $P_{\vartheta_0}^n$, say. Though our main interest is the case of independent, identically distributed observations, we present the results of this chapter in the following more general framework. For $\vartheta \in \Theta \subset \mathbb{R}^p$ and $n \in \mathbb{N}$, let $P_\vartheta^{(n)}$ be a p–measure on some measurable space $(X_n, \mathcal{A}_n)$. Let $p_n(\cdot, \vartheta)$ be a density of $P_\vartheta^{(n)}$ with respect to some σ–finite measure $\mu_n | \mathcal{A}_n$. Within this framework, the local behavior of $P_\vartheta^{(n)}$, $n \in \mathbb{N}$, can be described in a technically useful way by the density of $P_{\vartheta+c_n^{-1}a}^{(n)}$ with respect to $P_\vartheta^{(n)}$ if the p–measures are mutually absolutely continuous. The results of this chapter are based on the so–called

LAN Condition 8.1.1. For every $a \in \mathbb{R}^p$ and all sufficiently large $n \in \mathbb{N}$,

$$\log \frac{p_n(x, \vartheta + c_n^{-1}a)}{p_n(x, \vartheta)} = a^\top \Delta_n(x, \vartheta) - \frac{1}{2} a^\top D(\vartheta)a + R_n(x, \vartheta, a), \qquad (8.1.2)$$

where $D(\vartheta)$ is a nonsingular, symmetric $p \times p$–matrix, $\Delta_n(\cdot, \vartheta) : X_n \to \mathbb{R}^p$ a measurable function such that

$$P_\vartheta^{(n)} \circ \Delta_n(\cdot, \vartheta) \Rightarrow N_{(0, D(\vartheta))}, \qquad (8.1.3)$$

and $R_n(\cdot, \vartheta, a)$ a remainder term which converges to zero. For different purposes, different versions of this convergence to 0 are needed, namely

$$R_n(\cdot, \vartheta, a) \to 0 \quad (P_\vartheta^{(n)}) \text{ for every } a \in \mathbb{R}^p, \qquad (8.1.4')$$

$$R_n(\cdot, \vartheta, a_n) \to 0 \quad (P_\vartheta^{(n)}) \text{ for every bounded sequence } a_n, \ n \in \mathbb{N}, \qquad (8.1.4'')$$

$$\sup_{\|a\| \le M} |R_n(\cdot, \vartheta, a)| \to 0 \quad (P_\vartheta^{(n)}) \text{ for every } M > 0. \qquad (8.1.4''')$$

It was one of the great achievements of LeCam (1960) to see that the LAN condition grasps all of the local structure of $\{P_\vartheta^{(n)} : \vartheta \in \Theta\}$, $n \in \mathbb{N}$, which is needed to obtain asymptotic bounds for the accuracy of statistical procedures.

The LAN condition itself will be used in Section 8.4. In Sections 8.2 and 8.3 we need only a particular consequence, namely that

$$P_\vartheta^{(n)} \circ \log \frac{p_n(\cdot, \vartheta + c_n^{-1}a)}{p_n(\cdot, \vartheta)} \Rightarrow N_{(-\frac{1}{2}a^\top D(\vartheta)a, \ a^\top D(\vartheta)a)}. \qquad (8.1.5)$$

From (8.1.3) and (8.1.2) with a remainder term fulfilling (8.1.4') we obtain that

$$P_\vartheta^{(n)} \circ \frac{p_n(\cdot, \vartheta + c_n^{-1}a)}{p_n(\cdot, \vartheta)}$$

$$\Rightarrow N_{(0, D(\vartheta))} \circ \left(u \to \exp[a^\top u - \frac{1}{2} a^\top D(\vartheta)a] \right).$$

Since

$$\int \exp\left[a^\top u - \frac{1}{2} a^\top D(\vartheta)a\right] N_{(0, D(\vartheta))}(du) = 1,$$

this implies contiguity of $P_{\vartheta + c_n^{-1}a}^{(n)}$, $n \in \mathbb{N}$, with respect to $P_\vartheta^{(n)}$, $n \in \mathbb{N}$, by Corollary 8.1.15 (LeCam's 1st Lemma). Conversely, (8.1.3) follows from (8.1.2) and contiguity.

Proposition 8.1.6. *Assume that* $P_\vartheta^{(n)}$, $n \in \mathbb{N}$, *fulfills condition (8.1.2) with a remainder term (8.1.4'). If* $P_{\vartheta + c_n^{-1}a}^{(n)}$, $n \in \mathbb{N}$, *and* $P_\vartheta^{(n)}$, $n \in \mathbb{N}$, *are mutually contiguous for every* a *in some open subset of* $\mathbb{R}^p$, *then (8.1.3) follows.*

Proof. By Lemma 8.1.13 contiguity implies that $p_n(\cdot, \vartheta + c_n^{-1}a)/p_n(\cdot, \vartheta)$, $n \in \mathbb{N}$, is stochastically bounded and stochastically bounded away from 0 under $P_\vartheta^{(n)}$. Hence $a^\top \Delta_n(\cdot, \vartheta)$, $n \in \mathbb{N}$, is stochastically bounded under $P_\vartheta^{(n)}$, $n \in \mathbb{N}$, and every subsequence $\mathbb{N}_0$ contains a subsequence $\mathbb{N}_1$ such that $P_\vartheta^{(n)} \circ \Delta_n(\cdot, \vartheta)$, $n \in \mathbb{N}_1$, is weakly convergent to some p–measure, say M_ϑ. Using (8.1.2) we obtain that

$$P_\vartheta^{(n)} \circ p_n(\cdot, \vartheta + c_n^{-1}a)/p_n(\cdot, \vartheta) \underset{n \in \mathbb{N}_1}{\Longrightarrow} M_\vartheta \circ (u \to \exp[a^\top u - \frac{1}{2}a^\top D(\vartheta)a]).$$

By Corollary 8.1.15 this implies that

$$\int \exp[a^\top u - \frac{1}{2}a^\top D(\vartheta)a] M_\vartheta(du) = 1,$$

i.e.

$$\int \exp[a^\top u] M_\vartheta(du) = \exp[\frac{1}{2}a^\top D(\vartheta)a] \qquad \text{for } a \in U.$$

Since $a \to \int \exp[a^\top u] M_\vartheta(du)$ is holomorphic, this relation holds with a replaced by ia, whence $M_\vartheta = N_{(0,D(\vartheta))}$. Since $N_{(0,D(\vartheta))}$ is independent of the initial subsequence $\mathbb{N}_0$, (8.1.3) follows. $\qquad\square$

Observe that the formulation of the LAN Condition 8.1.1 is somewhat redundant. If (8.1.2) holds locally uniformly in ϑ with $-\frac{1}{2}a^\top D(\vartheta)a$ replaced by some function $A(\vartheta, a)$, then $A(\vartheta, a) = -\frac{1}{2}a^\top D(\vartheta)a$ follows under suitable regularity conditions from the fact that

$$\frac{p_n(x, \vartheta + c_n^{-1}(a + b))}{p_n(x, \vartheta)} \tag{8.1.7}$$
$$= \frac{p_n(x, (\vartheta + c_n^{-1}a) + c_n^{-1}b)}{p_n(x, \vartheta + c_n^{-1}a)} \cdot \frac{p_n(x, \vartheta + c_n^{-1}a)}{p_n(x, \vartheta)} .$$

(See LeCam and Yang, 1990, p. 78, Proposition 5.)

The following proposition (see Section 7.5 for the notation) gives a convenient sufficient condition for LAN in the i.i.d. case. For more general conditions see LeCam and Yang (1990, p. 103, Proposition 1) and Strassser (1985, p. 412, Example 80.8).

Proposition 8.1.8. *Assume that $\vartheta \to \ell(x, \vartheta)$ admits a 2nd derivative which is continuous in a neighborhood of ϑ_0 and fulfills conditions (7.5.1)–(7.5.3) and 7.5.5(i).*

Then the LAN Condition 8.1.1 holds at ϑ_0 for $P_\vartheta^{(n)} = P_\vartheta^n$ and $c_n = n^{1/2}$ with

$$\Delta_n(\underline{x}, \vartheta) = n^{-1/2} \sum_1^n \ell^{(\cdot)}(x_\nu, \vartheta)$$

$$D(\vartheta) = L(\vartheta) \tag{8.1.10}$$

and a remainder term fulfilling (8.1.4''').

Notice that, whatever the conditions which guarantee (8.1.2) and (8.1.4'), we necessarily have (8.1.10) according to Proposition 8.1.6.

Proof. A Taylor expansion yields

$$\sum_1^n \left(\ell(x_\nu, \vartheta_0 + n^{-1/2}a) - \ell(x_\nu, \vartheta_0) \right)$$

$$= n^{-1/2} a^\top \sum_1^n \ell^{(\cdot)}(x_\nu, \vartheta_0) - \frac{1}{2} a L(\vartheta_0) a^\top + R_n(x, \vartheta_0, a),$$

with $R_n(x, \vartheta, a) = n^{-1} \sum_1^n a^\top r_n(x_\nu, \vartheta, a) a$, where $r_n(x, \vartheta, a)$ is the $p \times p$–matrix with elements

$$r_n^{(ij)}(x, \vartheta, a) := \int_0^1 (1-u) \left(\ell^{(ij)}(x, \vartheta + n^{-1/2}au) - P_\vartheta(\ell^{(ij)}(\cdot, \vartheta)) \right) du.$$

By Lemma 6.7.5 for every $\varepsilon > 0$ there exists a neighborhood U_ε of ϑ_0 such that

$$P_{\vartheta_0} \left(\overline{\ell}^{(ij)}(\cdot, U_\varepsilon) \right) - \varepsilon < P_{\vartheta_0} \left(\ell^{(ij)}(\cdot, \vartheta_0) \right).$$

If n is sufficiently large, we have

$$\ell^{(ij)}(x, \vartheta_0 + n^{-1/2}au) \leq \overline{\ell}^{(ij)}(x, U_\varepsilon),$$

and therefore

$$r_n^{(ij)}(x, \vartheta_0, a) \leq \frac{1}{2} \left(\overline{\ell}^{(ij)}(x, U_\varepsilon) - P_{\vartheta_0}(\ell^{(ij)}(\cdot, \vartheta_0)) \right)$$

for $\|a\| \leq M$. This implies

$$P_{\vartheta_0}^n \left\{ \underline{x} \in X^n : \sup_{\|a\| \leq M} n^{-1} \sum_1^n r_n^{(ij)}(x_\nu, \vartheta_0, a) > \varepsilon \right\}$$

$$\leq P_{\vartheta_0}^n \left\{ \underline{x} \in X^n : n^{-1} \sum_1^n (\overline{\ell}^{(ij)}(x_\nu, U_\varepsilon) - P_{\vartheta_0}(\ell^{(ij)}(\cdot, \vartheta_0))) > 2\varepsilon \right\}$$

$$\leq P_{\vartheta_0}^n\{\underline{x}\in X^n : n^{-1}\sum_1^n(\bar{\ell}^{(ij)}(x_\nu,U_\varepsilon) - P_{\vartheta_0}(\bar{\ell}^{(ij)}(\cdot,U_\varepsilon))) > \varepsilon\} \to 0.$$

Together with the corresponding inequality using $\underline{\ell}^{(ij)}$, this implies the assertion. □

In Section 8.4 it will be shown that under the LAN condition, the optimal limit distribution for regular estimator sequences of ϑ is $N_{(0,D(\vartheta)^{-1})}$. According to Theorem 7.4.3, estimator sequences with this limit distribution can be obtained as (asymptotic) solutions to the estimating equations $\Delta_n(x,\vartheta) = 0$, provided (see (7.4.2))

$$\Delta_n(x,\vartheta + c_n^{-1}a) = \Delta_n(x,\vartheta) - D(\vartheta)a + r_n(x,\vartheta,a)a. \qquad (8.1.11)$$

(Notice that for the particular function $g_n(\cdot,\vartheta) = \Delta_n(\cdot,\vartheta)$, we have $G(\vartheta) = D(\vartheta)$.)

It is easy to see that (8.1.11) follows — almost — from the LAN condition, if the latter holds locally uniformly in ϑ. In this case we obtain from (8.1.7) and (8.1.2) that

$$b^\top\Delta_n(x,\vartheta + c_n^{-1}a) = b^\top\Delta_n(x,\vartheta) - b^\top D(\vartheta)a + \tilde{R}_n(x,\vartheta,a,b)$$

with

$$\tilde{R}_n(x,\vartheta,a,b) = R_n(x,\vartheta,a+b) - R_n(x,\vartheta,a)$$
$$- R_n(x,\vartheta + c_n^{-1}a,b) + \frac{1}{2}b^\top(D(\vartheta + c_n^{-1}a) - D(\vartheta))b.$$

If D is continuous and the remainder term converges to 0 in $P_\vartheta^{(n)}$–measure, locally uniformly in ϑ, then $\tilde{R}_n(\cdot,\vartheta,a,b) \to 0$ $(P_\vartheta^{(n)})$ by contiguity, which implies

$$\Delta_n(x,\vartheta + c_n^{-1}a) = \Delta_n(x,\vartheta) - D(\vartheta)a + \tilde{R}_n(x,\vartheta,a), \qquad (8.1.12)$$

with a remainder term $\tilde{R}_n(\cdot,\vartheta,a)$ converging stochastically to 0.

There remains, however, a slight difference concerning the remainder terms: To prove the convolution theorem, the LAN condition with a remainder term fulfilling (8.1.4') suffices. To establish the asymptotic distribution of solutions to the estimating equations $\Delta_n(x,\vartheta) = 0$, we need the slightly stronger conditions (7.4.4) on the remainder term, which follow if the error term in the version of (8.1.12) fulfills

$$\sup_{\|a\|\leq c_n M} \|\tilde{R}_n(\cdot,\vartheta,a)\| \to 0 \quad (P_\vartheta^{(n)}),$$

for every $M > 0$. The less restrictive condition $\sup_{\|a\|\leq M}\|\tilde{R}_n(\cdot,\vartheta,a)\| \to 0$ $(P_\vartheta^{(n)})$ suffices for the improvement procedure, specified in Proposition 7.4.13.

The reader interested in more theoretical results will realize that the uniform versions of these conditions can be replaced by corresponding conditions on sequences a_n, $n \in \mathbb{N}$, if a sequence of discretized estimators is used in the improvement procedure (see Remark 7.4.22).

Recall that the sequence P'_n, $n \in \mathbb{N}$, is *contiguous* with respect to the sequence P_n, $n \in \mathbb{N}$, if for any sequence $A_n \in \mathcal{A}_n$, $n \in \mathbb{N}$, $P_n(A_n) \to 0$ implies $P'_n(A_n) \to 0$.

Lemma 8.1.13. *Assume that for every $n \in \mathbb{N}$, $P'_n|\mathcal{A}_n$ and $P_n|\mathcal{A}_n$ are mutually absolutely continuous p–measures, and let p'_n be a P_n–density of P'_n. Then the following assertions are equivalent:*
(i) P'_n, $n \in \mathbb{N}$, is contiguous with respect to P_n, $n \in \mathbb{N}$.
(ii) p'_n, $n \in \mathbb{N}$, is stochastically bounded under P'_n, $n \in \mathbb{N}$.

Proof. (i) $P_n\{p'_n > c\} \leq 1/c$ implies that p'_n is stochastically bounded under P_n, $n \in \mathbb{N}$, hence also under P'_n, $n \in \mathbb{N}$, by contiguity.

(ii) $P'_n(A_n) = P'_n(A_n \cap \{p'_n \leq c\}) + P'_n(A_n \cap \{p'_n > c\}) \leq cP_n(A_n) + P'_n\{p'_n > c\}$. If p'_n, $n \in \mathbb{N}$, is stochastically bounded under P'_n, for every $\varepsilon > 0$, there exists $c_\varepsilon > 0$ such that $P'_n\{p'_n > c_\varepsilon\} < \varepsilon$ for $n \in \mathbb{N}$. Hence $P'_n(A_n) \leq c_\varepsilon P_n(A_n) + \varepsilon$, and $\lim_{n \to \infty} P_n(A_n) = 0$ implies $\lim_{n \to \infty} P'_n(A_n) = 0$. □

Lemma 8.1.14. *Let $q \geq 0$ be continuous. Assume that $Q_n \Rightarrow Q$ and $Q_n(q) \to 1$. Then $\int q(u)Q(du) = 1$ iff q is uniformly integrable under Q_n, $n \in \mathbb{N}$, $n \geq n_0$.*

Proof. Let q_0 be a bounded and continuous function such that $0 \leq q_0 \leq q$, say $q_0 = \min\{q, c_0\}$. Since $Q_n \Rightarrow Q$ and $Q_n(q) \to 1$, we have $\lim_{n \to \infty} Q_n(q - q_0) = 1 - Q(q_0)$. As a special consequence, $Q(q) \leq 1$.

Let $\varepsilon > 0$ be arbitrary. (i) If $Q(q) = 1$, q_0 can be chosen such that $Q(q_0) > 1 - \varepsilon$. This implies $\lim_{n \to \infty} Q_n(q - q_0) < \varepsilon$, hence uniform integrability of q under Q_n, $n \in \mathbb{N}$, $n \geq n_0$. (ii) If q is uniformly integrable under Q_n, $n \in \mathbb{N}$, $n \geq n_0$, q_0 can be chosen such that $Q_n(q - q_0) < \varepsilon$ for all sufficiently large $n \in \mathbb{N}$, which implies $1 - Q(q_0) < \varepsilon$. Together with $Q(q) \leq 1$ this implies $Q(q) = 1$. □

For a related result see Lemma 6.7.10. As an immediate consequence of Lemmas 8.1.13 and 8.1.14 we obtain LeCam's 1st lemma.

Corollary 8.1.15. *Assume that $P'_n|\mathcal{A}_n$ and $P_n|\mathcal{A}_n$ are mutually absolutely continuous p–measures, and let p'_n be a P_n–density of P'_n. Assume that $P_n \circ p'_n \Rightarrow Q$. Then P'_n, $n \in \mathbb{N}$, is contiguous with respect to P_n, $n \in \mathbb{N}$ iff $\int uQ(du) = 1$.*

Proof. Since

$$P_n(p'_n 1_{\{p'_n > c\}}) = P'_n\{p'_n > c\} \qquad \text{for } c > 0,$$

contiguity of P'_n with respect to P_n is equivalent to uniform P_n–integrability of p'_n, $n \in \mathbb{N}$, by Lemma 8.1.13. By Lemma 8.1.14, applied with $Q_n = P_n \circ p'_n$ and $q(u) = u$, uniform P_n–integrability of p'_n, $n \in \mathbb{N}$, is equivalent to $\int u Q(du) = 1$. $\square$

Exercise 8.1.16. Assume that P'_n, $n \in \mathbb{N}$, is contiguous with respect to P_n, $n \in \mathbb{N}$. Then $P_n(|f_n|) \to 0$ for some sequence of bounded functions f_n, $n \in \mathbb{N}$, implies $P'_n(|f_n|) \to 0$. (Hint: $P_n(|f_n|) \to 0$ implies $P_n(A_{n,\epsilon}) \to 0$ for $A_{n,\epsilon} := \{|f_n| > \epsilon\}$.)

8.2 Regular sequences of confidence bounds and median unbiased estimators

The asymptotic power function given in this section confirms that the asymptotic confidence bounds obtained in Section 7.3 on the basis of asymptotically efficient estimator sequences are asymptotically optimal. Moreover, it is shown that a confidence interval is asymptotically optimal if the probabilities of over– and underestimation are the same.

Let $\Theta \subset \mathbb{R}^p$ be open and $\kappa : \Theta \to \mathbb{R}$ a real valued functional. Proposition 7.3.7 gives conditions for the existence of a sequence of (upper) confidence bounds, say $\kappa_\beta^{(n)}$, $n \in \mathbb{N}$, with locally uniform asymptotic covering probability β, i.e. (see Definition 7.3.2)

$$P_\vartheta^{(n)}\{x \in X_n : \kappa(\vartheta) \leq \kappa_\beta^{(n)}(x)\}, \qquad n \in \mathbb{N},$$

converges to β, locally uniformly, in the sense that for every $\vartheta_0 \in \Theta$ there exists a neighborhood U of ϑ_0 such that

$$\lim_{n \to \infty} \sup_{\vartheta \in U} |P_\vartheta^{(n)}\{x \in X_n : \kappa(\vartheta) \leq \kappa_\beta^{(n)}(x)\} - \beta| = 0. \qquad (8.2.1)$$

In the present section we shall show that sequences of confidence bounds fulfilling (8.2.1) cannot have an asymptotic power function better than $\Phi(u_{1-\beta} - t/\sigma_*(\vartheta))$, with

$$\sigma_*^2(\vartheta) = \kappa^{(\cdot)}(\vartheta) H(\vartheta) \kappa^{(\cdot)}(\vartheta)^\top \quad \text{and} \quad H(\vartheta) = D(\vartheta)^{-1}. \qquad (8.2.2)$$

In fact, a condition weaker than (8.2.1) suffices for this purpose, namely

$$\varlimsup_{n\to\infty} P^{(n)}_{\vartheta_0+c_n^{-1}a}\{x\in X_n : \kappa(\vartheta_0+c_n^{-1}a) < \kappa_\beta^{(n)}(x)\} \le \beta \qquad (8.2.1')$$

$$\le \varliminf_{n\to\infty} P^{(n)}_{\vartheta_0+c_n^{-1}a}\{x\in X_n : \kappa(\vartheta_0+c_n^{-1}a) \le \kappa_\beta^{(n)}(x)\}$$

for every $a \in \mathbb{R}^p$.

In the absence of a better name, we call such sequences of confidence bounds "regular" (in analogy to (7.1.7)).

Even condition (8.2.1') is somewhat circumspect. Under the conditions of Theorem 8.2.3, we have

$$\lim_{n\to\infty} P^{(n)}_{\vartheta_0+c_n^{-1}a}\{x\in X_n : \kappa_\beta^{(n)}(\underline{x}) = \kappa(\vartheta_0+c_n^{-1}a)\} = 0 \quad \text{for every } a\in\mathbb{R}^p.$$

Hence (8.2.1') holds under these assumptions with $\le$ replaced by $=$, and $\varliminf$, $\varlimsup$ replaced by lim. (See also Proposition 7.1.11 which obtains a stronger result, based on the assumption that $\kappa_\beta^{(n)}$, $n \in \mathbb{N}$, is "regular".)

To see this, let $Q_a^{(n)} := P^{(n)}_{\vartheta_0+c_n^{-1}a} \circ \kappa_\beta^{(n)}$. Condition (8.2.1') now reads

$$\varlimsup_{n\to\infty} Q_a^{(n)}\big(\kappa(\vartheta_0+c_n^{-1}a),\infty\big) \le \beta \le \varliminf_{n\to\infty} Q_a^{(n)}\big[\kappa(\vartheta_0+c_n^{-1}a),\infty\big) \quad \text{for } a\in\mathbb{R}^p.$$

We have to show that $\lim_{n\to\infty} Q_a^{(n)}\{\kappa(\vartheta_0+c_n^{-1}a)\} = 0$ for $a \in \mathbb{R}^p$. Relation (8.2.4) implies $\lim_{n\to\infty} Q_0^{(n)}\{\kappa(\vartheta_0)\} = 0$. Let $\mathbb{N}_0 \subset \mathbb{N}$ be arbitrary. According to Bauer (1990a, p. 237, Korollar 31.3) there exists a subsequence $\mathbb{N}_1 \subset \mathbb{N}_0$ such that $Q_0^{(n)}$, $n \in \mathbb{N}_1$, converges vaguely to some sub–probability measure. Hence for every $\varepsilon > 0$ there exists an open neighborhood $U_\varepsilon \subset \mathbb{R}$ of $\kappa(\vartheta_0)$, such that $\varlimsup_{n\in\mathbb{N}_1} Q_0^{(n)}(U_\varepsilon) < \varepsilon$. Let $a \in \mathbb{R}^p$ be arbitrary. If κ is continuous at ϑ_0, we have $\kappa(\vartheta_0+c_n^{-1}a) \in U_\varepsilon$ for n sufficiently large. This implies $\varlimsup_{n\in\mathbb{N}_1} Q_0^{(n)}\{\kappa(\vartheta_0+c_n^{-1}a)\} < \varepsilon$, hence $\lim_{n\in\mathbb{N}_1} Q_0^{(n)}\{\kappa(\vartheta_0+c_n^{-1}a)\} = 0$. Since $\mathbb{N}_0 \subset \mathbb{N}$ was arbitrary, this implies $\lim_{n\to\infty} Q_0^{(n)}\{\kappa(\vartheta_0+c_n^{-1}a)\} = 0$. Since $Q_a^{(n)}$, $n \in \mathbb{N}$, is contiguous with respect to $Q_0^{(n)}$, $n \in \mathbb{N}$, this implies $\lim_{n\to\infty} Q_a^{(n)}\{\kappa(\vartheta_0+c_n^{-1}a)\} = 0$.

The proof of the following theorem is based on the Neyman–Pearson lemma. This idea goes back to C.R. Rao (1963, pp. 196/7) and Bahadur (1964, pp. 1548/9), who used the Neyman–Pearson lemma to obtain bounds for the asymptotic variance of estimator sequences.

Theorem 8.2.3. *For* $\Theta \subset \mathbb{R}^p$ *let* $P_\vartheta^{(n)}$, $n \in \mathbb{N}$, $\vartheta \in \Theta$, *be a family of sequences of p–measures fulfilling at* ϑ_0 *condition (8.1.5).*

Assume that the functional κ has continuous partial derivatives, say $\kappa^{(\cdot)}(\vartheta)$, in some neighborhood of ϑ_0.

If a sequence of upper confidence bounds fulfills the local uniformity condition (8.2.1') at ϑ_0, then

$$\varlimsup_{n\to\infty} P_{\vartheta_0}^{(n)}\{x \in X_n : \kappa(\vartheta_0) - c_n^{-1}t' < \kappa_\beta^{(n)}(x) < \kappa(\vartheta_0) + c_n^{-1}t''\} \quad (8.2.4)$$
$$\leq N_{(0,1)}(u_\beta - t'/\sigma_*(\vartheta_0), u_\beta + t''/\sigma_*(\vartheta_0)) \qquad \text{for } t', t'' \geq 0.$$

Corollary 8.2.5. *For asymptotically median unbiased estimator sequences $\kappa^{(n)}$, $n \in \mathbb{N}$ (fulfilling (8.2.1') at ϑ_0 with $\beta = 1/2$),*

$$\varlimsup_{n\to\infty} P_{\vartheta_0}^{(n)}\{x \in X_n : \kappa(\vartheta_0) - c_n^{-1}t' < \kappa^{(n)}(x) < \kappa(\vartheta_0) + c_n^{-1}t''\} \quad (8.2.6)$$
$$\leq N_{(0,\sigma_*^2(\vartheta))}(-t', t'') \qquad \text{for } t', t'' \geq 0.$$

Addendum. *For the particular case $\kappa(\vartheta_1, \ldots, \vartheta_p) = \vartheta_i$ and asymptotically median unbiased estimator sequences $\vartheta_i^{(n)}$, this implies*

$$\varlimsup_{n\to\infty} P_{\vartheta_0}^{(n)}\{x \in X_n : \vartheta_{0i} - c_n^{-1}t' < \vartheta_i^{(n)} < \vartheta_{0i} + c_n^{-1}t''\} \quad (8.2.7)$$
$$\leq N_{(0,H_{ii}(\vartheta_0))}(-t', t'') \qquad \text{for } t', t'' \geq 0.$$

Proof of Theorem 8.2.3. Let $\ell_n(x) := \log p_n(x, \vartheta_0)/p_n(x, \vartheta_0 + c_n^{-1}a)$ with ϑ_0 and a fixed. From assumption (8.1.5) and LeCam's 3rd Lemma (see Corollary 6.7.11),

$$P_{\vartheta_0+c_n^{-1}a}^{(n)} \circ \ell_n \Rightarrow N_{(-\frac{1}{2}\sigma^2,\sigma^2)}, \quad \text{with } \sigma^2 = a^\top D(\vartheta_0)a.$$

Let $A_n := \{x \in X_n : \kappa(\vartheta_0 + c_n^{-1}a) < \kappa_\beta^{(n)}(x)\}$. The lower inequality in (8.2.1') implies

$$\varlimsup_{n\to\infty} P_{\vartheta_0+c_n^{-1}a}^{(n)}(A_n) \leq \beta.$$

Hence Lemma 8.2.15, applied with $P_n = P_{\vartheta_0+c_n^{-1}a}^{(n)}$ and $P_n' = P_{\vartheta_0}^{(n)}$, implies

$$\varlimsup_{n\to\infty} P_{\vartheta_0}^{(n)}(A_n) \leq \Phi(u_{1-\beta} + \sigma).$$

Written explicitly, this is

$$\varlimsup_{n\to\infty} P_{\vartheta_0}^{(n)}\{x \in X_n : \kappa(\vartheta_0 + c_n^{-1}a) < \kappa_\beta^{(n)}(x)\} \quad (8.2.8)$$
$$\leq \Phi(u_{1-\beta} + (a^\top D(\vartheta_0)a)^{1/2}).$$

Since $\kappa^{(\cdot)}$ is continuous, we have

$$\kappa(\vartheta_0 + c_n^{-1}a) = \kappa(\vartheta_0) + c_n^{-1} \int_0^1 \kappa^{(\cdot)}(\vartheta_0 + c_n^{-1}au)a\,du,$$

and

$$\lim_{n\to\infty} \int_0^1 \kappa^{(\cdot)}(\vartheta_0 + c_n^{-1}au)\,du = \kappa^{(\cdot)}(\vartheta_0).$$

Hence $\kappa^{(\cdot)}(\vartheta_0)a < -t$ implies $\kappa(\vartheta_0 + c_n^{-1}a) < \kappa(\vartheta_0) - c_n^{-1}t$ for n sufficiently large. Therefore, (8.2.8) implies

$$\overline{\lim_{n\to\infty}} \, P_{\vartheta_0}^{(n)} \big\{ x \in X_n : \kappa(\vartheta_0) - c_n^{-1}t \le \kappa_\beta^{(n)}(x) \big\} \tag{8.2.9}$$
$$\le \Phi\big(u_{1-\beta} + (a^\mathsf{T} D(\vartheta_0)a)^{1/2}\big) \quad \text{if } \kappa^{(\cdot)}(\vartheta_0)a < -t.$$

To have the upper bound in (8.2.9) as small as possible, we determine the infimum of $a^\mathsf{T} D(\vartheta_0)a$, subject to the side condition $\kappa^{(\cdot)}(\vartheta_0)a < -t$. This infimum is $t/\sigma_*(\vartheta_0)$, attained for $a = -tH(\vartheta_0)\kappa^{(\cdot)}(\vartheta_0)^\mathsf{T}/\kappa^{(\cdot)}(\vartheta_0)H(\vartheta_0)\kappa^{(\cdot)}(\vartheta_0)^\mathsf{T}$. Hence

$$\overline{\lim_{n\to\infty}} \, P_{\vartheta_0}^{(n)} \big\{ x \in X_n : \kappa(\vartheta_0) - c_n^{-1}t \le \kappa_\beta^{(n)}(x) \big\} \tag{8.2.10$'$}$$
$$\le \Phi\big(u_{1-\beta} + t/\sigma_*(\vartheta_0)\big) \quad \text{for } t \ge 0.$$

Starting from the upper inequality in (8.2.1$'$) we obtain in the same way (now working with $A_n = \big\{ x \in X_n : \kappa(\vartheta_0 + c_n^{-1}a) > \kappa_\beta^{(n)}(x) \big\}$) the relation

$$\lim_{n\to\infty} P_{\vartheta_0}^{(n)} \big\{ x \in X_n : \kappa(\vartheta_0) + c_n^{-1}t \le \kappa_\beta^{(n)}(x) \big\} \tag{8.2.10$''$}$$
$$\ge \Phi\big(u_{1-\beta} - t/\sigma_*(\vartheta_0)\big) \quad \text{for } t \ge 0.$$

From (8.2.10$'$) and (8.2.10$''$)

$$\overline{\lim_{n\to\infty}} \, P_{\vartheta_0}^{(n)} \big\{ x \in X_n : \kappa(\vartheta_0) - c_n^{-1}t' \le \kappa_\beta^{(n)}(x) < \kappa(\vartheta_0) + c_n^{-1}t'' \big\}$$
$$\le \Phi\big(u_{1-\beta} + t'/\sigma_*(\vartheta_0)\big) - \Phi\big(u_{1-\beta} - t''/\sigma_*(\vartheta_0)\big)$$
$$= N_{(0,1)}\big(u_{1-\beta} - t''/\sigma_*(\vartheta_0),\, u_{1-\beta} + t'/\sigma_*(\vartheta_0)\big)$$
$$= N_{(0,1)}\big(u_\beta - t'/\sigma_*(\vartheta_0),\, u_\beta + t''/\sigma_*(\vartheta_0)\big). \qquad \square$$

Theorem 7.4.3 and Proposition 7.4.13, applied with $g_n(x,\vartheta) = \Delta_n(x,\vartheta)$ give conditions which ensure the existence of estimator sequences $\vartheta^{(n)}$, $n \in \mathbb{N}$, for ϑ with asymptotic distribution $N_{(0,H(\vartheta))}$. According to Proposition 7.2.1 the estimator sequence $\kappa^{(n)} = \kappa \circ \vartheta^{(n)}$ for $\kappa(\vartheta)$ has asymptotic distribution $N_{(0,\sigma_*^2(\vartheta))}$. From such estimator sequences, upper confidence bounds with asymptotic power function $t \to \Phi\big(u_{1-\beta} - t/\sigma_*(\vartheta)\big)$ can be obtained according to Proposition 7.3.7. If (8.1.5) holds true, this is, according to Theorem 8.2.3, the optimal asymptotic power function.

So far, our considerations have been concerned with upper confidence bounds. Corresponding results hold for lower confidence bounds. It suggests itself to construct a *confidence interval* by means of a lower and an upper confidence bound, say $\underline{\kappa}^{(n)}$ and $\overline{\kappa}^{(n)}$. Throughout the following we assume that $\underline{\kappa}^{(n)}(x) \le \overline{\kappa}^{(n)}(x)$ for $x \in X_n$. Under this assumption,

$$P_\vartheta^{(n)}\big\{x \in X_n : \underline{\kappa}^{(n)}(x) \le \kappa(\vartheta) \le \overline{\kappa}^{(n)}(x)\big\}$$
$$= P_\vartheta^{(n)}\big\{x \in X_n : \kappa(\vartheta) \le \overline{\kappa}^{(n)}(x)\big\} - P_\vartheta^{(n)}\big\{x \in X_n : \kappa(\vartheta) < \underline{\kappa}^{(n)}(x)\big\}.$$

If $\overline{\kappa}^{(n)}$ has asymptotic covering probability $\overline{\beta}$ and $\underline{\kappa}^{(n)}$ (as a lower confidence bound) asymptotic covering probability $\underline{\beta}$, i.e. if

$$\lim_{n\to\infty} P_\vartheta^{(n)}\big\{x \in X_n : \kappa(\vartheta) \le \overline{\kappa}^{(n)}(x)\big\} = \overline{\beta}$$

$$\lim_{n\to\infty} P_\vartheta^{(n)}\big\{x \in X_n : \underline{\kappa}^{(n)}(x) \le \kappa(\vartheta)\big\} = \underline{\beta},$$

then

$$\lim_{n\to\infty} P_\vartheta^{(n)}\big\{x \in X_n : \underline{\kappa}^{(n)}(x) \le \kappa(\vartheta) \le \overline{\kappa}^{(n)}(x)\big\} = \overline{\beta} + \underline{\beta} - 1. \quad (8.2.11)$$

If we choose $\underline{\beta}, \overline{\beta}$ such that $\overline{\beta} + \underline{\beta} - 1 = \beta$, we obtain a confidence interval $[\underline{\kappa}^{(n)}, \overline{\kappa}^{(n)}]$ with asymptotic covering probability β. How should $\underline{\beta}, \overline{\beta}$ be chosen (subject to the side condition $\overline{\beta} + \underline{\beta} - 1 = \beta$) to obtain a confidence interval with favorable asymptotic properties?

A natural evaluation of the confidence interval $[\underline{\kappa}^{(n)}, \overline{\kappa}^{(n)}]$ is based on

$$P_\vartheta^{(n)}\big\{x \in X_n : \kappa(\vartheta) + c_n^{-1}t \le \underline{\kappa}^{(n)}(x)\big\}$$

and

$$P_\vartheta^{(n)}\big\{x \in X_n : \overline{\kappa}^{(n)}(x) \le \kappa(\vartheta) - c_n^{-1}t\big\}.$$

Both probabilities should be as small as possible for $t \ge 0$. If we consider the underestimation of $\kappa(\vartheta)$ by an amount $c_n^{-1}t$ as equally important as the overestimation by the same amount, it is justified to add the probabilities of these (disjoint) events and to base the evaluation of the confidence procedure on

$$I_\vartheta^{(n)}(t) := P_\vartheta^{(n)}\big\{x \in X_n : \kappa(\vartheta) + c_n^{-1}t \le \underline{\kappa}^{(n)}(x)\big\} \qquad (8.2.12)$$
$$+ P_\vartheta^{(n)}\big\{x \in X_n : \overline{\kappa}^{(n)}(x) \le \kappa(\vartheta) - c_n^{-1}t\big\}.$$

If $\vartheta \in (0, \infty)$ is a scale parameter, it might be more natural to consider $\vartheta(1+\Delta)$ and $\vartheta/(1+\Delta)$ (with $\Delta > 0$) as being at equal distance from ϑ. In nonasymptotic theory this fits naturally with equivariant estimators. If $\Delta = c_n^{-1}t$, the difference between $\vartheta/(1+\Delta)$ and $\vartheta(1-\Delta)$, being in this case of the order c_n^{-2}, is irrelevant under asymptotic approximations of 1st order.

If both bounds keep their asymptotic covering probability locally uniformly in the sense of (8.2.1'), we have according to (8.2.4)

$$\lim_{n\to\infty} P_\vartheta^{(n)}\{x \in X_n : \kappa(\vartheta) + c_n^{-1}t \le \underline{\kappa}^{(n)}(x)\} \ge \Phi(u_{\underline{\beta}} - t/\sigma_*(\vartheta)),$$

and

$$\lim_{n\to\infty} P_\vartheta^{(n)}\{x \in X_n : \overline{\kappa}^{(n)}(x) \le \kappa(\vartheta) - c_n^{-1}t\} \ge \Phi(u_{\overline{\beta}} - t/\sigma_*(\vartheta)),$$

hence

$$\lim_{n\to\infty} I_\vartheta^{(n)}(t) \ge \Phi(u_{\underline{\beta}} - t/\sigma_*(\vartheta)) + \Phi(u_{\overline{\beta}} - t/\sigma_*(\vartheta)). \tag{8.2.13}$$

This lower asymptotic bound is attained if the sequences of confidence bounds $\underline{\kappa}^{(n)}, \overline{\kappa}^{(n)}$ are based on asymptotically efficient estimator sequences for ϑ (see Section 7.3 for details). For such confidence bounds equality holds in (8.2.13), with $\underline{\lim}$ replaced by lim. To have $\lim_{n\to\infty} I_\vartheta^{(n)}(t)$ as small as possible for every $t > 0$, one has to choose $\underline{\beta}, \overline{\beta}$ such that

$$\Phi(u_{\underline{\beta}} - t/\sigma_*(\vartheta)) + \Phi(u_{\overline{\beta}} - t\sigma_*(\vartheta)) = \min,$$

subject to the side condition $\overline{\beta} + \underline{\beta} - 1 = \beta$. According to Lemma 8.2.14, this is achieved for $\underline{\beta} = \overline{\beta} = (1+\beta)/2$. With this choice we obtain

$$\lim_{n\to\infty} I_\vartheta^{(n)}(t) = 2\Phi(u_{(1+\beta)/2} - t\sigma_*(\vartheta)).$$

Lemma 8.2.14. $\alpha \to \Phi(\Phi^{-1}(\alpha) + t)$ is $\begin{smallmatrix}concave\\convex\end{smallmatrix}$ on $(0,1)$ if $t \begin{smallmatrix}>\\<\end{smallmatrix} 0$, i.e.

$$\Phi(\Phi^{-1}(\alpha') + t) + \Phi(\Phi^{-1}(\alpha'') + t)$$
$$\begin{smallmatrix}\le\\\ge\end{smallmatrix} 2\Phi(\Phi^{-1}(\frac{\alpha' + \alpha''}{2}) + t) \quad for \ t \begin{smallmatrix}>\\<\end{smallmatrix} 0.$$

Proof. $\frac{\partial}{\partial\alpha}\Phi(\Phi^{-1}(\alpha) + t) = \varphi(\Phi^{-1}(\alpha) + t)/\varphi(\Phi^{-1}(\alpha))$. Since

$$u \to \frac{\varphi(u + t)}{\varphi(u)} \quad is \quad \begin{smallmatrix}decreasing\\increasing\end{smallmatrix} \quad for \quad t \begin{smallmatrix}>\\<\end{smallmatrix} 0,$$

this implies for every $\alpha \in (0, 1)$

$$\frac{\partial^2}{\partial \alpha^2} \Phi(\Phi^{-1}(\alpha) + t) \begin{smallmatrix} < \\ > \end{smallmatrix} 0 \quad \text{for} \quad t \begin{smallmatrix} > \\ < \end{smallmatrix} 0. \qquad \square$$

The following Lemma 8.2.15 is a convenient asymptotic version of the Neyman–Pearson lemma for LAN families.

Lemma 8.2.15. *Assume that P_n, P_n' are mutually absolutely continuous p–measures with $p_n' \in dP_n'/dP_n$ such that $P_n \circ \log p_n' \Rightarrow N_{(-\frac{1}{2}\sigma^2, \sigma^2)}$. For $n \in \mathbb{N}$ let $\varphi_n : X_n \to [0, 1]$ be A_n–measurable. Then the following holds true for any subsequence $\mathbb{N}_0 \subset \mathbb{N}$, and any $\alpha \in (0, 1)$.*

$$\overline{\lim_{n \in \mathbb{N}_0}} \, P_n(\varphi_n) \leq \alpha \quad \text{implies} \quad \overline{\lim_{n \in \mathbb{N}_0}} \, P_n'(\varphi_n) \leq \Phi(\Phi^{-1}(\alpha) + \sigma).$$

As an immediate consequence, we obtain the same implication with $\mathbb{N}_0 = \mathbb{N}$ and $\overline{\lim}$ replaced by $\underline{\lim}$.

Proof. With $\alpha_n := P_n(\varphi_n)$, let c_n be defined by

$$P_n\{\log p_n' > c_n\} \leq \alpha_n \leq P_n\{\log p_n' \geq c_n\}. \qquad (8.2.16)$$

Since $\left((P_n \circ \log p_n')(c_n, \infty) - N_{(-\frac{1}{2}\sigma^2, \sigma^2)}(c_n, \infty)\right) \to 0$, we obtain from (8.2.16) that

$$\overline{\lim_{n \in \mathbb{N}_0}} \, N_{(-\frac{1}{2}\sigma^2, \sigma^2)}(c_n, \infty) = \overline{\lim_{n \in \mathbb{N}_0}} \, P_n\{\log p_n' > c_n\}$$

$$\leq \overline{\lim_{n \in \mathbb{N}_0}} \, \alpha_n \leq \alpha = N_{(-\frac{1}{2}\sigma^2, \sigma^2)}\left(-\frac{1}{2}\sigma^2 - \sigma\Phi^{-1}(\alpha), \infty\right),$$

which implies

$$\underline{\lim_{n \in \mathbb{N}_0}} \, c_n \geq -\frac{1}{2}\sigma^2 - \sigma\Phi^{-1}(\alpha). \qquad (8.2.17)$$

By the Neyman–Pearson Lemma 4.3.3(ii'), relation (8.2.16) implies

$$P_n'(\varphi_n) \leq P_n'\{\log p_n' \geq c_n\} \quad \text{for } n \in \mathbb{N}.$$

Since $P_n' \circ \log p_n' \Rightarrow N_{(\frac{1}{2}\sigma^2, \sigma^2)}$ by Corollary 6.7.11 we obtain from (8.2.17)

$$\overline{\lim_{n \in \mathbb{N}_0}} \, P_n'(\varphi_n) \leq \overline{\lim_{n \in \mathbb{N}_0}} \, P_n'\{\log p_n' \geq c_n\}$$

$$= \overline{\lim_{n \in \mathbb{N}_0}} \, N_{(\frac{1}{2}\sigma^2, \sigma^2)}[c_n, \infty) \leq \Phi(\Phi^{-1}(\alpha) + \sigma). \qquad \square$$

8.3 Sequences of confidence bounds and median unbiased estimators with limit distributions

The results of Section 8.2 are based on the assumption that the asymptotic confidence bounds keep the covering probability locally uniformly. If the distributions of the (standardized) sequences of confidence bounds converge to a limit distribution, then this limit distribution cannot be more concentrated in the spread order than the normal distribution with minimal asymptotic variance (8.2.2). This yields, in particular, a bound for multidimensional normal limit distributions.

Theorem 8.3.1. *For $\Theta \subset \mathbb{R}^p$ let $P_{\vartheta}^{(n)}$, $n \in \mathbb{N}$, $\vartheta \in \Theta$, be a family of sequences of p–measures fulfilling at ϑ_0 condition (8.1.5). Let $\kappa : \Theta \to \mathbb{R}$ be a real functional with continuous derivatives in a neighborhood of ϑ_0, and $\kappa^{(n)} : X_n \to \mathbb{R}$, $n \in \mathbb{N}$, a sequence such that, for every $a \in \mathbb{R}^p$,*

$$P_{\vartheta_0+c_n^{-1}a}^{(n)} \circ \left(\kappa^{(n)} - \kappa(\vartheta_0 + c_n^{-1}a)\right) \Rightarrow M_{\vartheta_0}, \tag{8.3.2}$$

a p–measure on $\mathbb{B}$.

Then M_{ϑ_0} is more spread out (see Definition 2.3.3) than the normal distribution with variance $\sigma_*^2(\vartheta_0) := \kappa^{(\cdot)}(\vartheta_0)H(\vartheta_0)\kappa^{(\cdot)}(\vartheta_0)^{\top}$, where $H(\vartheta) = D(\vartheta)^{-1}$.

Relation (8.3.2) implies in particular that $\kappa^{(n)}$, $n \in \mathbb{N}$, fulfills (8.2.1'), i.e. it may be viewed as a sequence of "regular" upper confidence bounds with covering probability $M_{\vartheta_0}[0, \infty)$. For the purpose of illustration consider a sequence of asymptotically median unbiased estimators, i.e. the case $M_{\vartheta_0}[0, \infty) = 1/2$. Theorem 8.2.3 implies that

$$M_{\vartheta_0}(-t', t'') \leq N_{(0,\sigma_*^2(\vartheta_0))}(-t', t'') \qquad \text{for } t', t'' \geq 0.$$

The assertion of Theorem 8.3.1 is stronger: Not only is M_{ϑ_0} less concentrated about 0 than $N_{(0,\sigma_*^2(\vartheta_0))}$, but M_{ϑ_0} and $N_{(0,\sigma_*^2(\vartheta_0))}$ are even comparable in the spread order, i.e. M_{ϑ_0} is "everywhere" less concentrated than $N_{(0,\sigma_*^2(\vartheta_0))}$ (see Definition 2.3.3 and Proposition 2.3.10 for a precise formulation). Notice that this asymptotic property, expressed through the limit distribution, is stronger than what we can get — in favorable cases — in nonasymptotic theory: Even if there exists a maximally concentrated median unbiased estimator, this estimator is not necessarily comparable in the spread order with all other median unbiased estimators (see Example 2.7.7).

Proof of Theorem 8.3.1. Let $v \in \mathbb{R}$ be fixed. By (8.3.2),

$$\lim_{n\to\infty} P^{(n)}_{\vartheta_0 + c_n^{-1}a}\big\{x \in X_n : c_n\big(\kappa^{(n)}(x) - \kappa(\vartheta_0 + c_n^{-1}a)\big) > v\big\}$$
$$\geq M_{\vartheta_0}(v, \infty) \qquad \text{for } a \in \mathbb{R}^p.$$

By Theorem 8.2.3, applied with $\kappa^{(n)} - c_n^{-1}v$ in place of $\kappa^{(n)}$, and $\beta = M_{\vartheta_0}(v, \infty)$, $t' = -\infty$ and $t'' = t$ we obtain

$$\overline{\lim_{n\to\infty}} P^{(n)}_{\vartheta_0}\big\{x \in X_n : \kappa^{(n)}(x) - c_n^{-1}v < \kappa(\vartheta) + c_n^{-1}t\big\}$$
$$\leq N_{(0,1)}\big(-\infty, u_\beta + t/\sigma_*(\vartheta_0)\big),$$

hence, with Ψ denoting the distribution function of M_{ϑ_0},

$$\Psi(v + t) \leq \Phi\big(\Phi^{-1}(\Psi(v)) + t/\sigma_*(\vartheta_0)\big),$$

or

$$\Phi^{-1}\big(\Psi(v + t)\big) \leq \Phi^{-1}\big(\Psi(v)\big) + t/\sigma_*(\vartheta_0). \tag{8.3.3}$$

Relation (8.3.3) holds for arbitrary $v \in \mathbb{R}$ and $t \geq 0$.

For arbitrary $0 < \alpha < \beta < 1$, there exist $v \in \mathbb{R}$ and $t > 0$ such that $\Psi(v) = \alpha$ and $\Psi(v + t) = \beta$ (since M_{ϑ_0} is nonatomic by Proposition 7.1.11). Hence

$$\Phi^{-1}(\beta) - \Phi^{-1}(\alpha) \leq \frac{t}{\sigma_*(\vartheta_0)} = \frac{1}{\sigma_*(\vartheta_0)}\big[\Psi^{-1}(\beta) - \Psi^{-1}(\alpha)\big],$$

or

$$\sigma_*(\vartheta_0)\Phi^{-1}(\beta) - \sigma_*(\vartheta_0)\Phi^{-1}(\alpha) \leq \Psi^{-1}(\beta) - \Psi^{-1}(\alpha).$$

Since $\sigma_*(\vartheta)\Phi^{-1}$ is the inverse of the distribution function of $N_{(0,\sigma_*^2(\vartheta_0))}$, this implies the assertion. $\qquad\qquad\square$

Theorem 8.3.1 is based on the "technical" uniformity condition (8.3.2). If

$$P^{(n)}_\vartheta \circ c_n\big(\kappa^{(n)} - \kappa(\vartheta)\big) \Rightarrow M_\vartheta \quad \text{for every } \vartheta \in \Theta,$$

Bahadur's lemma (see Corollary 8.4.24) can be applied as in the proof of Theorem 8.4.14 to show that the conclusion of Theorem 8.3.1 holds for λ^p–a.a. $\vartheta \in \Theta$. It holds for all $\vartheta \in \Theta$ if M_ϑ and $\kappa^{(\cdot)}(\vartheta)H(\vartheta)\kappa^{(\cdot)}(\vartheta)^\top$ depend on ϑ continuously (see Section 8.4 for some more details). Without local uniformity, the assertion of Theorem 8.3.1 is not necessarily true (see Example 8.6.2).

Theorem 8.3.1 follows, in fact, from the Convolution Theorem 8.4.1, applied for 1–dimensional functionals. The proof given here is elementary and may, therefore, be of didactical interest. Moreover, it requires condition (8.1.5) only, and not the "full" LAN Condition 8.1.1.

Theorem 8.3.1 yields also a first result on the asymptotic concentration of multidimensional estimator sequences. This result refers to estimator sequences which are asymptotically normal and is, therefore, much weaker than the Convolution Theorem 8.4.1. It would be worthwhile to derive the (multidimensional) assertion (8.4.4) for *arbitrary* limit distributions from the 1–dimensional version given in Theorem 8.3.1, but that seems to be difficult.

Proposition 8.3.4. *For* $\Theta \subset \mathbb{R}^p$ *let* $P_\vartheta^{(n)}$, $n \in \mathbb{N}$, $\vartheta \in \Theta$, *be a family of sequences of p-measures fulfilling at* ϑ_0 *condition (8.1.5). Let* $\kappa : \Theta \to \mathbb{R}^q$ *with* $q \leq p$ *be a functional with the components* $\kappa = (\kappa_1, \ldots, \kappa_q)^\top$. *Assume that* κ_i *has continuous partial derivatives in a neighborhood of* ϑ_0. *Let* K *denote the* $q \times p$*-matrix with elements* $K_{ij} = \kappa_i^{(j)}$.
  If $\kappa^{(n)} : X_n \to \mathbb{R}^q$ *fulfills*

$$P_{\vartheta_0 + c_n^{-1} a}^{(n)} \circ c_n \big(\kappa^{(n)} - \kappa(\vartheta_0 + c_n^{-1} a) \big) \Rightarrow N_{(0, \Sigma(\vartheta_0))} \qquad \text{for } a \in \mathbb{R}^p,$$

then

$$\Sigma(\vartheta_0) \geq_L K(\vartheta_0) H(\vartheta_0) K(\vartheta_0)^\top. \tag{8.3.5}$$

Proof. With $u \in \mathbb{R}^q$ let $\kappa_0(\vartheta) := u^\top \kappa(\vartheta)$ and $\kappa_0^{(n)}(x) = u^\top \kappa^{(n)}(x)$. We have

$$P_{\vartheta_0 + c_n^{-1} a}^{(n)} \circ c_n \big(\kappa_0^{(n)} - \kappa_0(\vartheta_0 + c_n^{-1} a) \big) \Rightarrow N_{(0, u^\top \Sigma(\vartheta) u)}.$$

Since $\kappa_0^{(\cdot)}(\vartheta) = u^\top K(\vartheta)$, the minimal asymptotic variance pertaining to the functional κ_0 is (by Theorem 8.3.1)

$$\sigma_*^2(\vartheta_0) = u^\top K(\vartheta_0) H(\vartheta_0) K(\vartheta_0)^\top u.$$

Hence

$$u^\top \Sigma(\vartheta_0) u \geq u^\top K(\vartheta_0) H(\vartheta_0) K(\vartheta_0)^\top u.$$

Since $u \in \mathbb{R}^q$ was arbitrary, this implies (8.3.5).   □

8.4 The convolution theorem

According to Theorem 8.3.1 the limit distribution of a 1–dimensional estimator sequence cannot be more concentrated than $N_{(0, \kappa^{(\cdot)}(\vartheta) H(\vartheta) \kappa^{(\cdot)}(\vartheta)^\top)}$. The convolution theorem generalizes this result to arbitrary finite dimensional limit distributions.

Convolution Theorem 8.4.1. (i) *Let* $P_\vartheta^{(n)}$, $n \in \mathbb{N}$, $\vartheta \in \Theta \subset \mathbb{R}^p$, *be a family of sequences of p-measures fulfilling the LAN Condition 8.1.1 at* ϑ_0, *with a remainder term fulfilling (8.1.4′).*

(ii) *Let* $\kappa : \Theta \to \mathbb{R}^q$, $q \leq p$, *be a functional which has at* ϑ_0 *continuous partial derivatives. Let* $K(\vartheta)$ *denote the* $q \times p$-*matrix with elements* $K_{ij}(\vartheta) = \kappa_i^{(j)}(\vartheta)$, $i = 1, \ldots, q$, $j = 1, \ldots, p$. *Assume that* $K(\vartheta)$ *has rank* q.

(iii) *Let* $\kappa^{(n)} : X_n \to \mathbb{R}^q$, $n \in \mathbb{N}$, *be a sequence of estimators which are "regular" at* ϑ_0 *in the following sense: There exists a* p-*measure* $M_{\vartheta_0} | \mathbb{B}^q$ *such that, for every* $a \in \mathbb{R}^p$,

$$P^{(n)}_{\vartheta_0 + c_n^{-1} a} \circ c_n\big(\kappa^{(n)} - \kappa(\vartheta_0 + c_n^{-1} a)\big) \Rightarrow M_{\vartheta_0}. \tag{8.4.2}$$

Under these assumptions there exists a p-*measure* $R_{\vartheta_0} | \mathbb{B}^q$ *such that*

$$P^{(n)}_{\vartheta_0} \circ \big(\Delta_n(\cdot, \vartheta_0),\ c_n(\kappa^{(n)} - \kappa(\vartheta_0)) - K(\vartheta_0) H(\vartheta_0) \Delta_n(\cdot, \vartheta_0)\big) \tag{8.4.3}$$
$$\Rightarrow N_{(0, D(\vartheta_0))} \times R_{\vartheta_0}.$$

This implies

$$M_{\vartheta_0} = N_{(0, \Sigma_*(\vartheta_0))} * R_{\vartheta_0}, \tag{8.4.4}$$

with $\Sigma_*(\vartheta_0) = K(\vartheta_0) H(\vartheta_0) K(\vartheta_0)^{\top}$.

An essential point in (8.4.3) is that the limit distribution is a product measure. This justifies saying that the sequences $\Delta_n(\cdot, \vartheta_0)$ and $c_n\big(\kappa^{(n)} - \kappa(\vartheta_0)\big) - K(\vartheta_0) H(\vartheta_0) \Delta_n(\cdot, \vartheta_0)$, $n \in \mathbb{N}$, are asymptotically independent under $P^{(n)}_{\vartheta_0}$, $n \in \mathbb{N}$.

By (8.4.4), the limit distribution M_{ϑ_0} has a positive λ^q-density. Recall Proposition 7.1.11 according to which this holds under slightly more general conditions.

The statistical significance of Theorem 8.4.1 will be discussed in Section 8.5.

Proof. By assumptions (i) and (iii),

$$P^{(n)}_{\vartheta_0} \circ \Delta_n(\cdot, \vartheta_0) \Rightarrow N_{(0, D(\vartheta_0))} \tag{8.4.5'}$$

and

$$P^{(n)}_{\vartheta_0} \circ c_n\big(\kappa^{(n)} - \kappa(\vartheta_0)\big) \Rightarrow M_{\vartheta_0}. \tag{8.4.5''}$$

Since either of the two sequences of marginal distributions of

$$P^{(n)}_{\vartheta_0} \circ \big(\Delta_n(\cdot, \vartheta_0),\ c_n(\kappa^{(n)} - \kappa(\vartheta_0))$$

converges to a p-measure, there exists by Lemma 8.4.18 for any subsequence $\mathbb{N}_0 \subset \mathbb{N}$ a subsequence $\mathbb{N}_1 \subset \mathbb{N}_0$ and a p-measure $Q_0 | \mathbb{B}^p \times \mathbb{B}^q$ such that

$$P^{(n)}_{\vartheta_0} \circ \big(\Delta_n(\cdot, \vartheta_0),\ c_n(\kappa^{(n)} - \kappa(\vartheta_0))\big) \underset{n \in \mathbb{N}_1}{\Longrightarrow} Q_0. \tag{8.4.6}$$

Let $Q_a | \mathbb{B}^p \times \mathbb{B}^q$ be defined by its Q_0–density

$$(u, v) \rightarrow \exp\left[a^\top u - \frac{1}{2} a^\top D(\vartheta_0)a\right]. \tag{8.4.7}$$

Q_a is a p–measure by (8.4.5$'$).

From Lemma 8.4.26, applied with $P_n = P_{\vartheta_0}^{(n)}$, $P_n' = P_{\vartheta_0 + c_n^{-1}a}^{(n)}$, $f_n = (\Delta_n(\cdot, \vartheta_0)$, $c_n(\kappa^{(n)} - \kappa(\vartheta_0)))$, $q(u, v) = a^\top u - \frac{1}{2} a^\top D(\vartheta_0)a$, $(u, v) \in \mathbb{R}^p \times \mathbb{R}^q$, and $M = Q_0$, $M' = Q_a$, we obtain that

$$P_{\vartheta_0 + c_n^{-1}a}^{(n)}\left(\Delta_n(\cdot, \vartheta), \ c_n(\kappa^{(n)} - \kappa(\vartheta_0))\right) \underset{n \in \mathbb{N}_1}{\Longrightarrow} Q_a. \tag{8.4.8}$$

From (8.4.8),

$$P_{\vartheta_0 + c_n^{-1}a}^{(n)} \circ \left(c_n(\kappa^{(n)} - \kappa(\vartheta_0)) - K(\vartheta_0)a\right) \underset{n \in \mathbb{N}_1}{\Longrightarrow} Q_a \circ \left((u, v) \rightarrow v - K(\vartheta_0)a\right).$$

Since $c_n\left(\kappa(\vartheta_0 + c_n^{-1}a) - \kappa(\vartheta_0)\right) \rightarrow K(\vartheta_0)a$ by assumption (ii), this implies

$$P_{\vartheta_0 + c_n^{-1}a}^{(n)} \circ \left(c_n(\kappa^{(n)} - \kappa(\vartheta_0 + c_n^{-1}a))\right) \underset{n \in \mathbb{N}_1}{\Longrightarrow} Q_a \circ \left((u, v) \rightarrow v - K(\vartheta_0)a\right).$$

Together with (8.4.2) this implies

$$Q_a \circ \left((u, v) \rightarrow v - K(\vartheta_0)a\right) = M_{\vartheta_0}. \tag{8.4.9}$$

Now we derive properties of the family of limit distributions $\{Q_a : a \in \mathbb{R}^p\}$. From (8.4.9) we obtain

$$\int \exp\left[it^\top (v - K(\vartheta_0)a) Q_a(d(u, v))\right] = \Psi(t),$$

with $\Psi(t) := \int \exp[it^\top v] M_{\vartheta_0}(dv)$. Together with (8.4.7) this implies

$$\int \exp\left[it^\top (v - K(\vartheta_0)a) + a^\top u - \frac{1}{2} a^\top D(\vartheta_0)a\right] Q_0(d(u, v)) = \Psi(t).$$

Since the left hand side of this equation is a holomorphic function of a, it holds for all $a \in \mathbb{C}^p$, hence in particular for $a = i(s - H(\vartheta_0)K(\vartheta_0)^\top t)$. This leads to

$$\int \exp\left[i(s^\top u + t^\top (v - K(\vartheta_0)H(\vartheta_0)u))\right] Q_0(d(u, v)) \tag{8.4.10}$$

$$= \exp\left[-\frac{1}{2} s^\top D(\vartheta_0)s\right] \chi(t),$$

with

$$\chi(t) = \Psi(t) \exp\left[\frac{1}{2} t^\top K(\vartheta_0)H(\vartheta_0)K(\vartheta_0)^\top t\right]. \tag{8.4.11}$$

From (8.4.10), applied with $s = 0$, we obtain that χ is the characteristic function of $R_{\vartheta_0} := Q_0 \circ \left((u, v) \rightarrow v - K(\vartheta_0)H(\vartheta_0)u\right)$. Since $s \rightarrow \exp[-\frac{1}{2} s^\top D(\vartheta_0)s]$ is the characteristic function of $N_{(0,D(\vartheta_0))}$, (8.4.10) implies

$$Q_0 \circ ((u,v) \rightarrow (u, v - K(\vartheta_0)H(\vartheta_0)u)) = N_{(0\ D(\vartheta_0))} \times R_{\vartheta_0}.$$

Together with (8.4.6), this implies

$$P_{\vartheta_0}^{(n)} \circ (\Delta_n(\cdot, \vartheta_0),\ c_n(\kappa^{(n)} - \kappa(\vartheta_0)) - K(\vartheta_0)H(\vartheta_0)\Delta_n(\cdot, \vartheta_0)) \quad (8.4.12)$$

$$\underset{n \in \mathbb{N}_1}{\Longrightarrow} \quad N_{(0,D(\vartheta_0))} \times R_{\vartheta_0}.$$

According to (8.4.11), R_{ϑ_0} depends in no way on the subsequences $\mathbb{N}_1 \subset \mathbb{N}_0$. Hence the convergence in (8.4.12) holds with $\mathbb{N}_1$ replaced by $\mathbb{N}$. □

The asymptotic bound given in the convolution theorem refers to estimator sequences $\kappa^{(n)}$, $n \in \mathbb{N}$, which are "regular" in the sense of condition (8.4.2). According to Proposition 7.1.8, this condition is always fulfilled if the convergence of $P_\vartheta^{(n)} \circ c_n(\kappa^{(n)} - \kappa(\vartheta))$, $n \in \mathbb{N}$, to M_ϑ is locally uniform in ϑ, and if the limit distribution M_ϑ is continuous in ϑ with respect to weak convergence. Such properties are necessary for $\kappa^{(n)}$ to be suitable for the computation of asymptotic confidence bounds (see Examples 7.3.3 and 7.3.9). Notwithstanding this fact, it would be nicer to have the assertion of the convolution theorem under a less restrictive condition, like

$$P_\vartheta^{(n)} \circ c_n(\kappa^{(n)} - \kappa(\vartheta)) \Rightarrow M_\vartheta \quad \text{for every } \vartheta \in \Theta. \quad (8.4.13)$$

The following theorem shows that the assertion of the convolution theorem holds under condition (8.4.13) for λ^p–a.a. $\vartheta \in \Theta$.

Theorem 8.4.14. *Assume conditions (i) and (ii) of the convolution theorem for every $\vartheta \in \Theta$, and condition (8.4.13) for λ^p–a.a. $\vartheta \in \Theta$.*
   Then (8.4.4) holds for λ^p–a.a. $\vartheta \in \Theta$.

Addendum. *If $D(\vartheta)$ and M_ϑ depend on ϑ continuously, then (8.4.4) holds for every $\vartheta \in \Theta$. (This is not necessarily true for relation (8.4.3). See Exercise 8.4.15.)*

Proof. According to Theorem 6.6 in Parthasarathy (1967, p. 47) there exists a countable family of bounded and continuous functions h_i, $i \in \mathbb{N}$, such that for any sequence of p–measures $Q_n | \mathbb{B}^q$, $Q_n \Rightarrow Q_0$ iff $Q_n(h_i) \underset{n \in \mathbb{N}}{\longrightarrow} Q_0(h_i)$ for $i \in \mathbb{N}$.
   Let

$$f_{i,n}(\vartheta) := P_\vartheta^{(n)}(h_i \circ c_n(\kappa^{(n)} - \kappa(\vartheta))), \qquad n \in \mathbb{N},$$

and

$$f_{i,0}(\vartheta) := M_\vartheta(h_i).$$

By assumption there exists a λ^p–null set $N \in \mathbb{B}^p$ such that

$$f_{i,n}(\vartheta) \xrightarrow[n \in \mathbb{N}]{} f_{i,0}(\vartheta) \qquad \text{for } i \in \mathbb{N},\ \vartheta \in \Theta \cap N^c.$$

By Corollary 8.4.24 to Bahadur's lemma there exists a subsequence $\mathbb{N}_0 \subset \mathbb{N}$ and a λ^p–null set $N_0 \supset N$ such that

$$f_{i,n}(\vartheta + c_n^{-1}a) \xrightarrow[n \in \mathbb{N}_0]{} f_{i,0}(\vartheta) \quad \text{for } i \in \mathbb{N},\ a \in \mathbb{Q}^p,\ \vartheta \in \Theta \cap N_0^c.$$

Hence condition (8.4.2) holds along the subsequence $\mathbb{N}_0$ for all $a \in \mathbb{Q}^p$. Relation (8.4.6) holds with a subsequence $\mathbb{N}_1 \subset \mathbb{N}_0$, and one obtains (8.4.9) for $a \in \mathbb{Q}^p$. Since $\mathbb{Q}^p$ is dense in $\mathbb{R}^p$, this suffices to prove (8.4.12), which, in turn, implies (8.4.4).

To prove the addendum, let $\vartheta_0 \in N$ be arbitrary, and $\vartheta_n \in \Theta \cap N^c$, $n \in \mathbb{N}$, be a sequence converging to ϑ_0. Let $\Sigma(\vartheta) := K(\vartheta)H(\vartheta)K(\vartheta)^\top$. Since (8.4.4) holds for $\vartheta \in \Theta \cap N^c$, we have

$$M_{\vartheta_n} = N_{(0,\Sigma(\vartheta_n))} * R_n \qquad \text{for } n \in \mathbb{N}.$$

$\vartheta_n \to \vartheta_0$ implies $\Sigma(\vartheta_n) \to \Sigma(\vartheta_0)$, and therefore $N_{(0,\Sigma(\vartheta_n))} \Rightarrow N_{(0,\Sigma(\vartheta_0))}$. Moreover, we have $M_{\vartheta_n} \Rightarrow M_{\vartheta_0}$ by assumption. This implies that R_n, $n \in \mathbb{N}$, converges weakly to some p–measure, say R_0, and that $M_{\vartheta_0} = N_{(0,\Sigma(\vartheta_0))} * R_0$.

To see the latter, assume that, more generally, $M_n = P_n * Q_n$ and $M_n \Rightarrow M_0$, $P_n \Rightarrow P_0$. Hence

$$\int \exp[it^\top z] M_n(dz) = \int \exp[it^\top u] P_n(du) \int \exp[it^\top v] Q_n(dv)$$

for $n \in \mathbb{N}$.

Since $M_n \Rightarrow M_0$ and $P_n \Rightarrow P_0$, we have

$$\int e^{it^\top z} M_n(dz) \to \int e^{it^\top z} M_0(dz)$$

and

$$\int e^{it^\top u} P_n(du) \to \int e^{it^\top u} P_0(du).$$

Being characteristic functions, these limits are continuous. Hence

$$\int e^{it^\top v} Q_n(dv), \qquad n \in \mathbb{N},$$

converges to a function which is continuous at $t = 0$. This implies weak convergence of Q_n, $n \in \mathbb{N}$, to some p–measure Q_0 (see, e.g., Chow and Teicher, 1988, p. 271, Theorem 3), and $M_0 = P_0 * Q_0$. $\square$

Exercise 8.4.15. Continues Example 7.3.3. Show that, despite of the fact that $N_{(\vartheta,1)}^n \circ n^{1/2}(\vartheta^{(n)} - \vartheta) \Rightarrow N_{(0,1)}$ for every $\vartheta \in \mathbb{R}$, the estimator sequence $\vartheta^{(n)}$,

$n \in \mathbb{N}$, is not regular at $\vartheta = 0$, and $n^{1/2}\big(\vartheta^{(n)}(\underline{x}) - \vartheta\big) - n^{-1/2}\sum_1^n \ell^{(\cdot)}(x_\nu, \vartheta)$ is not asymptotically independent of $n^{-1/2}\sum_1^n \ell^{(\cdot)}(x_\nu, \vartheta)$. The joint distribution converges at $\vartheta = 0$ to $N_{(0,\Sigma)}$ with $\Sigma_{11} = 1$, $\Sigma_{12} = -1$, $\Sigma_{22} = 2$.

Example 8.4.16. *Minimal asymptotic variance for estimators of the β-quantile.*
For $\vartheta \in \Theta \subset \mathbb{R}^p$ let $P_\vartheta | \mathbb{B}$ be a p-measure with positive and continuous Lebesgue density $p(\cdot, \vartheta)$ and distribution function F_ϑ. Assume regularity conditions which guarantee that the minimal asymptotic variance for estimators of ϑ is $\Lambda(\vartheta)$.

To obtain the minimal asymptotic variance for the β-quantile $q_\beta(\vartheta)$, defined by $F_\vartheta\big(q_\beta(\vartheta)\big) = \beta$, we need the partial derivatives $q_\beta^{(i)}(\vartheta)$, $i = 1, \ldots, p$. Differentiating the identity $F_\vartheta\big(q_\beta(\vartheta)\big) \equiv \beta$ for $\vartheta \in \Theta$, we obtain

$$q_\beta^{(i)}(\vartheta) = -F_\vartheta^{(i)}\big(q_\beta(\vartheta)\big)/p\big(q_\beta(\vartheta), \vartheta\big).$$

Hence the minimal asymptotic variance for estimator sequences of the β-quantile becomes

$$\hat\sigma_\beta^2(\vartheta) := \frac{1}{p(q_\beta(\vartheta), \vartheta)^2} \sum_{i,j=1}^p F_\vartheta^{(i)}\big(q_\beta(\vartheta)\big) F_\vartheta^{(j)}\big(q_\beta(\vartheta)\big) \Lambda_{ij}(\vartheta).$$

A possible estimator sequence for the β-quantile is the β-quantile of the sample, $x_{[n\beta]:n}$, which is asymptotically normal with variance

$$\sigma_\beta^2(\vartheta) = \frac{1}{p(q_\beta(\vartheta), \vartheta)^2}\beta(1 - \beta).$$

From Theorem 8.4.14 it is clear that $\hat\sigma_\beta^2(\vartheta) \leq \sigma_\beta^2(\vartheta)$. The reader is invited to check that $x_{[n\beta]:n}$ is asymptotically linear with influence function

$$k(x, \vartheta) = \big(\beta - 1_{(-\infty, q_\beta(\vartheta)]}(x)\big)/p\big(q_\beta(\vartheta), \vartheta\big),$$

whereas asymptotically efficient estimator sequences have influence function

$$k_0(x, \vartheta) = \sum_{i,j=1}^p q_\beta^{(i)}(\vartheta) \Lambda_{ij}(\vartheta) \ell^{(j)}(x, \vartheta).$$

Specialized for a location- and scale parameter family $P_{a,c}$, $a \in \mathbb{R}$, $c > 0$, with densities $x \to c^{-1}p((x - a)/c)$ we have $q_\beta(a, c) = a + cq_\beta(0, 1)$, hence $q_\beta^{(1)}(a, c) = 1$ and $q_\beta^{(2)}(a, c) = q_\beta(0, 1)$. This leads to the minimal asymptotic variance

$$\hat\sigma_\beta^2(a, c) = c^2\big(\Lambda_{11} + 2q_\beta(0, 1)\Lambda_{12} + q_\beta(0, 1)^2\Lambda_{22}\big),$$

since $\Lambda(a, c) = c^2\Lambda$ (see Lemma 9.3.16).

The following figure shows the asymptotic relative efficiency of the sequence of sample quantiles with respect to the optimal estimator sequence for the β–quantile in the family of normal and Cauchy distributions as a function of β.

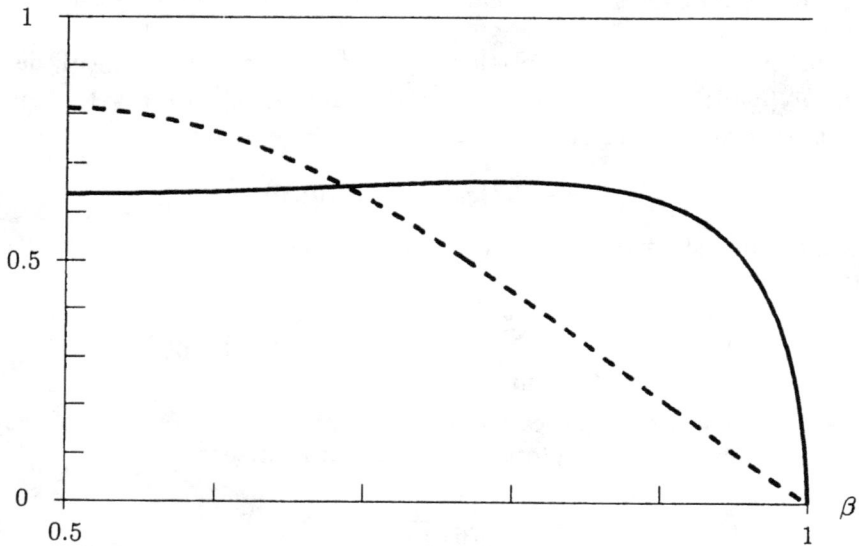

Figure 8.1   ——— normal;   - - - - Cauchy.

Historical Remark 8.4.17. The convolution theorem has a forerunner in results of Wolfowitz (1965, pp. 258/9, Theorem) for onedimensional parameters, and Kaufman (1966, p. 157, Theorem 2.1) for multidimensional parameters, who establish that the maximum likelihood estimator is asymptotically maximally concentrated on symmetric convex sets. An essential ingredient in these proofs is the asymptotic stochastic independence between an asymptotically efficient estimator sequence and the difference of the latter to an arbitrary estimator sequence, an idea going back to Fisher (1925, p. 706). In two independent papers, Hájek (1970, pp. 324/5, Theorem) and Inagaki (1970, p. 10, Theorem 3.1), the convolution theorem has been brought to its final form, saying that the limit distribution of an arbitrary estimator sequence can be represented as the convolution product of a normal distribution (which is the limit distribution of ML estimators) with some other distribution. In the meantime there are numerous proofs available. The proof given above is based on ideas of Bickel (see Roussas, 1972, p. 136).

Lemma 8.4.18. *For $i = 1, 2$ let $(X_i, \mathcal{A}_i)$ be Polish spaces. Let $P_n|\mathcal{A}_1 \times \mathcal{A}_2$, $n \in \mathbb{N}$, be a sequence of p–measures such that either of the two sequences of marginal distributions converges weakly to a p–measure. Then there exists a subsequence $\mathbb{N}_1 \subset \mathbb{N}$ and a p–measure $P_0|\mathcal{A}_1 \times \mathcal{A}_2$ such that $P_n \underset{n \in \mathbb{N}_1}{\Longrightarrow} P_0$.*

Proof. By Prohorov's Theorem II (see Billingsley, 1968, p. 37, Theorem 6.2), for every $\varepsilon > 0$ there exist compact sets $C_{i,\varepsilon}$ such that for the i-th marginal $P_{n,i}$ of P_n, $P_{n,i}(C_{i,\varepsilon}^c) < \varepsilon/2$ for $n \in \mathbb{N}$. Let $C_\varepsilon := C_{1,\varepsilon} \times C_{2,\varepsilon}$. Since $C_\varepsilon^c \subset (C_{1,\varepsilon}^c \times X_2) \cup (X_1 \times C_{2,\varepsilon}^c)$, we obtain $P_n(C_\varepsilon^c) \leq \varepsilon$ for $n \in \mathbb{N}$. Hence there exists a convergent subsequence by Prohorov's Theorem I (see Billingsley, 1968, p. 37, Theorem 6.1). □

The following lemma is a slightly improved version of Bahadur's lemma (Bahadur, 1964, p. 1549, Lemma 4), due to Droste and Wefelmeyer (1984, p. 140, Proposition 3.7).

Lemma 8.4.19 (Bahadur). *For $n \in \{0\} \cup \mathbb{N}$ let $f_n : \mathbb{R}^p \to \mathbb{R}$ be a measurable function such that*

$$\lim_{n \to \infty} f_n(u) = f_0(u) \qquad for \ \lambda^p\text{--}a.a. \ u \in \mathbb{R}^p .$$

Then the following is true: For every sequence $t_n \in \mathbb{R}^p$ with $\lim_{n \to \infty} t_n = 0$, there exists an infinite subsequence $\mathbb{N}_0 \subset \mathbb{N}$ and a λ^p–null set $N \in \mathbb{B}^p$ such that

$$\lim_{n \in \mathbb{N}_0} f_n(u + t_n) = f_0(u) \qquad for \ u \notin N. \tag{8.4.20}$$

An example of Rényi and Erdös (see Schmetterer, 1966, pp. 304/5) shows that $\mathbb{N}_0$ cannot be replaced by $\mathbb{N}$ in (8.4.20).

Proof. It suffices to prove that

$$f_n(\cdot + t_n) \underset{n \in \mathbb{N}}{\longrightarrow} f_0 \qquad (N_{(0,I)}). \tag{8.4.21}$$

By the Theorem of Riesz (see, e.g., Ash, 1972, p. 93, Theorem 2.5.3) this implies convergence $N_{(0,I)}$–a.e., hence also λ^p–a.e., for some subsequence.

In the following we use repeatedly that $d(N_{(t_n,I)}, N_{(0,I)}) \to 0$ by Scheffé's lemma, where d denotes the sup–metric.

At first we shall show that $g_n \to 0$ $(N_{(0,I)})$ for some measurable function $g_n : \mathbb{R}^p \to \mathbb{R}$ implies

$$g_n(\cdot + t_n) \to 0 \quad (N_{(0,I)}). \tag{8.4.22}$$

This follows from

$$N_{(0,I)}\{u \in \mathbb{R}^p : |g_n(u + t_n)| > \varepsilon\} = N_{(t_n,I)}\{u \in \mathbb{R}^p : |g_n(u)| > \varepsilon\}$$
$$\leq d(N_{(t_n,I)}, N_{(0,I)}) + N_{(0,I)}\{u \in \mathbb{R}^p : |g_n(u)| > \varepsilon\}.$$

Next we shall prove

$$f_0(\cdot + t_n) \to f_0 \quad (N_{(0,I)}). \tag{8.4.23}$$

Let $\varepsilon, \delta > 0$ be arbitrary. By Lusin's theorem (see, e.g., Ash, 1972, p. 186, Theorem 4.3.16) there exists a continuous function f_δ such that $N_{(0,I)}\{u \in \mathbb{R}^p : f_0(u) \neq f_\delta(u)\} < \delta$. Therefore,

$$N_{(0,I)}\{u \in \mathbb{R}^p : |f_0(u + t_n) - f_0(u)| > \varepsilon\}$$
$$\leq N_{(0,I)}\{u \in \mathbb{R}^p : f_0(u + t_n) \neq f_\delta(u + t_n)\}$$
$$+ N_{(0,I)}\{u \in \mathbb{R}^p : f_\delta(u) \neq f_0(u)\}$$
$$+ N_{(0,I)}\{u \in \mathbb{R}^p : |f_\delta(u + t_n) - f_\delta(u)| > \varepsilon\}.$$

Since

$$N_{(0,I)}\{u \in \mathbb{R}^p : f_0(u + t_n) \neq f_\delta(u + t_n)\}$$
$$= N_{(t_n,I)}\{u \in \mathbb{R}^p : f_0(u) \neq f_\delta(u)\} \leq d(N_{(t_n,I)}, N_{(0,I)}) + \delta,$$

and $f_\delta(u + t_n) \to f_\delta(u)$ for every $u \in \mathbb{R}^p$, relation (8.4.23) follows.

Relation (8.4.21) now follows from (8.4.23) and (8.4.22), applied for $g_n = f_n - f_0$.   □

Corollary 8.4.24. *For $i, n \in \mathbb{N}$ and $n = 0$ let $f_{i,n} : \mathbb{R}^p \to \mathbb{R}$ be measurable functions such that*

$$\lim_{n \to \infty} f_{i,n}(u) = f_{i,0}(u) \quad \text{for } i \in \mathbb{N} \text{ and } \lambda^p\text{-a.a. } u \in \mathbb{R}^p.$$

For $j \in \mathbb{N}$ let $t_{j,n} \in \mathbb{R}^p$, $n \in \mathbb{N}$, be a sequence with $\lim_{n \to \infty} t_{j,n} = 0$.

Then there exists an infinite subsequence $\mathbb{N}_0 \subset \mathbb{N}$ and a λ^p-null set $N \in \mathbb{B}^p$ such that

$$\lim_{n \in \mathbb{N}_0} f_{i,n}(u + t_{j,n}) = f_{i,0}(u) \quad \text{for } i, j \in \mathbb{N}, \ u \notin N. \tag{8.4.25}$$

Proof. According to Bahadur's Lemma 8.4.19, there exists a subsequence $\mathbb{N}_1 \subset \mathbb{N}$ and a λ^p-null set N_1 such that

$$\lim_{n \in \mathbb{N}_1} f_{1,n}(u + t_{1,n}) = f_{1,0}(u) \quad \text{for } u \notin N_1.$$

By Bahadur's lemma (applied to the sequence $t_{2,n}$, $n \in \mathbb{N}_1$) there exists a subsequence $\mathbb{N}_2 \subset \mathbb{N}_1$ and a λ^p-null set N_2 such that

$$\lim_{n \in \mathbb{N}_2} f_{1,n}(u + t_{2,n}) = f_{1,0}(u) \quad \text{for } u \notin N_2.$$

Proceeding in this way we obtain subsequences $\mathbb{N}_1 \supset \mathbb{N}_2 \supset \ldots \supset \mathbb{N}_j$ and λ^p–null sets $N_1, \ldots, N_j$ such that

$$\lim_{n \in \mathbb{N}_j} f_{1,n}(u + t_{j,n}) = f_{1,0}(u) \qquad \text{for } u \notin N_j.$$

Let $\hat{\mathbb{N}}_1$ denote the diagonal sequence determined by $\mathbb{N}_j$, $j \in \mathbb{N}$, and let $\hat{N}_1 := \bigcup_{j=1}^{\infty} N_j$. For $j \in \mathbb{N}$ fixed, all sufficiently large elements of $\hat{\mathbb{N}}_1$ are in $\mathbb{N}_j$. Moreover, $N_j \subset \hat{N}_1$. Hence

$$\lim_{n \in \hat{\mathbb{N}}_1} f_{1,n}(u + t_{j,n}) = f_{1,0}(u) \qquad \text{for } j \in \mathbb{N}, \ u \notin \hat{N}_1.$$

Applying the same procedure to the sequences $f_{i,n}$, $n \in \hat{\mathbb{N}}_1$ leads to the assertion. $\qquad\qquad\qquad\qquad\qquad\qquad\qquad\qquad\qquad\qquad\qquad\quad$ $\square$

For $n \in \mathbb{N}$, let P'_n be a p–measure with P_n–density p'_n. If $P_n \circ \log p'_n$, $n \in \mathbb{N}$, is weakly convergent, then it is possible to derive the limit distribution of $P'_n \circ f_n$, $n \in \mathbb{N}$, from the limit distribution of $P_n \circ f_n$, $n \in \mathbb{N}$, if p'_n is approximable by a contraction of f_n. This is made precise in Lemma 8.4.26. (Applied with $f_n = (\log p'_n, S_n)$, $q(u, v) = u$ and M the normal distribution with parameter $\left(-\frac{1}{2}\sigma_1^2, \mu_2, \sigma_1^2, \sigma_2^2, \varrho\right)$, this yields that M' is the normal distribution with parameter $\left(\frac{1}{2}\sigma_1^2, \mu_2 + \varrho\sigma_1\sigma_2, \sigma_1^2, \sigma_2^2, \varrho\right)$, hence $N_{(\mu_2 + \varrho\sigma_1\sigma_2, \sigma_2^2)}$ the limit distribution of $P'_n \circ S_n$. This is LeCam's 3rd Lemma.)

Lemma 8.4.26. *Let $(X, \mathcal{A})$ and $(Y, \mathcal{B})$ be Hausdorff spaces, endowed with their Borel algebra. Let $P_n|\mathcal{A}$, $n \in \mathbb{N}$, be a sequence of p–measures, and $f_n : (X, \mathcal{A}) \to (Y, \mathcal{B})$ a measurable function such that*

$$P_n \circ f_n \Rightarrow M. \tag{8.4.27}$$

Let $P'_n|\mathcal{A}$ be a p–measure with P_n–density p'_n. Assume that

$$\log p'_n = q \circ f_n + r_n, \quad \text{with } r_n \to 0 \quad (P_n), \tag{8.4.28}$$

where $q : Y \to \mathbb{R}$ is a continuous function fulfilling

$$\int \exp[q(v)] M(dv) = 1.$$

Then

$$P'_n \circ f_n \Rightarrow M', \tag{8.4.29}$$

the p–measure with M–density $v \to \exp[q(v)]$.

Proof. Relation (8.4.27) implies

$$P_n \circ (q \circ f_n, f_n) \Rightarrow Q := M \circ (v \to (q(v), v)).$$

By Slutzky's Lemma 7.7.8, relation (8.4.28) implies

$$P_n \circ (\log p_n',\ f_n) \Rightarrow Q.$$

Since $P_n(p_n') = 1$ and $\int \exp[u]Q(d(u,v)) = \int \exp[q(v)]M(dv) = 1$, we obtain from Corollary 6.7.11, applied with $Q_n = P_n \circ (\log p_n', f_n)$ and $q'(u,v) = \exp[u]$, that

$$\int g(v)\exp[u]Q_n(d(u,v)) \to \int g(v)\exp[u]Q(d(u,v))$$

for every bounded and continuous function $g : Y \to \mathbb{R}$.

Since

$$\int g(v)\exp[u]Q_n(d(u,v)) = \int g(f_n(x))p_n'(x)P_n(dx)$$

$$= \int g(f_n(x))P_n'(dx) = \int g(v)P_n' \circ f_n(dv),$$

and

$$\int g(v)\exp[u]Q(d(u,v)) = \int g(v)\exp[q(v)]M(dv),$$

this implies (8.4.29).   □

8.5 Maximally concentrated limit distributions

Presents various results on concentration bounds for limit distributions. Discusses the relation between componentwise asymptotic optimality and joint asymptotic optimality.

The Convolution Theorem 8.4.1 gives conditions on the family of sequences $P_\vartheta^{(n)}$, $n \in \mathbb{N}$, $\vartheta \in \Theta$, the functional κ, and the estimator sequence $\kappa^{(n)}$, $n \in \mathbb{N}$, under which the limit distribution M_ϑ of $P_\vartheta^{(n)} \circ c_n(\kappa^{(n)} - \kappa(\vartheta))$ can be represented as the convolution product of $N_{(0,\Sigma_*(\vartheta))}$, with $\Sigma_*(\vartheta) = K(\vartheta)H(\vartheta)K(\vartheta)^\top$, and some other p–measure. As a consequence of Anderson's theorem (see Corollary 2.4.13(ii) and Exercise 2.4.16) such a representation entails

$$M_\vartheta(C) \le N_{(0,\Sigma_*(\vartheta))}(C) \tag{8.5.1}$$

for every set $C \in \mathbb{B}^q$ which is convex and symmetric about 0. To say that $N_{(0,\Sigma_*(\vartheta))}$ is the optimal asymptotic distribution for "regular" estimator sequences has, therefore, a clear operational meaning.

Relation (8.5.1) does not extend to *arbitrary* convex sets C containing 0. Even if M_ϑ is a normal distribution with mean 0, relation (8.5.1) fails for some convex sets containing 0, unless the covariance matrix of M_ϑ is proportional to $\Sigma_*(\vartheta)$ (see Proposition 2.4.18). There is, however, an important exception: If the functional κ is 1–dimensional, M_ϑ is more spread out than a normal distribution with variance $\sigma_*^2(\vartheta)$ by Proposition 2.3.21(i) and therefore less concentrated on *arbitrary* intervals containing 0. In Theorem 8.3.1, this result was obtained by a more elementary proof.

Theorem 7.4.3 and Proposition 7.4.13, applied with $g_n(\cdot,\vartheta) = \Delta_n(\cdot,\vartheta)$, give regularity conditions under which there exist estimator sequences $\vartheta^{(n)}$, $n \in \mathbb{N}$, for ϑ with limit distribution $N_{(0,H(\vartheta))}$. These estimator sequences are "regular" according to Proposition 7.1.9. According to Proposition 7.2.1, $\kappa^{(n)} := \kappa \circ \vartheta^{(n)}$, $n \in \mathbb{N}$, is a regular estimator sequence with limit distribution $N_{(0,\Sigma_*(\vartheta))}$. Under the conditions of the convolution theorem this is the optimal limit distribution for "regular" estimator sequences of the functional κ.

The convolution theorem gives not only the optimal limit distribution, but also the stochastic approximation for estimator sequences attaining this optimal limit distribution.

Proposition 8.5.2. *Under regularity conditions (i)–(iii) of the Convolution Theorem 8.4.1, any "regular" estimator sequence $\kappa^{(n)}$, $n \in \mathbb{N}$, with the optimal limit distribution has the stochastic expansion*

$$c_n\big(\kappa^{(n)} - \kappa(\vartheta_0)\big) = K(\vartheta_0)H(\vartheta_0)\Delta_n(\cdot,\vartheta_0) + r_n(\cdot,\vartheta_0) \qquad (8.5.3)$$

with $r_n(\cdot,\vartheta_0) \to 0 \quad (P_{\vartheta_0}^{(n)})$.

Proof. By assumption, $P_{\vartheta_0}^{(n)} \circ c_n\big(\kappa^{(n)} - \kappa(\vartheta_0)\big) \Rightarrow N_{(0,\Sigma_*(\vartheta_0))}$, hence $R_{\vartheta_0}\{0\} = 1$ by Lemma 2.4.19.

Since R_{ϑ_0} is the weak limit of

$$P_{\vartheta_0}^{(n)} \circ \big(c_n(\kappa^{(n)} - \kappa(\vartheta_0)) - K(\vartheta_0)H(\vartheta_0)\Delta_n(\cdot,\vartheta_0)\big), \qquad n \in \mathbb{N},$$

$R_{\vartheta_0}\{0\} = 1$ implies

$$c_n\big(\kappa^{(n)} - \kappa(\vartheta_0)\big) - K(\vartheta_0)H(\vartheta_0)\Delta_n(\cdot,\vartheta_0) \to 0 \quad (P_{\vartheta_0}^{(n)}). \qquad \square$$

Corollary 8.5.4. *Let $\hat{\kappa}^{(n)}$, $n \in \mathbb{N}$, and $\kappa^{(n)}$, $n \in \mathbb{N}$, be "regular" estimator sequences fulfilling condition (iii) of the Convolution Theorem 8.4.1. If $\kappa^{(n)}$, $n \in \mathbb{N}$, is asymptotically efficient at ϑ_0, then $c_n(\hat{\kappa}^{(n)} - \kappa^{(n)})$, $n \in \mathbb{N}$, is asymptotically stochastically independent of $c_n(\kappa^{(n)} - \kappa(\vartheta_0))$, $n \in \mathbb{N}$.*

Proof. By the Convolution Theorem 8.4.1

$$c_n\big(\hat{\kappa}^{(n)} - \kappa(\vartheta_0)\big) - K(\vartheta_0)H(\vartheta_0)\Delta_n(\cdot,\vartheta_0), \qquad n \in \mathbb{N},$$

is asymptotically stochastically independent of $\Delta_n(\cdot, \vartheta_0)$, $n \in \mathbb{N}$. If $\kappa^{(n)}$, $n \in \mathbb{N}$, is asymptotically efficient, it fulfills (8.5.3). This implies the assertion.   □

Exercise 8.5.5. As a special consequence of the convolution theorem we obtained in Corollary 8.5.4 that $c_n(\hat{\kappa}^{(n)} - \kappa^{(n)})$ and $c_n(\kappa^{(n)} - \kappa(\vartheta))$ are asymptotically independent, if $\kappa^{(n)}$ is asymptotically optimal. Here is an elementary way to a comparable result: If

$$P_\vartheta^{(n)} \circ \left(c_n(\kappa^{(n)} - \kappa(\vartheta)), \; c_n(\hat{\kappa}^{(n)} - \kappa(\vartheta)) \right) \Rightarrow N_{(0, \Sigma(\vartheta))},$$

then $\kappa_\alpha^{(n)} = (1 - \alpha)\kappa^{(n)} + \alpha\hat{\kappa}^{(n)}$ is asymptotically normal with variance

$$\sigma_\alpha^2(\vartheta) = \sigma_{11}(\vartheta) + 2\alpha(\sigma_{12}(\vartheta) - \sigma_{11}(\vartheta)) + \alpha^2(\sigma_{11}(\vartheta) - 2\sigma_{12}(\vartheta) + \sigma_{22}(\vartheta)).$$

Since $\sigma_\alpha^2(\vartheta) \geq \sigma_{11}(\vartheta)$ for every $\alpha \in \mathbb{R}$, this implies $\sigma_{12}(\vartheta) = \sigma_{11}(\vartheta)$, hence asymptotic independence between $c_n(\hat{\kappa}^{(n)} - \kappa^{(n)})$ and $c_n(\kappa^{(n)} - \kappa(\vartheta))$.

The relation $\sigma_{12}(\vartheta) = \sigma_{11}(\vartheta)$ implies that the asymptotic efficiency of $\hat{\kappa}^{(n)}$, $n \in \mathbb{N}$, (relative to the asymptotically optimal estimator sequence $\kappa^{(n)}$, $n \in \mathbb{N}$), is equal to $\varrho^2(\vartheta)$ (where $\varrho(\vartheta)$ denotes the correlation coefficient of $N_{(0, \Sigma(\vartheta))}$).

Now we consider the relation between the efficiency of a multivariate estimator sequence and the efficiency of its components.

If $P_\vartheta^{(n)} \circ c_n(\kappa^{(n)} - \kappa(\vartheta)) \Rightarrow M_\vartheta$, then the distribution of any subvector, say $c_n(\kappa_{i_r}^{(n)} - \kappa_{i_r}(\vartheta))$, $r = 1, \ldots, m$ (with $\{i_1, \ldots, i_m\}$ a subset of $\{1, \ldots, p\}$), converges weakly to the marginal distribution of M_ϑ, defined by

$$M_\vartheta \circ \left((u_1, \ldots, u_p) \to (u_{i_1}, \ldots, u_{i_m}) \right).$$

In case $M_\vartheta = N_{(0, \Sigma(\vartheta))}$, this is the normal distribution $N_{(0, \hat{\Sigma}(\vartheta))}$, where $\hat{\Sigma}(\vartheta)$ is the $m \times m$-matrix consisting of the rows and columns of $\Sigma(\vartheta)$ with subscripts $i_1, \ldots, i_m$.

If $\kappa^{(n)}$, $n \in \mathbb{N}$, is asymptotically optimal, then $\Sigma(\vartheta) = \Sigma_*(\vartheta)$, i.e. $\Sigma_{ij}(\vartheta) = \kappa_i^{(\cdot)}(\vartheta) H(\vartheta) \kappa_j^{(\cdot)}(\vartheta)^\top$, and therefore

$$\hat{\Sigma}_{i_r i_s}(\vartheta) = \kappa_{i_r}^{(\cdot)}(\vartheta) H(\vartheta) \kappa_{i_s}^{(\cdot)}(\vartheta)^\top, \qquad r, s = 1, \ldots, m.$$

This, however, is the "minimal" covariance matrix for regular estimator sequences of the functional $\hat{\kappa}(\vartheta) = \left(\kappa_{i_1}(\vartheta), \ldots, \kappa_{i_m}(\vartheta) \right)^\top$.

To summarize: *If $(\kappa_1^{(n)}, \ldots, \kappa_p^{(n)})$ is asymptotically optimal for $(\kappa_1, \ldots, \kappa_p)$, then $(\kappa_{i_1}^{(n)}, \ldots, \kappa_{i_m}^{(n)})$ is asymptotically optimal for $(\kappa_{i_1}, \ldots, \kappa_{i_p})$, for any subset $\{i_1, \ldots, i_m\}$ of $\{1, \ldots, p\}$.* This refutes what appears not unreasonable at first sight, namely: That one can be the more accurate the fewer functionals are to be estimated. In particular: That one can obtain an asymptotically superior estimator for ϑ_1 (if $\vartheta_2, \ldots, \vartheta_p$ are treated as unknown nuisance parameters), than if one has to estimate all parameters simultaneously.

One would hesitate to mention a trivial result like this, were it not the case that a renowned journal like Sankhyā gave room to an author insisting on a deviating opinion about the minimal asymptotic variance in the presence of nuisance parameters. (See the comment from Katti to Bar–Lev and Reiser, 1983, p. 302.)

The following Proposition 8.5.6 provides a converse: Asymptotically efficient regular estimator sequences for the components of $\kappa : \Theta \to \mathbb{R}^q$, taken together, constitute an asymptotically efficient regular multivariate estimator sequence for κ. This results from the fact that asymptotically efficient regular estimator sequences have a unique stochastic expansion (by Proposition 8.5.2).

Proposition 8.5.6. *Assume regularity conditions (i) and (ii) of the Convolution Theorem 8.4.1. For $i \in \{1, \ldots, q\}$ let $\kappa_i^{(n)}$, $n \in \mathbb{N}$, be a "regular" estimator sequence for κ_i which is asymptotically efficient at ϑ_0.*
Then $\kappa^{(n)} = (\kappa_1^{(n)}, \ldots, \kappa_q^{(n)})$, $n \in \mathbb{N}$, is asymptotically efficient at ϑ_0 for the functional $\kappa = (\kappa_1, \ldots, \kappa_q)$.

Proof. Since $\kappa_i^{(n)}$, $n \in \mathbb{N}$, is asymptotically efficient for κ_i at ϑ_0, we obtain from Proposition 8.5.2, applied for the case $q = 1$, that

$$c_n\big(\kappa_i^{(n)} - \kappa_i(\vartheta_0)\big) = \kappa_i^{(\cdot)}(\vartheta_0)H(\vartheta_0)\Delta_n(\cdot, \vartheta_0) + r_{i,n}(\cdot, \vartheta_0),$$

with $r_{i,n}(\cdot, \vartheta_0) \to 0 \quad (P_{\vartheta_0}^{(n)})$.

By assumption, this holds for every $i \in \{1, \ldots, q\}$. Hence (8.5.3) holds with $r_n(\cdot, \vartheta_0) = (r_{1,n}(\cdot, \vartheta_0), \ldots, r_{q,n}(\cdot, \vartheta_0))$, and $r_n(\cdot, \vartheta_0) \to 0 \quad (P_{\vartheta_0}^{(n)})$. This implies

$$P_{\vartheta_0}^{(n)} \circ c_n\big(\kappa^{(n)} - \kappa(\vartheta_0)\big) \Rightarrow N_{(0, K(\vartheta_0)H(\vartheta_0)K(\vartheta_0)^\top)}. \qquad \Box$$

The intuitive interpretation of Proposition 8.5.6 is straightforward: *componentwise asymptotic optimality* implies *joint asymptotic optimality*. That this is not as obvious as it might appear at a first view can be seen from Example 8.5.7 which shows that "componentwise better" does not imply "jointly better", even if both distributions are normal. (See in this connection also Example 2.4.3.)

Example 8.5.7. Assume that (x_1, x_2, x_2) is normally distributed with mean vector $(\vartheta_1, \vartheta_2, \vartheta_1 + \vartheta_2)$ and covariance matrix

$$\begin{pmatrix} 1 & 0 & 0 \\ 0 & 1 & 0 \\ 0 & 0 & \frac{1}{2} \end{pmatrix}.$$

The problem is to estimate $(\vartheta_1, \vartheta_2)$. With $\vartheta_i^{(n)}(\underline{x}) = \bar{x}_{i,n}$, the distribution of $\left(n^{1/2}(\vartheta_1^{(n)} - \vartheta_1), n^{1/2}(\vartheta_2^{(n)} - \vartheta_2)\right)$ is normal with mean vector 0 and covariance matrix Σ with elements $\Sigma_{11} = \Sigma_{22} = 1$ and $\Sigma_{12} = 0$.

As an alternative, we consider the estimators

$$\hat{\vartheta}_1^{(n)}(\underline{x}) = \bar{x}_{3,n} - \bar{x}_{2,n} \quad \text{and} \quad \hat{\vartheta}_2^{(n)}(\underline{x}) = (\bar{x}_{1,n} + 6\bar{x}_{2,n} - \bar{x}_{3,n})/5.$$

$\left(n^{1/2}(\hat{\vartheta}_1^{(n)} - \vartheta_1), n^{1/2}(\hat{\vartheta}_2^{(n)} - \vartheta_2)\right)$ is jointly normal with mean vector 0 and covariance matrix $\hat{\Sigma}$ with elements $\hat{\Sigma}_{11} = \hat{\Sigma}_{22} = 3/2$ and $\hat{\Sigma}_{12} = -13/10$.

Hence $n^{1/2}(\vartheta_i^{(n)} - \vartheta_i)$ is more concentrated than $n^{1/2}(\hat{\vartheta}_i^{(n)} - \vartheta_i)$, for $i = 1, 2$: We have

$$N_{(0,\Sigma)}\{u \in \mathbb{R}^2 : |u_i| \le t\} > N_{(0,\hat{\Sigma})}\{u \in \mathbb{R}^2 : |u_i| \le t\} \quad \text{for } t > 0.$$

In spite of this we have for some $t > 0$ (e.g. for all $t \in (0,1)$)

$$N_{(0,\Sigma)}\{u \in \mathbb{R}^2 : |u_i| \le t \text{ for } i = 1,2\} < N_{(0,\hat{\Sigma})}\{u \in \mathbb{R}^2 : |u_i| \le t \text{ for } i = 1,2\}.$$

Hence $\vartheta_i^{(n)}$ is better than $\hat{\vartheta}_i^{(n)}$ for $i = 1, 2$, but $(\vartheta_1^{(n)}, \vartheta_2^{(n)})$ is not better than $(\hat{\vartheta}_1^{(n)}, \hat{\vartheta}_2^{(n)})$.

We return now to the problem of estimating *some* of the parameters $(\vartheta_1, \ldots \ldots, \vartheta_p)$, say $(\vartheta_1, \ldots, \vartheta_q)$, considering $(\vartheta_{q+1}, \ldots, \vartheta_p)$ as nuisance parameters. How does the optimal limit distribution change if the nuisance parameters are known? (The realistic question is, of course, the opposite one: How does the optimal limit distribution change if we use — to be on the safe side — a more general model including additional (nuisance) parameters?)

Estimators of $(\vartheta_1, \ldots, \vartheta_q)$ may be obtained by two different strategies. We may ignore the knowledge of $\vartheta_{q+1}, \ldots, \vartheta_p$, determine an asymptotically optimal estimator sequence $(\vartheta_1^{(n)}, \ldots, \vartheta_p^{(n)})$, and use $(\vartheta_1^{(n)}, \ldots, \vartheta_q^{(n)})$ to estimate the unknown parameters $(\vartheta_1, \ldots, \vartheta_q)$. Alternatively, we may determine an asymptotically optimal estimator sequence $(\hat{\vartheta}_1^{(n)}, \ldots, \hat{\vartheta}_q^{(n)})$, based on the knowledge of $\vartheta_{q+1}, \ldots, \vartheta_p$. Since $(\hat{\vartheta}_1^{(n)}, \ldots, \hat{\vartheta}_q^{(n)})$, $n \in \mathbb{N}$, is asymptotically optimal, $(\vartheta_1^{(n)}, \ldots \ldots, \vartheta_q^{(n)})$ cannot be asymptotically better. In the following we discuss the relationship between the pertaining limit distributions.

For this purpose we decompose the matrix $D(\vartheta)$ according to the distinction between $\vartheta_1, \ldots, \vartheta_q$ and $\vartheta_{q+1}, \ldots, \vartheta_p$:

$$D = \begin{pmatrix} D_{(1,1)} & D_{(1,2)} \\ D_{(2,1)} & D_{(2,2)} \end{pmatrix}$$

with

$$D_{(1,1)} = (D_{ij})_{i,j=1,\ldots,q}, \quad D_{(1,2)} = (D_{ij})_{\substack{i=1,\ldots,q \\ j=q+1,\ldots,p}}, \quad D_{(2,2)} = (D_{ij})_{i,j=q+1,\ldots,p}.$$

The corresponding decomposition of the matrix $H = D^{-1}$ is

$$H = \begin{pmatrix} H_{(1,1)} & H_{(1,2)} \\ H_{(2,1)} & H_{(2,2)} \end{pmatrix}.$$

We obtain

$$D_{(1,1)}H_{(1,1)} + D_{(1,2)}H_{(2,1)} = I, \tag{8.5.8$'$}$$
$$D_{(1,1)}H_{(1,2)} + D_{(1,2)}H_{(2,2)} = 0. \tag{8.5.8$''$}$$

(8.5.8$''$) implies

$$D_{(1,2)} = -D_{(1,1)}H_{(1,2)}H_{(2,2)}^{-1}.$$

Inserting this in (8.5.8$'$) yields

$$D_{(1,1)}\left(H_{(1,1)} - H_{(1,2)}H_{(2,2)}^{-1}H_{(2,1)}\right) = I.$$

Hence

$$D_{(1,1)}^{-1} = H_{(1,1)} - H_{(1,2)}H_{(2,2)}^{-1}H_{(2,1)}. \tag{8.5.9}$$

If

$$P_{\vartheta}^{(n)} \circ \left(c_n(\vartheta_i^{(n)} - \vartheta_i)\right)_{i=1,\ldots,p} \Rightarrow N_{(0,H(\vartheta))},$$

we obtain

$$P_{\vartheta}^{(n)} \circ \left(c_n(\vartheta_i^{(n)} - \vartheta_i)\right)_{i=1,\ldots,q} \Rightarrow N_{(0,H_{(1,1)}(\vartheta))}.$$

On the other hand,

$$P_{\vartheta}^{(n)} \circ \left(c_n(\hat{\vartheta}_i^{(n)} - \vartheta_i)\right)_{i=1,\ldots,q} \Rightarrow N_{(0,D_{(1,1)}(\vartheta)^{-1})}.$$

Relation (8.5.9) implies

$$D_{(1,1)}(\vartheta)^{-1} \leq_L H_{(1,1)}(\vartheta),$$

a relationship which was clear in advance from the convolution theorem. From this relationship we may infer under what conditions the knowledge of $\vartheta_{q+1},\ldots,$ $\ldots,\vartheta_p$ is — asymptotically — of no help for the estimation of $\vartheta_1,\ldots,\vartheta_q$: This is the case if $D_{(1,2)} = 0$. Since $D_{(2,1)} = D_{(1,2)}^{\top}$, this implies $D_{(2,1)} = 0$, hence also $H_{(1,2)} = 0$ and $H_{(2,1)} = 0$, so that, by (8.5.9), $D_{(1,1)}^{-1} = H_{(1,1)}$. (This follows from $D_{(2,1)}H_{(1,1)} + D_{(2,2)}H_{(2,1)} = 0$.)

Parameters ϑ_i, ϑ_j are called "independent" if $D_{ij}(\vartheta) = 0$ (a condition which reduces to $P_{\vartheta}\left(\ell^{(i)}(\cdot,\vartheta)\ell^{(j)}(\cdot,\vartheta)\right) = 0$ in the i.i.d. case). Hence one can say that the knowledge of $\vartheta_{q+1},\ldots,\vartheta_p$ is asymptotically irrelevant for the estimation of $\vartheta_1,\ldots,\vartheta_q$ iff for $i \in \{1,\ldots,q\}$ and $j \in \{q+1,\ldots,p\}$ the parameters ϑ_i and ϑ_j are independent.

This explains, in particular, why the knowledge of the location parameter is asymptotically irrelevant for the optimal estimation of the scale parameter — and vice versa — if the underlying p–measure is symmetric (see Section 9.4, p. 329, for details).

The convolution theorem has as a particular consequence that the optimal limit distribution of regular estimator sequences is necessarily normal. For the i.i.d. case, the normality of optimal limit distributions follows also by an elementary argument. To simplify the presentation, we assume that the functional $\kappa|\Theta$ is real valued. The extension to multivariate functionals is straightforward.

Proposition 8.5.10. *Let K be a class of estimator sequences for κ which is closed under convex combinations. Assume that $M_\vartheta|\mathbb{B}$ is an optimal limit distribution for estimator sequences in K. That means*

(i) M_ϑ is an asymptotic concentration bound in the sense that

$$\lim_{n\to\infty} P_\vartheta^n\{\underline{x} \in X^n : -t' < n^{1/2}\big(\kappa^{(n)}(\underline{x}) - \kappa(\vartheta)\big) < t''\} \qquad (8.5.11)$$
$$\leq M_\vartheta(-t',t'') \qquad \text{for } t', t'' \geq 0,$$

for any estimator sequence $\kappa^{(n)}$, $n \in \mathbb{N}$, in K for which the limit on the left hand side of (8.5.11) exists;

(ii) M_ϑ is attainable, i.e. there exists an estimator sequence in K such that equality holds in (8.5.11);

(iii) $\int u M_\vartheta(du) = 0$ and $\int u^2 M_\vartheta(du) < \infty$.

Then M_ϑ is a normal distribution.

Proof. Let $\hat{\kappa}^{(n)}$, $n \in \mathbb{N}$, be an estimator sequence for which equality holds in (8.5.11). For $\alpha \in (0,1)$ define

$$\kappa_\alpha^{(n)}(x_1, \ldots, x_n) = \alpha \hat{\kappa}^{(n_\alpha)}(x_1, \ldots, x_{n_\alpha})$$
$$+ (1-\alpha)\hat{\kappa}^{(n-n_\alpha)}(x_{n_\alpha+1}, \ldots, x_n),$$

with $n_\alpha := [\alpha n]$. We have

$$n^{1/2}\big(\kappa_\alpha^{(n)}(x_1, \ldots, x_n) - \kappa(\vartheta)\big)$$
$$= \alpha\big(\frac{n}{n_\alpha}\big)^{1/2} n_\alpha^{1/2}\big(\hat{\kappa}^{(n_\alpha)}(x_1, \ldots, x_{n_\alpha}) - \kappa(\vartheta)\big)$$
$$+ (1-\alpha)\big(\frac{n}{n-n_\alpha}\big)^{1/2}(n-n_\alpha)^{1/2}\big(\hat{\kappa}^{(n-n_\alpha)}(x_{n_\alpha+1}, \ldots, x_n) - \kappa(\vartheta)\big).$$

The assumption on $\hat{\kappa}^{(n)}$ implies that $n^{1/2}\big(\kappa_\alpha^{(n)} - \kappa(\vartheta)\big)$, $n \in \mathbb{N}$, converges weakly to $M_\vartheta^2 \circ \big((u,v) \to \alpha^{1/2}u + (1-\alpha)^{1/2}v\big)$.

Since $\kappa_\alpha^{(n)}$, $n \in \mathbb{N}$, is in $\mathcal{K}$, the optimality of M_ϑ among the limit distributions of estimator sequences in $\mathcal{K}$ implies

$$M_\vartheta^2\{(u,v) \in \mathbb{R}^2 : -t' < \alpha^{1/2}u + (1-\alpha)^{1/2}v < t''\} \tag{8.5.12}$$
$$\leq M_\vartheta(-t', t'') \qquad \text{for } t', t'' > 0.$$

As a consequence of (iii),

$$\int \left(\alpha^{1/2}u + (1-\alpha)^{1/2}v\right)^2 M_\vartheta(du)M_\vartheta(dv) = \int w^2 M_\vartheta(dw).$$

Hence equality holds in (8.5.12), i.e.

$$M_\vartheta^2 \circ \left((u,v) \to (\alpha^{1/2}u + (1-\alpha)^{1/2}v)\right) = M_\vartheta.$$

Since $\alpha \in (0,1)$ was arbitrary, this implies that M_ϑ is stable. Since M_ϑ has a finite 2nd moment, M_ϑ is a normal distribution (see Chow and Teicher, 1988, p. 449, Theorem 2). □

Now we will show that *the minimal asymptotic covariance matrix for estimators of a functional κ is independent of how the family of p–measures is parametrized.*

Assume that the conditions (i)–(iii) of the Convolution Theorem 8.4.1 are fulfilled for the family $P_\vartheta^{(n)}$, $n \in \mathbb{N}$, $\vartheta \in \Theta$, with $\Theta \subset \mathbb{R}^p$, and the functional $\kappa : \Theta \to \mathbb{R}^q$, $q \leq p$. Then the minimal asymptotic covariance matrix is $K(\vartheta)H(\vartheta)K(\vartheta)^\top$. For $T \subset \mathbb{R}^p$ assume that $g : T \to \Theta$ is injective with continuous partial derivatives, say G. For $\tau \in T$ let $\hat{P}_\tau := P_{g(\tau)}$, and $\hat{\kappa}(\tau) = \kappa(g(\tau))$. We have to show that the minimal asymptotic covariance matrix, obtained from the model $\{\hat{P}_\tau : \tau \in T\}$ and the functional $\hat{\kappa} : T \to \mathbb{R}^q$, is equal to $K(g(\tau))H(g(\tau))K(g(\tau))^\top$.

If the LAN condition holds in its uniform version, we obtain

$$\log \frac{p_n(x, g(\tau + c_n^{-1}a))}{p_n(x, g(\tau))} = a^\top \hat{\Delta}(x, \tau) - \frac{1}{2}a^\top \hat{D}(\tau)a + r_n(x, \tau)$$

with

$$\hat{\Delta}(x, \tau) = G(\tau)\Delta(x, g(\tau)),$$

and

$$\hat{D}(\tau) = G(\tau)D(g(\tau))G(\tau)^\top.$$

Moreover, $\hat{K}(\tau) = K(g(\tau))G(\tau)$. Therefore,

$$\hat{K}(\tau)\hat{D}(\tau)^{-1}\hat{K}(\tau)^\top = K(g(\tau))H(g(\tau))K(g(\tau))^\top,$$

as was to be shown.

In the i.i.d. case, there is yet another *invariance condition* that asymptotic bounds have to meet. If we consider a sample of size $n = km$, i.e. km independent realizations x_ν, $\nu = 1, \ldots, km$, of P_ϑ, as m independent realizations $\big(x_{k(\nu-1)+1}, \ldots, x_{k\nu}\big)$, $\nu = 1, \ldots, m$, of P_ϑ^k, we may compute the asymptotic bound for $P_\vartheta^n \circ n^{1/2}(\vartheta^{(n)} - \vartheta)$, $n \in \mathbb{N}$, also as asymptotic bound for $(P_\vartheta^k)^m \circ k^{1/2} m^{1/2}(\tilde{\vartheta}^{(m)} - \vartheta)$, $m \in \mathbb{N}$, if we write $\vartheta^{(km)}(x_1, x_2, \ldots, x_{km})$ as

$$\tilde{\vartheta}^{(m)}\big((x_1, \ldots, x_k), (x_{k+1}, \ldots, x_{2k}), \ldots, (x_{k(m-1)+1}, \ldots, x_{km})\big).$$

Since $(x_1, \ldots, x_k) \to \prod_{\alpha=1}^k p(x_\alpha, \vartheta)$ is a density of P_ϑ^k (with respect to μ^k), the matrix $L^{(k)}$ for the "concentrated" model has elements

$$L_{ij}^{(k)}(\vartheta) = \int \sum_{\alpha=1}^k \ell^{(i)}(x_\alpha, \vartheta) \sum_{\beta=1}^k \ell^{(j)}(x_\beta, \vartheta) P_\vartheta^k\big(d(x_1, \ldots, x_k)\big)$$

$$= k L_{ij}(\vartheta).$$

Hence the asymptotic bound for $(P_\vartheta^k)^m \circ m^{1/2}(\tilde{\vartheta}^{(m)} - \vartheta)$, $m \in \mathbb{N}$, comes out as $k^{-1}\Lambda(\vartheta)$, and the asymptotic bound for $P_\vartheta^{km} \circ (km)^{1/2}(\vartheta^{(km)} - \vartheta) = (P_\vartheta^k)^m \circ k^{1/2} m^{1/2}(\tilde{\vartheta}^{(m)} - \vartheta)$, $m \in \mathbb{N}$, is $\Lambda(\vartheta)$, in agreement with the original model.

8.6 Superefficiency

Superefficiency at a certain parameter value brings on subefficiency in a neighborhood of order c_n^{-1}.

Recall that an estimator sequence $\kappa^{(n)}$, $n \in \mathbb{N}$, is "regular" at ϑ if (see (8.4.2))

$$P_{\vartheta+c_n^{-1}a}^{(n)} \circ c_n\big(\kappa^{(n)} - \kappa(\vartheta + c_n^{-1}a)\big) \Rightarrow M_\vartheta \quad \text{for } a \in \mathbb{R}^p. \tag{8.6.1}$$

Under conditions (i) and (ii) of the Convolution Theorem 8.4.1, the limit distribution M_ϑ cannot be more concentrated (on convex sets, symmetric about the origin) than $N_{(0, \Sigma_*(\vartheta))}$. What happens for parameter values ϑ at which the estimator sequence is not regular, say: $P_{\vartheta+c_n^{-1}a}^{(n)} \circ c_n\big(\kappa^{(n)} - \kappa(\vartheta + c_n^{-1}a)\big)$, $n \in \mathbb{N}$, converges, but the limit depends on a?

Example 8.6.2 presents an estimator sequence which is "superefficient" at a single point. The same idea, due to Hodges Jr. (see LeCam, 1953, p. 280) can be used to obtain superefficiency on a countable subset of Θ. LeCam (1953, p. 291, Example 4) indicates how this technique can be applied to obtain superefficiency on an uncountable subset of Θ (of λ^p–measure 0, of course). But this construction has never been carried through in detail.

Example 8.6.2. For the family $\{N_{(\vartheta,1)} : \vartheta \in \mathbb{R}\}$ the optimal limit distribution of regular estimator sequences for ϑ is $N_{(0,1)}$.

For $n \in \mathbb{N}$, let

$$\vartheta^{(n)}(\underline{x}) := \begin{cases} \overline{x}_n \\ \frac{1}{2}\overline{x}_n \end{cases} \quad \text{if} \quad |\overline{x}_n| \begin{matrix} > \\ \leq \end{matrix} n^{-1/4}.$$

This estimator sequence is superefficient at $\vartheta = 0$; $N^n_{(\vartheta,1)} \circ n^{1/2}(\vartheta^{(n)} - \vartheta)$, $n \in \mathbb{N}$, converges weakly to the optimal limit distribution $N_{(0,1)}$ for $\vartheta \neq 0$. For $\vartheta = 0$ it converges to a nondegenerate limit distribution which is more concentrated than $N_{(0,1)}$, namely $N_{(0,1/4)}$. The limit distribution pretends that something has been gained at $\vartheta = 0$, and nothing has been lost for $\vartheta \neq 0$. To demonstrate that the limit distribution renders, in this case, no adequate description of the performance of the estimator sequence for large sample sizes, we consider $\vartheta \to N^n_{(\vartheta,1)}\{\underline{x} \in \mathbb{R}^n : n^{1/2}|\vartheta^{(n)}(\underline{x}) - \vartheta| < 1\}$. Figure 8.2 shows this function for two different sample sizes.

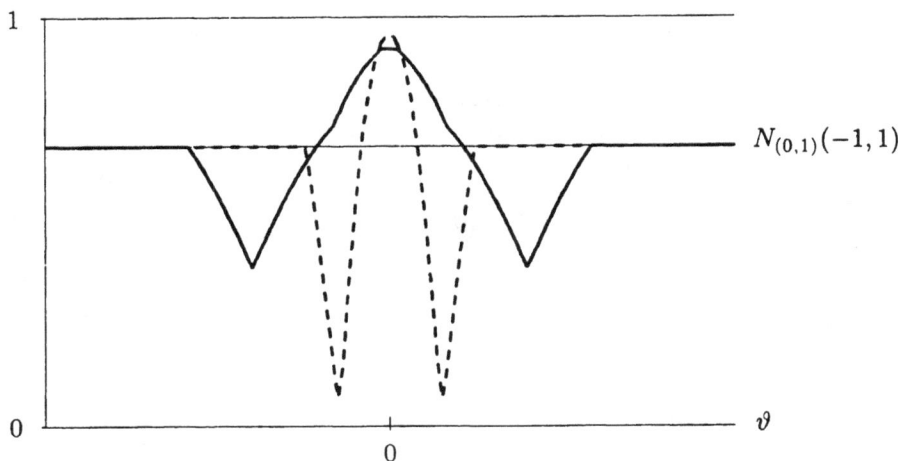

Figure 8.2   —— $n = 10$;   - - - - $n = 150$

However large n may be, there are always parameter values ϑ for which the "superefficient" estimator $\vartheta^{(n)}$ is inferior. But this "zone of inferiority" shrinks to 0 as $n^{-1/2}$ and will, therefore, be disjoint from any fixed $\vartheta \neq 0$ if n is sufficiently large.

What we observe in this particular example is true in general: If an estimator sequence is "supereffificient" at a certain point ϑ_0, then it is necessarily "subeffificient" in the immediate neighborhood of ϑ_0. This phenomenon was first observed by LeCam (1953, p. 327, Theorem 14). The following theorem illustrates this phenomenon from a different point of view.

Theorem 8.6.3. *Let* $\Theta \subset \mathbb{R}^p$ *be an open set. Let* $P_\vartheta^{(n)}$, $n \in \mathbb{N}$, $\vartheta \in \Theta$, *be a family of sequences of p–measures fulfilling the following condition.*
For every sequence ϑ_n, *with* $\sup_{n \in \mathbb{N}} c_n |\vartheta_n - \vartheta_0| < \infty$, *and every sequence* $a_n \to a$,

$$P_{\vartheta_n}^{(n)} \circ \log \frac{p_n(\cdot, \vartheta_n + c_n^{-1} a_n)}{p_n(\cdot, \vartheta_n)} \Rightarrow N_{(-\frac{1}{2} a^\top D(\vartheta_0) a, \, a^\top D(\vartheta_0) a)}. \tag{8.6.4}$$

(This condition holds true if the family $P_\vartheta^{(n)}$, $n \in \mathbb{N}$, $\vartheta \in \Theta$, *fulfills the LAN Condition 8.1.1 locally uniformly at* ϑ_0.)
Let $\kappa : \Theta \to \mathbb{R}$ *be a functional with continuous partial derivatives, say* $\kappa^{(\cdot)}$. *According to the Convolution Theorem 8.4.1,* $N_{(0, \sigma_*^2(\vartheta))}$, *with* $\sigma_*^2(\vartheta) :=$ $\kappa^{(\cdot)}(\vartheta) H(\vartheta) \kappa^{(\cdot)}(\vartheta)^\top$, *is the optimal limit distribution for regular estimator sequences. This implies in particular that*

$$\lim_{n \to \infty} P_\vartheta^{(n)} \{ x \in X_n : c_n |\kappa^{(n)}(x) - \kappa(\vartheta)| < t \} \leq N_{(0, \sigma_*^2(\vartheta))}(-t, t)$$

for every $t > 0$ *if* $\kappa^{(n)}$, $n \in \mathbb{N}$, *is regular at* ϑ.
Let now $\kappa^{(n)}$, $n \in \mathbb{N}$, *be an estimator sequence such that for some* $\vartheta_0 \in \Theta$, *some subsequence* $\mathbb{N}_0 \subset \mathbb{N}$ *and some* $t > 0$,

$$\lim_{n \in \mathbb{N}_0} P_{\vartheta_0}^{(n)} \{ x \in X_n : c_n |\kappa^{(n)}(x) - \kappa(\vartheta_0)| < t \} > N_{(0, \sigma_*^2(\vartheta_0))}(-t, t). \tag{8.6.5}$$

Then

$$\inf_{M > 0} \overline{\lim_{n \in \mathbb{N}_0}} \inf_{\|\vartheta - \vartheta_0\| \leq c_n^{-1} M} P_\vartheta^{(n)} \{ x \in X_n : c_n |\kappa^{(n)}(x) - \kappa(\vartheta)| < t \} \tag{8.6.6}$$
$$< N_{(0, \sigma_*^2(\vartheta_0))}(-t, t).$$

Remark 8.6.7. The implication from (8.6.5) to (8.6.6) is equivalent to the following implication.
For every subsequence $\mathbb{N}_0 \subset \mathbb{N}$,

$$\lim_{n \in \mathbb{N}_0} \inf_{\|\vartheta - \vartheta_0\| \leq c_n^{-1} M} P_\vartheta^{(n)} \{ x \in X_n : c_n |\kappa^{(n)}(x) - \kappa(\vartheta)| < t \} \tag{8.6.8}$$
$$\geq N_{(0, \sigma_*^2(\vartheta_0))}(-t, t) \qquad \text{for every } M > 0$$

implies

$$\lim_{n\in\mathbb{N}_0} \inf_{\|\vartheta-\vartheta_0\|\le c_n^{-1}M} P_\vartheta^{(n)}\{x\in X_n : c_n|\kappa^{(n)}(x) - \kappa(\vartheta)| < t\} \qquad (8.6.9)$$
$$= N_{(0,\sigma_*^2(\vartheta_0))}(-t,t) \qquad \text{for every } M > 0.$$

Roughly speaking, this means that asymptotic efficiency for all sequences ϑ_n, $n\in\mathbb{N}$, converging to ϑ_0 at a rate c_n^{-1} excludes superefficiency at ϑ_0.

As a particular consequence, we obtain from relation (8.6.9) that, for any estimator sequence $\kappa^{(n)}$, $n\in\mathbb{N}$,

$$\inf_{M>0} \overline{\lim_{n\in\mathbb{N}}} \inf_{\|\vartheta-\vartheta_0\|\le c_n^{-1}M} P_\vartheta^{(n)}\{x\in X_n : c_n|\kappa^{(n)}(x) - \kappa(\vartheta)| < t\} \quad (8.6.10)$$
$$\le N_{(0,\sigma_*^2(\vartheta_0))}(-t,t) \qquad \text{for every } t > 0.$$

These relations follow from Lemma 8.6.24, applied with

$$D_n(M) := \inf_{\|\vartheta-\vartheta_0\|\le c_n^{-1}M} [P_\vartheta^{(n)}\{x\in X_n : c_n|\kappa^{(n)}(x) - \kappa(\vartheta)| < t\}$$
$$- N_{(0,\sigma_*^2(\vartheta_0))}(-t,t)].$$

Relation (8.6.10) corresponds to the usual minimax theorem. It is slightly weaker than the assertion of Theorem 8.6.3, since it does not imply that *superefficiency* at ϑ_0 (in the sense of (8.6.5)) entails *subefficiency* in the neighborhood of ϑ_0 (in the sense of (8.6.6)) and is therefore compatible with

$$\lim_{n\in\mathbb{N}_0} \inf_{\|\vartheta-\vartheta_0\|\le c_n^{-1}M} P_\vartheta^{(n)}\{x\in X_n : c_n|\kappa^{(n)}(x) - \kappa(\vartheta)| < t\}$$
$$> N_{(0,\sigma_*^2(\vartheta_0))}(-t,t) \qquad \text{for } M > 0, \ t > 0.$$

Proof of Theorem 8.6.3. To simplify our notations, we write N for $N_{(0,1)}$, $\sigma(\vartheta_0)$ for $\sigma_*(\vartheta_0)$, and we replace t by $t\sigma(\vartheta_0)$. Let

$$\alpha := \lim_{n\in\mathbb{N}_0} P_{\vartheta_0}^{(n)}\{x\in X_n : c_n|\kappa^{(n)}(x) - \kappa(\vartheta_0)| < t\sigma(\vartheta_0)\} \qquad (8.6.11)$$
$$- N(-t,t),$$

and

$$\beta := N(-t,t) \qquad (8.6.12)$$
$$- \inf_{M>0} \overline{\lim_{n\in\mathbb{N}_0}} \inf_{\|\vartheta-\vartheta_0\|\le c_n^{-1}M} P_\vartheta^{(n)}\{x\in X_n : c_n|\kappa^{(n)}(x) - \kappa(\vartheta)| < t\sigma(\vartheta_0)\}.$$

We have $\alpha > 0$ according to condition (8.6.5). Relation (8.6.6) asserts that $\beta > 0$. We shall show that $\beta \ge \alpha^2 t/4\varphi(t)$. This will be carried through by showing that the assumption

$$\Delta := \alpha^2 t/4\varphi(t) - \beta > 0 \qquad (8.6.13)$$

leads to a contradiction.

Let $a_n \in \mathbb{R}^p$, $n \in \mathbb{N}$, be a sequence such that

$$a_n^\top D(\vartheta_0) a_n \to 1, \tag{8.6.14}$$

and

$$\inf_{\|\vartheta - \vartheta_0\| \leq c_n^{-1} M} \kappa^{(\cdot)}(\vartheta) a_n > \sigma(\vartheta_0) \quad \text{for every } M > 0 \text{ and all } n \geq n(M). \tag{8.6.15}$$

To see that such a sequence a_n, $n \in \mathbb{N}$, exists, let

$$\delta_n := \inf_{\|\vartheta - \vartheta_0\| \leq c_n^{-1/2}} \kappa^{(\cdot)}(\vartheta) H(\vartheta_0) \kappa^{(\cdot)}(\vartheta_0)^\top / \sigma^2(\vartheta_0).$$

Since $\kappa^{(\cdot)}$ is continuous at ϑ_0, we have $\delta_n \uparrow 1$. The sequence

$$a_n := \Big(1 + \frac{1}{n}\Big) \delta_n^{-1} H(\vartheta_0) \kappa^{(\cdot)}(\vartheta_0)^\top / \sigma(\vartheta_0)$$

has the properties required by (8.6.14) and (8.6.15).

For $n \in \mathbb{N}$ and $k \in \{0\} \cup \mathbb{N}$ let

$$\vartheta_{n,k} := \vartheta_0 + c_n^{-1} 2kt a_n,$$
$$Q_{n,k} := P^{(n)}_{\vartheta_{n,k}},$$
$$A^+_{n,k} := \{x \in X_n : \kappa^{(n)}(x) \geq \kappa(\vartheta_{n,k}) + c_n^{-1} t\sigma(\vartheta_0)\},$$
$$A^-_{n,k} := \{x \in X_n : \kappa^{(n)}(x) \leq \kappa(\vartheta_{n,k}) - c_n^{-1} t\sigma(\vartheta_0)\},$$
$$\alpha^\pm_{n,k} := Q_{n,k}(A^\pm_{n,k}).$$

Relation (8.6.15) implies

$$\kappa(\vartheta_{n,k+1}) > \kappa(\vartheta_{n,k}) + c_n^{-1} 2t\sigma(\vartheta_0) \quad \text{for } n \geq n_k \text{ (say)},$$

hence

$$(A^-_{n,k+1})^c \subset A^+_{n,k} \quad \text{for } n \geq n_k. \tag{8.6.16}$$

According to (8.6.25) there exists a subsequence $\mathbb{N}_1 \subset \mathbb{N}_0$ such that

$$\inf_{M>0} \varliminf_{n \in \mathbb{N}_1} \inf_{\|\vartheta - \vartheta_0\| \leq c_n^{-1} M} P^{(n)}_\vartheta \{x \in X_n : c_n |\kappa^{(n)}(x) - \kappa(\vartheta)| < t\sigma(\vartheta_0)\}$$

$$= \inf_{M>0} \varlimsup_{n \in \mathbb{N}_0} \inf_{\|\vartheta - \vartheta_0\| \leq c_n^{-1} M} P^{(n)}_\vartheta \{x \in X_n : c_n |\kappa^{(n)}(x) - \kappa(\vartheta)| < t\sigma(\vartheta_0)\}.$$

Together with (8.6.12) this implies

$$\inf_{M>0} \varliminf_{n \in \mathbb{N}_1} \inf_{\|\vartheta - \vartheta_0\| \leq c_n^{-1} M} P^{(n)}_\vartheta \{x \in X_n : c_n |\kappa^{(n)}(x) - \kappa(\vartheta)| < t\sigma(\vartheta_0)\}$$

$$= N(-t, t) - \beta,$$

hence, for every $k \in \mathbb{N}$,

$$\varlimsup_{n \in \mathbb{N}_1} (\alpha_{n,k}^- + \alpha_{n,k}^+) \leq 2\Phi(-t) + \beta. \tag{8.6.17}$$

Using (8.6.16) and (8.6.17) we obtain that

$$\lim_{n \in \mathbb{N}_1} \alpha_{n,k}^+ \leq \Phi(-t) - \delta \qquad \text{for some } \delta \in \big(0, \Phi(-t)\big) \tag{8.6.18}$$

implies

$$\lim_{n \in \mathbb{N}_1} \alpha_{n,k+1}^+ < \Phi(-t) - \delta - (\delta^2 t/\varphi(t) - \beta). \tag{8.6.19}$$

To prove (8.6.19) we proceed as follows. Interpreting $A_{n,k}^+$ as a critical region of level $\alpha_{n,k}^+$ for testing the hypothesis $Q_{n,k}$ against the alternative $Q_{n,k+1}$, we obtain

$$\varliminf_{n \in \mathbb{N}_1} Q_{n,k+1}(A_{n,k}^+) \leq \Phi\big(\Phi^{-1}(\Phi(-t) - \varepsilon) + 2t\big). \tag{8.6.20}$$

This can be seen as follows. Since $Q_{n,k} = P_{\vartheta_{n,k}}^{(n)}$ and $Q_{n,k+1} = P_{\vartheta_{n,k}+c_n^{-1}2ta_n}^{(n)}$, condition (8.6.4) implies (by (8.6.14)) that

$$Q_{n,k} \circ \log q_{n,k} \Rightarrow N_{(-\frac{1}{2}\sigma^2, \sigma^2)}$$

with $\sigma^2 = 4t^2$, where $q_{n,k} \in dQ_{n,k+1}/dQ_{n,k}$. Hence (8.6.20) follows from Lemma 8.2.15 (applied with $\varphi_n = 1_{A_{n,k}^+}$).

Because of (8.6.16), relation (8.6.20) implies

$$\varliminf_{n \in \mathbb{N}_1} \alpha_{n,k+1}^- \geq \Phi\big(-\Phi^{-1}(\Phi(-t) - \delta) - 2t\big)$$
$$> \Phi(-t) + \delta + \delta^2 t/\varphi(t),$$

with the last inequality following from Lemma 8.6.23. Together with (8.6.17) this implies (8.6.19):

$$\lim_{n \in \mathbb{N}_1} \alpha_{n,k+1}^+ \leq 2\Phi(-t) + \beta - \varliminf_{n \in \mathbb{N}_1} \alpha_{n,k+1}^- < \Phi(-t) - \delta - (\delta^2 t/\varphi(t) - \beta).$$

From (8.6.11) we obtain

$$\varlimsup_{n \in \mathbb{N}_0} (\alpha_{n,0}^- + \alpha_{n,0}^+) = 2\Phi(-t) - \alpha, \tag{8.6.21}$$

hence

$$\lim_{n \in \mathbb{N}_1} \alpha_{n,0}^- \leq \Phi(-t) - \alpha/2 \tag{8.6.21'}$$

or

$$\lim_{n \in \mathbb{N}_1} \alpha_{n,0}^+ \leq \Phi(-t) - \alpha/2. \tag{8.6.21''}$$

W.l.g. we assume (8.6.21''). In this case, (8.6.18) holds true for $k = 0$ with $\delta = \alpha/2$. Assume now that Δ (see (8.6.13)) is positive. Proceeding by induction over k, we obtain from the implication between (8.6.18) and (8.6.19) that

$$\varliminf_{n \in \mathbb{N}_1} \alpha_{n,k}^+ < \Phi(-t) - \alpha/2 - k\Delta \qquad (8.6.22)$$

for $k \in \mathbb{N}$ with $(k-1)\Delta < \Phi(-t) - \alpha/2$.

Let now k_0 denote the largest integer $k \in \mathbb{N}$ fulfilling $k\Delta < \Phi(-t) - \alpha/2$. Hence (8.6.22) holds with k replaced by k_0+1. Since $\Phi(-t) - \alpha/2 - (k_0+1)\Delta \leq 0$, the assumption $\Delta > 0$ leads to a contradiction. $\square$

Lemma 8.6.23. $\Phi\big(-\Phi^{-1}(\Phi(-t) - \delta) - 2t\big) > \Phi(-t) + \delta + \delta^2 t/\varphi(t)$ *for* $t > 0$ *and* $0 < \delta < \Phi(-t)$.

Proof. For $\delta \in \big[0, \Phi(-t)\big)$ let

$$f(\delta) := \Phi\big(-\Phi^{-1}(\Phi(-t) - \delta) - 2t\big).$$

We have

$$f'(\delta) = \frac{\varphi(\Phi^{-1}(\Phi(-t) - \delta) + 2t)}{\varphi(\Phi^{-1}(\Phi(-t) - \delta))},$$

and

$$f''(\delta) = 2t \frac{\varphi(\Phi^{-1}(\Phi(-t) - \delta) + 2t)}{\varphi(\Phi^{-1}(\Phi(-t) - \delta))^2}.$$

Since $f(0) = \Phi(-t)$, $f'(0) = 1$ and $f''(\delta) \geq 2t/\varphi(t)$ for $\delta \in \big[0, \Phi(-t)\big)$, we have

$$f(\delta) \geq \Phi(-t) + \delta + \delta^2 t/\varphi(t) \qquad \text{for } 0 \leq \delta < \Phi(-t),$$

with strict inequality if $\delta > 0$. $\square$

Lemma 8.6.24. *For any sequence of antitone functions* $D_n : [0, \infty) \to [-1, 1]$ *the following holds true:*

(i) *There exists a subsequence* $\mathbb{N}_0 \subset \mathbb{N}$ *such that*

$$\inf_{c>0} \varliminf_{n \in \mathbb{N}_0} D_n(c) = \inf_{c>0} \varlimsup_{n \in \mathbb{N}} D_n(c). \qquad (8.6.25)$$

(ii) *Assertions A and B are equivalent.*

(A) *For every subsequence* $\mathbb{N}_0 \subset \mathbb{N}$,

$$\varliminf_{n \in \mathbb{N}_0} D_n(0) > 0 \quad \text{implies} \quad \inf_{c>0} \varlimsup_{n \in \mathbb{N}_0} D_n(c) < 0.$$

(B) *For every subsequence* $\mathbb{N}_0 \subset \mathbb{N}$,

$$\inf_{c>0} \varliminf_{n \in \mathbb{N}_0} D_n(c) \geq 0 \quad \text{implies} \quad \varliminf_{n \in \mathbb{N}_0} D_n(c) = 0 \quad \text{for every } c \geq 0.$$

(iii) *B implies*

$$\inf_{c>0} \overline{\lim_{n\to\infty}} \, D_n(c) \le 0. \tag{8.6.26}$$

Proof. (i) Let $I := \inf_{c>0} \overline{\lim}_{n\in\mathbb{N}} D_n(c)$. For every $m \in \mathbb{N}$ there exists $n_m \in \mathbb{N}$, $n_m > n_{m-1}$, such that $D_{n_m}(m) > I - 1/m$. This implies $\underline{\lim}_{m\to\infty} D_{n_m}(m) \ge I$, hence $\underline{\lim}_{m\to\infty} D_{n_m}(c) \ge I$ for every $c > 0$. Therefore, (8.6.25) holds with $\mathbb{N}_0 = \{n_m : m \in \mathbb{N}\}$.

(ii) A implies B. Assume that $\inf_{c>0} \underline{\lim}_{n\in\mathbb{N}_0} D_n(c) \ge 0$, and $\overline{\lim}_{n\in\mathbb{N}_0} D_n(c) > 0$ for some $c \ge 0$, which implies $\overline{\lim}_{n\in\mathbb{N}_0} D_n(0) > 0$. Hence there exists a subsequence $\mathbb{N}_1 \subset \mathbb{N}_0$ such that $\lim_{n\in\mathbb{N}_1} D_n(0) > 0$. Because of A, this implies $\inf_{c>0} \overline{\lim}_{n\in\mathbb{N}_1} D_n(c) < 0$, which is contradictory.

B implies A. Assume that $\underline{\lim}_{n\in\mathbb{N}_0} D_n(0) > 0$, and $\inf_{c>0} \overline{\lim}_{n\in\mathbb{N}_0} D_n(c) \ge 0$. By (i) there exists a subsequence $\mathbb{N}_1 \subset \mathbb{N}_0$ such that $\inf_{c>0} \underline{\lim}_{n\in\mathbb{N}_1} D_n(c) \ge 0$. Because of B this implies $\lim_{n\in\mathbb{N}_1} D_n(0) = 0$, which is contradictory.

(iii) If $\inf_{c>0} \overline{\lim}_{n\to\infty} D_n(c) > 0$, then there exists by (i) a subsequence $\mathbb{N}_0 \subset \mathbb{N}$ such that $\inf_{c>0} \underline{\lim}_{n\in\mathbb{N}_0} D_n(c) > 0$, which is impossible according to B.

$\square$

Chapter 9
Miscellaneous results on asymptotic distributions

9.1 Examples of ML estimators

Contains various examples illustrating the determination of asymptotically efficient estimator sequences by the ML method and by the improvement procedure.

Example 9.1.1. The problem is to estimate ϑ in the family $\{N_{(\vartheta, a^2\vartheta^2)} : \vartheta > 0\}$ of normal distributions with known coefficient of variation $a > 0$. It will be convenient to consider $\{N_{(\vartheta, a^2\vartheta^2)} : \vartheta > 0\}$ as a scale parameter family, generated by $N_{(1,a^2)}$, and to write the density of $N_{(\vartheta, a^2\vartheta^2)}$ as

$$x \to \frac{1}{\vartheta} q_a\left(\frac{x}{\vartheta}\right), \quad \text{with } q_a(x) = \frac{1}{\sqrt{2\pi}a} \exp\left[-(x-1)^2/2a^2\right]. \tag{9.1.2}$$

It is straightforward to see that $(\sum_1^n x_\nu, \sum_1^n x_\nu^2)$ is a minimal sufficient statistic for the sample size n. An equivalent statistic is $(\bar{x}_n, s_n^2(\underline{x}))$, with $s_n^2(\underline{x}) := n^{-1} \sum_1^n (x_\nu - \bar{x}_n)^2$.

The ML estimator for ϑ can be written as

$$\vartheta^{(n)}(\underline{x}) := \left((4a^2 s_n^2(\underline{x}) + (1 + 4a^2)\bar{x}_n^2)^{1/2} - \bar{x}_n\right)/2a^2. \tag{9.1.3}$$

$\vartheta^{(n)}$ is consistent for ϑ, since $\bar{x}_n$ is consistent for ϑ, and $s_n^2(\underline{x})$ consistent for $a^2\vartheta^2$. Since $\vartheta^{(n)}$ is equivariant under dilations, the consistency is locally uniform.

Using

$$\ell^{(1)}(x, \vartheta) = \frac{1}{\vartheta a^2}\left(\left(\frac{x}{\vartheta}\right)^2 - \frac{x}{\vartheta} - a^2\right)$$

$$\ell^{(1,1)}(x, \vartheta) = \frac{1}{\vartheta^2 a^2}\left(-3\left(\frac{x}{\vartheta}\right)^2 + 2\frac{x}{\vartheta} + a^2\right)$$

it is easy to check conditions (7.5.1), (7.5.2) and the other regularity conditions of Theorem 7.5.5.

Since

$$N_{(\vartheta,a^2\vartheta^2)}\left(\big(\ell^{(1)}(\cdot,\vartheta)\big)^2\right) = \frac{1}{\vartheta^2}\frac{1+2a^2}{a^2},$$

Theorem 7.5.5 implies

$$N^n_{(\vartheta,a^2\vartheta^2)} \circ n^{1/2}(\vartheta^{(n)} - \vartheta)/\vartheta \Rightarrow N_{(0,a^2/(1+2a^2))}. \qquad (9.1.4)$$

Since $\vartheta^{(n)}$ is equivariant under dilations, the distribution of $\vartheta^{(n)}/\vartheta$ under $N^n_{(\vartheta,a^2\vartheta^2)}$ is independent of ϑ. According to Exercise 9.3.6 there exists a factor c_n such that $c_n\vartheta^{(n)}$ is mean unbiased. Since $c_n - 1 = o(n^{-1/2})$ (see Exercise 9.3.12), the estimator sequence $c_n\vartheta^{(n)}$, $n \in \mathbb{N}$, is asymptotically efficient in the class of all regular estimator sequences.

Because c_n has to be computed numerically, another approach to an asymptotically efficient sequence of mean unbiased estimators may be of interest. The starting point is that $\vartheta_0^{(n)}(\underline{x}) = \bar{x}_n$ and $\vartheta_1^{(n)}(\underline{x}) = d_n s_n(\underline{x})/a$ with

$$d_n = \sqrt{n/2}\,\Gamma((n-1)/2)/\Gamma(n/2),$$

are mean unbiased. Hence

$$\vartheta_\alpha^{(n)} := (1-\alpha)\vartheta_0^{(n)} + \alpha\vartheta_1^{(n)} \qquad (9.1.5)$$

is mean unbiased for $\alpha \in [0,1]$.

(i) Show that $\vartheta_{\alpha_n}^{(n)}$ minimizes the quadratic risk in the class of all estimators (9.1.5) if

$$\alpha_n = \frac{a^2}{n\left(\frac{n-1}{2}(\Gamma(\frac{n-1}{2})/\Gamma(\frac{n}{2}))^2 - 1\right) + a^2}.$$

Since $\bar{x}_n$ and $s_n(\underline{x})$ are stochastically independent, the joint asymptotic distribution of $\big(n^{1/2}(\vartheta_0^{(n)} - \vartheta)/\vartheta, n^{1/2}(\vartheta_1^{(n)} - \vartheta)/\vartheta\big)$ is normal with covariance matrix

$$\begin{pmatrix} a^2 & 0 \\ 0 & 1/2 \end{pmatrix}.$$

Hence $n^{1/2}(\vartheta_{\beta_n}^{(n)} - \vartheta)/\vartheta$ is asymptotically normal with variance $((1-\beta)^2a^2 + \beta^2/2)$, for any sequence $\beta_n \to \beta$. Using

$$\lim_{n\to\infty} n\left(\frac{n-1}{2}(\Gamma(\frac{n-1}{2})/\Gamma(\frac{n}{2}))^2 - 1\right) = \frac{1}{2}$$

we obtain $\alpha_n \to 2a^2/(1+2a^2)$. Hence $n^{1/2}(\vartheta_{\alpha_n}^{(n)} - \vartheta)/\vartheta$ is asymptotically normal with variance $a^2/(1+2a^2)$. This is the minimal asymptotic variance. The asymptotic performance of the estimator sequence (9.1.5) with minimal quadratic risk corresponds to the fact that $(1-\beta)^2a^2 + \beta^2/2$ attains its minimum for $\beta = 2a^2/(1 + 2a^2)$. Another sequence of mean unbiased and asymptotically optimal estimators is $\vartheta_\alpha^{(n)}$ with $\alpha = 2a^2/(1 + 2a^2)$, i.e. $\left(\bar{x}_n + 2ac_n s_n(\underline{x})\right)/(1 + 2a^2)$.

Notice that the optimum property of $\vartheta_{\alpha_n}^{(n)}$ for finite sample sizes is an optimality within the class of estimators defined by (9.1.5), and with respect to the quadratic loss function, whereas the asymptotic optimum property means maximal asymptotic concentration on arbitrary intervals containing ϑ within the class of "regular" estimator sequences.

Since the sufficient statistic is not complete (recall that $N_{(\vartheta, a^2 \vartheta^2)}^n (\vartheta_0^{(n)} - \vartheta_1^{(n)}) = 0$) there may be mean unbiased estimators which are asymptotically equivalent to $\vartheta_{\alpha_n}^{(n)}$, but superior for finite sample sizes. Kariya (1989, p. 924) gives an equivariant estimator of minimal risk for the loss function $L(u, \vartheta) = (u - \vartheta)^2/\vartheta^2$, for $a = 1$. See also Gleser and Healy (1976, p. 979).

The equivariance of $\vartheta^{(n)}$ can also be used to obtain sequences of median unbiased estimators, of confidence bounds, and of tests for ϑ which are asymptotically efficient.

Whether the conditional tests for ϑ discussed in literature (see Hinkley, 1977, or Lehmann, 1986, p. 549, Example 7), based on the conditional distribution of s_n, given $s_n/\bar{x}_n$, share the property of asymptotic efficiency, remains to be shown.

Readers interested in the estimation of the mean if the coefficient of variation is known may consult Joshi and Sathe (1982) and the references cited there. Multidimensional generalizations of this problem can be found in Kariya, Giri and Perron (1988).

Exercise 9.1.6. By Example 3.3.2, the estimators $\mu^{(n)}(\underline{x}) = \bar{x}_n$, $\lambda^{(n)}(\underline{x}) = (n - 3)/\sum_1^n (x_\nu^{-1} - \bar{x}_n^{-1})$ are mean unbiased with minimal convex risk among all mean unbiased estimators in the family of inverse normal distributions.

(i) Use the relations

$$\int x P_{\mu,\lambda}(dx) = \mu, \quad \int x^{-1} P_{\mu,\lambda}(dx) = \lambda^{-1} + \mu^{-1},$$

$$\int (x - \mu)^2 P_{\mu,\lambda}(dx) = \mu^3/\lambda, \quad \int \left(x^{-1} - (\lambda^{-1} + \mu^{-1})\right)^2 P_{\mu,\lambda}(dx) = \frac{2}{\lambda^2} + \frac{1}{\lambda\mu},$$

and observe that $\lambda \sum_1^n (x_\nu^{-1} - \bar{x}_n^{-1})$ is distributed as χ_{n-1}^2, independently of $\bar{x}_n$, to show that $n^{1/2}\left((\mu^{(n)}, \lambda^{(n)}) - (\mu, \lambda)\right)$ is asymptotically normal with minimal covariance matrix

$$\Lambda(\mu, \lambda) = \begin{pmatrix} \mu^3/\lambda & 0 \\ 0 & 2\lambda^2 \end{pmatrix}.$$

(ii) Use the relation

$$\frac{\lambda}{2\mu^2} \sum_1^n \frac{(x_\nu - \mu)^2}{x_\nu} = \frac{\lambda}{2} \sum_1^n (x_\nu^{-1} - \bar{x}_n^{-1}) + \frac{n\lambda(\bar{x}_n - \mu)^2}{2\mu^2 \bar{x}_n}$$

to show that $(\bar{x}_n, n/\sum_1^n(x_\nu^{-1} - \bar{x}_n))$ is the ML estimator for (μ, λ).

The following example illustrates the use of various estimators for Λ in the improvement procedure.

Example 9.1.7. For $\vartheta \in [0,1]$ let

$$P_\vartheta = (1 - \vartheta)N_{(0,\sigma^2)} + \vartheta N_{(1,\sigma^2)},$$

with σ^2 known. The problem is to estimate ϑ.

Since $\int x P_\vartheta(dx) = \vartheta$, we obtain $\vartheta_0^{(n)}(\underline{x}) := \bar{x}_n$ as a preliminary $\sqrt{n}$–consistent estimator sequence for ϑ. The estimator sequence $\vartheta_0^{(n)}$, $n \in \mathbb{N}$, is asymptotically normal with variance $\vartheta(1 - \vartheta) + \sigma^2$.

Since P_ϑ has Lebesgue density

$$p(x, \vartheta) = \frac{1}{\sigma} \left[(1 - \vartheta)\varphi(\frac{x}{\sigma}) + \vartheta\varphi(\frac{x-1}{\sigma}) \right], \tag{9.1.8}$$

we have

$$\ell^{(1)}(x, \vartheta) = \left(\varphi(\frac{x-1}{\sigma}) - \varphi(\frac{x}{\sigma}) \right) \Big/ \left((1 - \vartheta)\varphi(\frac{x}{\sigma}) + \vartheta\varphi(\frac{x-1}{\sigma}) \right),$$

and

$$\ell^{(1,1)}(x, \vartheta) = -\ell^{(1)}(x, \vartheta)^2,$$

from which conditions (7.5.1), (7.5.2) and the other regularity conditions of Theorem 7.5.9 follow easily for every $\vartheta \in \Theta$.

By Exercise 7.5.11 the estimators

$$L_1^{(n)}(\underline{x}) := L(\bar{x}_n)$$

and

$$L_2^{(n)}(\underline{x}) := n^{-1} \sum_1^n \ell^{(1)}(x_\nu, \bar{x}_n)^2$$

are consistent for $L(\vartheta)$, locally uniformly on $(0, 1)$. Since $L(\bar{x}_n)$ has to be computed by numerical integration, $L_2^{(n)}$ seems to be a convenient alternative. According to Theorem 7.5.9, both estimator sequences

$$\hat{\vartheta}_i^{(n)}(\underline{x}) := \bar{x}_n + n^{-1} \sum_1^n \ell^{(1)}(x_\nu, \bar{x}_n)/L_i^{(n)}(\underline{x}), \qquad i = 1, 2, \tag{9.1.9}$$

are asymptotically normal with minimal variance $L(\vartheta)^{-1}$, locally uniformly on $(0, 1)$.

The following table shows the performance of the estimators $\vartheta_i^{(n)}$, $i = 0, 1, 2$, in a simulation experiment.

Table 9.1. Performance of the estimators $\vartheta_i^{(n)}$ for $\vartheta = 0.1$, $\sigma^2 = 0.1$ and sample size $n = 100$, based on $N = 10000$ simulations.

estimator	bias	mean deviation theor.	mean deviation empir.	e.m.	coverage
$\vartheta_0^{(n)}$	0.001	1.263	1.249	0.024	0.891
$\vartheta_1^{(n)}$	0.006	1	1.017	0.022	0.879
$\vartheta_2^{(n)}$	−0.061	1	1.115	0.022	0.821

The performance of each estimator $\vartheta_i^{(n)}$ is measured by its bias, $\vartheta_i^{(n)} - \vartheta$, and its standardized mean deviation, $n^{1/2}|\vartheta_i^{(n)} - \vartheta|$. The average of these quantities over the N simulation experiments is to be set against the theoretical values.

In the case of bias, the empirical value is $N^{-1} \sum_{j=1}^{N} \vartheta_i^{(n)}(\underline{x}_j) - \vartheta$ (where $\underline{x}_j$ is the sample obtained in the j-th simulation experiment). The theoretical value is zero. To evaluate the relevance of the bias, it has to be seen in relation to the random error of the estimator. Therefore, we give in the column "bias" not the bias of $\vartheta_i^{(n)}$ as estimated by the simulation experiment, but its relation to the 99%–error bound of $\vartheta_i^{(n)}$, computed as $2.58 s_{i*}$, where s_{i*}^2 is the variance between $\vartheta_i^{(n)}(\underline{x}_j)$, $j = 1, \ldots, N$.

In case of the standardized mean deviation, the empirical value is

$$N^{-1} \sum_{j=1}^{N} n^{1/2}|\vartheta_i^{(n)}(\underline{x}_j) - \vartheta|.$$

The theoretical value is computed as $\sqrt{2/\pi}\sigma_i(\vartheta)$, with $\sigma_i^2(\vartheta)$ denoting the asymptotic variance of $\vartheta_i^{(n)}$.

To assess the deviation of the empirical mean deviation from its theoretical value, we give, in the column "e.m.", the 99%–error margin for the empirical values, computed from the variance between the N simulation results (i.e. $2.58 s_i/\sqrt{N}$, where s_i^2 is the variance between $n^{1/2}|\vartheta_i^{(n)}(\underline{x}_j) - \vartheta|$, $j = 1, \ldots, N$).

To make these results transparent, the numbers under the headline "mean deviation" are not given in their absolute value, but in their relation to the mean deviation corresponding to the minimal asymptotic variance.

Finally, we present under "coverage" the relative frequencies with which the symmetric 0.9–confidence interval covers the true value ϑ_0.

Now follow examples concerning the family of gamma distributions $\{\Gamma_{a,b} : a, b > 0\}$, with density

$$x \to \frac{1}{a^b \Gamma(b)} x^{b-1} \exp[-x/a], \qquad x > 0.$$

Example 9.1.10. To represent the ML estimator we use the function $H(u) := \log u - \psi(u)$, $u > 0$, where $\psi(u) := \Gamma'(u)/\Gamma(u)$ is the Gaussian ψ–function. Using these notations, the ML estimator for (a, b) may be written as

$$a^{(n)}(\underline{x}) := \bar{x}_n / b^{(n)}(\underline{x}) \tag{9.1.11'}$$

$$b^{(n)}(\underline{x}) := H^{-1}\left(\log\left(\bar{x}_n / \left(\prod_1^n x_\nu\right)^{1/n}\right)\right). \tag{9.1.11''}$$

Approximations to $b^{(n)}$ are given in Bowman and Shenton (1983) and the literature cited there.

It is easily seen from $\bar{x}_n \to ab$ and $n^{-1} \sum_1^n \log x_\nu \to \psi(b) + \log a$ $\Gamma_{a,b}^n$–a.e. that $(a^{(n)}, b^{(n)})$, $n \in \mathbb{N}$, is consistent for (a, b).

According to Theorem 7.5.5, $n^{1/2}\big((a^{(n)}, b^{(n)}) - (a, b)\big)$ is asymptotically normal with covariance matrix

$$\Lambda(a, b) = \frac{1}{b\psi'(b) - 1} \begin{pmatrix} a^2 \psi'(b) & -a \\ -a & b \end{pmatrix}. \tag{9.1.12}$$

To prove that $b\psi'(b) - 1 > 0$, use $\psi'(x) = \sum_0^\infty (x + \nu)^{-2}$ (see Whittaker and Watson, 1958, p. 241, equation 12.16) and $x^{-1} = \sum_0^\infty \left(\frac{1}{x+\nu} - \frac{1}{x+\nu+1}\right)$.

Since the estimators $a^{(n)}, b^{(n)}$ are equivariant in the sense that $a^{(n)}(\alpha\underline{x}) = \alpha a^{(n)}(\underline{x})$, $b^{(n)}(\alpha\underline{x}) = b^{(n)}(\underline{x})$, the distribution of $\big(n^{1/2}(a^{(n)} - a)/a, n^{1/2}(b^{(n)} - b)\big)$ is independent of $a > 0$. Correspondingly, the asymptotic distribution has covariance matrix $\Lambda(1, b)$.

Since the ML estimator for b, given by (9.1.11''), is difficult to compute, the improvement procedure (7.5.10) offers an attractive alternative.

As a preliminary $\sqrt{n}$–consistent estimator sequence one can use

$$a_0^{(n)}(\underline{x}) = s_n^2(\underline{x})/\bar{x}_n \qquad (9.1.13')$$
$$b_0^{(n)}(\underline{x}) = \bar{x}_n^2/s_n^2(\underline{x}). \qquad (9.1.13'')$$

The estimators (9.1.13) can be obtained as *moment estimators* from

$$\int x\Gamma_{a,b}(dx) = ab, \quad \int x^2\Gamma_{a,b}(dx) = a^2 b(1+b).$$

(Using $\int \log x\Gamma_{a,b}(dx)$ instead of $\int x^2\Gamma_{a,b}(dx)$ leads to the ML estimators (9.1.11).)

$n^{1/2}((a_0^{(n)}, b_0^{(n)}) - (a,b))$, $n \in \mathbb{N}$, is asymptotically normal with covariance matrix

$$\Sigma(a,b) = \begin{pmatrix} a^2(2b+3)/b & -2a(b+1) \\ -2a(b+1) & 2b(b+1) \end{pmatrix}. \qquad (9.1.14)$$

To check the regularity conditions for the application of Theorem 7.5.9 we use the following relations.

$$\ell^{(1)}(x,a,b) = \frac{1}{a}\left(\frac{x}{a} - b\right)$$
$$\ell^{(2)}(x,a,b) = \log\frac{x}{a} - \psi'(b)$$
$$\ell^{(1,1)}(x,a,b) = -\frac{1}{a^2}\left(2\frac{x}{a} - b\right) \qquad (9.1.15)$$
$$\ell^{(2,1)}(x,a,b) = \ell^{(1,2)}(x,a,b) = -1/a$$
$$\ell^{(2,2)}(x,a,b) = -\psi'(b),$$

$$\int x^k\Gamma_{a,b}(dx) = a^k b(b+1)\cdot\ldots\cdot(b+k-1) \quad \text{for } k \in \mathbb{N}$$
$$\int x^k \log x\Gamma_{a,b}(dx) = (ab)^k\Gamma_{1,b+k}(\log) + (ab)^k \log a, \quad k = 0,1,$$

$$(9.1.16)$$

and

$$\frac{\partial^k}{\partial b^k}\Gamma(b) = \Gamma(b)\Gamma_{1,b}((\log)^k), \quad k \in \mathbb{N}. \qquad (9.1.17)$$

To estimate $\Lambda(a,b)$, we use $\Lambda_1^{(n)}(\underline{x}):=\Lambda(a_0^{(n)}(\underline{x}), b_0^{(n)}(\underline{x}))$ and $\Lambda_2^{(n)}(\underline{x})$, the inverse of the matrix with elements

$$n^{-1}\sum_1^n \ell^{(i)}(x_\nu, a_0^{(n)}(\underline{x}), b_0^{(n)}(\underline{x}))\ell^{(j)}(x_\nu, a_0^{(n)}(\underline{x}), b_0^{(n)}(\underline{x})), \quad i,j = 1,2.$$

In principle one could also consider to estimate $L_{ij}(a, b)$ by

$$-n^{-1} \sum_1^n \ell^{(ij)}\big(x_\nu, a_0^{(n)}(\underline{x}), b_0^{(n)}(\underline{x})\big),$$

but these values agree with $L_{ij}\big(a_0^{(n)}(\underline{x}), b_0^{(n)}(\underline{x})\big)$.

Since $\sum_1^n \ell^{(1)}\big(x_\nu, a_0^{(n)}(\underline{x}), b_0^{(n)}(\underline{x})\big) = 0$, we obtain from Theorem 7.5.9, the following improved estimators

$$\hat{a}^{(n)}(\underline{x}) := a_0^{(n)}(\underline{x}) + n^{-1} \sum_1^n \Lambda_{12}^{(n)}(\underline{x})\ell^{(2)}\big(x_\nu, a_0^{(n)}(\underline{x}), b_0^{(n)}(\underline{x})\big) \quad (9.1.18')$$

$$\hat{b}^{(n)}(\underline{x}) := b_0^{(n)}(\underline{x}) + n^{-1} \sum_1^n \Lambda_{22}^{(n)}(\underline{x})\ell^{(2)}\big(x_\nu, a_0^{(n)}(\underline{x}), b_0^{(n)}(\underline{x})\big). \quad (9.1.18'')$$

In the following, $(\hat{a}_i^{(n)}, \hat{b}_i^{(n)})$ denotes the improved estimator, computed according to (9.1.18) with the matrix $\Lambda_i^{(n)}(\underline{x})$.

We obtain

$$\hat{a}_1^{(n)}(\underline{x}) = a_0^{(n)}(\underline{x})\big(1 + f_n(\underline{x})\big) \qquad (9.1.19')$$

$$\hat{b}_1^{(n)}(\underline{x}) = b_0^{(n)}(\underline{x})\big(1 - f_n(\underline{x})\big) \qquad (9.1.19'')$$

with

$$f_n(\underline{x}) = \frac{\psi(b_0^{(n)}(\underline{x})) + \log a_0^{(n)}(\underline{x}) - n^{-1} \sum_1^n \log x_\nu}{b_0^{(n)}(\underline{x})\psi'(b_0^{(n)}(\underline{x})) - 1}.$$

Furthermore

$$\hat{a}_2^{(n)}(\underline{x}) = a_0^{(n)}(\underline{x}) \qquad (9.1.20')$$

$$+ g_n(\underline{x})\frac{a_0^{(n)}(\underline{x})^2 h_n(\underline{x})}{s_n^2(\underline{x})(g_n(\underline{x})^2 + n^{-1}\sum_1^n(\log x_\nu)^2 - (n^{-1}\sum_1^n \log x_\nu)^2) - h_n(\underline{x})^2},$$

$$\hat{b}_2^{(n)}(\underline{x}) = b_0^{(n)}(\underline{x}) \qquad (9.1.20'')$$

$$- g_n(\underline{x})\frac{s_n^2(\underline{x})}{s_n^2(\underline{x})(g_n(\underline{x})^2 + n^{-1}\sum_1^n(\log x_\nu)^2 - (n^{-1}\sum_1^n \log x_\nu)^2) - h_n(\underline{x})^2},$$

with

$$g_n(\underline{x}) = \psi'\big(b_0^{(n)}(\underline{x})\big) + \log a_0^{(n)}(\underline{x}) - n^{-1} \sum_1^n \log x_\nu,$$

$$h_n(\underline{x}) = n^{-1} \sum_1^n x_\nu \log x_\nu - \bar{x}_n n^{-1} \sum_1^n \log x_\nu.$$

The following table shows the behavior of the estimators $(a_i^{(n)}, b_i^{(n)})$, $i = 0, 1, 2$, in a simulation experiment. The content of this table is described in Example 9.1.7, pp. 303/4.

Since a is a scale parameter, it suffices to present the results for $a = 1$.

Table 9.2   Performance of the estimators for $b = 2$ and sample size $n = 100$, based on $N = 10000$ simulations.

estimator	bias	mean deviation theor.	empir.	e.m.	coverage
$a_0^{(n)}$	-0.034	1.254	1.230	0.024	0.880
$\hat{a}_1^{(n)}$	-0.026	1	0.996	0.019	0.884
$\hat{a}_2^{(n)}$	-0.074	1	1.015	0.020	0.860
$b_0^{(n)}$	0.099	1.318	1.347	0.028	0.911
$\hat{b}_1^{(n)}$	0.043	1	1.040	0.022	0.892
$\hat{b}_2^{(n)}$	0.116	1	1.082	0.024	0.901

In Examples 5.6.9 and 5.6.13, upper confidence bounds for a and b were obtained which are maximally concentrated (in a nonasymptotic sense). Since these confidence bounds can be computed numerically only, asymptotic confidence bounds with asymptotic optimum properties are still of interest. Such ones can be obtained according to Proposition 7.3.7. This leads to the following confidence bounds with asymptotic covering probability β for a and b, respectively (see (9.1.12))

$$a_\beta^{(n)}(\underline{x}) := a^{(n)}(\underline{x})\big(1 + n^{-1/2} u_{1-\beta} \sigma_1(b^{(n)}(\underline{x}))\big) \qquad (9.1.21')$$
$$\text{with } \sigma_1^2(b) = \psi'(b)/\big(b\psi'(b) - 1\big)$$
$$b_\beta^{(n)}(\underline{x}) := b^{(n)}(\underline{x}) + n^{-1/2} u_{1-\beta} \sigma_2(b^{(n)}(\underline{x})), \qquad (9.1.21'')$$
$$\text{with } \sigma_2^2(b) = b/\big(b\psi'(b) - 1\big),$$

where $(a^{(n)}, b^{(n)})$, $n \in \mathbb{N}$, is an asymptotically efficient estimator sequence.

Exercise 9.1.22. The problem is to find an estimator for the β–quantile in the family $\{\Gamma_{a,b} : a, b > 0\}$. If $q_\beta(b)$ denotes the β–quantile of $\Gamma_{1,b}$, the β–quantile of $\Gamma_{a,b}$ is $q_\beta(a,b) = aq_\beta(b)$.

(i) Show that the minimal asymptotic variance for $q_\beta(a,b)$ is

$$\sigma_0^2(a,b) = a^2\left(q_\beta(b)^2/b + \left(q_\beta(b)/\sqrt{b} - f(b)\right)^2/\left(b\psi'(b) - 1\right)\right)$$

with

$$f(b) = \sqrt{b}\,\Gamma(b)\exp\left[q_\beta(b)\right]q_\beta(b)^{1-b}\left(\beta\psi(b) - \int_0^{q_\beta(b)} \log t\,\Gamma_{1,b}(dt)\right).$$

An estimator sequence with minimal asymptotic variance can be obtained as $q_\beta^{(n)}(x) = a^{(n)}(x)q_\beta\left(b^{(n)}(x)\right)$, if $(a^{(n)}, b^{(n)})$, $n \in \mathbb{N}$, is asymptotically efficient.

Another estimator is the sample quantile $\hat{q}_\beta^{(n)} = x_{[\beta n]:n}$.

(ii) Show that $\hat{q}_\beta^{(n)}$ is asymptotically normal with variance

$$\hat{\sigma}^2(a,b) = a^2\Gamma(b)^2\exp\left[2q_\beta(b)\right]\beta(1 - \beta)/q_\beta(b)^{2(b-1)}.$$

Hint: Use that $n^{1/2}\left(x_{[n\beta]:n} - q_\beta(P)\right)$ is asymptotically normal with variance $\beta(1 - \beta)/p\left(q_\beta(P)\right)^2$, provided the Lebesgue density p of P is positive and continuous at $q_\beta(P)$ (see Reiß, 1989, p. 109, Example 4.1.1).

The following figure shows the asymptotic relative efficiency of $\hat{q}_\beta^{(n)}$ with respect to $q_\beta^{(n)}$ as a function of β for $b = 1$.

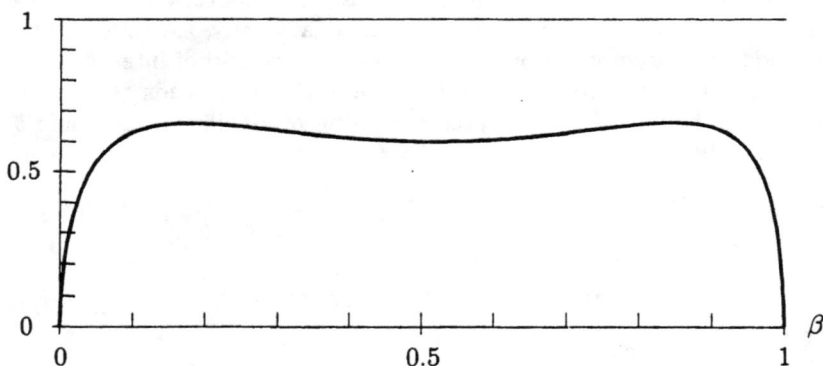

Figure 9.1

Example 9.1.23. Let $(X, \mathcal{A})$ be a measurable space and $\mu|\mathcal{A}$ a σ–finite measure. For $\vartheta \in \Theta$ (an open subset of $\mathbb{R}$) and $\eta \in H$ let $P_{\vartheta,\eta}|\mathcal{A}$ be a p–measure equivalent to μ. Assume that there exists a statistic S from $(X, \mathcal{A})$ to some measurable space $(Y, \mathcal{B})$ which is sufficient for each of the families $\{P_{\vartheta,\eta} : \eta \in H\}$, $\vartheta \in \Theta$. (Notice that S is assumed to be the same for every $\vartheta \in \Theta$.) According to the Factorization Theorem 1.2.10 the μ–density of $P_{\vartheta,\eta}$ can be written as

$$p(x, \vartheta, \eta) = q(x, \vartheta)p_0\big(S(x), \vartheta, \eta\big), \qquad x \in X.$$

Throughout the following we assume that this factorization is standardized such that $p_0(\cdot, \vartheta, \eta)$ is the density of $P_{\vartheta,\eta} \circ S$ with respect to some σ–finite measure $\nu|\mathcal{B}$.

Under this standardization, we have

$$P_{\vartheta,\eta}\big(q(\cdot, \delta)/q(\cdot, \vartheta)\big) = 1 \qquad \text{for all } \vartheta, \delta \in \Theta, \ \eta \in H.$$

This can be seen from

$$
\begin{aligned}
1 = \nu\big(p_0(\cdot, \vartheta, \eta)\big) &= P_{\delta,\eta} \circ S\big(p_0(\cdot, \vartheta, \eta)/p_0(\cdot, \delta, \eta)\big) \\
&= P_{\delta,\eta}\big(p_0(S(\cdot), \vartheta, \eta)/p_0(S(\cdot), \delta, \eta)\big) \\
&= \mu\big(q(\cdot, \delta)p_0(S(\cdot), \vartheta, \eta)\big) = P_{\vartheta,\eta}\big(q(\cdot, \delta)/q(\cdot, \vartheta)\big).
\end{aligned}
$$

Since the function "log" is strictly concave, Jensen's inequality implies

$$P_{\vartheta,\eta}\big(\big[\log q(\cdot, \delta) - \log q(\cdot, \vartheta)\big]\big) < 0 \qquad \text{for } \delta \neq \vartheta, \ \eta \in H$$

unless $q(\cdot, \vartheta)/q(\cdot, \vartheta) = 1$ μ–a.e. Hence $\delta \to \log q(\cdot, \delta)$ fulfills condition (6.3.2). This motivates the definition of "conditional" ML estimators, by

$$\sum_1^n \log q(x_\nu, \delta) = \max.$$

If the assumptions of Theorem 7.5.5 are fulfilled, the sequence of conditional ML estimators, say $\vartheta_*^{(n)}$, $n \in \mathbb{N}$, is asymptotically normal with asymptotic variance

$$\sigma_*^2(\vartheta, \eta) = 1/P_{\vartheta,\eta}\big((q^{(\cdot)}(\cdot, \vartheta)/q(\cdot, \vartheta))^2\big).$$

As a consequence of the asymptotic optimality of the ML estimator, it is clear that $\vartheta_*^{(n)}$ cannot be asymptotically better than $\vartheta^{(n)}$, the so–called "marginal" ML estimator of ϑ, i.e. the 1st component of the ML estimator $(\vartheta^{(n)}, \eta^{(n)})$ for (ϑ, η). It will, in fact, be inferior, in general. To show this we assume that $H \subset \mathbb{R}^q$. Denoting the derivative with respect to ϑ by (0), and the derivative with respect to η_i by (i), we have

$$L_{00}(\vartheta,\eta) = P_{\vartheta,\eta}\Big(\big(\frac{q^{(\cdot)}(\cdot,\vartheta)}{q(\cdot,\vartheta)}\big)^2\Big) + P_{\vartheta,\eta} \circ S\Big(\big(\frac{p_0^{(\cdot)}(\cdot,\vartheta,\eta)}{p_0(\cdot,\vartheta,\eta)}\big)^2\Big)$$

$$L_{ij}(\vartheta,\eta) = P_{\vartheta,\eta} \circ S\Big(\frac{p_0^{(i)}}{p_0} \cdot \frac{p_0^{(j)}}{p_0}\Big) \quad \text{unless } i = j = 0.$$

This follows from

$$\mu\big(q(\cdot,\delta)p_0(S(\cdot),\vartheta,\eta)\big) = 1,$$

if differentiation with respect to δ, ϑ, η may be interchanged with μ–integration.

Let $\hat{\vartheta}^{(n)}$ denote the marginal ML estimator for ϑ in the family $P_{\vartheta,\eta} \circ S$, $\vartheta \in \Theta$, $\eta \in \mathrm{H}$. Then $\hat{\vartheta}^{(n)}$ has asymptotic variance

$$\hat{\sigma}^2(\vartheta,\eta) = \frac{\det L'_{00}(\vartheta,\eta)}{\det L(\vartheta,\eta) - P_{\vartheta,\eta}((q^{(\cdot)}(\cdot,\vartheta)/q(\cdot,\vartheta))^2)\det L'_{00}(\vartheta,\eta)},$$

where L'_{00} denotes the matrix obtained by crossing out the 0–th row and 0–th column in the matrix L.

The estimators $\vartheta_*^{(n)}$ and $\hat{\vartheta}^{(n)}$ are asymptotically independent, since their score functions,

$$q^{(\cdot)}(\cdot,\vartheta)/q(\cdot,\vartheta) \quad \text{and} \quad p_0\big(S(\cdot),\vartheta,\eta\big)/p_0\big(S(\cdot),\vartheta,\eta\big),$$

are orthogonal under $P_{\vartheta,\eta}$. Hence,

$$\alpha\vartheta_*^{(n)} + (1-\alpha)\hat{\vartheta}^{(n)}, \qquad \alpha \in [0,1],$$

is asymptotically normal with variance $\alpha^2\sigma_*^2(\vartheta,\eta) + (1-\alpha)^2\hat{\sigma}^2(\vartheta,\eta)$. The minimal value of this asymptotic variance, attained for

$$\alpha(\vartheta,\eta) = \hat{\sigma}^2(\vartheta,\eta)/\big(\sigma_*^2(\vartheta,\eta) + \hat{\sigma}^2(\vartheta,\eta)\big),$$

is equal to

$$\sigma_*^2(\vartheta,\eta)\hat{\sigma}^2(\vartheta,\eta)/\big(\sigma_*^2(\vartheta,\eta) + \hat{\sigma}^2(\vartheta,\eta)\big) = \Lambda_{00}(\vartheta,\eta).$$

The optimal linear combination of the conditional ML estimator, given S, and the marginal ML estimator, based on the distribution of S, is asymptotically efficient.

9.2 Tolerance bounds

Introduces the concept of tolerance bounds and gives an asymptotic bound for the concentration of tolerance bounds.

Throughout the following X is a measurable subset of $\mathbb{R}$ (usually $X = \mathbb{R}$ or $X = \mathbb{R}_+$). Let $\{P_\vartheta : \vartheta \in \Theta\}$ be a family of p–measures on $\mathbb{B} \cap X$, and $\underline{x} \in X^n$ a sample from P_ϑ^n. Wanted an estimator for a subset of $\mathbb{R}$, which contains the amount β of P_ϑ. In a slightly different interpretation this is a subset of $\mathbb{R}$ which will contain the next observation with probability β.

For $\underline{x} \in X^n$ let $I_\beta^{(n)}(\underline{x})$ be a nonempty, measurable subset of X.

$I_\beta^{(n)}$ is a β–*expectation tolerance set* if

$$\int P_\vartheta\big(I_\beta^{(n)}(\underline{x})\big) P_\vartheta^n(d\underline{x}) = \beta \qquad \text{for } \vartheta \in \Theta. \tag{9.2.1}$$

Motivated by the intended applications, we shall require that $I_\beta^{(n)}(\underline{x})$ is an interval or a ray. In the following we concentrate our attention on rays, say $I_\beta^{(n)}(\underline{x}) = \big(-\infty, T_\beta^{(n)}(\underline{x})\big]$. The map $T_\beta^{(n)} : X^n \to X$ is an *upper β–expectation tolerance bound* if

$$\int F_\vartheta\big(T_\beta^{(n)}(\underline{x})\big) P_\vartheta^n(d\underline{x}) = \beta \qquad \text{for } \vartheta \in \Theta, \tag{9.2.1'}$$

where F_ϑ denotes the distribution function of P_ϑ.

We start with special conditions under which tolerance bounds fulfilling (9.2.1') can easily be obtained.

If Θ is a group of transformations $\vartheta : X \to X$, and $P_\vartheta = P \circ (x \to \vartheta x)$, one may try to find sets $I_\beta^{(n)}(\underline{x})$ which are equivariant in the sense that $I_\beta^{(n)}(\vartheta x_1, \dots, \vartheta x_n) = \vartheta I_\beta^{(n)}(x_1, \dots, x_n)$, and which fulfill (9.2.1) for P. This implies (9.2.1) for arbitrary $\vartheta \in \Theta$, since the distribution of $\underline{x} \to P_\vartheta(I_\beta^{(n)}(\underline{x}))$ under P_ϑ^n does not depend on ϑ. We have

$$P_\vartheta^n \circ \big(\underline{x} \to P_\vartheta(I_\beta^{(n)}(\underline{x}))\big) = P^n \circ \big(\underline{x} \to P_\vartheta(I_\beta^{(n)}(\vartheta\underline{x}))\big)$$
$$= P^n \circ \big(\underline{x} \to P_\vartheta(\vartheta I_\beta^{(n)}(\underline{x}))\big) = P^n \circ \big(\underline{x} \to P(I_\beta^{(n)}(\underline{x}))\big).$$

If $I_\beta^{(n)}(\underline{x}) = \big(-\infty, T_\beta^{(n)}(\underline{x})\big]$, equivariance of the function $T_\beta^{(n)}$ implies equivariance of the set $I_\beta^{(n)}$.

In Section 9.3, pp. 318/9, this will be carried through for families with location and scale parameters.

Another method for the construction of tolerance bounds (suggested by Faulkenberry, 1973) requires a statistic $S^{(1+n)} : X^{1+n} \to Y$ which is sufficient for $\{P_\vartheta^{1+n} : \vartheta \in \Theta\}$. According to Proposition 1.3.1 there exists a Markov kernel $M^{(1+n)}|Y \times \mathbb{B}$ such that $M^{(1+n)}(\cdot, B)$ is a conditional expectation of $(x_0, x_1, \dots, x_n) \to 1_B(x_0)$, given $S^{(1+n)}$, with respect to $\{P_\vartheta^{1+n} : \vartheta \in \Theta\}$. This implies, in particular, that

$$\int M^{(1+n)}(y, B) P_\vartheta^{1+n} \circ S^{(1+n)}(dy) = P_\vartheta(B) \quad \text{for } B \in \mathbb{B}, \ \vartheta \in \Theta.$$

For $y \in Y$ we determine a set $B_\beta^{(n)}(y) \in \mathbb{B} \cap X$ such that

$$M^{(1+n)}(y, B_\beta^{(n)}(y)) = \beta, \tag{9.2.2}$$

and such that $(x_0, y) \to 1_{B_\beta^{(n)}(y)}(x_0)$ is $(\mathbb{B} \cap X) \times \mathcal{B}$–measurable. Let

$$I_\beta^{(n)}(x_1, \ldots, x_n) := \left\{ x_0 \in X : x_0 \in B_\beta^{(n)}(S^{(1+n)}(x_0, x_1, \ldots, x_n)) \right\}.$$

Then $I_\beta^{(n)}$ fulfills (9.2.1), since

$$\int P_\vartheta(I_\beta^{(n)}(\underline{x})) P_\vartheta^n(d\underline{x})$$

$$= P_\vartheta^{1+n} \left\{ (x_0, x_1, \ldots, x_n) \in X^{1+n} : x_0 \in B_\beta^{(n)}(S^{(1+n)}(x_0, x_1, \ldots, x_n)) \right\}$$

$$= \int M^{(1+n)}(y, B_\beta^{(n)}(y)) P_\vartheta^{1+n} \circ S^{(1+n)}(dy) = \beta.$$

(Hint: Apply Proposition 1.10.26 with $f(x_0, y) = 1_{B_\beta^{(n)}(y)}(x_0)$.)

If $M^{(1+n)}(y, \cdot)| \mathbb{B} \cap X$ is nonatomic, there exists a function $u_\beta^{(n)} : Y \to X$ such that (9.2.2) holds with $B_\beta^{(n)}(y) = (-\infty, u_\beta^{(n)}(y)]$, leading to

$$I_\beta^{(n)}(x_1, \ldots, x_n) = \left\{ x_0 \in X : x_0 \leq u_\beta^{(n)}(S^{(1+n)}(x_0, x_1, \ldots, x_n)) \right\}.$$

Observe that the set $I_\beta^{(n)}(x_1, \ldots, x_n)$ thus defined is not necessarily an interval.

Exercise 9.2.3. For $\vartheta > 0$ let $E_\vartheta| \mathbb{B}_+$ denote the *exponential distribution*. According to Example 3.4.17 (with $(x_1, \ldots, x_n)$ replaced by $(x_0, x_1, \ldots, x_n)$), $x \to \frac{n}{y}(1 - \frac{x_0}{y})^{n-1}$, $0 < x_0 < y$, is the Lebesgue density of the conditional distribution of x_0, given $\sum_0^n x_\nu = y$. Show that $u_\beta^{(n)}(y) = (1 - (1 - \beta)^{1/n})y$. This leads to

$$I_\beta^{(n)}(x_1, \ldots, x_n) = \left(0, ((1 - \beta)^{-1/n} - 1) \sum_1^n x_\nu \right)$$

as a β–expectation tolerance interval.

Now we turn to an evaluation of tolerance bounds. After having observed $\underline{x}$, we use $(-\infty, T_\beta^{(n)}(\underline{x})]$ as a half–ray which contains approximately the amount β of P_ϑ. Hence it is desirable that $P_\vartheta^n \circ (F_\vartheta \circ T_\beta^{(n)})$, i.e. the distribution of

$\underline{x} \to F_\vartheta\big(T_\beta^{(n)}(\underline{x})\big)$ under P_ϑ^n, should be concentrated as closely as possible about β. So far, there seems to be no general optimal solution to this problem for every $n \in \mathbb{N}$. This is, perhaps, due to the fact that the machinery working with "sufficiency" and "completeness" does not apply here. Following Wilks (1941), many authors restrict their consideration to tolerance bounds depending on the observations through a sufficient statistic only, without showing that the class of tolerance bounds thus obtained is complete in the sense that for any tolerance bound there is one depending on the observations through the sufficient statistic only, which is at least as good. (This class is complete, of course, if one admits randomized tolerance bounds, based on the Markov kernel depending on x through $S(x)$ only. See Sections 1.1 and 1.3.)

It is clear that there cannot be a tolerance bound, say $\hat{T}_\beta^{(n)}$, which is optimal in the strong sense that for any other β–expectation tolerance bound, say $T_\beta^{(n)}$,

$$P_\vartheta^n\big\{\underline{x} \in X^n : F_\vartheta\big(\hat{T}_\beta^{(n)}(\underline{x})\big) \in (\beta',\beta'')\big\}$$
$$\geq P_\vartheta^n\big\{\underline{x} \in X^n : F_\vartheta\big(T_\beta^{(n)}(\underline{x})\big) \in (\beta',\beta'')\big\} \quad \text{for all } \beta' < \beta < \beta''.$$

This would require that

$$P_\vartheta^n\big\{\underline{x} \in X^n : F_\vartheta\big(\hat{T}_\beta^{(n)}(\underline{x})\big) \leq \beta\big\}$$
$$= P_\vartheta^n\big\{\underline{x} \in X^n : F_\vartheta\big(T_\beta^{(n)}(\underline{x})\big) \leq \beta\big]$$

(see p. 75) a condition which is, in general, incompatible with (9.2.1'). Guttman (1970, Section 3) obtains for the normal distribution tolerance regions which he claims to be "minimax" and "most stringent" (see p. 38), but these optimum properties are based on some "desirability concept" for tolerance regions, the operational significance of which seems to be questionable. Guenther (1975, p. 33, Theorem 1) states that a tolerance bound which is a contraction of a complete sufficient statistic is optimal, but the "proof" of his theorem is beyond the point.

In the absence of an optimal solution for finite sample sizes, we are content with an asymptotic solution.

Proposition 9.2.4. *Assume that P_ϑ has a continuous and positive Lebesgue density on $\mathbb{R}$. If*

$$P_\vartheta^n \circ n^{1/2}\big(F_\vartheta \circ T_\beta^{(n)} - \beta\big) \Rightarrow Q_\vartheta \qquad \text{for } \vartheta \in \Theta \subset \mathbb{R}^p, \tag{9.2.5}$$

then the following holds true for λ^p–a.a. $\vartheta \in \Theta$: $N_{(0,\sigma_\beta^2(\vartheta))} <_s Q_\vartheta$, *with*

$$\sigma_\beta^2(\vartheta) = F_\vartheta^{(\cdot)}\big(q_\beta(\vartheta)\big)\Lambda(\vartheta)F_\vartheta^{(\cdot)}\big(q_\beta(\vartheta)\big)^\top, \tag{9.2.6}$$

where $q_\beta(\vartheta)$ denotes the β–quantile of P_ϑ.

Proof. Since $p(\cdot, \vartheta)$, the Lebesgue density of P_ϑ, is continuous and positive on $\mathbb{R}$, $t \to F_\vartheta(t)$ is increasing. According to Proposition 7.2.1 (applied with $h = F_\vartheta^{-1}$), relation (9.2.5) implies

$$P_\vartheta^n \circ n^{1/2}\big(T_\beta^{(n)} - \kappa_\beta(\vartheta)\big) \Rightarrow Q_\vartheta \circ \big(u \to u/\hat{p}_\beta(\vartheta)\big) \qquad \text{for } \vartheta \in \Theta,$$

with $\hat{p}_\beta(\vartheta) := p\big(q_\beta(\vartheta), \vartheta\big)$.

By Theorem 8.3.1, for λ^P–a.a. $\vartheta \in \Theta$, the limit distribution $Q_\vartheta \circ \big(u \to u/\hat{p}_\beta(\vartheta)\big)$ is more spread out than $N_{(0,\hat{\sigma}_\beta^2(\vartheta))}$ with

$$\hat{\sigma}_\beta^2(\vartheta) = \frac{1}{\hat{p}_\beta(\vartheta)^2} F_\vartheta^{(\cdot)}\big(q_\beta(\vartheta)\big) \Lambda(\vartheta) F_\vartheta^{(\cdot)}\big(q_\beta(\vartheta)\big)^\top. \tag{9.2.7}$$

Hence, for λ^P–a.a. $\vartheta \in \Theta$, Q_ϑ is more spread out than

$$N_{(0,\hat{\sigma}_\beta^2(\vartheta))} \circ \big(u \to \hat{p}_\beta(\vartheta)u\big) = N_{(0,\sigma_\beta^2(\vartheta))}. \qquad \square$$

Recall that (9.2.5) implies $F_\vartheta \circ T_\beta^{(n)} \to \beta$ (P_ϑ^n), hence $\int F_\vartheta\big(T_\beta^{(n)}(\underline{x})\big) P_\vartheta^n(d\underline{x}) \to \beta$. What one expects from (9.2.5) and $\int u Q_\vartheta(du) = 0$ is

$$n^{1/2}\Big(\int F_\vartheta(T_\beta^{(n)}(\underline{x})) P_\vartheta^n(d\underline{x}) - \beta\Big) \to 0,$$

but this follows only under additional conditions (like uniform integrability of $n^{1/2}(F_\vartheta \circ T_\beta^{(n)} - \beta)$ under P_ϑ^n, $n \in \mathbb{N}$; see Section 7.2).

9.3 Probability measures with location– and scale parameters

The results of Sections 7.3 and 7.5 are applied to obtain estimates and confidence bounds for the location– and the scale parameter, and tolerance bounds for the p–measure. Contains sufficient conditions for consistency and asymptotic normality of ML estimators.

Throughout this section, $P|\mathbb{B}$ is a p–measure with positive Lebesgue density p. For $a \in \mathbb{R}$, $c > 0$, let $P_{a,c} := P \circ (x \to a + cx)$, the p–measure generated from P by the transformation $x \to a + cx$.

$$x \to \frac{1}{c}p\big(\frac{x-a}{c}\big), \qquad x \in \mathbb{R},$$

is a Lebesgue density of $P_{a,c}$.

Straightforward modifications apply to families with only a location parameter or a scale parameter.

Notice that certain results of this section apply also to two–parameter families which can be transformed into a location– and scale parameter family, such as the family of lognormal distributions, or the family of Weibull distributions.

Throughout the following, the functions $a^{(n)} : \mathbb{R}^n \to \mathbb{R}$ and $c^{(n)} : \mathbb{R}^n \to \mathbb{R}_+$ are *equivariant* in the following sense: For $\alpha \in \mathbb{R}$ and $\gamma > 0$,

$$a^{(n)}(\alpha + \gamma x_1, \ldots, \alpha + \gamma x_n) = \alpha + \gamma a^{(n)}(x_1, \ldots, x_n) \qquad (9.3.1')$$
$$c^{(n)}(\alpha + \gamma x_1, \ldots, \alpha + \gamma x_n) = \gamma c^{(n)}(x_1, \ldots, x_n). \qquad (9.3.1'')$$

Typical examples of statistics which are equivariant in this sense are

$$\left(\bar{x}_n, \; n^{-1} \sum_{1}^{n} |x_\nu - \bar{x}_n|\right) \quad \text{or} \quad \left(\bar{x}_n, \; \left(n^{-1} \sum_{1}^{n} (x_\nu - \bar{x}_n)^2\right)^{1/2}\right).$$

Proposition 9.3.2. *Assume that $(a^{(n)}, c^{(n)})$ are equivariant. Then the following holds true if $n \geq 2$:*
(i) The joint distribution of $((a^{(n)} - a)/c, \; c^{(n)}/c)$ under $P_{a,c}^n$, say

$$Q^{(n)} := P_{a,c}^n \circ ((a^{(n)} - a)/c, \; c^{(n)}/c), \qquad (9.3.3)$$

does not depend on a and c.
(ii) $Q^{(n)} | \mathbb{B} \times \mathbb{B}_+$ has a positive λ^2–density on $\mathbb{R} \times (0, \infty)$.

Proof. (i) follows immediately from the equivariance.

(ii) Since P has a positive Lebesgue density, the measures $P_{a,c}$, $a \in \mathbb{R}$, $c > 0$, are mutually absolutely continuous. Hence $P_{a,c}^n \circ (a^{(n)}, c^{(n)})$, $a \in \mathbb{R}$, $c > 0$, are mutually absolutely continuous. This implies that the measures $Q^{(n)} \circ ((u, v) \to (a + cu, cv))$, $a \in \mathbb{R}$, $c > 0$, on $\mathbb{B} \times \mathbb{B}_+$ are mutually absolutely continuous. Hence the assertion follows from Lemma 1.9.6, applied for the topological group $G = \mathbb{R} \times (0, \infty)$ with the group operation $(a', c') \circ (a'', c'') = (a' + c'a'', c'c'')$, and the right invariant Haar measure $\nu | \mathbb{B} \times \mathbb{B}_+$ with $\lambda^2 | \mathbb{B} \times \mathbb{B}_+$–density $(u, v) \to v^{-1}$. $\qquad \square$

Example 9.3.4. If $P = N_{(0,1)}$ and $a^{(n)}(\underline{x}) := \bar{x}_n$, $c^{(n)}(\underline{x}) := \left(\frac{1}{n-1} \sum_{1}^{n}(x_\nu - \bar{x}_n)^2\right)^{1/2}$, then

$$Q^{(n)} = N_{(0,1/n)} \times V_n,$$

with

$$V_n = \chi_{n-1}^2 \circ (u \to (u/(n-1))^{1/2}).$$

Exercise 9.3.5. Show that $\int uv Q^{(n)}(d(u,v)) = 0$, if this integral exists and if P is symmetric and $a^{(n)}$, $c^{(n)}$ are equivariant under shifts (i.e. they fulfill (9.3.1) with $\gamma = 1$) and symmetric in the sense that

$$a^{(n)}(-x_1, \ldots, -x_n) = -a^{(n)}(x_1, \ldots, x_n),$$

and

$$c^{(n)}(-x_1, \ldots, -x_n) = c^{(n)}(x_1, \ldots, x_n).$$

To distinguish $(a^{(n)}, c^{(n)})$ as an estimator for (a, c), we have to require that $(a^{(n)}, c^{(n)})$ is centered around (a, c) in some appropriate sense, say mean- or median unbiased. To achieve this in a way which preserves equivariance, we consider the family of functions $(a^{(n)} + uc^{(n)}, vc^{(n)})$, with $u \in \mathbb{R}$ and $v > 0$. By choosing u and v appropriately (say $u = u_n$, $v = v_n$), $a^{(n)} + u_n c^{(n)}$ and $v_n c^{(n)}$ can be made (mean- or median-) unbiased.

Exercise 9.3.6. Show that $a^{(n)} + u_n^* c^{(n)}$ and $v_n^* c^{(n)}$ are mean unbiased for a and c if

$$u_n^* = -\int r Q^{(n)}(d(r,s)) \Big/ \int s Q^{(n)}(d(r,s)),$$

and

$$v_n^* = 1 \Big/ \int s Q^{(n)}(d(r,s)).$$

The same idea can be used to obtain confidence bounds with *exact* covering probability.

Proposition 9.3.7. *For every* $\beta \in (0,1)$ *there exist* $u_{n,\beta} \in \mathbb{R}$ *and* $v_{n,\beta} > 0$ *such that* $a^{(n)} + u_{n,\beta} c^{(n)}$ *and* $v_{n,\beta} c^{(n)}$ *are upper confidence bounds for* a *and* c, *respectively, with covering probability* β, *i.e., for* $a \in \mathbb{R}$, $c > 0$ *and* $n \in \mathbb{N}$, $n \geq 2$,

$$P_{a,c}^n \big\{ \underline{x} \in \mathbb{R}^n : a \leq a^{(n)}(\underline{x}) + u_{n,\beta} c^{(n)}(\underline{x}) \big\} = \beta,$$

and

$$P_{a,c}^n \big\{ \underline{x} \in \mathbb{R}^n : c \leq v_{n,\beta} c^{(n)}(\underline{x}) \big\} = \beta.$$

As a particular case we mention $a^{(n)} + u_{n,1/2} c^{(n)}$ and $v_{n,1/2} c^{(n)}$ as median unbiased estimators for a and c.

Proof. (i) Equivariance of $a^{(n)}$ and $c^{(n)}$ implies

$$P_{a,c}^n \big\{ \underline{x} \in \mathbb{R}^n : a \leq a^{(n)}(\underline{x}) + u c^{(n)}(\underline{x}) \big\}$$

$$= P^n\Big\{\underline{x} \in \mathbb{R}^n : \frac{a^{(n)}(\underline{x}) - a}{c^{(n)}(\underline{x})} \geq -u\Big\}$$

$$= Q^{(n)}\Big\{(r, s) \in \mathbb{R} \times (0, \infty) : \frac{r}{s} \geq -u\Big\}.$$

Since $Q^{(n)}$ has a positive λ^2–density on $\mathbb{R} \times (0, \infty)$, there exists a unique $u \in \mathbb{R}$ such that

$$Q^{(n)}\Big\{(r, s) \in \mathbb{R} \times (0, \infty) : \frac{r}{s} \geq -u\Big\} = \beta.$$

(ii) The existence of $v_{n,\beta}$ is proved in the same way.   □

There are numerous equivariant statistics $a^{(n)}$, $c^{(n)}$. In general, none of these leads to optimal procedures (except for the case $P = N_{(0,1)}$ and $a^{(n)}(\underline{x}) = \bar{x}_n$, $c^{(n)}(\underline{x}) = s_n(\underline{x})$; see Examples 4.7.12, 4.7.13 and 5.6.2, 5.6.3). Hence asymptotic optimality comes into focus. Since ML estimators are among the equivariant statistics, we expect that the resulting estimators and confidence procedures will be asymptotically optimal if Exercise 9.3.6 and Proposition 9.3.7 are applied with ML estimators.

Assume now that we are given a sequence $(a^{(n)}, c^{(n)})$ such that

$$P_{a,c}^n \circ \big(n^{1/2}(a^{(n)} - a), n^{1/2}(c^{(n)} - c)\big), \qquad n \in \mathbb{N},$$

is asymptotically normal with mean vector 0. In other words: $(a^{(n)}, c^{(n)})$ are asymptotically normal estimator sequences, which are asymptotically median unbiased anyway. The following considerations show that the transformations to exact mean– or median unbiasedness have no influence on the asymptotic distribution. Hence asymptotically better estimator sequences for (a, c) lead to asymptotically better sequences of unbiased estimators, and sequences of ML estimators lead to asymptotically optimal ones.

If $P_{a,c}^n \circ \big(n^{1/2}(a^{(n)} - a), n^{1/2}(c^{(n)} - c)\big)$, $n \in \mathbb{N}$, is asymptotically normal with mean vector 0, equivariance of $(a^{(n)}, c^{(n)})$ implies that the covariance matrix can be written as $c^2\Sigma$, whence

$$P_{a,c}^n \circ \big(n^{1/2}(a^{(n)} - a)/c, n^{1/2}(c^{(n)} - c)/c\big) \Rightarrow N_{(0,\Sigma)}.$$

Together with (9.3.3) this implies

$$Q^{(n)} \circ \big((r, s) \to (n^{1/2}r, n^{1/2}(s - 1))\big) \Rightarrow N_{(0,\Sigma)}. \tag{9.3.8}$$

From the proof of Proposition 9.3.7, we know that $u_{n,\beta}$ and $v_{n,\beta}$ are defined by

$$Q^{(n)}\big\{(r, s) \in \mathbb{R} \times (0, \infty) : r + u_{n,\beta}s \geq 0\big\} = \beta \tag{9.3.9$'$}$$

and

$$Q^{(n)}\big\{(r, s) \in \mathbb{R} \times (0, \infty) : v_{n,\beta}s \geq 1\big\} = \beta. \tag{9.3.9$''$}$$

Using the fact that weak convergence to a limit distribution with Lebesgue density implies uniform convergence on convex sets, we obtain from (9.3.8) and (9.3.9) that

$$\lim_{n \to \infty} N_{(0,\Sigma)}\{(r, s) \in \mathbb{R} \times (0, \infty) : r + u_{n,\beta} s \geq -n^{1/2} u_{n,\beta}\} = \beta \quad (9.3.10')$$

and

$$\lim_{n \to \infty} N_{(0,\Sigma)}\{(r, s) \in \mathbb{R} \times (0, \infty) : v_{n,\beta} s \geq n^{1/2}(1 - v_{n,\beta})\} = \beta. \quad (9.3.10'')$$

These relations imply

$$u_{n,\beta} = n^{-1/2} u_{1-\beta} \sigma_1 + o(n^{-1/2}) \quad (9.3.11')$$

and

$$v_{n,\beta} = 1 + n^{-1/2} u_{1-\beta} \sigma_2 + o(n^{-1/2}) \quad (9.3.11'')$$

(where σ_1^2 and σ_2^2 are the elements in the main diagonal of Σ).

Hence the confidence bounds with exact covering probability β given in Proposition 9.3.7 have the same asymptotic power function as the confidence bounds with asymptotic covering probability β, based on the limit distributions of the estimator sequence $(a^{(n)}, c^{(n)})$, $n \in \mathbb{N}$. In particular:

The confidence bounds with exact covering probability obtained from the sequence of ML estimators are asymptotically optimal.

Exercise 9.3.12. Show that the sequence of mean unbiased estimators described in Exercise 9.3.6 has the same limit distribution $N_{(0,\Sigma)}$ as the original estimator sequence $(a^{(n)}, c^{(n)})$, $n \in \mathbb{N}$, provided

$$\lim_{n \to \infty} n \int r^2 Q^{(n)}(d(r, s)) = \sigma_1^2$$

and

$$\lim_{n \to \infty} n\left[\int s^2 Q^{(n)}(d(r, s)) - 1\right] = \sigma_2^2.$$

(Hint: Use Exercise 9.3.6.)

Now we derive similar results for *tolerance bounds*. For tolerance regions of the type $(-\infty, a^{(n)} + tc^{(n)}]$, the distribution of $P_{a,c}(-\infty, a^{(n)} + tc^{(n)}]$ under $P_{a,c}^n$ does not depend on a and c, since

$$P_{a,c}^n \circ \left(P_{a,c}(-\infty, a^{(n)} + tc^{(n)}]\right)$$

$$= P_{a,c}^n \circ \left(P\left(-\infty, \frac{a^{(n)} - a}{c} + t\frac{c^{(n)}}{c}\right]\right)$$

$$= Q^{(n)} \circ \left((r, s) \to P(-\infty, r + ts]\right).$$

Since

$$\lim_{t\to\infty} P(-\infty, r+ts] = 1 \quad \text{and} \quad \lim_{t\to-\infty} P(-\infty, r+ts] = 0,$$

for every $\beta \in (0,1)$ there exists $t_{n,\beta} \in \mathbb{R}$ such that

$$\int P(-\infty, r+t_{n,\beta}s]Q^{(n)}(d(r,s)) = \beta. \tag{9.3.13}$$

Hence $(-\infty, a^{(n)} + t_{n,\beta}c^{(n)}]$ is a β–expectation tolerance region.

According to Proposition 9.2.4, the optimal limit distribution of $P(-\infty, T_\beta^{(n)}]$ is $N_{(0,\sigma_\beta^2(a,c))}$, with

$$\sigma_\beta^2(a,c) = c^2 p(q_\beta)^2(\Lambda_{11} + 2q_\beta\Lambda_{12} + q_\beta^2\Lambda_{22}).$$

(For the definition of Λ see Lemma 9.3.16.) We shall show that this limit distribution is attained for $T_\beta^{(n)} = a^{(n)} + t_{n,\beta}c^{(n)}$, if $(a^{(n)}, c^{(n)})$ is an estimator sequence with the optimal limit distribution, $N_{(0,c^2\Lambda)}$.

By assumption (see (9.3.8)),

$$Q_*^{(n)} := Q^{(n)} \circ \left((r,s) \to (n^{1/2}r, n^{1/2}(s-1))\right) \Rightarrow N_{(0,\Sigma)}. \tag{9.3.14}$$

Hence (9.3.13) may be rewritten as

$$\int P\left(-\infty, n^{-1/2}(r + t_{n,\beta}s) + t_{n,\beta}\right]Q_*^{(n)}(d(r,s)) = \beta. \tag{9.3.15}$$

From (9.3.14) it follows easily that $t_{n,\beta} \to q_\beta$. However, we have to show that $t_{n,\beta} = q_\beta + o(n^{-1/2})$. This follows if p is bounded and continuous, and if $(x,y) \to |x|$, $(x,y) \to |y|$ are uniformly integrable under $Q_*^{(n)}$, $n \in \mathbb{N}$.

Under such conditions,

$$|n^{1/2}\left(P(-\infty, n^{-1/2}(x + t_{n,\beta}y) + t_{n,\beta}] - P(-\infty, t_{n,\beta}]\right) - (x + t_{n,\beta}y)p(q_\beta)|$$

$$\leq \varepsilon_n(x,y)|x + t_{n,\beta}y|, \quad \text{with a bounded sequence } \varepsilon_n \to 0 \quad (Q_*^{(n)}).$$

Using $\int x N_{(0,\Sigma)}(d(x,y)) = 0$, $\int y N_{(0,\Sigma)}(d(x,y)) = 0$, this implies by (9.3.15) that

$$|P(-\infty, t_{n,\beta}] - \beta| = o(n^{-1/2}), \quad \text{hence } t_{n,\beta} = q_\beta + o(n^{-1/2}).$$

Therefore,

$$n^{1/2}\left(P(-\infty, a^{(n)}+t_{n,\beta}c^{(n)}] - P(-\infty, a^{(n)}+q_\beta c^{(n)}]\right) \to 0 \quad (P^n).$$

Since

$$P^n \circ n^{1/2}\left(P(-\infty, a^{(n)} + q_\beta c^{(n)}] - \beta\right) \Rightarrow N_{(0,\sigma_\beta^2(a,c))},$$

this proves the assertion.   □

Now we turn to the problem of obtaining an asymptotically optimal sequence of equivariant estimators for (a, b). Throughout the following, we denote $\ell(x) = \log p(x)$.

Lemma 9.3.16. *Assume that p has a continuous derivative, and that the integrals occurring below exist.*

Then the minimal asymptotic covariance matrix, defined by (7.5.7), is equal to $c^2 \Lambda$, where Λ is the inverse of the matrix L with elements

$$L_{11} = \int \ell'(x)^2 P(dx).$$

$$L_{12} = \int x\ell'(x)^2 P(dx)$$

$$L_{22} = \int \left(1 + x\ell'(x)\right)^2 P(dx)$$

If P is symmetric, then

$$\Lambda = \begin{pmatrix} 1/L_{11} & 0 \\ 0 & 1/L_{22} \end{pmatrix}. \tag{9.3.17}$$

Proof. Follows immediately from

$$\ell^{(1)}(x, a, c) = -\frac{1}{c}\ell'\left(\frac{x-a}{c}\right)$$

$$\ell^{(2)}(x, a, c) = -\frac{1}{c}\left(1 + \frac{x-a}{c}\ell'\left(\frac{x-a}{c}\right)\right).$$

$\square$

The most convenient way to asymptotically efficient sequences of equivariant estimators is the improvement procedure (7.5.10) which leads to:

$$\hat{a}^{(n)}(\underline{x}) = a^{(n)}(\underline{x}) - c^{(n)}(\underline{x})n^{-1}\sum_{1}^{n} h_1\left(\frac{x_\nu - a^{(n)}(\underline{x})}{c^{(n)}(\underline{x})}\right)$$

$$\hat{c}^{(n)}(\underline{x}) = c^{(n)}(\underline{x})\left(1 - n^{-1}\sum_{1}^{n} h_2\left(\frac{x_\nu - a^{(n)}(\underline{x})}{c^{(n)}(\underline{x})}\right)\right)$$

with

$$h_1(x) = \Lambda_{11}\ell'(x) + \Lambda_{12}(1 + x\ell'(x))$$
$$h_2(x) = \Lambda_{21}\ell'(x) + \Lambda_{22}(1 + x\ell'(x)).$$

It is straightforward to see that $\hat{a}^{(n)}, \hat{c}^{(n)}$, $n \in \mathbb{N}$, are equivariant if the preliminary estimators $a^{(n)}, c^{(n)}$ have this property.

The estimator sequence $(\hat{a}^{(n)}, \hat{c}^{(n)})$ is asymptotically normal with covariance matrix $c^2 \Lambda$ if the preliminary estimator sequence is $\sqrt{n}$–consistent and if p fulfills the following regularity conditions.

$$\int p'(x)dx = 0 \quad \text{and} \quad \int xp'(x)dx = -1 \quad \text{(to guarantee (7.5.1))},$$

and

$$\int x^k p''(x)dx = 0 \quad \text{for } k = 0, 1 \text{ and } \int x^2 p''(x)dx = 2 \quad \text{(to guarantee (7.5.2))}.$$

These conditions are, in particular, fulfilled if

$$\lim_{|x| \to \infty} xp(x) = 0 \quad \text{and} \quad \lim_{|x| \to \infty} x^2 p'(x) = 0.$$

Condition 7.5.5(i) holds true if ℓ'' is continuous and if the function

$$x \to \sup_{\substack{|\alpha| < \epsilon \\ |\gamma - 1| < \epsilon}} \left(1 + \left(\frac{x - \alpha}{\gamma}\right)^2\right) \left| \ell''\left(\frac{x - \alpha}{\gamma}\right)\right|$$

is P–integrable. Condition 7.5.5(ii) holds if $x \to (1 + x^2)\ell'(x)^2$ is P–integrable [locally uniformly] at $(0, 1)$.

Provided $\int x^2 P(dx) < \infty$, a convenient pair of equivariant $\sqrt{n}$–consistent estimator sequences is $a^{(n)}(\underline{x}) = \bar{x}_n$, $c^{(n)}(\underline{x}) = n^{-1} \sum_1^n |x_\nu - \bar{x}_n|$, if P is standardized such that $\int xP(dx) = 0$ and $\int |x|P(dx) = 1$. If $x^2 P(dx) = \infty$, functions of order statistics may be used for this purpose.

Compared with the improvement procedure, ML estimators are much more difficult to compute in general. In spite of this, the following results may be of interest.

Exercise 9.3.18. For $n \geq 2$ the ML estimators can be chosen equivariant on the complement of the λ^n–null set $\{\underline{x} \in \mathbb{R}^n : x_1 = \ldots = x_n\}$. Hence the ML estimator is automatically equivariant, if it is unique.

Hint: Starting from an arbitrary ML estimator $(a^{(n)}, c^{(n)})$, define

$$\hat{a}^{(n)}(\underline{x}) := \bar{x}_n + a^{(n)}(\underline{y})n^{-1} \sum_{\mu=1}^n |x_\mu - \bar{x}_n|$$

$$\hat{c}^{(n)}(\underline{x}) := c^{(n)}(\underline{y})n^{-1} \sum_{\mu=1}^n |x_\mu - \bar{x}_n|,$$

with

$$y_\nu := (x_\nu - \bar{x}_n)/n^{-1} \sum_{\mu=1}^n |x_\mu - \bar{x}_n|.$$

That $(\hat{a}^{(n)}, \hat{c}^{(n)})$ is a ML estimator, follows from

$$\sup\{\frac{1}{c^n}\prod_1^n p(\frac{(\alpha + \gamma x_\nu) - a}{c}) : a \in \mathbb{R}, \ c > 0\}$$

$$= \gamma^{-n} \sup\{\frac{1}{c^n}\prod_1^n p(\frac{x_\nu - a}{c}) : a \in \mathbb{R}, \ c > 0\}.$$

The consistency theorem for ML estimators requires regularity conditions (6.3.1) on the density, and the Covering Condition 6.3.8. Exercise 9.3.19 shows that the covering condition is never fulfilled on the parameter space $\Theta = \mathbb{R} \times (0, \infty)$ for $f(x, a, c) = \log \frac{1}{c}p(\frac{x-a}{c})$. As a consequence of Lemma 6.3.9, it is fulfilled for

$$f(x_1, \dots, x_k, a, c) = \frac{1}{k}\sum_{\nu=1}^k \log \frac{1}{c}p(\frac{x_\nu - a}{c})$$

for *some* $k \in \mathbb{N}$, $k > 2$, under condition (9.3.22).

That the covering condition fails on $\Theta = \mathbb{R} \times (0, \infty)$ finds an intuitive explanation: $\frac{1}{c}p(\frac{x-a}{c})$ becomes arbitrarily large for $a = x$ and $c \to 0$ (provided $p(0) > 0$). This is made precise in Exercise 9.3.19. The same effect cannot occur with

$$\frac{1}{c}p(\frac{x_1 - a}{c}) \cdot \frac{1}{c}p(\frac{x_2 - a}{c}) \quad \text{if} \quad x_1 \neq x_2.$$

Exercise 9.3.19. Let U_0 be a bounded neighborhood of $(a_0, c_0) \in \mathbb{R} \times (0, \infty)$, and $V_1, \dots, V_m$ a cover of U_0^c. Let $\ell(x, a, c) := \log \frac{1}{c}p(\frac{x-a}{c})$. Show that

$$P_{a_0, c_0}(\bar{\ell}(\cdot, V_i)^+) = \infty \quad \text{for some } i \in \{1, \dots, m\}. \tag{9.3.20}$$

If $\int p(x) \log p(x)dx \in \mathbb{R}$, then $\bar{\ell}(\cdot, V_i) - \ell(\cdot, a_0, c_0)$ is not P_{a_0, c_0}–integrable.

Hint: Since U_0 is bounded, $|x|$ sufficiently large implies $(x, c) \in U_0^c$ for every $c > 0$. For any sequence $c_n > 0$, $n \in \mathbb{N}$, there is $k \in \{1, \dots, m\}$ such that $(x, c_n) \in V_k$, hence

$$\bar{\ell}(x, V_k)^+ \geq \ell(x, x, c_n) = -\log c_n + \log p(0),$$

for infinitely many $n \in \mathbb{N}$. Thus $c_n \to 0$ leads to (9.3.20).   □

The proof of the following proposition is taken from an unpublished manuscript of Reiß. See also Kiefer and Wolfowitz (1956, pp. 897/9), and Pitman (1979, p. 71, Theorem 3).

Proposition 9.3.21. *Let* $p : \mathbb{R} \to (0, \infty)$ *be a continuous p–density, fulfilling*

$$\lim_{|x| \to \infty} |x|^{1+\epsilon} p(x) = 0 \quad \text{for some } \epsilon > 0. \tag{9.3.22}$$

Denote $\ell_k(x_1, \ldots, x_k, a, c) := \frac{1}{k} \sum_1^k \ell(x_\nu, a, c)$.
The following holds true for every $k > 1 + 1/\epsilon$.
(i) $P^k_{a_0, c_0} \big(\bar{\ell}_k(\cdot, \Theta) - \ell_k(\cdot, a_0, c_0) \big) < \infty$, *and*
(ii) *there exists a compact neighborhood* C *of* (a_0, c_0) *such that*

$$P^k_{a_0, c_0} \big(\bar{\ell}_k(\cdot, C^c) - \ell_k(\cdot, a_0, c_0) \big) < 0.$$

Proof. (i) At first we shall show that (9.3.22) implies

$$\int p(x) \log p(x) dx > -\infty. \tag{9.3.23}$$

As a consequence of (9.3.22), there exists $r \geq e$ such that

$$p(x) \leq |x|^{-(1+\epsilon)} \quad \text{for } |x| \geq r.$$

Since $u \log u$ is antitone on $(0, 1/e)$, this implies

$$p(x) \log p(x) \geq |x|^{-(1+\epsilon)} \log |x|^{-(1+\epsilon)} \quad \text{for } |x| \geq r,$$

hence

$$\int_{|x| \geq r} p(x) \log p(x) dx > -\infty.$$

Moreover, $\int_{|x| < r} p(x) \log p(x) dx > -\infty$, since $u \log u \geq -1/e$. Hence (9.3.23) follows.
(ii) For $(x_1, \ldots, x_k) \in \mathbb{R}^k$ let

$$m(x_1, \ldots, x_k) := \min \Big\{ \frac{|x_i - x_j|}{2} : \ i, j = 1, \ldots, k; \ i \neq j \Big\}.$$

For any $a \in \mathbb{R}$, we have $|x_i - a| \geq m(x_1, \ldots, x_k)$ for all $i = 1, \ldots, k$ with at most one exception.
This implies

$$\inf_{a \in \mathbb{R}} \prod_{i=1}^k \Big(1 + \big(\frac{|x_i - a|}{c} \big)^{1+\epsilon} \Big) \geq \Big(1 + \big(\frac{m(x_1, \ldots, x_k)}{c} \big)^{1+\epsilon} \Big)^{k-1}. \tag{9.3.24}$$

Since p is continuous, it is bounded on every finite interval. Hence (9.3.22) implies the existence of $M > 0$ such that

$$p(x) \leq M/(1 + |x|^{1+\epsilon}) \quad \text{for } x \in \mathbb{R}.$$

Together with (9.3.24) this implies

$$\sup_{a \in \mathbf{R}} \prod_{i=1}^{k} \frac{1}{c} p\left(\frac{x_i - a}{c}\right) \le \frac{M^k}{c^k} \left(1 + \left(\frac{m(x_1, \ldots, x_k)}{c}\right)^{1+\varepsilon}\right)^{-(k-1)}. \tag{9.3.25}$$

Furthermore, $k > 1 + 1/\varepsilon$ implies $1 + \varepsilon > k/(k-1)$, so that

$$1 + u^{1+\varepsilon} > u^{k/(k-1)} \qquad \text{for every } u > 0.$$

Applied with $u = m(x_1, \ldots, x_k)/c$ this yields

$$\inf_{c > 0} c^k \left(1 + \left(\frac{m(x_1, \ldots, x_k)}{c}\right)^{1+\varepsilon}\right)^{k-1} \ge m(x_1, \ldots, x_k)^k.$$

Together with (9.3.25) this implies

$$\sup_{c > 0} \sup_{a \in \mathbf{R}} \prod_{i=1}^{k} \frac{1}{c} p\left(\frac{x_i - a}{c}\right) \le M^k m(x_1, \ldots, x_k)^{-k}. \tag{9.3.26}$$

Next we shall show that

$$\int \log m(x_1, \ldots, x_k) P^k \left(d(x_1, \ldots, x_k)\right) > -\infty. \tag{9.3.27}$$

Together with (9.3.23) and (9.3.26) this implies assertion (i).
    It suffices to prove

$$\int_{0 < m < 1} \log m(x_1, \ldots, x_k) P^k \left(d(x_1, \ldots, x_k)\right) > -\infty.$$

Since $\log m(x_1, \ldots, x_m) < 0$ implies

$$\log m(x_1, \ldots, x_m) \ge -\sum \left(\log \frac{|x_i - x_j|}{2}\right)^{-}$$

(with the summation extending over all $i, j = 1, \ldots, k$; $i \ne j$), it suffices to
prove that

$$\int \left(\log \frac{|x - y|}{2}\right)^{-} P(dx) P(dy) < \infty.$$

Since P has a bounded density, $P \times P \circ \left((x, y) \to \frac{|x-y|}{2}\right)$ has a bounded density,
say q. Together with $\int_0^1 \log z \, dz = -1$, this implies

$$\int_0^1 (\log z) q(z) dz > -\infty.$$

(iii) Now we shall show that

$$C_r(x_1,\ldots,x_k) := \{(a,c) \in \mathbb{R} \times (0,\infty) : \prod_1^k \frac{1}{c}p(\frac{x_i - a}{c}) \geq r\}$$

is bounded, with the second component bounded away from 0 for every $(x_1,\ldots$
$\ldots,x_k) \in \mathbb{R}^k$ with $m(x_1,\ldots,x_k) > 0$. By Lemma 9.3.28 this implies that for every $K > 0$ there is a compact set C_K such that

$$P^k\big(\overline{\ell}_k(\cdot,C_K^c)\big) < -K.$$

Together with (9.3.23), this implies assertion (ii).

From (9.3.25), $C_r(x_1,\ldots,x_k)$ is contained in the set

$$\{(a,c) \in \mathbb{R} \times (0,\infty) : \frac{M^k}{c^k}\Big(1 + (\frac{m(x_1,\ldots,x_k)}{c})^{1+\varepsilon}\Big)^{-(k-1)} \geq r\}.$$

If $m(x_1,\ldots,x_k) > 0$, then $k > 1 + 1/\varepsilon$ implies that this set is of the type $\mathbb{R} \times [c_0,c_1]$, with $0 < c_0 < c_1 < \infty$.

Moreover,

$$\prod_{i=1}^k \frac{1}{c}p(\frac{x_i - a}{c}) \leq \frac{M^k}{c^k} \cdot \prod_{i=1}^k \Big(1 + (\frac{|x_i - a|}{c})^{1+\varepsilon}\Big)^{-1}.$$

Hence

$$C_r(x_1,\ldots,x_k) \subset \{(a,c) \in \mathbb{R} \times [c_0,c_1] : \frac{M}{c_0^k}\prod_{i=1}^k \Big(1 + (\frac{|x_i - a|}{c_1})^{1+\varepsilon}\Big)^{-1} \geq r\}$$

which is bounded. $\square$

Lemma 9.3.28. *Let* Θ *be* σ-*locally compact, but not compact. Assume that* $f : X \times \Theta \to \mathbb{R}$ *fulfills conditions (6.3.1).*

Assume that $P_\vartheta\big(\overline{f}(\cdot,\Theta)\big) < \infty$, *and that* $\{\vartheta \in \Theta : f(x,\vartheta) \geq r\}$ *is compact or empty for every* $x \in X$, $r \in \mathbb{R}$.

Then for every $K > 0$ *there exists a compact set* $C_K \subset \Theta$ *such that*

$$P_\vartheta\big(\overline{f}(\cdot,C_K^c)\big) < -K.$$

(Since Θ *is not compact,* $C_K^c \neq \emptyset$.)

Proof. By assumption, $C_r(x) := \{\vartheta \in \Theta : f(x,\vartheta) \geq r\}$ is compact or empty for every $r \in \mathbb{R}$. Hence $\inf\{\overline{f}(x,C^c) : C \subset \Theta \text{ compact}\} \leq \overline{f}(x,C_r^c(x)) \leq r$. Since $r \in \mathbb{R}$ was arbitrary, this implies $\inf\{\overline{f}(x,C^c) : C \subset \Theta \text{ compact}\} = -\infty$. Let C_n be an increasing sequence of compact sets such that $\Theta = \bigcup_1^\infty C_n^\circ$. Since $P_\vartheta\big(\overline{f}(\cdot,\Theta)\big) < \infty$, $\inf\{\overline{f}(x,C_n^c) : n \in \mathbb{N}\} = -\infty$ implies $\lim_{n\to\infty} P_\vartheta\big(\overline{f}(x,C_n^c)\big) = -\infty$. $\square$

9.4 Miscellaneous results on estimators

Collects a few scattered remarks related to the computation of estimators, e.g., remarks on moment estimators as preliminary estimators for the improvement procedure, and so–called "quasi ML estimators".

Theorem 7.4.3, which leads to asymptotically efficient estimator sequences under suitable regularity conditions, can be successfully applied if the Covering Condition 6.3.8 is fulfilled, or if a consistent preliminary estimator sequence is available. This theorem applies under less restrictive conditions on the estimating equation if the preliminary estimator sequence is c_n–consistent. If this is the case, the improvement procedure offers an attractive alternative.

In many instances, $\sqrt{n}$–consistent estimator sequences can be obtained in a comparatively simple way.

Let $\Theta \subset \mathbb{R}^p$ be an open subset, and $\{P_\vartheta : \vartheta \in \Theta\}$ a family of p–measures. For $k = 1, \ldots, p$ let $f_k : X \to \mathbb{R}$ be an $\mathcal{A}$–measurable function such that f_k^2 is P_ϑ–integrable [locally uniformly on Θ]. If the functions f_k, $k = 1, \ldots, p$, are appropriately chosen, there exist functions $h_i : \mathbb{R}^p \to \mathbb{R}$, $i = 1, \ldots, p$, such that

$$\vartheta_i = h_i\big(P_\vartheta(f_1), \ldots, P_\vartheta(f_p)\big).$$

Since $n^{-1} \sum_{\nu=1}^n f_k(x_\nu)$, $n \in \mathbb{N}$, converges to $P_\vartheta(f_k)$ for $P_\vartheta^{\mathbb{N}}$–a.a. $\underline{x} \in X^{\mathbb{N}}$,

$$\vartheta_i^{(n)}(\underline{x}) := h_i\Big(n^{-1} \sum_{\nu=1}^n f_1(x_\nu), \ldots, n^{-1} \sum_{\nu=1}^n f_p(x_\nu)\Big)$$

defines an estimator sequence which is strongly consistent, provided h_i is continuous. This estimator sequence will be [locally uniformly] $\sqrt{n}$–consistent, if the functions h_i have continuous partial derivatives. This follows immediately from Propositions 7.6.8 and 7.2.1, applied with $\kappa_k^{(n)}(\underline{x}) = n^{-1} \sum_1^n f_k(x_\nu)$, $\kappa_k(\vartheta) = P_\vartheta(f_k)$, $k = 1, \ldots, p$, and

$$\Sigma_{ij}(\vartheta) = \int \big(f_i(x) - P_\vartheta(f_i)\big)\big(f_j(x) - P_\vartheta(f_j)\big) P_\vartheta(dx), \quad i, j = 1, \ldots, p.$$

If $f_1, \ldots, f_p$ are linearly P_ϑ–independent, the matrix Σ is nonsingular.

This procedure is a straightforward generalization of K. Pearson's "method of moments" which is based on the functions $f_k(x) = x^k$ (for $X = \mathbb{R}$). There are some papers about how to choose the functions f_k to obtain an estimator sequence of high efficiency. The practical relevance of these papers seems doubtful. Starting from a $\sqrt{n}$–consistent estimator sequence obtained as indicated above, an asymptotically efficient estimator sequence can be obtained immediately by the improvement procedure (see Proposition 7.4.13). As an application we mention Example 9.1.10 (see (9.1.13)).

Now we consider the following problem: In some cases it is comparatively easy to obtain

(i) an asymptotically efficient estimator sequence for ϑ_1 under the assumption that $(\vartheta_2, \ldots, \vartheta_p)$ are known, say $\vartheta_1^{(n)}(\cdot, \vartheta_2, \ldots, \vartheta_p)$, $n \in \mathbb{N}$, and

(ii) an estimator sequence for $(\vartheta_2, \ldots, \vartheta_p)$, say $(\vartheta_2^{(n)}, \ldots, \vartheta_p^{(n)})$, $n \in \mathbb{N}$.

Under which conditions is $\underline{x} \to \vartheta_1^{(n)}(\underline{x}, \vartheta_2^{(n)}(\underline{x}), \ldots, \vartheta_p^{(n)}(\underline{x}))$ an asymptotically efficient estimator sequence for ϑ_1 in the full family $\{P_\vartheta^n : \vartheta \in \Theta\}$?

Example 9.4.1. Let $P|\mathbb{B}$ be a p–measure fulfilling $\int x P(dx) = 0$ and $\int |x| P(dx) = 1$. Let $\{P_{a,c} : a \in \mathbb{R}, c > 0\}$ denote the pertaining location– and scale parameter family. Since $\int |x - a| P_{a,c}(dx) = c$, $n^{-1} \sum_1^n |x_\nu - a|$ is a reasonable (though not necessarily asymptotically optimal) estimator for c, if a is known. Is it possible to replace a by an estimator, say by $\bar{x}_n$?

The following considerations refer to a family of sequences $P_\vartheta^{(n)}$, $n \in \mathbb{N}$, $\vartheta \in \Theta \subset \mathbb{R}^p$, fulfilling the LAN Condition 8.1.1 with remainder term (8.1.4'). To simplify our notations we assume that $\Theta = \overset{p}{\underset{i=1}{\times}} \Theta_i$. Assume that $\vartheta_1^{(n)}(\cdot, \vartheta_2, \ldots, \vartheta_p)$, $n \in \mathbb{N}$, is an estimator sequence for ϑ_1, which is regular and asymptotically efficient for the family $P_{(\vartheta_1, \vartheta_2, \ldots, \vartheta_p)}^{(n)}$, $n \in \mathbb{N}$, $\vartheta_1 \in \Theta_1$ (with $\vartheta_2, \ldots, \vartheta_p$ fixed). As a consequence of the convolution theorem (see Proposition 8.5.2) we have

$$c_n\big(\vartheta_1^{(n)}(\cdot, \vartheta_2, \ldots, \vartheta_p) - \vartheta_1\big) = \Delta_{n1}(\cdot, \vartheta)/D_{11}(\vartheta) + r_n(\cdot, \vartheta), \qquad (9.4.2)$$

with $r_n(\cdot, \vartheta) \to 0$ $(P_\vartheta^{(n)})$.

Moreover, let $(\vartheta_2^{(n)}, \ldots, \vartheta_p^{(n)})$ be an estimator sequence which is regular for $(\vartheta_2, \ldots, \vartheta_p)$ in the full family $P_\vartheta^{(n)}$, $n \in \mathbb{N}$, $\vartheta \in \Theta$.

Specialized to the i.i.d. case (see Proposition 8.1.8), the following Proposition 9.4.3 gives conditions under which $n^{-1/2} \sum_1^n \ell^{(1)}(x_\nu, \vartheta_1, \vartheta_2, \ldots, \vartheta_p)$ is asymptotically independent of $(n^{1/2}(\vartheta_2^{(n)} - \vartheta_2), \ldots, n^{1/2}(\vartheta_p^{(n)} - \vartheta_p))$, if $(\vartheta_2^{(n)}, \ldots, \vartheta_p^{(n)})$, $n \in \mathbb{N}$, is a regular estimator sequence for $(\vartheta_2, \ldots, \vartheta_p)$ in the family $\{P_\vartheta : \vartheta \in \Theta\}$. This implies asymptotic independence between $(n^{1/2}(\vartheta_2^{(n)} - \vartheta_2), \ldots, n^{1/2}(\vartheta_p^{(n)} - \vartheta_p))$ and $n^{1/2}(\vartheta_1^{(n)}(\cdot, \vartheta_2, \ldots, \vartheta_p) - \vartheta_1)$, for any regular estimator sequence $\vartheta_1^{(n)}(\cdot, \vartheta_2, \ldots, \vartheta_p)$, $n \in \mathbb{N}$, which is asymptotically efficient for ϑ_1 in the family $\{P_{(\vartheta_1, \vartheta_2, \ldots, \vartheta_p)} : \vartheta_1 \in \Theta_1\}$. Considered as a "surprising fact", these relations were established by Parke (1986, p. 356) under unclear conditions.

Proposition 9.4.3. *Assume that $P_\vartheta^{(n)}$, $n \in \mathbb{N}$, $\vartheta \in \Theta$, fulfills the LAN Condition 8.1.1 with a remainder term fulfilling (8.1.4').*

If $(\vartheta_2^{(n)}, \ldots, \vartheta_p^{(n)})$, $n \in \mathbb{N}$, is a regular estimator sequence for $(\vartheta_2, \ldots, \vartheta_p)$, and if $P_\vartheta^{(n)} \circ (\Delta_{n1}(\cdot, \vartheta), c_n(\vartheta_2^{(n)} - \vartheta_2), \ldots, c_n(\vartheta_p^{(n)} - \vartheta_p))$, $n \in \mathbb{N}$, is asymptotically normal, then Δ_{n1} and $(c_n(\vartheta_2^{(n)} - \vartheta_2), \ldots, c_n(\vartheta_p^{(n)} - \vartheta_p))$, $n \in \mathbb{N}$, are asymptotically independent.

Proof. With $H(\vartheta) := D(\vartheta)^{-1}$ let $h_{ni}(x, \vartheta) = \sum_{j=1}^p H_{ij}(\vartheta) \Delta_{nj}(x, \vartheta)$, $i = 1, \ldots, p$ (where Δ_{ni}, $i = 1, \ldots, p$, are the components of Δ_n). The relation $P_\vartheta^{(n)} \circ \Delta_n(\cdot, \vartheta) \Rightarrow N_{(0, D(\vartheta))}$ implies $P_\vartheta^{(n)} \circ (\Delta_{n1}, h_{n2}, \ldots, h_{np}) \Rightarrow N_{(0, \Sigma(\vartheta))}$, with $\Sigma_{1i}(\vartheta) = 0$ for $i = 2, \ldots, p$. (Hint: Apply (2.4.15) with $A_{1j} = \delta_{1j}$ for $j = 1, \ldots, p$, and $A_{ij} = H_{ij}(\vartheta)$ for $i = 2, \ldots, p$, $j = 1, \ldots, p$. As a consequence of (8.4.3), applied for the functional $\kappa : \Theta \to \mathbb{R}^{p-1}$, defined by $\kappa_i(\vartheta_1, \ldots, \vartheta_p) = \vartheta_i$, $i = 2, \ldots, p$, the function Δ_{n1} is asymptotically independent of $(c_n(\vartheta_2^{(n)} - \vartheta_2) - h_{n2}, \ldots, c_n(\vartheta_p^{(n)} - \vartheta_p) - h_{np})$. Hence the assertion follows from Lemma 9.4.4, applied for $P_n = P_\vartheta^{(n)} \circ (\Delta_{n1}, h_{n2}, \ldots, h_{np}, c_n(\vartheta_2^{(n)} - \vartheta_2) - h_{n2}, \ldots, c_n(\vartheta_p^{(n)} - \vartheta_p) - h_{np})$, with $q = p - 1$. $\qquad \square$

Lemma 9.4.4. *Let P_n, $n \in \mathbb{N}$, be a sequence of p-measures on $\mathbb{B}^{1+2q}$ with the following properties (with $u \in \mathbb{R}$ and $v, w \in \mathbb{R}^q$).*

(i) $P_n \circ ((u, v, w) \to (u, v)) \Rightarrow N_{(0, \Sigma)}$ *with* $\Sigma_{0i} = 0$ *for* $i = 1, \ldots, q$,

(ii) $P_n \circ ((u, v, w) \to (u, w)) \Rightarrow N_{(0, \Sigma_{00})} \times Q$ *for some p-measure Q.*

(iii) $P_n \circ ((u, v, w) \to (u, v + w)) \Rightarrow N_{(0, \hat{\Sigma})}$.

Then $\hat{\Sigma}_{0i} = 0$ for $i = 1, \ldots, q$, i.e. u and $v + w$ are asymptotically independent under P_n, $n \in \mathbb{N}$.

Proof. By Lemma 8.4.18, there exists a subsequence $\mathbb{N}_0$ such that P_n, $n \in \mathbb{N}_0$, converges weakly to some p-measure $P_0 | \mathbb{B}^{1+2q}$. Since $P_0 \circ ((u, v, w) \to (u, v + w)) = N_{(0, \hat{\Sigma})}$ by (iii), this implies the existence of $\hat{\Sigma}_{0i} = \int u(v_i + w_i) P_0(d(u, v, w))$. Assumption (i) implies $\int u v_i P_0(d(u, v, w)) = \Sigma_{0i} = 0$. Hence $\int u w_i P_0(d(u, v, w))$ exists, and is equal to 0 by assumption (ii). This implies $\hat{\Sigma}_{0i} = 0$, for $i = 1, \ldots, q$. $\qquad \square$

Throughout the following we assume that the remainder term in (9.4.2) fulfills $\sup_{\|a\| \le c_n M} |r_n(\cdot, \vartheta + c_n^{-1}a)| \to 0$ $(P_\vartheta^{(n)})$ for $M > 0$, and that $\Delta_n(\cdot, \vartheta)$ fulfills (see (8.1.12))

$$\Delta_n(x, \vartheta + c_n^{-1}a) = \Delta_n(x, \vartheta) - D(\vartheta)a + R_n(x, \vartheta, a) \qquad (9.4.5)$$

with

$$\sup_{\|a\|\le c_n M} \|R_n(\cdot,\vartheta,a)\| \to 0 \quad (P_\vartheta^{(n)}) \quad \text{for } M > 0.$$

Under these assumptions,

$$\Delta_{n1}\big(x,\vartheta_1,\vartheta_2^{(n)}(x),\ldots,\vartheta_p^{(n)}(x)\big)$$

$$= \Delta_{n1}(x,\vartheta_1,\vartheta_2,\ldots,\vartheta_p) - \sum_{i=2}^{p} D_{1i}(\vartheta)c_n\big(\vartheta_i^{(n)}(x) - \vartheta_i\big) + \bar{r}_n(x,\vartheta),$$

with $\bar{r}_n(\cdot,\vartheta) \to 0 \ (P_\vartheta^{(n)})$.

This implies (see (9.4.2))

$$c_n\big(\vartheta_1^{(n)}\big(x,\vartheta_2^{(n)}(x),\ldots,\vartheta_p^{(n)}(x)\big) - \vartheta_1\big)$$

$$= \Big[\Delta_{n1}(x,\vartheta) - \sum_{i=2}^{p} D_{1i}(\vartheta)c_n(\vartheta_i^{(n)}(x) - \vartheta_i)\Big]\Big/ D_{11}(\vartheta) + \hat{r}_n(x,\vartheta),$$

with $\hat{r}_n(x,\vartheta) \to 0 \ (P_\vartheta^{(n)})$.

Using the stochastic independence stated in Proposition 9.4.3, we obtain that $c_n\big(\vartheta_1^{(n)}(\cdot,\vartheta_2^{(n)}(\cdot),\ldots,\vartheta_p^{(n)}(\cdot)) - \vartheta_1\big)$, $n \in \mathbb{N}$, is asymptotically normal with variance

$$\sigma_1^2(\vartheta) = \Big[D_{11}(\vartheta) + \sum_{i=2}^{p}\sum_{j=2}^{p} D_{1i}(\vartheta)D_{1j}(\vartheta)\hat{\Sigma}_{ij}(\vartheta)\Big]\Big/ D_{11}(\vartheta)^2, \qquad (9.4.6)$$

where $\hat{\Sigma}(\vartheta)$ denotes the asymptotic covariance matrix of $(\vartheta_2^{(n)},\ldots,\vartheta_p^{(n)})$.

The minimal asymptotic variance for regular estimator sequences of ϑ_1 in the full family $P_\vartheta^{(n)}$, $n \in \mathbb{N}$, $\vartheta \in \Theta$, is $H_{11}(\vartheta)$. Using the relation

$$\sum_{i=2}^{p}\sum_{j=2}^{p} D_{1i}D_{1j}H_{ij} = -D_{11} + H_{11}D_{11}^2,$$

we obtain

$$\sigma_1^2(\vartheta) - H_{11}(\vartheta) = \sum_{i=2}^{p}\sum_{j=2}^{p} D_{1i}(\vartheta)D_{1j}(\vartheta)\big(\hat{\Sigma}_{ij}(\vartheta) - H_{ij}(\vartheta)\big)\Big/ D_{11}(\vartheta)^2.$$

Hence $\vartheta_1^{(n)}(\cdot,\vartheta_2^{(n)}(\cdot),\ldots,\vartheta_p^{(n)}(\cdot))$, $n \in \mathbb{N}$, is asymptotically efficient if $\hat{\Sigma}_{ij}(\vartheta)$ $= H_{ij}(\vartheta)$, i.e. if $(\vartheta_2^{(n)},\ldots,\vartheta_p^{(n)})$, $n\in \mathbb{N}$, is asymptotically efficient for $(\vartheta_2,\ldots,\vartheta_p)$ in the full family $P_\vartheta^{(n)}$, $n \in \mathbb{N}$, $\vartheta \in \Theta$. If $p = 2$, the asymptotic efficiency of $\vartheta_2^{(n)}$, $n \in \mathbb{N}$, is even necessary for the asymptotic efficiency of $\vartheta_1^{(n)}(\cdot,\vartheta_2^{(n)}(\cdot))$, $n \in \mathbb{N}$, unless $D_{12}(\vartheta) = 0$. (In the latter case, any c_n–consistent estimator sequence for ϑ_2 renders $\vartheta_1^{(n)}(\cdot,\vartheta_2^{(n)}(\cdot))$ asymptotically efficient.)

Now we restrict our considerations to the i.i.d. case. For solid believers in the ML method, it might be tempting to determine $\vartheta_1^{(n)}(\underline{x})$ by maximizing $\vartheta_1 \to \prod_{\nu=1}^n p(x_\nu, \vartheta_1, \vartheta_2^{(n)}(\underline{x}), \ldots, \vartheta_p^{(n)}(\underline{x}))$. Under the usual regularity conditions, this leads to the estimating equation

$$\sum_{\nu=1}^n \ell^{(1)}(x_\nu, \vartheta_1, \vartheta_2^{(n)}(\underline{x}), \ldots, \vartheta_p^{(n)}(\underline{x})) = 0. \tag{9.4.7}$$

The results obtained above imply that the estimator sequence for ϑ_1 thus obtained is asymptotically efficient if $(\vartheta_2^{(n)}, \ldots, \vartheta_p^{(n)})$ is asymptotically efficient (in the full family). This result occurs in Gong and Samaniego (1981, pp. 864/5; see Theorem 2.2 and the Remark following this theorem).

If one thinks of obtaining the estimator sequence for ϑ_1 as the solution of an estimating equation, say

$$\sum_1^n g(x_\nu, \vartheta_1, \vartheta_2^{(n)}(\underline{x}), \ldots, \vartheta_p^{(n)}(\underline{x})) = 0,$$

there is no good reason for taking $g = \ell^{(1)}$. One could as well take $g = \ell^{(i)}$ for any $i \in \{2, \ldots, p\}$ (provided $L_{1i}(\vartheta) \neq 0$), or any linear combination $\sum_1^p a_i(\vartheta)\ell^{(i)}(\cdot, \vartheta)$ with $\sum_1^p a_i(\vartheta)L_{1i}(\vartheta) \neq 0$. Since (by Proposition 8.5.2) $\hat{\Lambda}_1(\cdot, \vartheta) = \sum_{i=1}^p \Lambda_{1i}(\vartheta)\ell^{(i)}(\cdot, \vartheta)$ is the influence function of the asymptotically efficient estimator sequence in the full family $\{P_\vartheta : \vartheta \in \Theta\}$, it suggests itself to try the estimating equation

$$\sum_1^n \hat{\Lambda}_1(x_\nu, \vartheta_1, \vartheta_2^{(n)}(\underline{x}), \ldots, \vartheta_p^{(n)}(\underline{x})) = 0. \tag{9.4.8}$$

Under suitable regularity conditions, we have $P_\vartheta(\hat{\Lambda}_1^{(i)}(\cdot, \vartheta)) = -\delta_{1i}$ (see (7.6.10)). Therefore (see Proposition 7.4.23 and Theorem 7.4.3), the sequence of solutions $\vartheta_1^{(n)}$, $n \in \mathbb{N}$, of (9.4.8) has the influence function $\hat{\Lambda}_1(\cdot, \vartheta)$, i.e. $\vartheta_1^{(n)}$, $n \in \mathbb{N}$, is asymptotically efficient for the full family $\{P_\vartheta : \vartheta \in \Theta\}$. For this, $\sqrt{n}$–consistency of $(\vartheta_2^{(n)}, \ldots, \vartheta_p^{(n)})$, $n \in \mathbb{N}$, suffices; asymptotic efficiency of this estimator sequence is not required.

Chapter 10
Asymptotic test theory

10.1 Introduction

Discusses the applicability of asymptotic test theory as compared with nonasymptotic theory.

The chapter on asymptotic test theory is restricted to the following simple problem: Given a family of p–measures $\{P_\vartheta : \vartheta \in \Theta\}$ with $\Theta \subset \mathbb{R}^p$, and a real valued functional $\kappa : \Theta \to \mathbb{R}$, the problem is to test the hypothesis $\kappa(\vartheta) \leq r_0$, based on a sample of size n. More precisely, this is the hypothesis $\{P_\vartheta^n : \vartheta \in \Theta, \kappa(\vartheta) \leq r_0\}$. The problem is to obtain for given $\alpha \in (0,1)$ a critical function φ_n such that $P_\vartheta^n(\varphi_n) \leq \alpha$ if $\kappa(\vartheta) \leq r_0$, and for which $P_\vartheta^n(\varphi_n)$ is as large as possible if $\kappa(\vartheta) > r_0$.

Even in the simplest case, i.e. $\Theta \subset \mathbb{R}$ and $\kappa(\vartheta) \equiv \vartheta$, this test problem is not yet uniquely defined unless one specifies the value of the alternative, say ϑ_1, for which the power, $P_{\vartheta_1}^n(\varphi_n)$, has to be maximized. Unless the family $\{P_\vartheta : \vartheta \in \Theta\}$ is exponential, the optimal critical function φ_n will depend on the alternative (see Section 4.5).

This dependence on the alternative may be so marked that a critical function φ_n which maximizes $P_{\vartheta_1}(\varphi_n)$ under the side condition "$P_\vartheta(\varphi_n) \leq \alpha$ for $\vartheta \leq \vartheta_0$" has poor power for values $\vartheta > \vartheta_1$ (see Example 4.1.4). In such a case it will be preferable to be content with critical functions which have a reasonably high power for all $\vartheta \geq \vartheta_1$, without any precisely defined optimum property. Asymptotic methods, leading to asymptotically optimal critical functions φ_n, are a practicable tool for this aim.

This is all the more true if $\Theta \subset \mathbb{R}^p$ with $p > 1$, and the hypothesis is $\kappa(\vartheta) \leq r_0$. The natural choice of an alternative, namely $\kappa(\vartheta) = r_1$, is not feasible; one has to choose a particular $\vartheta_1 \in \Theta$ to obtain a uniquely determined maximization problem. The resulting optimal critical function may be useless for practical purposes (see Example 4.4.5(ii)). It will be shown in Section 10.2 that asymptotic methods lead to practically useful results also in this case.

Some test problems are so vaguely defined that there is no starting point for an adequate mathematical treatment, asymptotic or not. As a trivial example we mention the problem (see p. 125, paragraph (ii)) of testing the hypothesis $\{P_0^n\}$ against all alternatives $\{P^n\}$ with $P \neq P_0$ (perhaps with P restricted to a certain family $\mathfrak{P}$).

As a rough approximation to the truth one could use the idea that different tests distribute their power differently about different alternatives. Without knowing the consequences which different deviations from the hypothesis have, and which deviations are likely to occur, there is no basis for evaluating the power function of a test. Hence there is no basis for formulating a mathematical optimization problem. Yet an approximate information about the power function might be of some help for an intuitive evaluation. As an example of such a service which asymptotic theory can render we mention the following problem. Tests for the hypothesis $\{P_0\}$ based on $\sup_{t \in \mathbf{R}} |Q_{\pm}^{(n)}(-\infty, t] - P_0(-\infty, t]|$ or $\int \left(Q_{\pm}^{(n)}(-\infty, t] - P_0(-\infty, t]\right)^2 P_0(dt)$ are usually considered as some sort of "omnibus"–test which should be used if one has no particular alternative in mind: Milbrodt and Strasser (1990) have shown that, in contrast to the general opinion, the power of such tests is concentrated to a surprising extent on alternatives deviating from P_0 in a particular direction.

In view of the somewhat problematic nature of the general test problem, we restrict the following consideration to the simplest case: Tests for a hypothesis based on a real valued functional $\kappa : \Theta \to \mathbb{R}$.

10.2 Tests for a real valued functional

Critical regions are obtained by converting confidence rays. The asymptotic power functions of such critical regions are derived from the asymptotic power functions of confidence bounds.

Let Θ be an open subset of $\mathbb{R}^p$. For $\vartheta \in \Theta$ and $n \in \mathbb{N}$ let $P_{\vartheta}^{(n)}$ be a p–measure on $(X_n, \mathcal{A}_n)$. The problem is to test a hypothesis about the value of a functional $\kappa : \Theta \to \mathbb{R}$, say $\kappa(\vartheta) \leq r_0$, against alternatives ϑ with $\kappa(\vartheta) > r_0$.

In Section 7.3 it was shown how to construct asymptotic confidence bounds for κ. Since critical regions for the hypothesis $\kappa(\vartheta) \leq r_0$ are related to lower confidence bounds, we state the results of Proposition 7.3.7 (referring to upper confidence bounds) for the case of lower confidence bounds, for easier reference.

Starting from an estimator sequence $\kappa^{(n)}$, $n \in \mathbb{N}$, with

$$P_{\vartheta}^{(n)} \circ c_n\left(\kappa^{(n)} - \kappa(\vartheta)\right) \Rightarrow Q_{\vartheta}, \tag{10.2.1}$$

one obtains the lower confidence bound

$$\kappa_\beta^{(n)}(x) := \kappa^{(n)}(x) - c_n^{-1} t_\beta^{(n)}(x), \tag{10.2.2}$$

for which

$$\lim_{n\to\infty} P_\vartheta^{(n)}\{x \in X_n : \kappa_\beta^{(n)}(x) \le \kappa(\vartheta) + c_n^{-1} t\} = Q_\vartheta(-\infty, t + t_\beta(\vartheta)]. \tag{10.2.3}$$

Desirable is a power function $t \to Q_\vartheta(-\infty, t + t_\beta(\vartheta)]$ which is as steep as possible.

If $\kappa_\beta^{(n)}$ is a lower confidence bound fulfilling

$$\lim_{n\to\infty} P_\vartheta^{(n)}\{x \in X_n : \kappa_\beta^{(n)}(x) \le \kappa(\vartheta)\} = \beta,$$

the confidence ray $[\kappa_\beta^{(n)}(x), \infty)$ can be converted into a critical region for the hypothesis $\kappa(\vartheta) \le r_0$ (against alternatives with $\kappa(\vartheta) > r_0$) by

$$C_{r_0}^{(n)} := \{x \in X_n : \kappa_\beta^{(n)}(x) > r_0\}.$$

$C_{r_0}^{(n)}$ consists of all $x \in X_n$ for which r_0 is not in the confidence ray. We have

$$\overline{\lim_{n\to\infty}} P_\vartheta^{(n)}(C_{r_0}^{(n)}) \le 1 - \beta \qquad \text{for } \kappa(\vartheta) \le r_0.$$

If the convergence in (10.2.3) holds locally uniformly in ϑ (conditions for this are given in Proposition 7.3.7), one may replace ϑ by $\vartheta + c_n^{-1} a$. Using the continuity of $\vartheta \to Q_\vartheta$ and $\vartheta \to t_\beta(\vartheta)$, one obtains

$$\lim_{n\to\infty} P_{\vartheta + c_n^{-1} a}^{(n)}\{x \in X_n : \kappa_\beta^{(n)}(x) \le \kappa(\vartheta + c_n^{-1} a) + c_n^{-1} t\}$$
$$= Q_\vartheta(-\infty, t + t_\beta(\vartheta)] \qquad \text{for } a \in \mathbb{R}^p.$$

Since $\kappa(\vartheta + c_n^{-1} a) = \kappa(\vartheta) + c_n^{-1} t_n$ with $t_n \to \kappa^{(\cdot)}(\vartheta) a$, this implies

$$\lim_{n\to\infty} P_{\vartheta + c_n^{-1} a}^{(n)}\{x \in X_n : \kappa_\beta^{(n)}(x) \le \kappa(\vartheta) + c_n^{-1}(\kappa^{(\cdot)}(\vartheta) a + t)\}$$
$$= Q_\vartheta(-\infty, t + t_\beta(\vartheta)].$$

Applied with $t = -\kappa^{(\cdot)}(\vartheta) a$, we obtain for $\vartheta \in \Theta$ and $a \in \mathbb{R}^p$

$$\lim_{n\to\infty} P_{\vartheta + c_n^{-1} a}^{(n)}\{x \in X_n : \kappa_\beta^{(n)}(x) \le \kappa(\vartheta)\}$$
$$= Q_\vartheta(-\infty, -\kappa^{(\cdot)}(\vartheta) a + t_\beta(\vartheta)].$$

For $\vartheta \in \Theta$ with $\kappa(\vartheta) = r_0$ this leads to

$$\lim_{n\to\infty} P_{\vartheta + c_n^{-1} a}^{(n)}(C_{r_0}^{(n)}) = Q_\vartheta[-\kappa^{(\cdot)}(\vartheta) a + t_\beta(\vartheta), \infty). \tag{10.2.4}$$

Hence

$$a \to Q_\vartheta \big[-\kappa^{(\cdot)}(\vartheta)a + t_\beta(\vartheta), \infty \big) \tag{10.2.5}$$

is the asymptotic power function of the sequence of critical regions $C_{r_0}^{(n)}$, $n \in \mathbb{N}$. Applied for $a = 0$ this confirms that $\lim_{n \to \infty} P_\vartheta^{(n)}(C_{r_0}^{(n)}) = 1 - \beta$ if $\kappa(\vartheta) = r_0$.

If relation (10.2.1) holds with $Q_\vartheta = N_{(0,\sigma^2(\vartheta))}$, then the asymptotic power function (10.2.5) becomes

$$a \to \Phi\big(u_\beta + \kappa^{(\cdot)}(\vartheta)a/\sigma(\vartheta) \big). \tag{10.2.6}$$

The *relative efficiency* for test sequences is defined in the spirit as the relative efficiency for estimator sequences in Section 2.6. Given two sequences $\varphi_i^{(n)}$, $n \in \mathbb{N}$, of critical functions for the hypothesis $\kappa(\vartheta) \le r_0$ fulfilling $P_\vartheta^{(n)}(\varphi_i^{(n)}) \le \alpha$ for $\kappa(\vartheta) \le r_0$, we choose a sequence of alternatives ϑ_n. Of interest are alternatives for which the rejection probability under $\varphi_1^{(n)}$ is reasonably high, say 0.8 at least. Then we determine the smallest sample size $m_n = \min\{m \in \mathbb{N} : P_{\vartheta_n}^{(m)}(\varphi_2^{(m)}) \ge P_{\vartheta_n}^{(n)}(\varphi_1^{(n)})\}$, for which $\varphi_2^{(m_n)}$ is, at ϑ_n, at least as good as $\varphi_1^{(n)}$. (This definition presumes that $m \to P_\vartheta^{(m)}(\varphi_2^{(m)})$ is isotone.) The *relative efficiency* of $\varphi_2^{(n)}$ with respect to $\varphi_1^{(n)}$ is defined as n/m_n. What makes this concept practically useful is the fact that in regular cases the ratio n/m_n is almost independent of the particular alternative if the sample sizes are large.

Assume now that $\lim_{n \to \infty} P_{\vartheta+n^{-1/2}a_n}^{(n)}(\varphi_i^{(n)}) = \Phi\big(u_{1-\alpha} + \kappa^{(\cdot)}(\vartheta)a/\sigma_i(\vartheta) \big)$ for $a_n \to a$. Writing

$$P_{\vartheta+n^{-1/2}a}^{(m)}(\varphi_2^{(m)}) = P_{\vartheta+m^{-1/2}(m/n)^{1/2}a}^{(m)}(\varphi_2^{(m)})$$

we obtain that

$$\big| P_{\vartheta+n^{-1/2}a}^{(m_n)}(\varphi_2^{(m_n)}) - P_{\vartheta+n^{-1/2}a}^{(n)}(\varphi_1^{(n)}) \big| \to 0$$

iff $n/m_n \to \sigma_1^2(\vartheta)/\sigma_2^2(\vartheta)$. Hence the *asymptotic relative efficiency* is independent of a, i.e. it is the same for all local alternatives $\vartheta + n^{-1/2}a$, $a \in \mathbb{R}^p$.

If two critical regions are derived from two asymptotically normal estimator sequences in the way indicated above, we obtain tests with asymptotic power function (10.2.6). Hence the asymptotic relative efficiency of these critical regions is the same as the asymptotic relative efficiency of the underlying estimator sequences.

Under suitable regularity conditions, the asymptotically optimal estimator sequence for $\kappa(\vartheta)$ is asymptotically normal with variance

$$\sigma_*^2(\vartheta) = \kappa^{(\cdot)}(\vartheta)H(\vartheta)\kappa^{(\cdot)}(\vartheta)^\mathsf{T}$$

(see Section 8.4). Hence the asymptotic power function of the sequence of critical regions $C_{r_0}^{(n)}$, $n \in \mathbb{N}$, derived from an asymptotically optimal estimator sequence for $\kappa(\vartheta)$ is given by (10.2.6) with $\sigma = \sigma_*$. Considering the fact that these critical regions were obtained by a relatively simple device, it is not unreasonable to expect that other methods might lead to critical regions with a better asymptotic power function. It will be shown in Section 10.3 that this is not the case.

10.3 The asymptotic envelope power function for tests for a real valued functional

Obtains asymptotic bounds for the power of arbitrary critical functions, and verifies that these bounds are attained by critical regions based on asymptotically efficient estimator sequences.

The following theorem gives the envelope power function for sequences of critical functions which keep the prescribed type one error locally uniformly, i.e. which fulfill for some neighborhood U of ϑ_0

$$\varlimsup_{n\to\infty} \sup_{\vartheta \in U} \{P_\vartheta^{(n)}(\varphi_n) : \kappa(\vartheta) \leq r_0\} \leq 1 - \beta.$$

For technical reasons, we need slightly less: Uniformity for $\vartheta = \vartheta_0 + c_n^{-1}b$, as expressed by condition (10.3.3) suffices.

In the following let $p_n(\cdot, \vartheta)$ be a density of $P_\vartheta^{(n)}$ with respect to some σ–finite measure $\mu_n | A_n$.

Theorem 10.3.1. *Let Θ be an open subset of $\mathbb{R}^p$. Assume that the family of sequences $P_\vartheta^{(n)}$, $n \in \mathbb{N}$, $\vartheta \in \Theta$, fulfills the following condition: For every $a, b \in \mathbb{R}^p$,*

$$P_{\vartheta_0+c_n^{-1}b}^{(n)} \circ \log \frac{p_n(\cdot, \vartheta_0 + c_n^{-1}a)}{p_n(\cdot, \vartheta_0 + c_n^{-1}b)} \Rightarrow N_{(-\frac{1}{2}\sigma^2, \sigma^2)} \qquad (10.3.2')$$

with

$$\sigma^2 = (a-b)^\top D(\vartheta_0)(a-b). \qquad (10.3.2'')$$

Recall that (10.3.2) follows if condition (8.1.5) holds locally uniformly at ϑ_0.

Assume that $\kappa : \Theta \to \mathbb{R}$ has a derivative which is continuous at ϑ_0. Then the following is true.

If a sequence φ_n, $n \in \mathbb{N}$, of critical functions fulfills

$$\varlimsup_{n\to\infty} P^{(n)}_{\vartheta_0+c_n^{-1}b}(\varphi_n) \le 1 - \beta \tag{10.3.3}$$

$$\text{for } b \in \mathbb{R}^p \text{ with } \kappa(\vartheta_0 + c_n^{-1}b) \le r_0,$$

then

$$\varlimsup_{n\to\infty} P^{(n)}_{\vartheta_0+c_n^{-1}a}(\varphi_n) \le \Phi\big(u_\beta + \kappa^{(\cdot)}(\vartheta_0)a/\sigma_*(\vartheta_0)\big) \tag{10.3.4'}$$

$$\text{for } a \in \mathbb{R}^p \text{ with } \kappa^{(\cdot)}(\vartheta_0)a > 0,$$

where

$$\sigma_*^2(\vartheta_0) = \kappa^{(\cdot)}(\vartheta_0)H(\vartheta_0)\kappa^{(\cdot)}(\vartheta_0)^\top. \tag{10.3.4''}$$

Proof. Except for certain technical details, the proof is the same as that of Theorem 8.2.3.

Let $\vartheta_0 \in \Theta$ with $\kappa(\vartheta_0) = r_0$ be fixed. We consider parameter values $\vartheta_0 + c_n^{-1}b$ with $\kappa^{(\cdot)}(\vartheta_0)b < 0$. This implies $\kappa(\vartheta_0 + c_n^{-1}b) < \kappa(\vartheta_0) = r_0$ for n sufficiently large. For any sequence of critical functions φ_n fulfilling condition (10.3.3), we have

$$\varlimsup_{n\to\infty} P^{(n)}_{\vartheta_0+c_n^{-1}b}(\varphi_n) \le 1 - \beta \quad \text{if } \kappa^{(\cdot)}(\vartheta_0)b < 0. \tag{10.3.5}$$

We determine a bound for $P^{(n)}_{\vartheta_0+c_n^{-1}a}(\varphi_n)$ under alternatives $\vartheta_0 + c_n^{-1}a$ with $\kappa^{(\cdot)}(\vartheta_0)a > 0$ (which relation implies $\kappa(\vartheta_0 + c_n^{-1}a) > r_0$ for n sufficiently large). Let

$$\ell_n(x) := \log p_n(x, \vartheta_0 + c_n^{-1}a)/p_n(x, \vartheta_0 + c_n^{-1}b).$$

By assumption (10.3.2),

$$P^{(n)}_{\vartheta_0+c_n^{-1}b} \circ \ell_n \quad \Rightarrow \quad N_{(-\frac{1}{2}\sigma^2,\sigma^2)}$$

with $\sigma^2 = (a-b)^\top D(\vartheta_0)(a-b)$. By Lemma 8.2.15, applied with $P_n = P_{\vartheta_0+c_n^{-1}b}$ and $P'_n = P_{\vartheta_0+c_n^{-1}a}$, this implies

$$\varlimsup_{n\to\infty} P^{(n)}_{\vartheta_0+c_n^{-1}a}(\varphi_n) \le \Phi(\sigma - u_{1-\beta}). \tag{10.3.6}$$

So far, $b \in \mathbb{R}^p$ is arbitrary, subject to the condition $\kappa^{(\cdot)}(\vartheta_0)b < 0$. To obtain a sharp upper bound, we choose b such that $(a-b)^\top D(\vartheta_0)(a-b) = \min$ under the side condition $\kappa^{(\cdot)}(\vartheta_0)b < 0$.

Notice that this is the choice of a *least favorable* element among the p–measures in the hypothesis. (See also Section 4.4.) Since

$$\inf\big\{(a-b)^\top D(\vartheta_0)(a-b) : b \in \mathbb{R}^p, \ \kappa^{(\cdot)}(\vartheta_0)b < 0\big\} = \big(\kappa^{(\cdot)}(\vartheta_0)a\big)^2/\sigma_*^2(\vartheta_0),$$

this leads to

$$\overline{\lim_{n \to \infty}} \, P^{(n)}_{\vartheta_0 + c_n^{-1}a}(\varphi_n) \leq \Phi\big(u_\beta + \kappa^{(\cdot)}(\vartheta_0)a/\sigma_*(\vartheta_0)\big).$$

$\square$

If (10.3.3) is replaced by the two–sided condition

$$\overline{\lim_{n \to \infty}} \, \big| P^{(n)}_{\vartheta_0 + c_n^{-1}b}(\varphi_n) - (1 - \beta) \big| = 0$$

for $b \in \mathbb{R}^p$ with $\kappa(\vartheta_0 + c_n^{-1}b) \leq r_0$, then we obtain in addition to (10.3.4') the relation

$$\lim_{n \to \infty} \, P^{(n)}_{\vartheta_0 + c_n^{-1}a}(\varphi_n) \geq \Phi\big(u_\beta + \kappa^{(\cdot)}(\vartheta_0)a/\sigma_*(\vartheta_0)\big)$$

for $\kappa^{(\cdot)}(\vartheta_0)a < 0$.

The asymptotic envelope power function given by (10.3.4) is attained by the critical regions given in Section 10.2, if these are based on asymptotically efficient estimator sequences. We remark that these critical regions fulfill the condition

$$\lim_{n \to \infty} P^{(n)}_\vartheta(C^{(n)}_{r_0}) = 1 - \beta \qquad \text{for } \kappa(\vartheta) = r_0$$

(and not only $\lim_{n \to \infty} P^{(n)}_\vartheta(\varphi_n) \leq 1 - \beta$ for $\kappa(\vartheta) \leq r_0$). Hence they are *asymptotically similar tests* of asymptotic level $1 - \beta$, and asymptotically optimal among *all* tests of asymptotic level $1 - \beta$.

Corollary 10.3.7. *Assume that $\{P^{(n)}_\vartheta : \vartheta \in \Theta, \, n \in \mathbb{N}\}$ fulfills conditions (10.3.2).*

Then the asymptotic envelope power function at ϑ_0 of tests for the hypothesis $\{(\vartheta_1, \ldots, \vartheta_p) \in \Theta : \vartheta_1 \leq \vartheta_{01}\}$ is

$$(a_1, \ldots, a_p) \to \Phi\big(u_\beta + a_1/H_{11}(\vartheta_0)^{1/2}\big).$$

Example 10.3.8. Let P be a p–measure with Lebesgue density p, symmetric about 0. Let $P_{a,c}$ be the p–measure with Lebesgue density $x \to \frac{1}{c}p\big(\frac{x-a}{c}\big)$.

According to Lemma 9.3.16, the minimal covariance matrix is given by (9.3.17). Presuming that $p(0) > 0$, the sequence of sample medians, $\tilde{x}_n, n \in \mathbb{N}$, is $\sqrt{n}$–consistent. Hence (see (7.5.10))

$$a^{(n)}(\underline{x}) := \tilde{x}_n + c^{(n)}(\underline{x})\Lambda_{11}n^{-1}\sum_1^n \ell'\Big(\frac{x_\nu - \tilde{x}_n}{c^{(n)}(\underline{x})}\Big), \qquad n \in \mathbb{N},$$

is asymptotically efficient for a, if $c^{(n)}, n \in \mathbb{N}$, is a consistent estimator sequence for c.

Notice that $a^{(n)}$ is equivariant:

$$a^{(n)}(\alpha + \gamma x_1, \ldots, \alpha + \gamma x_n) = \alpha + \gamma a^{(n)}(x_1, \ldots, x_n),$$

if $c^{(n)}$ is equivariant. A critical region for the hypothesis $a \leq a_0$, $c > 0$ is

$$C_{a_0}^{(n)} := \{\underline{x} \in \mathbb{R}^n : a^{(n)}(\underline{x}) > a_0 + n^{-1/2}u_{1-\beta}c^{(n)}(\underline{x})\sqrt{\Lambda_{11}}\}.$$

We have

$$\lim_{n \to \infty} P^n_{a_0+n^{-1/2}t, c+n^{-1/2}s}(C_{a_0}^{(n)}) = \Phi(u_\beta + t/c\sqrt{\Lambda_{11}}).$$

References

Anderson, T.W. (1955). The integral of a symmetric unimodal function over a symmetric convex set and some probability inequalities. Proc. Amer. Math. Soc. 6, 170–176.

Anderson, T.W. (1984). An Introduction to Multivariate Statistical Analysis, 2nd ed. Wiley, New York.

Anscombe, F.J. (1948). The validity of comparative experiments (with discussion). J. Roy. Statist. Soc. Ser. A 111, 181–211.

Arbuthnot, J. (1710). An argument for divine providence, taken from the constant regularity observ'd in the births of both sexes. Philos. Trans. Roy. Soc. London 27, 186–190. (Reprinted in Kendall and Plackett, 1977, 30–34.)

Ash, R.B. (1972). Real Analysis and Probability. Academic Press, New York.

Asrabadi, B.R. (1985). The exact confidence interval for the scale parameter and the MVUE of the Laplace distribution. Comm. Statist. A — Theory Methods 14, 713–733.

Bahadur, R.R. (1955). A characterization of sufficiency. Ann. Math. Statist. 26, 286–293.

Bahadur, R.R. (1957). On unbiased estimates of uniformly minimum variance. Sankhyā 18, 211–224.

Bahadur, R.R. (1958). Examples of inconsistency of maximum likelihood estimates. Sankhyā 20, 207–210.

Bahadur, R.R. (1964). On Fisher's bound for asymptotic variances. Ann. Math. Statist. 35, 1545–1552.

Bahadur, R.R. (1971). Some Limit Theorems in Statistics. Conference Board of the Mathematical Sciences, Regional Conference Series in Applied Mathematics, no. 4. SIAM, Philadelphia.

Bahadur, R.R. and Lehmann, E.L. (1955). Two comments on "Sufficiency and statistical decision functions". Ann. Math. Statist. 26, 139–142.

Bain, L.J. and Weeks, D.L. (1964). A note on the truncated exponential distribution. Ann. Math. Statist. 35, 1366–1367.

Barankin, E.W. (1949). Locally best unbiased estimates. Ann. Math. Statist. 20, 477–501.

Barankin, E.W. (1950). Extension of a theorem of Blackwell. Ann. Math. Statist. 21, 280–284.

Bar-Lev, S.K. and Reiser, B. (1983). A note on maximum conditional likelihood estimation for the gamma distribution. Sankhyā Ser. B 45, 300–302.

Barndorff-Nielsen, O. (1969). Levy homeomorphic parametrization and exponential families. Z. Wahrscheinlichkeitstheorie verw. Gebiete 12, 56–58.

Barndorff-Nielsen, O. (1978). Information and Exponential Families in Statistical Theory. Wiley, New York.

Barton, D.E. (1961). Unbiased estimation of a set of probabilities. Biometrika 48, 227–229.

Basu, A.P. (1964). Estimates of reliability for some distributions useful in life testing. Technometrics 6, 215–219.

Basu, D. (1955a). On statistics independent of a complete sufficient statistic. Sankhyā 15, 377–380.

Basu, D. (1955b). A note on the theory of unbiased estimation. Ann. Math. Statist. 26, 345–348.

Basu, D. (1958). On statistics independent of sufficient statistics. Sankhyā 20, 223–226.

Basu, D. (1978). On partial sufficiency: A review. J. Statist. Plann. Inference 2, 1–13.

Bauer, H. (1990a). Maß– und Integrationstheorie. De Gruyter, Berlin.

Bauer, H. (1990b). Wahrscheinlichkeitstheorie, 4. Auflage. De Gruyter, Berlin.

Berk, R.H. (1972). A note on invariance and sufficiency. Ann. Math. Statist. 43, 647–650.

Berk, R.H. and Bickel, P.J. (1968). On invariance and almost invariance. Ann. Math. Statist. 39, 1573–1576.

Bernoulli, D. (1734). Quelle est la cause physique de l'inclinaison des plans des orbites des planètes par rapport au plan de l'équateur de la revolution du soleil autour de son axe. Recueil des Pièces qui ont Remporté le Prix de l'Académie Royale des Sciences 3, 95–122.

Bernoulli, D. (1777). Dijudicatio maxime probabilis plurium observationum discrepantium atque verisimillima inductio inde formanda. Acta Acad. Petropolitanae 1, 3–23. (Reprinted with translation in Kendall, 1961.)

Bhattacharya, R.N. and Rao, R.R. (1976). Normal Approximation and Asymptotic Expansions. Wiley, New York.

Bickel, P.J. and Lehmann, E.L. (1969). Unbiased estimation in convex families. Ann. Math. Statist. 40, 1523–1535.

Bickel, P.J. and Lehmann, E.L. (1979). Descriptive statistics for nonparametric models. IV. Spread. In: Contributions to Statistics. Jaroslav Hájek Memorial Volume (J. Jurečková, ed.), 33–40, Academia, Prague.

Billingsley, P. (1968). Convergence of Probability Measures. Wiley, New York.

Birnbaum, A. (1961). A unified theory of estimation, I. Ann. Math. Statist. 32, 112–135.

Birnbaum, A. (1964). Median-unbiased estimators. Bull. Math. Statist. 11, 25–34.

Blackwell, D. (1947). Conditional expectation and unbiased sequential estimation. Ann. Math. Statist. 18, 105–110.

Blackwell, D. (1951). Comparison of experiments. In: Proceedings of the Second Berkeley Symposium on Mathematical Statistics and Probability, 93–102, University of California Press, Berkeley.

Blackwell, D. and Girshick, M.A. (1954). Theory of Games and Statistical Decisions. Wiley, New York.

Blackwell, D. and Ryll-Nardzewski, C. (1963). Non-existence of everywhere proper conditional distributions. Ann. Math. Statist. 34, 223–225.

Bomze, I.M. (1990). A Functional Analytic Approach to Statistical Experiments. Pitman Research Notes in Mathematics Series, 237. Longman, Harlow.

Bondesson, L. (1975). Uniformly minimum variance estimation in location parameter families. Ann. Statist. 3, 637–660.

Borges, R. and Pfanzagl, J. (1963). A characterization of the one parameter exponential family of distributions by monotonicity of likelihood ratios. Z. Wahrscheinlichkeitstheorie verw. Gebiete 2, 111–117.

Bourbaki, N. (1958). Fonctions d'une Variable Réelle, Chap. I–III. Hermann, Paris.

Bowman, K.O. and Shenton, L.R. (1983). Maximum likelihood estimators for the gamma distribution revisited. Comm. Statist. B – Simulation Comput. 12, 697–710.

Brown, G.W. (1947). On small-sample estimation. Ann. Math. Statist. 18, 582–585.

Brown, L.D. (1967). The conditional level of Student's t-test. Ann. Math. Statist. 38, 1068–1071.

Brown, L.D. (1986). Fundamentals of Statistical Exponential Families with Applications in Statistical Decision Theory. IMS Lecture Notes - Monograph Series, 9.

Brown, L.D., Cohen, A. and Strawderman, W.E. (1976). A complete class theorem for strict monotone likelihood ratio with applications. Ann. Statist. 4, 712–722.

Chhikara, R.S. and Folks, J.L. (1989). The Inverse Gaussian Distribution. Dekker, New York.

Chow, Y.S. and Teicher, H. (1988). Probability Theory, 2nd ed. Springer, New York.

Chung, K.L. (1951). The strong law of large numbers. In: Proceedings of the Second Berkeley Symposium on Mathematical Statistics and Probability, 341–352. University of California Press.

Clopper, C.J. and Pearson, E.S. (1934). The use of confidence or fiducial limits illustrated in the case of the binomial. Biometrika 26, 404–413.

Copas, J.B. (1975). On the unimodality of the likelihood for the Cauchy distribution. Biometrika 62, 701–704.

Cournot, A.A. (1843). Exposition de la théorie des chances et des probabilités. Paris.

Cramér, H. (1946). Mathematical Methods of Statistics. Princeton University Press.

Czuber, E. (1891). Theorie der Beobachtungsfehler. Teubner, Leipzig.

Dantzig, G.B. and Wald, A. (1951). On the fundamental lemma of Neyman and Pearson. Ann. Math. Statist. 22, 87–93.

Darmois, G. (1935). Sur les lois de probabilité à estimation exhaustive. C. R. Acad. Sci. Paris 260, 1265–1266.

Deemer, W.L. and Votaw, D.F. (1955). Estimation of parameters of truncated or censored exponential distributions. Ann. Math. Statist. 26, 498–504.

De Lury, D.B. (1938). Note on correlations. Ann. Math. Statist. 9, 149–151.

De Morgan, A. (1864). On the theory of errors of observation. Trans. Cambridge Philos. Soc. 10, 409–427.

Denny, J.L. (1964). A continuous real–valued function on E^n almost everywhere $1-1$. Fund. Math. 55, 95–99.

Doksum, K. (1969). Starshaped transformations and the power of rank tests. Ann. Math. Statist. 40, 1167–1176.

Doob, J.L. (1934). Probability and statistics. Trans. Amer. Math. Soc. 36, 759–775.

Doob, J.L. (1953). Stochastic Processes. Wiley, New York.

Droste, W. and Wefelmeyer, W. (1984). On Hájek's convolution theorem. Statist. Decisions 2, 131–144.

Droste, W. and Wefelmeyer, W. (1985). A note on strong unimodality and dispersivity. J. Appl. Probab. 22, 235–239.

Dudley, R.M. (1989). Real Analysis and Probability. Wadsworth & Brooks/Cole, Pacific Grove.

Dugué, D. (1936). Sur le maximum de précision des estimations gaussiennes à la limite. C. R. Acad. Sci. Paris 202, 193–195.

Dvoretzky, A., Wald, A. and Wolfowitz, J. (1950). Elimination of randomization in certain problemes of statistics and of the theory of games. Proc. Nat. Acad. Sci. U.S.A. 36, 256–260.

Dvoretzky, A., Wald, A. and Wolfowitz, J. (1951). Elimination of randomization in certain statistical decision procedures and zero–sum games. Ann. Math. Statist. 22, 1–21.

Dynkin, E.B. (1951). Necessary and sufficient statistics for a family of probability distributions. Uspeki Matem. Nauk (N. S.) 6, 68–90. (English translation in Select. Transl. Math. Statist. Probab. 1 (1961), 17–40.)

Eaton, M.L. and Morris, C.N. (1970). The application of invariance to unbiased estimation. Ann. Math. Statist. 41, 1708–1716.

Edgeworth, F.Y. (1883). The method of least squares. Philos. Mag. (Fifth Series) 16, 360–375.

Edgeworth, F.Y. (1908, 1909). On the probable errors of frequency-constants. J. Roy. Statist. Soc. 71, 381–397, 499–512, 651–678; Addendum 72, 81–90.

Eisenhart, Ch. (1949). Probability center lines for standard deviation and range charts. Industrial Quality Control 6, 24–26.

Eisenhart, Ch. and Martin, C.S. (1948). The relative frequencies with which certain estimators of the standard deviation of a normal population tend to underestimate its value. Ann. Math. Statist. 19, 600, Abstract.

Encke, J.F. (1832). Über die Begründung der Methode der kleinsten Quadrate. Berliner Astron. Jahrb. für 1834, 248–304.

Engelhardt, M. and Bain, L.J. (1977). Uniformly most powerful unbiased tests on the scale parameter of a gamma distribution with a nuisance shape parameter. Technometrics 19, 77–81.

Erdélyi, A. et al. (1954). Tables of Integral Transforms, Vol. I. McGraw-Hill, New York.

Eudey, M.W. (1949). On the treatment of discontinuous random variables. Thesis at the Univ. of California.

Fabian, V. (1970). On uniform convergence of measures. Z. Wahrscheinlichkeitstheorie verw. Gebiete 15, 139–143.

Faulkenberry, G.D. (1973). A method of obtaining prediction intervals. J. Amer. Statist. Assoc. 68, 433–435.

Ferguson, T.S. (1962, 1963). Location and scale parameters in exponential families of distributions. Ann. Math. Statist. 33, 986–1001; Correction 34, 1603.

Fisher, R.A. (1915). Frequency distribution of the values of the correlation coefficient in samples from an indefinitly large population. Biometrika 10, 507–521.

Fisher, R.A. (1920). A mathematical examination of the methods of determining the accuracy of an observation by the mean error, and by the mean square error. Monthly Not. Roy. Astr. Soc. 80, 758–770. (Reprinted in Fisher, 1950.)

Fisher, R.A. (1922). On the mathematical foundations of theoretical statistics. Philos. Trans. Roy. Soc. London Ser. A 222, 309–368. (Reprinted in Fisher, 1950.)

Fisher, R.A. (1925). Theory of statistical estimation. Proc. Cambridge Philos. Soc. 22, 700–725. (Reprinted in Fisher, 1950.)

Fisher, R.A. (1930). Inverse probability. Proc. Cambridge Philos. Soc. 26, 528–535. (Reprinted in Fisher, 1950.)

Fisher, R.A. (1934). Two new properties of mathematical likelihood. Proc. Roy. Soc. London Ser. A 144, 285–307. (Reprinted in Fisher, 1950.)

Fisher, R.A. (1935). The logic of inductive inference (with discussion). J. Roy. Statist. Soc. 98, 39–82. (Reprinted in Fisher, 1950.)

Fisher, R.A. (1950). Contributions to Mathematical Statistics. Wiley, New York.

Fourier, J.B.J. (1826). Recherches Statistiques sur la Ville de Paris et le Département de la Seine, Vol. 3.

Foutz, R.V. (1977). On the unique consistent solution to the likelihood equations. J. Amer. Statist. Assoc. 72, 147–148.

Fraser, D.A.S. (1954). Non-parametric theory: Scale and location parameters. Canad. J. Math. 6, 46–68.

Fraser, D.A.S. (1957). Nonparametric Methods in Statistics. Wiley, New York.

Fraser, D.A.S. (1966). On sufficiency and conditional sufficiency. Sankhyā Ser. A 28, 145–150.

Galambos, J. (1978). The Asymptotic Theory of Extreme Order Statistics. Wiley, New York.

Gauß, C.F. (1809). Theoria motus corporum coelestium in sectionibus conicis solem ambientum. Hamburg. (Reprinted in Gauß' collected works, Vol. 7, 1–280.)

Gauß, C.F. (1816). Bestimmung der Genauigkeit der Beobachtungen. Z. Astron. und verw. Wiss. 1. (Reprinted in Gauß' collected works, Vol. 4, 109–119.)

Gauß, C.F. (1821). Theoria combinationis observationum erroribus minimis obnoxiae, Part 1. Commentationes societatis regiae scientarum Gottingensis recentiores, Vol. 5. (Reprinted in Gauß' collected works, Vol. 4, 1–26.)

Gauß, C.F. (1825). Letter to W. Olbers. In: C. Schilling: W. Olbers, Sein Leben und sein Werk, Bd. 2. Briefwechsel zwischen Gauß und Olbers, p. 425.

Gauß, C.F. (1839). Letter to F.W. Bessel. In: Briefwechsel zwischen Gauß und Bessel, p. 523, Berlin, 1880.

Geary, R.C. (1942). The estimation of many parameters. J. Roy. Statist. Soc. 105, 213–217.

Ghurye, S.G. (1958). Note on sufficient statistics and two-stage procedures. Ann. Math. Statist. 29, 155–166.

Ghurye, S.G. and Wallace, D.L. (1959). A convolution class of monotone likelihood ratio families. Ann. Math. Statist. 30, 1158–1164.

Glaisher, J.W.L. (1872). On the law of facility of errors of observations and on the methods of least squares. Mem. Roy. Astr. Soc. 39(2), 75–124.

Glasser, G.J. (1962). Minimum variance unbiased estimators for Poisson probabilities. Technometrics 4, 409–418.

Gleser, L.J. and Healy, J.D. (1976). Estimating the mean of a normal distribution with known coefficient of variation. J. Amer. Statist. Assoc. 71, 977–981.

Gong, G. and Samaniego, F.J. (1981). Pseudo maximum likelihood estimation: Theory and applications. Ann. Statist. 9, 861–869.

Guenther, W.C. (1975). Best β–expectation tolerance limits a problem in estimation. Metron 33, 31–40.

Guttman, I. (1970). Statistical Tolerance Regions. Griffin, London.

Hájek, J. (1970). A characterization of limiting distributions of regular estimates. Z. Wahrscheinlichkeitstheorie verw. Gebiete 14, 323–330.

Hall, W.J., Wijsman, R.A. and Ghosh, J.K. (1965). The relationship between sufficiency and invariance with applications in sequential analysis. Ann. Math. Statist. 36, 575–614.

Halmos, P.R. (1946). The theory of unbiased estimation. Ann. Math. Statist. 17, 34–43.

Halmos, P.R. and Savage, L.J. (1949). Application of the Radon-Nikodym theorem to the theory of sufficient statistics. Ann. Math. Statist. 20, 225–241.

Heyer, H. (1969). Erschöpftheit und Invarianz beim Vergleich von Experimenten. Z. Wahrscheinlichkeitstheorie verw. Gebiete 12, 21–55.

Heyer, H. (1982). Theory of Statistical Experiments. Springer, New York.

Hinkley, D.V. (1977). Conditional inference about a normal mean with known coefficient of variation. Biometrika 64, 105–108.

Hipp, C. (1974). Sufficient statistics and exponential families. Ann. Statist. 2, 1283–1292.

Hipp, C. (1975). Note on the paper "Transformation groups and sufficient statistics" by J. Pfanzagl. Ann. Statist. 3, 478–482.

Hodges, J.L., Jr., and Lehmann, E.L. (1950). Some problems in minimax point estimation. Ann. Math. Statist. 21, 182–197.

Hodges, J.L., Jr., and Lehmann, E.L. (1963). Estimates of location based on rank tests. Ann. Math. Statist. 34, 598–611.

Hoeffding, W. (1977). Some incomplete and boundedly complete families of distributions. Ann. Statist. 5, 278–291.

Holla, M.S. (1967). Reliability estimation of the truncated exponential model. Technometrics 9, 332–336.

Hotelling, H. (1931). The generalization of Student's ratio. Ann. Math. Statist. 2, 360–378.

Hsu, C.T. (1940). On samples from a normal bivariate population. Ann. Math. Statist. 11, 410–426.

Ibragimov, I.A. (1956). On the composition of unimodal distributions. Theory Probab. Appl. 1, 255–260.

Ibragimov, I.A. and Has'minskii, R.Z. (1981). Statistical Estimation: Asymptotic Theory. Springer, New York.

Inagaki, N. (1970). On the limiting distribution of a sequence of estimators with uniformity property. Ann. Inst. Statist. Math. 22, 1–13.

Ioffe, A.D. (1978). Survey of measurable selection theorems: Russian literature supplement. SIAM J. Control Opt. 16, 728–732.

Isenbeck, M. and Rüschendorf, L. (1992). Completeness in location families. Prob. Math. Statist. 13, 321–343.

Joshi, S. and Sathe, Y.S. (1982). On approximating the equivariant estimator of the mean of a normal distribution with known coefficient of variation. Comm. Statist. A – Theory Methods 11, 209–215.

Kale, B.K. (1961). On the solution of the likelihood equation by iteration processes. Biometrika 48, 452–456.

Kariya, T. (1989). Equivariant estimation in a model with an ancillary statistic. Ann. Statist. 17, 920–928.

Kariya, T., Giri, N. and Perron, F. (1988). Invariant estimation of mean vector μ of $N_{(\mu,\Sigma)}$ with $\mu'\Sigma^{-1}\mu = 1$ or $\Sigma^{-1/2}\mu = C$ or $\Sigma = \delta^2\mu'\mu I$. J. Multivariate Anal. 27, 270–283.

Karlin, S. (1957). Pólya type distributions, II. Ann. Math. Statist. 28, 281–308.

Karlin, S. and Rubin, H. (1956). The theory of decision procedures for distributions with monotone likelihood ratio. Ann. Math. Statist. 27, 272–299.

Karlin, S. and Studden, W.J. (1966). Tchebycheff Systems: With Applications in Analysis and Statistics. Interscience Publishers, New York.

Kaufman, S. (1966). Asymptotic efficiency of the maximum likelihood estimator. Ann. Inst. Statist. Math. 18, 155–178.

Kelley, J.L. (1955). General Topology. Van Nostrand, New York.

Kendall, M.G. (1961). Daniel Bernoulli on maximum likelihood. Biometrika 48, 1–18.

Kendall, M.G. and Plackett, R.L. (1977). Studies in the History of Statistics and Probability, Vol. 2. Griffin, London.

Kendall, M.G. and Stuart, A.S. (1961). The Advanced Theory of Statistics, Vol. 2, 3rd. ed. Griffin, London.

Kiefer, J. and Wolfowitz, J. (1956). Consistency of the maximum likelihood estimator in the presence of infinitely many incidental parameters. Ann. Math. Statist. 27, 887–906.

Kitagawa, T. (1956). The operational calculus and the estimation of functions of parameters admitting sufficient statistics. Bull. Math. Statist. 6, 95–108.

Klaassen, C.A.J. (1985). Strong unimodality. Adv. in Appl. Probab. 17, 905–907.

Klaassen, C.A.J. (1987). Consistent estimation of the influence function of locally asymptotically linear estimators. Ann. Math. Statist. 15, 1548–1562.

Kolmogorov, A.N. (1950). Unbiased estimates. Izvestiya Akad. Nauk SSSR Ser. Mat. 14, 303-326. (Amer. Math. Soc. Transl. No. 98, reprinted in Translations Series 1, Amer. Math. Soc. 11 (1962), 144–170.)

Koopman, B.O. (1936). On distributions admitting a sufficient statistic. Trans. Amer. Math. Soc. 39, 399–409.

Kozek, A. (1977). On the theory of estimation with convex loss functions. In: Proceedings of the Symposium to Honour Jerzy Neyman (R. Bartoszynski, E. Fidelis and W. Klonecki, eds.) 177–202, PWN-Polish Scientific Publishers, Warszawa.

Krafft, O. and Witting, H. (1967). Optimale Tests und ungünstigste Verteilungen. Z. Wahrscheinlichkeitstheorie verw. Gebiete 7, 289–302.

Laha, R.G. (1954). On a characterization of the gamma distributions. Ann. Math. Statist. 25, 784–787.

Lambert, J.H. (1760). Photometria, sive de mensura et gradilus luminis colorum et umbrae. Augustae Vindelicorum.

Landers, D. (1972). Existence and consistency of modified minimum contrast estimates. Ann. Math. Statist. 43, 74–83.

Landers, D. and Rogge, L. (1973). On sufficiency and invariance. Ann. Statist. 1, 543–544.

Landers, D. and Rogge, L. (1976). A note on completeness. Scand. J. Statist. 3, 139.

Laplace, P.S. (1774). Mémoire sur la probabilité des causes par les événements. Mémoires de l'Académie royale des Sciences de Paris (Savants étrangers) 6, 621–656. (Œuvres Complètes, Vol. 8 (1891), 27–65, Paris.)

Laplace, P.S. (1781). Mémoire sur les probabilités. Mémoires de l'Académie royale des Sciences de Paris, 1778, 227–332. (Œuvres Complètes, Vol. 9 (1893), 381–485, Paris.)

Laplace, P.S. (1812). Théorie Analytique des Probabilités, Paris. (Œuvres Complètes, Vol. 7 (1847), Paris.)

LeCam, L. (1953). On some asymptotic properties of maximum likelihood estimates and related Bayes' estimates. Univ. of Calif. Publ. in Statist. 1, 277–330.

LeCam, L. (1956). On the asymptotic theory of estimation and testing hypotheses. In: Proceedings of the Third Berkeley Symposium of Mathematical Statistics and Probability 1, (J. Neyman, ed.), 129–156, University of California Press, Berkeley.

LeCam, L. (1960). Locally asymptotically normal families of distributions. Univ. of Calif. Publ. in Statist. 3, 37–98.

LeCam, L. and Yang, G.L. (1990). Asymptotics in Statistics. Some Basic Concepts. Springer, New York.

Lehmann, E.L. (1951). A general concept of unbiasedness. Ann. Math. Statist. 22, 587–592.

Lehmann, E.L. (1952). On the existence of least favorable distributions. Ann. Math. Statist. 23, 408–416.

Lehmann, E.L. (1959). Testing Statistical Hypotheses. Wiley, New York.

Lehmann, E.L. (1983). Theory of Point Estimation. Wiley, New York.

Lehmann, E.L. (1986). Testing Statistical Hypotheses, 2nd ed. Wiley, New York.

Lehmann, E.L. and Scheffé, H. (1947). On the problem of similar regions. Proc. Nat. Acad. Sci. U.S.A. 33, 382–386.

Lehmann, E.L. and Scheffé, H. (1950, 1955, 1956). Completeness, similar regions, and unbiased estimation. Sankhyā 10, 305–340; 15, 219–236; Correction 17, 250.

Lehmann, E.L. and Scholz, F.W. (1992). Ancillarity. In: Current Issues in Statistical Inference: Essays in Honor of D. Basu (M. Ghosh and P.K. Pathak, eds.), 32–51, IMS Lecture Notes — Monograph Series, 17.

Lehmann, E.L. and Stein, C. (1948). Most powerful tests of composite hypotheses. Ann. Math. Statist. 19, 495–516.

Lewis, T. and Thompson, J.W. (1981). Dispersive distributions, and the connection between dispersivity and strong unimodality. J. Appl. Probab. 18, 76–90.

Lexis, W. (1875). Einleitung in die Theorie der Bevölkerungsstatistik . Straßburg.

Lieberman, G.J. and Resnikoff, G.J. (1955). Sampling plans for inspection by variables. J. Amer. Statist. Assoc. 50, 457–516.

Loève, M. (1977). Probability Theory I, 4th ed. Springer, Berlin.

Lukacs, E. (1956). Characterization of populations by properties of suitable statistics. In: Proceedings of the Third Berkeley Symposium on Mathematical Statistics and Probability 2, (J. Neyman, ed.), 195–214, University of California Press, Berkeley.

Lukacs, E. (1970). Characteristic functions, 2nd ed. Griffin, London.

Mäkeläinen, T., Schmidt, K. and Styan, G. (1981). On the existence and uniqueness of the maximum likelihood estimate of a vector-valued parameter in fixed-size samples. Ann. Statist. 9, 758–767.

Mattner, L. (1992). Completeness of location families, translated moments, and uniqueness of charges. Probab. Theory Related Fields 92, 137–149.

Mattner, L. (1993). Some incomplete but boundedly complete location families. Ann. Statist. 21, 2158–2162.

McCord, J.R. (1964). On asymptotic moments of extreme statistics. Ann. Math. Statist. 35, 1738–1745.

Milbrodt, H. and Strasser, H. (1990). On the asymptotic power of the two–sided Kolmogorov–Smironov test. J. Statist. Plann. Inference 26, 1–23.

Mittal, M.M. and Dahiya, R.C. (1989). Estimating the parameters of a truncated Weibull distribution. Commun. Statist. A — Theory Methods 18(6), 2027-2042.

Morgan, W.A. (1939). A test for the significance of the difference between the two variances in a sample from a normal bivariate population. Biometrika 31, 13–19.

Mussmann, D. (1987). On a characterization of monotone likelihood ratio experiments. Ann. Inst. Statist. Math. 39, 263–274.

Nachbin, L. (1965). The Haar Integral. Van Nostrand, Princeton, N.J.

Neyman, J. (1934). On the two different aspects of the representative method of stratified sampling and the method of purposive selection. J. Roy. Statist. Soc. 97, 558–606. Discussion, 607–625.

Neyman, J. (1935). Sur un teorema concernente le cosidette statistiche sufficienti. Giorn. Ist. Ital. Attuari 6, 320–334.

Neyman, J. (1937). Outline of a theory of statistical estimation based on the classical theory of probability. Philos. Trans. Roy. Soc. London Ser. A 236, 333–380.

Neyman, J. (1938). L'estimation statistique traitée comme un problème classique de probabilité. Actualités Sci. Ind. 739, 25–57.

Neyman, J. (1941). Fiducial argument and the theory of confidence intervals. Biometrika 32, 128–150.

Neyman, J. and Pearson, E.S. (1928). On the use and interpretation of certain test criteria for purposes of statistical inference. Biometrika 20A, 175–240, 263–295. (Reprinted in Neyman and Pearson, 1967, 1–66 and 67–98.)

Neyman, J. and Pearson, E.S. (1933a). On the problem of the most efficient tests of statistical hypotheses. Philos. Trans. Roy. Soc. London Ser. A 231, 289–337. (Reprinted in Neyman and Pearson, 1967, 140–185.)

Neyman, J. and Pearson, E.S. (1933b). The testing of statistical hypotheses in relation to probabilities a priori. Proc. Cambridge Philos. Soc. 24, 492–510. (Reprinted in Neyman and Pearson, 1967, 186–202.)

Neyman, J. and Pearson, E.S. (1936). Contributions to the theory of testing statistical hypotheses I. Unbiased critical regions of type A and type A_1. Statist. Res. Mem. London 1, 1–37.

Neyman, J. and Pearson, E.S. (1967). Joint Statistical Papers. University of California Press, Berkeley.

Nöbeling, G. (1978). Integralsätze der Analysis. De Gruyter, Berlin.

Olkin, I. and Pratt, J.W. (1958). Unbiased estimation of certain correlation coefficients. Ann. Math. Statist. 29, 201–211.

Padmanabhan, A.R. (1970). Some results on minimum variance unbiased estimation. Sankhyā Ser. A 32, 107–114.

Parke, W.R. (1986). Pseudo maximum likelihood estimation: The asymptotic distribution. Ann. Statist. 14, 355–357.

Parthasarathy, K.R. (1967). Probability Measures on Metric Spaces. Academic Press, New York.

Patil, G.P. and Wani, J.K. (1966). Minimum variance unbiased estimation of the distribution function admitting a sufficient statistic. Ann. Inst. Statist. Math. 18, 39–47.

Pearson, K. (1896). Contributions to the mathematical theory of evolution. III. Philos. Trans. Roy. Soc. London Ser. A 187, 253–318.

Pearson, K. and Filon, L.N.G. (1898). Mathematical contributions to the theory of evolution. IV. On the probable errors of frequency constants and on the influence of random selection and correlation. Philos. Trans. Roy. Soc. London Ser. A 191, 229–311.

Perlman, M.D. (1972). On the strong consistency of approximate maximum likelihood estimators. In: Proceedings of the Sixth Berkeley Symposium on Mathematical Statistics and Probability 1, (L.M. LeCam, J. Neyman and E.L. Scott, eds.), 263–281, University of California Press, Berkeley.

Pfaff, T. (1982). Quick consistency of quasi maximum likelihood estimators. Ann. Statist. 10, 990–1005.

Pfanzagl, J. (1960). Über die Existenz überall trennscharfer Tests. Metrika 3, 169–176.

Pfanzagl, J. (1962). Überall trennscharfe Tests und monotone Dichtequotienten. Z. Wahrscheinlichkeitstheorie verw. Gebiete 1, 109–115.

Pfanzagl, J. (1967). A technical lemma for monotone likelihood ratio families. Ann. Math. Statist. 38, 611–613.

Pfanzagl, J. (1968). A characterization of the one parameter exponential family by existence of uniformly most powerful tests. Sankhyā Ser. A 30, 147–156.

Pfanzagl, J. (1969a). Further remarks on topology and convergence in some ordered families of distributions. Ann. Math. Statist. 40, 51–65.

Pfanzagl, J. (1969b). On the measurability and consistency of minimum contrast estimators. Metrika 14, 249–272.

Pfanzagl, J. (1972). Transformation groups and sufficient statistics. Ann. Math. Statist. 43, 553–568.

Pfanzagl, J. (1974). A characterization of sufficiency by power functions. Metrika 21, 197–199.

Pfanzagl, J. (1979). On optimal median unbiased estimators in the presence of nuisance parameters. Ann. Statist. 7, 187–193.

Pfanzagl, J. (1993). Sequences of optimal unbiased estimators need not be asymptotically optimal. Scand. J. Statist. 20, 73–76.

Pitman, E.J.G. (1936). Sufficient statistics and intrinsic accuracy. Proc. Cambridge Philos. Soc. 32, 567–579.

Pitman, E.J.G. (1939). A note on normal correlation. Biometrika 31, 9-12.

Pitman, E.J.G. (1979). Some Basic Theory for Statistical Inference. Chapman and Hall, London.

Pizzetti, P. (1892). I fundamenti matematici per la critica dei resultati sperimentali. Atti d'Univ. Genova.

Pratt, J.W. (1961). Length of confidence intervals. J. Amer. Statist. Assoc. 56, 549–567.

Pratt, J.W. (1976). F.Y. Edgeworth and R.A. Fisher on the efficiency of maximum likelihood estimation. Ann. Statist. 4, 501–514.

Pugh, E.L. (1963). The best estimate of reliability in the exponential case. Oper. Res. 11, 57–61.

Rao, C.R. (1945). Information and accuracy attainable in the estimation of statistical parameters. Bull. Calcutta Math. Soc. 37, 81–91.

Rao, C.R. (1947). Minimum variance and the estimation of several parameters. Proc. Cambridge Philos. Soc. 43, 280–283.

Rao, C.R. (1952). Some theorems on minimum variance estimation. Sankhyā 12, 27–42.

Rao, C.R. (1963). Criteria for estimation in large samples. Sankhyā Ser. A 25, 189–206.

Rao, M.M. (1965). Existence and determination of optimal estimators relative to convex loss. Ann. Inst. Statist. Math. 17, 133–147.

Rao, R.R. (1962). Relations between weak and uniform convergence of measures with applications. Ann. Math. Statist. 33, 659–680.

Reeds, J. (1985). Asymptotic number of roots of Cauchy location likelihood equations. Ann. Statist. 13, 775–784.

Reiß, R.-D. (1989). Approximate Distributions of Order Statistics. Springer, New York.

Roussas, G.R. (1972). Contiguity of Probability Measures: Some Applications in Statistics. Cambridge University Press.

Sacksteder, R. (1967). A note on statistical equivalence. Ann. Math. Statist. 38, 787–794.

Sathe, Y.S. and Varde, S.D. (1969). Minimum variance unbiased estimation of reliability for the truncated exponential distribution. Technometrics 11, 609–612.

Saunders, I.W. and Moran, P.A.P. (1978). On the quantiles of the gamma and the F distributions. J. Appl. Probab. 15, 426–432.

Scheffé, H. (1943). On a measure–problem arising in the theory of non–parametric tests. Ann. Math. Statist. 14, 227–233.

Schmetterer, L. (1960). On a problem of J. Neyman and E. Scott. Ann. Math. Statist. 31, 656–661.

Schmetterer, L. (1966). On the asymptotic efficiency of estimates. In: Research Papers in Statistics. Festschrift for J. Neyman (F.N. David, ed.), 301–317, Wiley, New York.

Schmetterer, L. (1974). Introduction to Mathematical Statistics, 2nd ed. Springer, Berlin.

Schmetterer, L. and Strasser, H. (1974). Zur Theorie der erwartungstreuen Schätzungen. Anz. Österreich. Akad. Wiss. Math.-Natur. Kl. 1974, no. 6, 59–66.

Shimizu, K. (1988). Point estimation. In: Lognormal Distributions. Theory and Applications (E.L. Crow and K. Shimizu, eds.), 27–86, Dekker, New York.

Smith, W.L. (1957). A note on truncation and sufficient statistics. Ann. Math. Statist. 28, 247–252.

Stein, C. (1964). Inadmissibility of the usual estimate for the variance of the normal distribution with unknown mean. Ann. Inst. Statist. Math. 16, 155–160.

Strasser, H. (1985). Mathematical Theory of Statistics. De Gruyter, Berlin.

Sverdrup, E. (1953). Similarity, unbiasedness, minimaxibility and admissibility of statistical test procedures. Skand. Aktuarietidskr. 36, 64–86.

Tate, R.F. (1959). Unbiased estimation: Functions of location and scale parameters. Ann. Math. Statist. 30, 341–366.

Tate, R.F. and Klett, G.W. (1959). Optimal confidence intervals for the variance of a normal distribution. J. Amer. Statist. Assoc. 54, 674–682.

Tocher, K.D. (1950). Extension of the Neyman–Pearson theory of tests to discontinuous variates. Biometrika 37, 130–144.

Torgersen, E. (1981). On complete sufficient statistics and uniformly minimum variance unbiased estimators. Sympos. Math. 25, 137–153.

Torgersen, E. (1991). Comparison of Statistical Experiments. Cambridge University Press.

Tweedie, M.C.K. (1957). Statistical properties of inverse Gaussian distributions I. Ann. Math. Statist. 28, 362–377.

van der Vaart, H.R. (1961). Some extensions of the idea of bias. Ann. Math. Statist. 32, 436–447.

Vogt, H. (1968). Zur Parameter- und Prozentpunktschätzung von Lebensdauerverteilungen bei kleinem Stichprobenumfang. Metrika 14, 117–131.

Voinov, V.G. (1985). Unbiased estimation of powers of the inverse of mean and related problems. Sankhyā Ser. B 47, 354–364.

von Weizsäcker, H. (1974). Zur Gleichwertigkeit zweier Arten von Randomisierung. Manuscripta Math. 11, 91–94.

Wagner, D.A. (1977). Survey of measurable selection theorems. SIAM J. Control Opt. 15, 859–903.

Wagner, D.A. (1980). Survey of measurable selection theorems: An update. In: Measure Theory, Oberwolfach 1979 (D. Kölzov, ed.), 176–219, Lecture Notes in Math. 794, Springer, Berlin.

Wald, A. (1939). Contributions to the theory of statistical estimation and hypothesis testing. Ann. Math. Statist. 10, 299–326.

Wald, A. (1943). Tests of statistical hypotheses concerning several parameters when the number of observations is large. Trans. Amer. Math. Soc. 54, 426–482.

Wald, A. (1945). Statistical decision functions which minimize the maximum risk. Ann. of Math. 46, 265–280.

Wald, A. (1950). Statistical Decision Functions. Wiley, New York.

Washio, Y., Morimoto, H. and Ikeda, N. (1956). Unbiased estimation based on sufficient statistics. Bull. Math. Statist. 6, 69–94.

Wertz, W. (1978). Statistical Density Estimation. A Survey. Vandenhoeck and Ruprecht, Göttingen.

Whittaker, E.T. and Watson, G.N. (1958). A Course of Modern Analysis, 4th ed. Cambridge University Press.

Wijsman, R.A. (1958). Incomplete sufficient statistics and similar tests. Ann. Math. Statist. 29, 1028–1045.

Wilks, S.S. (1932). Certain generalizations in the analysis of variance. Biometrika 24, 471–494.

Wilks, S.S. (1941). Determination of sample sizes for setting tolerance limits. Ann. Math. Statist. 12, 91–96.

Wilson, E.B. (1927). Probable inference, the law of succession, and statistical inference. J. Amer. Statist. Assoc. 22, 209–212.

Witting, H. (1985). Mathematische Statistik I. Teubner, Stuttgart.

Wolfowitz, J. (1965). Asymptotic efficiency of the maximum likelihood estimator. Theory Probab. Appl. 10, 247–260.

Working, H. and Hotelling, H. (1929). Applications of the theory of error to the interpretation of trends. J. Amer. Statist. Assoc. 24, 73–85.

Zacks, S. (1971). The Theory of Statistical Inference. Wiley, New York.

Zermelo, E. (1929). Die Berechnung der Turnier-Ergebnisse als ein Maximumproblem der Wahrscheinlichkeitsrechnung. Math. Z. 29, 436–460.

Author Index

Subject Index

Notation Index

antitone	nonincreasing
$B(x,y)$	beta function
$B_{n,p}$	binomial distribution
$\mathcal{B}_S$	$\{B \subset Y : S^{-1}B \in \mathcal{A}\}$ if $S : (X, \mathcal{A}) \to Y$
$\mathbb{B}$	Borel algebra of the real line
$\mathbb{B}_0$	$\mathbb{B} \cap [0,1]$ or $\mathbb{B} \cap (0,1)$
$\mathbb{B}_+$	$\mathbb{B} \cap [0, \infty)$ or $\mathbb{B} \cap (0, \infty)$
$\overline{\mathbb{B}}$	Borel algebra of the extended real line
contraction of S	a function which depends on x through $S(x)$ only
C_y	$\{x \in X : (x,y) \in C\}$ if $C \subset X \times Y$
$\mathbb{C}$	set of all complex numbers
$\mathcal{C}$	class of all convex Borel sets 226
$\mathfrak{C}$	class of all bounded and continuous functions 255
χ_m^2	χ^2–distribution with m degrees of freedom
$d(x,y)$	distance between x and y
$D(\vartheta)$	241, 264
δ_{ij}	Kronecker symbol
E_ϑ	exponential distribution 2
φ	Lebesgue density of $N_{(0,1)}$
Φ	distribution function of $N_{(0,1)}$
Φ_α	class of all level–α–tests 124
$G(\vartheta)$	240
$\Gamma(b)$	gamma function 154
$\Gamma_{a,b}$	gamma distribution 27
$H(\vartheta)$	$D(\vartheta)^{-1}$ 241, 269
iff	if and only if
i.i.d.	independent and identically distributed

$\mathfrak{P} \circ S$	$\{P \circ S : P \in \mathfrak{P}\}$
Π_a	Poisson distribution
ψ_S	139
$q_\beta(\vartheta)$	β–quantile of P_ϑ
$\mathbb{Q}$	set of all rational numbers
$Q_{\underline{x}}^{(n)}$	empirical distribution 189
$r_n(\underline{x}, \underline{y})$	sample correlation coefficient 152
$\mathbb{R}$	set of all real numbers
$\mathbb{R}_+$	$\mathbb{R} \cap [0, \infty)$
sgn	signum
sup–metric	$d(P, Q) = \sup_{A \in \mathcal{A}} |P(A) - Q(A)|$
$s_n^2(\underline{x})$	$n^{-1} \sum_1^n (x_\nu - \bar{x}_n)^2$ or $(n-1)^{-1} \sum_1^n (x_\nu - \bar{x}_n)^2$
$S(y, \delta)$	open ball with center y and radius δ
$\sigma_*^2(\vartheta)$	minimal asymptotic variance 269
$\Sigma_*(\vartheta)$	minimal asymptotic covariance matrix 279
t_m	t–distribution with m degrees of freedom
$t_{m,\beta}$	upper β–quantile of t_m
u_β	upper β–quantile of $N_{(0,1)}$
U	uniform distribution on $\big((0,1), \mathbb{B}_0\big)$
w.l.g.	without loss of generality
$\bar{x}_n$	sample mean
$\tilde{x}_n$	sample median 97
$x_{i:n}$	i-th order statistic
$\mathbb{Z}$	set of all integers
$P \circ f$	distribution of $f : (X, \mathcal{A}) \to (Y, \mathcal{B})$, i.e. $P \circ f(B) = P(f^{-1}B)$
$\mu * \nu$	convolution product
$\nu \ll \mu$	ν is dominated by μ
$\Sigma_1 \leq_L \Sigma_2$	Löwner order; $\Sigma_2 - \Sigma_1$ is positive semidefinite
$Q_1 <_s Q_2$	spread order 75
$f \underset{(P)}{=} g$	$f = g$ P–a.e.
$\alpha A + \beta B$	$\{\alpha a + \beta b : a \in A, \ b \in B\}$

$A = B$ (P)	$P(A \triangle B) = 0$
$A \subset B$ (P)	$P(B - A) = 0$
$A \in \mathcal{A}_0$ (P)	11
$\mathcal{A}_0 \subset \mathcal{A}$ (P)	16
C^b	boundary of C
C^c	complement of C
C°	interior of C
$a_n \to a$	convergence of a_n to a
$f_n \to f$ $(P^{(n)})$	stochastic convergence
$P^{(n)} \Rightarrow P$	weak convergence 255
$f^{(i)}(x, \vartheta_1, \ldots, \vartheta_p)$	partial derivative with respect to ϑ_i
$f^{(\cdot)}(x, \vartheta)$	row vector with components $f^{(i)}(x, \vartheta)$, $i = 1, \ldots, p$ (unless $f(x, \vartheta) = \ell(x, \vartheta)$)
$\overline{f}(x, V)$	$\sup\{f(x, y) : y \in V\}$
$\underline{f}(x, V)$	$\inf\{f(x, y) : y \in V\}$
$\lvert \cdot \rvert$	absolute value
$\lVert \cdot \rVert$	Euclidean norm
$[x]$	largest integer $\leq x$
$[g_1, \ldots, g_p]$	linear space spanned by the functions g_i, $i = 1, \ldots, p$
1_A	indicator function

www.ingramcontent.com/pod-product-compliance
Lightning Source LLC
Chambersburg PA
CBHW050657190326
41458CB00008B/2602